Chemical Reaction Engineering

Chemical Reaction Engineering

Edited by
Elsie Perkins

WILLFORD PRESS

www.willfordpress.com

Published by Willford Press,
118-35 Queens Blvd., Suite 400,
Forest Hills, NY 11375, USA

ISBN: 978-1-64728-330-8

Cataloging-in-Publication Data

Chemical reaction engineering / edited by Elsie Perkins.
    p. cm.
Includes bibliographical references and index.
ISBN 978-1-64728-330-8
1. Chemical engineering. 2. Chemical reactions. 3. Chemical reactors. I. Perkins, Elsie.
TP155 .C44 2022
660--dc23

For information on all Willford Press publications
visit our website at www.willfordpress.com

WILLFORD PRESS

Contents

Preface

This book was inspired by the evolution of our times; to answer the curiosity of inquisitive minds. Many developments have occurred across the globe in the recent past which has transformed the progress in the field.

Chemical reaction engineering is a sub-field of chemical engineering or industrial chemistry which deals with chemical reactors. It aims at the optimization of chemical reactions so as to determine the best reactor design. Various factors such as heat transfer, reaction kinetics, mass transfer and flow phenomena are studied to relate reactor performance with feed composition and operating conditions. Chemical reaction engineering is applied across the petroleum and petrochemical industries as well as in systems that require the engineering or modelling of reactions. This book is a valuable compilation of topics, ranging from the basic to the most complex advancements in the field of chemical reaction engineering. It presents this complex subject in the most comprehensible and easy to understand language. For all readers who are interested in chemical reaction engineering, the case studies included in this book will serve as an excellent guide to develop a comprehensive understanding.

This book was developed from a mere concept to drafts to chapters and finally compiled together as a complete text to benefit the readers across all nations. To ensure the quality of the content we instilled two significant steps in our procedure. The first was to appoint an editorial team that would verify the data and statistics provided in the book and also select the most appropriate and valuable contributions from the plentiful contributions we received from authors worldwide. The next step was to appoint an expert of the topic as the Editor-in-Chief, who would head the project and finally make the necessary amendments and modifications to make the text reader-friendly. I was then commissioned to examine all the material to present the topics in the most comprehensible and productive format.

I would like to take this opportunity to thank all the contributing authors who were supportive enough to contribute their time and knowledge to this project. I also wish to convey my regards to my family who have been extremely supportive during the entire project.

Editor

Pt-Co and Pt-Ni Catalysts of Low Metal Content for H_2 Production by Reforming of Oxygenated Hydrocarbons and Comparison with Reported Pt-Based Catalysts

Liza A. Dosso, Carlos R. Vera ⓘD, and Javier M. Grau ⓘD

Instituto de Investigaciones en Catálisis y Petroquímica "Ing. José Miguel Parera" (INCAPE), FIQ, UNL-CONICET, CCT CONICET Santa Fe, "Dr. Alberto Cassano," Colec. Ruta Nac. No. 168, KM 0, Paraje El Pozo, S3000AOJ Santa Fe, Argentina

Correspondence should be addressed to Javier M. Grau; jgrau@fiq.unl.edu.ar

Academic Editor: Eric Guibal

New catalysts of Pt, PtNi, PtCo, and NiCo supported on Al_2O_3 were developed for producing hydrogen by aqueous phase reforming (APR) of oxygenated hydrocarbons. The urea matrix combustion technique was used for loading the metal on the support in order to improve several aspects: increase both the metal-support interaction and the metal dispersion and decrease the metal load. The catalysts were characterized by MS/ICP, N_2 adsorption, XRD, TPR, CO chemisorption, and the test of cyclohexane dehydrogenation (CHD). The APR of a solution of 10% mass ethylene glycol (EG), performed in a tubular fixed bed reactor at 498 K, 22 bar, WHSV = 2.3 h^{-1}, was used as the main reaction test. After 10 h on-stream, the catalysts prepared by UMC had better hydrogen yield and catalytic stability than common catalysts prepared by IWI. The UMC/IWI H_2 yield ratio was 23.5/15.2 for Pt, 24.0/17.0 for PtCo, 26.6/21.0 for PtNi, and 8.0/3.9 for NiCo. Ni or Co addition to Pt increased the carbon conversion while keeping the H_2 turnover high. Cobalt also improves stability. Reports of several authors were revised for a comparison. The analysis indicated that the developed catalysts are a viable and cheaper alternative for H_2 production from a renewable resource.

1. Introduction

Current methods of industrial hydrogen production used in petroleum refineries require high temperatures, about 800 K in the case of naphtha reforming [1] and 1200 K for methane steam reforming [2]. These two processes make use of nonrenewable, fossil fuel raw materials and are carried out in the gas phase (GPR).

Nowadays, a growing need exists for processing heavy crudes with a high content of heteroatoms and polyaromatic molecules. Large amount of hydrogen is needed for the hydrotreatment of biomass-derived oxygenates. Many works about the hydrodeoxygenation of lignin-derived phenolic have been reported [3–5]. This makes the use of extensive hydrocracking, hydrodesulphurization, dearomatization, hydrodeoxygenation, and so on, a common refining and biorefining practice. Alternative sources of hydrogen supply are thus becoming necessary.

The burning of fossil fuels has a major impact on the increase of the concentration of CO_2 in the atmosphere. On the contrast, energy derived from biomass releases carbon with a carbon-energy ratio similar to that of coal. However, as indicated by Wuebbles and Jain [6], biomass has already absorbed an equal amount of carbon from the atmosphere before its emission, and therefore the net carbon emissions of biomass fuels are zero during their life cycle. As explained by Nozawa et al. [7], the production of hydrogen from biorenewable sources is a promising way of minimizing environmental problems associated with the combustion of fossil fuels.

The aqueous phase reforming (APR) of oxygenated hydrocarbons derived from biomass is considered a promising alternative process for supplying great amounts of hydrogen at a conveniently low cost. Biomass reforming also has a neutral CO_2 life cycle, making it convenient from an environmental point of view. APR of oxygenated hydrocarbons can also be performed at low reaction temperatures, for example, 500 K, thermodynamically favoring hydrogen formation with low carbon monoxide content. An overview of aqueous-phase catalytic processes for production of hydrogen and alkanes in a bio refinery has been shown by Huber and Dumesic [8]. In this sense, APR has been recently considered a promising route to upgrade organic compounds found in bio refinery water fractions [9].

One of the key aspects of the APR technology is the design of a suitable catalyst. Many reports study catalysts based on supported Pt. These reports highlight the influence of the amount of available metal surface on the overall activity and the selectivity to hydrogen. Thus, best-performing catalysts have high Pt surface areas per unit catalyst mass. Due to the difficulties of achieving and maintaining a high metal dispersion, most current Pt catalysts for APR have great Pt loads and high associated costs. One way of reducing the catalyst cost is to partially or totally replace Pt by another transition metal with similar properties. For instance, Ni provides good hydrogen yields in gas phase reforming. However, Davda et al. [10] have reported that the performance of Ni in APR is lower than that of pure Pt and that also Ni suffers from intense sintering.

Shabaker et al. [11] used Raney Ni for APR of oxygenated hydrocarbons and got good catalytic activity. However, preparation and handling of pyrophoric Raney Ni are cumbersome and hazardous. Co is also effective for breaking the C-C bond, key reaction step for the generation of H_2. However, it has too high selectivity to methane, an undesired final product that consumes much of the produced hydrogen [7]. Supported Co catalysts for APR are also unstable and sinter to a great extent. Their selectivity to coke is also high unless used in the form of Co-Pt alloys [12]. Best performing Co catalysts are those having small metal particles of easy reducibility [13]. This high metal dispersion is however difficult to achieve by common preparation techniques.

González-Cortés et al. [14, 15] described an alternative method for dispersing transition metals on Al_2O_3 by means of the coimpregnation of the metal salts with urea, followed by drying and fast calcination (combustion) of the formed urea matrix (UMC method). Combustion forms structural defects on the support that act as additional transition metal anchoring sites [16], increasing the number of metal particles and decreasing their average size. This preparation method reportedly allows the preparation of supported metal catalysts of higher dispersion and lower metal load, decreasing the catalysts fabrication costs.

Another issue is that of the choice of the support. Different groups have obtained catalysts with high hydrogen yields, but they not only use high Pt loads but also use special supports [10]. It would be desirable to use alumina as support since it has an affordable low price and good textural and mechanical properties, such as surface area, pore volume, pore size, crush, and sintering resistance.

Most APR catalysts have high fabrication costs. In this sense, the focus of the present work is put on the comparison of the catalysts of this work, which have relatively low metal content, with those prepared by conventional methods that

have been reported in the open literature. The aim is to obtain similar properties as those of these reported catalysts.

The catalytic properties are compared of reference Pt, PtNi, PtCo, and NiCo catalysts supported on alumina, prepared by a classical method (incipient wetness impregnation of the precursor solutions on gamma alumina) with catalysts of similar composition prepared by the UMC technique. Properties studied are activity, hydrogen yield, selectivity, hydrogen turn-over frequency TOF_{H_2}, and stability. Particularly, the stability is assessed by measuring both the velocity of coke fouling during the reaction test and the leaching of the metal phase. The catalytic properties are assessed by means of the reaction of aqueous phase reforming of ethylene glycol. Finally, the properties of the best performing catalysts are compared with the properties of catalysts reported by other authors.

2. Materials and Methods

2.1. Catalyst Preparation. Two different methods were used to incorporate the metals to the support: (i) the common simultaneous incipient wetness impregnation (IWI) of the metal precursors, followed by slow calcination; (ii) the method of simultaneous wet impregnation of the metal precursors with urea, followed by the fast combustion of the urea matrix (UMC) [14, 15, 17]. In both cases, the support used was γ-Al_2O_3 (Sasol Alumina spheres 2.5/210), thermally stabilized for 2 h at 873 K, 0.06 $cm^3 \cdot g^{-1}$ pore volumes, with 208.8 $m^2 \cdot g^{-1}$ specific surface area, and ground to a particle size of 35–80 mesh. Salt metal precursors were Pt $(NH_3)_4(NO_3)_2$, $Ni(NO_3)_2$, and $Co(NO_3)_2$ (all Sigma Aldrich, R.P.A). Aqueous solutions of adequate concentration were prepared and then used in the impregnation, in order to obtain final catalysts with 1% Pt and 3% Ni or Co on a mass basis.

In the case of the IWI route, the support was dried in a stove at 383 K for 12 h after impregnation and was then slowly heated from room temperature to 723 K (10 $K \cdot min^{-1}$), then kept and calcined at this temperature for 3 h in flowing air (30 $cm^3 \cdot min^{-1}$). In the case of the UMC route, a solution was prepared that contained urea and the metal precursors with a ratio of 10 mol urea per mol of metal (all metals). The pH of the solution was adjusted to 7, and then the support was immersed in the solution. The system was stirred gently for 3 h at 323 K. The solution excess was evaporated at 323 K until a slurry was formed, and then it was quickly calcined for 10 min at 773 K [14, 15]. Finally, the oxides formed by both routes were reduced in a stream of hydrogen, 1 h at 773 K in situ, in the same steel tubular reactor used for the reaction test.

Mono and bimetallic catalysts were identified according to the following description: Pt-IWI; PtNi-IWI; PtCo-IWI, NiCo-IWI, Pt-UMC; PtNi-UMC; PtCo-UMC; and NiCo-UMC. The UMC and IWI acronyms indicate the preparation route used.

2.2. Catalyst Characterization. The concentration of the supported metals (Pt, Ni, and Co) in the catalysts was determined by mass spectroscopy with inductively coupled plasma (MS/ICP). The equipment used was an ARL 3410 with argon as gas for the plasma. The solid sample was dissolved in a sulphuric acid solution (50% vol.), and then an aliquot was placed in the nebulizer of the ICP, and the concentration of the metal cations was determined from the mass spectrum of the ion source, as measured by means of a quadrupole mass spectrometer.

The concentration of exposed metal sites was determined using CO as a molecular probe. CO selectively chemisorbs on the surface metal atoms of the metal particles. For this test, an amount of 0.05 g of reduced catalyst was used. Calibrated pulses of a $CO:N_2$ mixture (1.46% CO in N_2, molar basis) were sent to the reactor cell until the metallic surface became saturated. The nonadsorbed CO was converted to CH_4 with H_2 over a Ni/Kieselguhr catalyst and detected by a flame ionization detector connected to the exhaust of the reactor. The adsorbed CO was determined from a mass balance.

The specific surface area was measured in a volumetric system from the N_2 adsorption isotherm obtained at 77 K. The specific surface area (BET method) and porosity measurements were performed in a model ASAP 2020 Micromeritics apparatus using 0.3 g of sample. Samples were outgassed by heat under vacuum ($1.333\,10^{-9}$ bar) at 523 K for 30 h before the nitrogen adsorption.

X-ray diffraction tests were performed in an XD-D1 Shimadzu diffractometer. A sample of about 0.3 g was dried in an oven and then ground to a powder. Then, it was placed on a sample holder and irradiated with Cu Kα monochromatic radiation of a wavelength of about 1.54 Å, filtered with Ni, and operated at 40 kV and 40 mA, at a scan rate of 2° min^{-1} and scanning the 20–80° 2θ range.

The temperature-programmed reduction technique (TPR) allows the study of the reducibility of the surface species on the solid support and the degree of interaction between them, especially metal-metal, metal-promoter, and metal-support interactions. Reducibility was measured by H_2 consumption as the sample was subjected to a heating schedule. An Ohkura TP 2002s equipment with a thermal conductivity detector (TCD) was used for these experiments. A known mass of catalyst was first treated in flowing air at 723 K for 1 h and then brought to room temperature. Then the sample was flushed with flowing Ar for 15 min, and the reducing mixture (5% H_2 in Ar) was passed over the sample at room temperature. Once the system was stabilized, the temperature was raised linearly from room temperature to 973 K at a heating rate of 10 K min^{-1}.

2.3. Catalytic Activity Tests. The cyclohexane dehydrogenation (CHD) reaction test allows evaluating the metallic phase of the catalyst. This reaction is known to proceed with a rate that is strictly proportional to the number of surface metal sites, with no regards to metal particle size or surface atom location. This means this reaction is not sensitive to the catalyst structure [18]. The reaction was performed in a glass reactor. The catalyst mass used was 0.1 g, the reaction temperature 573 K, pressure 1 bar, hydrogen flow rate

80 cm$^3 \cdot$min^{-1}, and cyclohexane flow rate 1.61 cm$^3 \cdot$h^{-1} (99.9% Merck). Before the reaction, the catalyst was reduced at 773 K for 1 h with hydrogen. The products were analyzed on-line in a gas chromatograph, with a copper capillary column 100 m long, 0.5 mm internal diameter, and a squalene coating phase.

Aqueous-phase reforming of ethylene glycol was performed in a tubular stainless steel AISI 316L reactor of 9.5 mm internal diameter and of 1.5 mm thickness, heated by an electric oven. The system had a similar configuration as that used by Shabaker et al. [11]. In the experiments, a mass of 0.5 g of catalyst was previously reduced in situ at 773 K for 1 h in hydrogen flow. Then the system was purged with He, and the reaction pressure was adjusted to the desired value by means of a back pressure regulator. The feed, an aqueous solution of ethylene glycol (10%, mass basis), was injected to the reactor by an HPLC pump at a flow rate of 0.02 cm$^3 \cdot$min^{-1}. The reaction was performed at 498 K, 22 bar, WHSV = 2.34 h^{-1} (LHSV = 0.86 h^{-1}). The reaction temperature and pressure were controlled to a narrow margin ($\pm$5 K, $\pm$0.1 bar). The products of the reaction were cooled down in a condenser connected on-line and downstream the reactor. The flow rate of the exit gases was 3 cm$^3 \cdot$min^{-1}. The gases were analyzed on-line with a Shimadzu 8A gas chromatograph using a thermal conductivity detector and a Restek Shin Carbon Micropacked ST column. Total time-on-stream was 10 h. Yields to hydrogen, methane, and other products were determined from compositional data, according to the following formula [10]:

$$H_2 \text{ yield} = \frac{H_2 \text{ in the gas products}}{H_2 \text{ theoretically calculated}} \times 100. \quad (1)$$

This is not the common definition of yield but is that used by Dumesic and coworkers. It has been adopted here for the sake of comparison. "H_2 *in the gas products*" is the molar hydrogen flow rate measured at the reactor outlet. As no hydrogen is fed to the reactor, this is equal to the hydrogen produced. "H_2 *theoretically calculated*" is the hypothetical hydrogen flow rate produced if all the EG feedstock was reformed according to

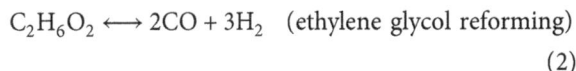

$$C_2H_6O_2 \longleftrightarrow 2CO + 3H_2 \quad \text{(ethylene glycol reforming)} \quad (2)$$

$$2CO + 2H_2O \longleftrightarrow 2CO_2 + 2H_2 \quad \text{(water gas shift)} \quad (3)$$

For calculation both (2) and (3) are considered to be irreversible.

$$\text{Conversion}_{C \text{ to gas}} = \frac{C \text{ in the gas products}}{C \text{ fed to the reactor}} \times 100, \quad (4)$$

$$H_2 \text{ selectivity} = \frac{H_2 \text{ in the gas products}}{C \text{ in the gas products}} \times \frac{1}{R} \times 100,$$

where R is the H_2/CO_2 ratio for the reforming, equal to 5/2 for ethylene glycol.

Selectivity to carbon products (CP) in the gas products (CO, CO_2, CH_4):

$$CP \text{ selectivity} = \frac{CP \text{ in the gas products}}{C \text{ fed to the reactor}} \times 100. \quad (5)$$

Other parameter calculated was the turn-over frequency for hydrogen production (TOF_{H_2}), in units of mols of H_2 per unit time and surface metal site. For calculating the rates of H_2 formation expressed as TOF_{H_2}, the number of surface metal sites was assumed to be equal to the amount of CO molecules irreversibly chemisorbed at 300 K. This is also the treatment used by Davda et al. [19].

$$TOF_{H_2} = \frac{H_2 \text{ in the gas product}}{CO \text{ chemisorbed on surface metal atoms} \times \min}. \quad (6)$$

2.4. Deactivation Study. The stability of the APR catalytic system is limited by three main factors: fouling by coke, sintering, and leaching of the metallic phase. The deactivation by coking was assessed by temperature programmed oxidation (TPO) of the coke deposit of the spent catalysts. An amount of 0.05 g of deactivated catalyst was first loaded into a quartz reactor, and then an oxidizing gas mixture (2% O_2 in N_2, molar basis, 30 $cm^3 \cdot min^{-1}$ flow rate) was forced through the sample. The reactor was heated at a rate of 10 $K \cdot min^{-1}$ from room temperature up to 973 K. In the presence of oxygen, the deposits were combusted to CO and CO_2. Both gases were transformed into methane over a Ni/kieselguhr catalyst in a methanation reactor and then sent to a flame ionization detector. The TPO trace was thus obtained by registering the FID voltage as a function of the cell temperature. The area under the signal is proportional to the amount of deposited coke. The sintering of the metal phase affects directly the WGS reaction, and it was indirectly assessed by the hydrogen yield obtained at a fixed reaction time. The leaching of active sites was determined by chemical analysis of the used catalyst at the end of each experiment. The analysis was performed by inductively coupled plasma with mass spectroscopy detection (ICP-MS).

To compare the stability of the catalysts, the fouling deactivation rate (after a stabilization period of 3 h) and leaching deactivation rate were calculated as follows:

$$r_{df}\left(\frac{g_{coke}}{100g_{cat} \cdot h}\right) = \frac{\%\left([Coke_T]_{t_f} - [Coke_T]_{t_s}\right)}{(t_f - t_s)},$$

$$r_{dl}\left(\frac{g_{metal\ leached}}{100g_{cat} \cdot h}\right) = \frac{\%\left([Metal_T]_{t_i} - [Metal_T]_{t_f}\right)}{(t_f - t_i)}, \quad (7)$$

where $\%[Coke_T]_t$ is the percentage of accumulated coke on catalyst and $\%[Metal_T]_t$ is the percentage of total metal charge on catalyst after t hours on-stream. t_i is the initial reaction time, 0 h; t_s is the stabilization time, 3 h; and t_f is the final reaction time.

2.5. Literature Data Comparison. The technology of aqueous phase reforming (APR) has been specially developed by Randy Cortright and James Dumesic of the University of Wisconsin [20]. Their first pioneering works indicated that it was possible to convert carbohydrates (ethylene glycol, sorbitol, fructose, etc.) in aqueous solution to hydrogen and nonoxygenated hydrocarbons by means of suitable heterogeneous catalysts and relatively mild reaction conditions. These papers were followed by intensive work by other groups. Most catalysts studied had noble metals, for example, Pt, in high loads.

In order to make a comparison, a literature survey was made and data collected related to APR over Pt/Al_2O_3 catalysts. Catalysts with a similar or higher loading of metal phase were considered (in comparison with the catalysts of this work). Data were also collected corresponding to APR catalysts that used similar metal contents but different supports.

3. Results and Discussion

3.1. Catalysts Characterization. Table 1 shows some properties of the metal function of the catalysts: chemical composition, CO chemisorption capacity, reducibility (hydrogen consumption in the TPR experiment), and activity in cyclohexane dehydrogenation (CHD). This table also shows values of specific area of the catalysts after their final activation.

It can be seen that the Ni, Co, and Pt mass contents are similar to the nominal contents. The metal charge does not substantially modify the specific surface of the catalysts. Nitrogen adsorption measurements indicate that the BET surface area of the catalysts varies no more than 8% in comparison to the original value of the metal-free support. The metal addition step is then supposed to occur without blocking of the pore mouths of the catalysts. This points out to a high efficiency of both methods (UMC and IWI) for loading the metals to the support, with negligible loss of metal mass or support-specific area. The results of chemisorption of CO have double importance. First, the CO chemisorption capacity is proportional to the dispersion of the metal. Secondly, CO adsorption is the first step for the water gas shift reaction that results in an increased production of hydrogen. The results indicate that the monometallic Pt catalysts, prepared either by IWI or UMC, have the highest capacity for CO chemisorption, that is, the highest metal particle dispersion. The CO chemisorption capacity of the bimetallic PtNi and PtCo catalysts is lower than the capacity of the monometallic Pt catalysts. The order of CO chemisorption capacity of the catalysts is Pt > PtCo > PtNi. These results coincide with the reports of Ko et al. [21] which indicate that the addition of Ni to Pt/Al_2O_3 catalysts decreases the number of accessible sites for CO adsorption, though it improves the activity for preferential CO oxidation in the presence of H_2. The CO chemisorption capacity of the NiCo-UMC catalysts is similar to that of the monometallic Pt catalysts, but the metal concentration of the latter is six times greater. On the contrast, if we observe the effect of the technique of incorporation of the metal, it is confirmed that in the case of the bimetallic catalysts prepared by UMC, an increase of dispersion of 40 and 45% (capacity of CO chemisorption) is obtained. The dehydrogenation capacity of each catalyst, as

TABLE 1: Chemical composition, CO chemisorption, H_2 consumption in TPR test, and specific surface area (Sg).

Catalyst	Pt (%)	Co (%)	Ni (%)	$\mu mol\ CO/g_{cat}$	$\mu mol\ H_2/g_{cat}$	Sg (m^2/g)	Conversion$_{CH\ to\ Bz}$ (%)
Pt-IWI	0.98	—	—	48.7	335.1	215	93.4
Pt-UMC	0.99	—	—	49.4	460.8	210	95.4
PtCo-IWI	1.02	3.05	—	25.0	496.8	193	67.3
PtCo-UMC	0.97	2.87	—	34.0	846.1	201	79.1
PtNi-IWI	1.01	—	3.10	18.6	668.4	200	52.0
PtNi-UMC	0.99	—	2.96	27.1	803.9	206	79.5
NiCo-IWI	—	3.05	2.95	35.2	755.4	197	20.3
NiCo-UMC	—	2.99	2.87	50.9	993.1	192	34.1

Cyclohexane to benzene conversion (Conversion$_{CH\ to\ Bz}$) after 1 h on-stream in a dehydrogenation test.

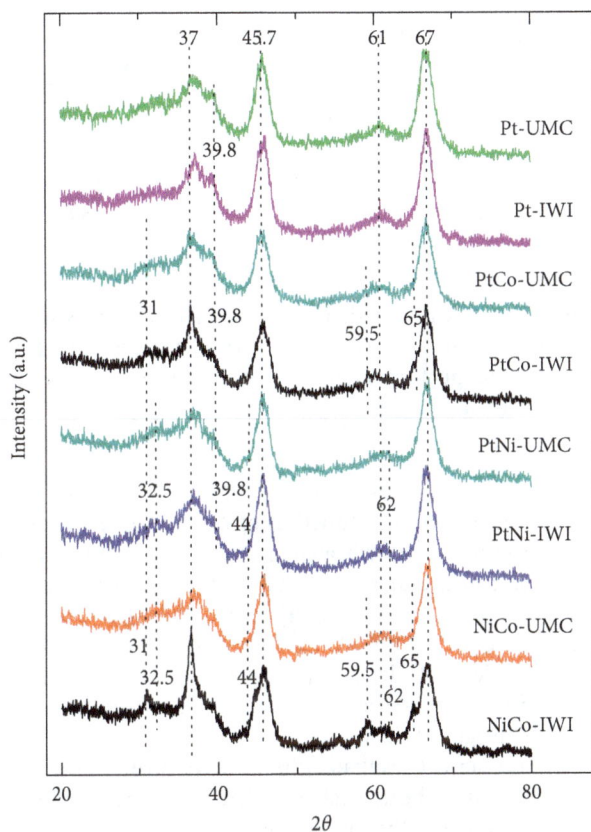

FIGURE 1: X-ray diffractograms of the studied catalysts.

indicated by the conversion of CH to benzene, is proportional to its CO adsorption capacity. However, its increase is higher for the catalysts containing Pt. For catalysts with the same metal load, the higher the amount of CO chemisorbed, the higher the conversion of cyclohexane. Finally Table 1 also shows values of the consumption of H_2 during the temperature-programmed reduction tests. Bimetallic catalysts have a higher consumption than the monometallic Pt, indicating the coreduction of Ni or Co atoms together or with Pt in each case.

The X-ray powder diffraction patterns of the prepared catalysts are shown in Figure 1.

We can see that all catalysts exhibited the peaks of gamma alumina at 2θ angles of 37°; 39.2°; 45.7°; 61°, and 67° (Joint Committee on Powder Diffraction Standards (JCPDS)) [22]. Low metal loads make it difficult to see bigger differences. If

we expand the scale, we can observe that all catalysts that contain Pt show differences at angles of 39.8°, a peak associated to metallic (zero-valent) platinum [23]. These modifications are more remarkable for the catalysts prepared by IWI than those prepared by UMC, indicating a higher size of the metal crystals, that is, a smaller dispersion. For the NiCo-UMC sample, with 3% Ni and 3% Co, small signals of NiO and $NiAl_2O_4$ are observed at 32.5°, 44°, and 62°, and there are no peaks indicating the presence of cobalt oxides [24]. This points to cobalt being in high dispersion or in strong interaction with Ni or with the support. The same catalyst prepared by IWI, NiCo-IWI, shows new peaks at 31°, 59.5°, and 65°, corresponding to Co_3O_4 and $CoAl_2O_4$. The same occurs in the case of the PtCo-IWI catalyst. In addition to the peak corresponding to Pt^0, other signals appear that correspond to cobalt oxides and cobalt aluminate [24]. The XRD spectrum is an evidence of the high degree of dispersion, the strong interaction between metals, or the strong metal-support interaction, in the catalysts prepared by the UMC technique.

TPR traces of the tested catalysts can be seen in Figure 2. The Pt-IWI catalyst has peaks at 500 K and at 680 K while the Pt-UMC catalyst has peaks at 435 K, 500 K, and 680 K. Peaks at temperatures over 640 K are related to Pt species with strong support interaction, like Pt_xAlO_x or Pt_3Al [25]. On the contrast, the 435 K peak of the Pt-UMC catalyst could be related to PtO_x species of low interaction with the support surface [26]. In the case of the PtCo-IWI and PtCo-UMC catalysts, peaks at temperatures below 475 K can be found that could be related to the reduction of Pt species. The shift of the peaks in comparison to the reduction of monometallic Pt could be due to the interference of Co ions. Hydrogen consumption in the region around 300–773 K has been attributed to the reduction of cobalt oxide, as Co_3O_4, along with the reduction of Co^{+3} surface ions [27]. The increase of the size of the first Pt reduction peak could thus be due to the coreduction of Co and Pt species. Above 570 K, the H_2 consumption is related to the reduction of cobalt oxide (Co_3O_4), which involves the reduction of Co_3O_4 to CoO occurring at around 609–640 K. The shift to lower temperatures could be related to H_2 spillover over Pt. Neither PtCo-IWI nor PtCo-UMC has peaks above 1100 K, and a peak appears at 778 K. This means that there are Co species with very high support interaction [28, 29]. As in the case of the monometallic Pt catalyst, the bimetallic catalyst prepared by UMC had a shift of the Pt reduction peak to lower temperatures, indicating the presence of smaller particles.

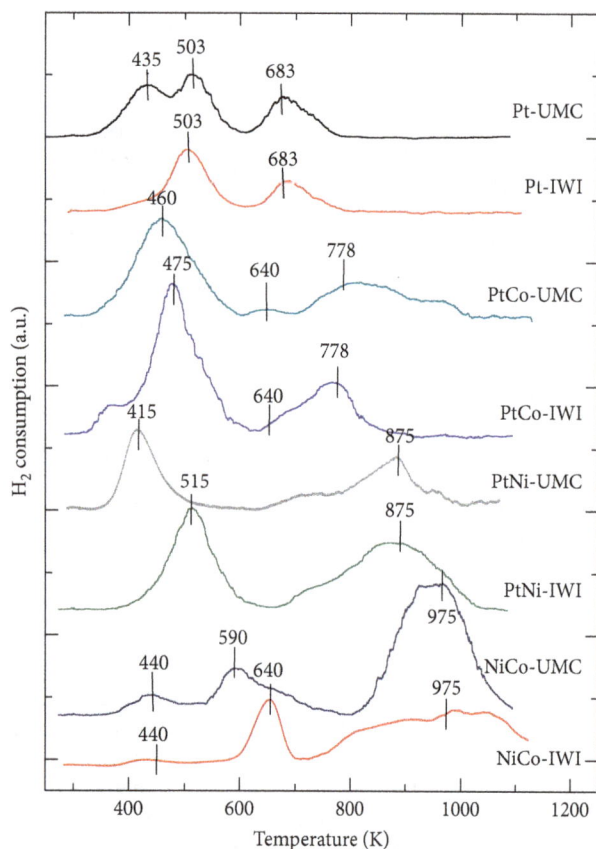

FIGURE 2: TPR traces of the tested catalysts.

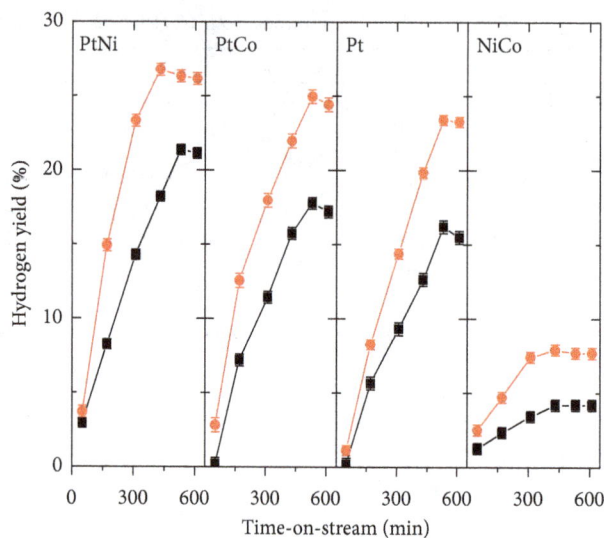

FIGURE 3: Hydrogen yield of the different catalysts. APR of ethylene glycol (10% aqueous solution). Fixed bed reactor, 498 K 22 bar, WHSV = 2.34 h^{-1}, LHSV = 0.86 h^{-1}. • UMC; ■ IWI.

For the PtNi-IWI and PtNi-UMC catalysts, two different peaks are seen. The first one, located at 415–515 K could be related to the reduction of PtOx species interacting with the oxygen atoms of Al_2O_3 and NiO, reduced by the spillover effect of H_2 over Pt. In the catalyst prepared by UMC, a shift to lower temperatures in the reduction peak of Pt species is found. On the contrast, the peak at 875 K is related to the reduction of nickel species. This could be NiO weakly interacting with the Al_2O_3 support ($T < 820$ K) or nickel aluminates, also reduced by the dissociation of H_2 to H• on Pt that penetrated into alumina and reduced more nickel at lower temperatures [30]. The lack of reduction peaks at temperatures above 1000 K could indicate the absence of Ni species with strong interaction with the support as a consequence of the promotion of the reduction of nickel oxides through hydrogen spillover on Pt [31].

The TPR traces of NiCo-IWI and NiCo-UMC catalysts had peaks at 440 K, 590–640 K, and 975 K. Reduction peaks at temperatures below 773 K are related to the reduction of nickel and cobalt ions to metallic Ni and Co. It is suggested by Nabgan et al. [32] that the 975 K peak corresponds to the reduction of a Ni-Co phase in high interaction with the Al_2O_3 support. In the catalyst obtained by UMC, a shift of the intermediate reduction peak is again found.

Summarizing the characterization results, we can observe that for the same metal loading, the UMC technique permits obtaining catalysts with a higher CO-chemisorption capacity, with more dispersed metal particles and with a greater specific area. A higher metal activity for

dehydrogenation was also obtained. These results would confirm the possibility of obtaining catalysts of similar catalytic performance but with lower noble metal concentration.

3.2. Catalytic Properties in APR of Ethylene Glycol. A plot of the hydrogen yield as a function of time-on-stream can be seen in Figure 3. The hydrogen production has an induction period of 6–8 h, and then it stabilizes to almost constant values. The order of hydrogen yield is PtNi-UMC > PtCo-UMC ≈ Pt-UMC > PtNi-IWI > PtCo-IWI ≈ Pt-IWI > NiCo-UMC > NiCo-IWI. It can be seen that the Pt catalyst prepared by the classical IWI technique displays one of the lowest hydrogen yield values.

The numerical results of the catalytic properties at the end of the run indicate (Table 2) that the highest hydrogen selectivity is that of the monometallic Pt catalysts. However, the catalytic conversion on these catalysts is not so high and therefore the hydrogen yield is not the highest.

The total time-on-stream of the APR tests was 10 h. On average PtNi-UMC, PtCo-UMC, and Pt-UMC had higher H_2 yields and TOF_{H2} than Pt-IWI. The catalyst PtNi-UMC had the best hydrogen yield. PtCo-UMC had better yields up to 400 min of time-on-stream, its activity level then becoming similar to that of Pt-UMC. All catalysts show growing hydrogen yields along the run, then stabilizing after 500 min on-stream. This is coincident with the report of Luo et al. [24]. Their catalysts had a stable activity after 10 h on-stream. Afterwards a drop in conversion was registered.

It can also be seen that once the catalysts reach a pseudo steady state, the hydrogen yield values become higher than that of the Pt-IWI catalyst. The difference is 50% for Pt-UMC and PtCo-UMC and 75% for PtNi-UMC. Also for an equal yield to hydrogen, the PtCo-UMC catalyst has a lower selectivity to methane than the Pt-UMC catalyst. Bimetallic PtNi catalysts are the most active and show the highest

TABLE 2: Results of APR of aqueous ethylene glycol (10% w).

Catalyst	Conversion$_{C\,to\,gas}$, (%)	Selectivity (%)		H$_2$ yield (%)	TOF$_{H_2}$ (min^{-1})
		CH$_4$	H$_2$		
Pt-IWI	42.9	2.0	35.4	15.2	1.01
Pt-UMC	65.0	1.7	36.2	23.5	1.53
PtCo-IWI	64.0	0.7	26.6	17.0	2.19
PtCo-UMC	86.0	0.6	27.9	24.0	2.27
PtNi-IWI	80.0	6.7	26.3	21.0	3.13
PtNi-UMC	86.0	4.6	30.9	26.6	3.16
NiCo-IWI	24.0	11.4	16.3	3.9	0.36
NiCo-UMC	48.0	12.7	16.7	8.0	0.51

Results at 10 h on-stream. Reaction conditions: 498 K, 22 bar, WHSV = 2.34 h^{-1}, and LHSV = 0.86 h^{-1}.

TABLE 3: Stability of the catalysts in APR of aqueous ethylene glycol (10%w).

Catalyst	Metal$_T$]$_t$ (%)		$10^4 \times r_{dl}$ ($g_{metal}/g_{cat} \cdot h$)	Coke$_T$]$_t$ (%)		$10^4 \times r_{df}$ ($g_{coke}/g_{cat} \cdot h$)
	0 h	10 h		3 h	10 h	
Pt-IWI	0.98	0.97	0.1	1.3	1.9	8.6
Pt-UMC	0.99	0.98	0.1	0.9	1.2	4.3
PtCo-IWI	4.07	4.01	0.6	1.2	1.4	2.8
PtCo-UMC	3.84	3.82	0.2	1.1	1.2	1.4
PtNi-IWI	4.11	4.03	0.8	2.8	3.6	11.4
PtNi-UMC	3.95	3.90	0.5	2.8	3.4	8.6
NiCo-IWI	6.00	5.92	0.8	2.4	3.4	14.3
NiCo-UMC	5.86	5.83	0.3	2.5	2.7	2.9

Reaction conditions: 498 K, 22 bar, WHSV = 2.34 h^{-1}, LHSV = 0.86 h^{-1}, and TOS = 10 (h); Metal$_T$]$_t$: total metal content on coke-free catalyst at (t) hour on-stream; Coke$_T$]$_t$: coke deposited on the catalyst at (t) hour on-stream; r_{dl}: leaching deactivation rate; r_{df}: fouling deactivation rate.

values of TOF$_{H2}$. However, their selectivity to methane is high, that is, they consume part of the H$_2$ produced.

Table 3 shows the percentages of total metal deposited in each catalyst before and after a period of 10 h of reaction and the amounts of carbon deposited on the catalysts after 3 h and 10 h on-stream. From these data, the rates of deactivation by leaching of the metal phase and fouling by coke are calculated.

Regarding leaching, although the variations of the results are within the experimental error of the analysis technique, there is a trend that confirms a stronger anchorage of the metallic particles in the catalysts prepared by UMC with respect to those obtained by the conventional IWI technique. Regarding fouling, the Pt-UMC and the PtCo-UMC catalysts had the lowest coke content. Coke content was below 2% after 10 h on-stream. Pt-UMC had the lowest amount of coke at 3 h on-stream. The Ni-containing catalysts had the highest amount of coke at short and long reaction times. For any of the catalysts, not less than 60% of the total coke was formed in the first 3 h of reaction. This indicates that after the catalyst is stabilized, the coking reactions become slower, probably because of deactivation of the sites of higher coking activity.

After 3 h of reaction, the coke on the metal is stabilized, and the deactivation rate due to fouling is minimal in the samples promoted with cobalt, both in PtCo and in NiCo. The fouling rate increases in the samples with nickel. This trend is more pronounced in the catalysts prepared by UMC.

Figure 4 shows the coke oxidation profiles of the catalysts after 10 h on-stream in the APR test.

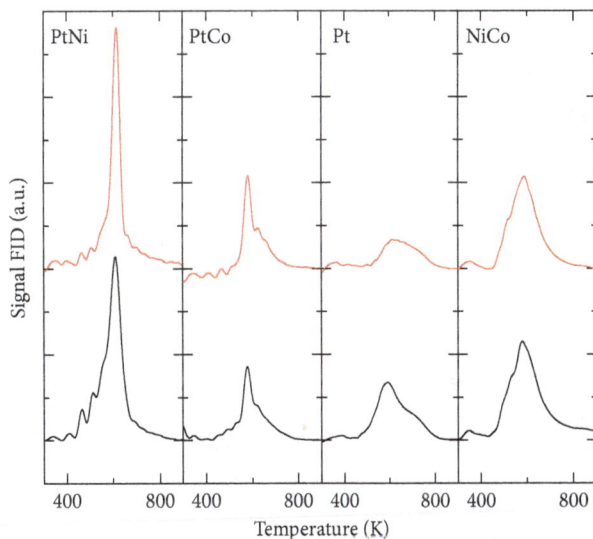

FIGURE 4: Temperature-programmed oxidation traces of different catalysts. Catalyst coked during a test of 10 h (APR of aqueous ethylene glycol, 10%). – UMC; – IWI.

It can be seen in all cases that coke is burned at temperatures lower than 873 K. The incorporation of Ni into Pt generates a more hydrogenated deposit giving a very sharp coke burning peak. Despite having accumulated the largest amount of coke, its steady-state activity is the largest. The incorporation of Co to Pt reduces the amount of coke and reduces the degree of polymerization, coke being eliminated in a lower temperature range. If we compare the coke

TABLE 4: Results of APR of ethylene glycol over different catalysts.

Catalyst	EG (% w)	T (K)	P (bar)	WHSV (h⁻¹)	LHSV (h⁻¹)	Conversion C to gas (%)	Selectivity (%)		H₂ yield (%)	TOF$_{H_2}$ (min⁻¹)	Reference
							C_5^-	H₂			
Pt(3)/Al$_2$O$_3$	1	498	29.3		0.6	90.0	4.0	96.0	86.4	5.3	Shabaker et al. [11]
R-Ni(14)Sn	1	498	25.8		5.1	93.0	4.0	95.0	88.4	1.4	Shabaker et al. [11]
Pt(1)	10	498	29.3		72.0	5.4	1.2	87.1	4.7	1.9	Huber et al. [12]
Pt(1)Ni(1)	10	498	29.3		70.2	5.9	0.0	91.0	5.4	5.2	Huber et al. [12]
Pt(1)Ni(5)	10	498	29.3		138.5	3.6	0.9	89.5	3.2	3.0	Huber et al. [12]
Pt(1)Ni(8)	10	498	29.3		128.6	3.5	2.0	90.2	3.2	2.8	Huber et al. [12]
Pt(1)Co(5)	10	498	29.3		57.4	8.4	0.5	88.2	7.4	5.1	Huber et al. [12]
Pt(1.5)/Al$_2$O$_3$	1	548	200.0	1.2		78	0.3				De Vlieger et al. [33]
Pt(1.5)/Al$_2$O$_3$	20	723	250.0	12		74	16.9				De Vlieger et al. [33]
Ni(19)/SiO2	10	498	22.0		n/r	2.3	13.0	57.0	1.3	14.0	Davda et al. [19]
Pd(5)/SiO$_2$	10	498	22.0		n/r	3.1	0.0	98.5	3.0	30.0	Davda et al. [19]
Pt(6)/SiO$_2$	10	498	22.0		n/r	21.0	13.0	77.9	16.4	275.0	Davda et al. [19]
Ru(6)/SiO$_2$	10	498	22.0		n/r	42.0	58.0	7.0	2.9	20.0	Davda et al. [19]
Pt(1)/CMK-3	10	523	45.6	2.0		25.4	4.5	n/r	26.6	103	Kim et al. [38]
Pt(3)/CMK-3	10	523	45.6	2.0		46.0	n/r	n/r	49.3	46.4	Kim et al. [36]
Pt(7)/CMK-3	10	523	45.6	2.0		69.8	7.1	n/r	72.1	31.2	Kim et al. [35]
Pt(3) CMK-9	10	523	45.6	2.0		44.9	4.2	n/r	49.3	n/r	Kim et al. [37]
Fe(3) CMK-9	10	523	45.6	2.0		12.1	8.9	n/r	27.1	n/r	Kim et al. [37]
Pt(3)Fe(3) (1:1)	10	523	45.6	2.0		60.8	3.3	n/r	64.9	n/r	Kim et al. [37]
Pt(3)Fe(6) (1:2)	10	523	45.6	2.0		63.2	3.1	n/r	66.7	n/r	Kim et al. [37]
Pt(3)Fe(9) (1:3)	10	523	45.6	2.0		68.4	2.8	n/r	71.1	~70-80	Kim et al. [37]
Pt(3)Fe(12) (1:4)	10	523	45.6	2.0		60.1	2.6	n/r	70.7	n/r	Kim et al. [37]
Pt(3)Fe(15) (1:5)	10	523	45.6	2.0		58.9	2.7	n/r	68.8	n/r	Kim et al. [37]

Steady state conditions, fixed bed reactor.

deposits obtained on catalysts of the same composition but synthesized by different methods, a reduction of the coke content on the UMC catalysts can be seen. The degree of metal dispersion obtained by each preparation route evidently modifies the stability of the catalyst.

3.3. Comparison of These Results with Those Obtained by Other Research Groups. In 2002, the group of Professor Dumesic of the University of Wisconsin developed a catalytic process that generated H_2 from the aqueous-phase reforming (APR) of biomass-derived oxygenated compounds such as methanol, ethylene glycol, glycerol, sugars, and sugar-alcohols [20]. Davda et al. [19] tested some Group VIII metals for the APR of 10% ethylene glycol at 483 and 498 K and at a total pressure of 22 bar. They found that the overall catalytic activity for ethylene glycol reforming, as measured by the rate of CO_2 production at 483 K, decreased in the following order for silica-supported metals: Pt~Ni > Ru > Rh~Pd > Ir. Silica supported Rh, Ru, and Ni showed a low selectivity for production of H_2 and a high selectivity for alkane production. In addition, Ni/SiO$_2$ showed significant deactivation at the higher temperature of 498 K. Silica-supported Pt and Pd catalysts exhibited relatively high selectivities for production of H_2, with low rates of alkane production. It seemed evident that catalysts based on Pt and Pd were promising materials for the selective production of hydrogen.

We can see in Table 4 a summary of results of the APR of solutions of ethylene glycol of varying concentration, at varying temperature, pressure, and flow rate conditions, as published by different authors.

Shabaker et al. [11] reported results of the aqueous phase reforming of solutions of 1% ethylene glycol using an industrial Raney NiSn catalyst and a Pt/Al$_2$O$_3$ catalyst with 3% Pt. While conversion and yield to hydrogen were quite high, the selectivity to alkanes was too high, almost 4% for both catalysts. Huber and Dumesic [12] reported results corresponding to a catalyst of 3% Pt on alumina for the APR of aqueous ethylene glycol (10%). Included in the table are also the results corresponding to bimetallic PtNi and PtCo catalysts and metal atomic ratios 1:1, 1:5, and 1:8 (Pt/second metal). None of the catalysts had a conversion above 10% even after reaching the steady state. Supported Pt catalysts over alumina had better results when the reaction conditions were modified [33]. Table 4 shows the conversion results of APR of ethylene glycol over a Pt/Al$_2$O$_3$ catalyst with 1.5% Pt. Although the conversion to gas was high, it was necessary to increase the reaction pressure by one order of magnitude. This involves a rise of operation costs. Results for other metals and supports can be inspected in Table 4 [19]. It can be seen that Ni and Pd loaded catalysts have low conversion values when supported on silica, even at high metal loadings, while Pt, Ru, and Rh give high values of conversion but a poor selectivity to hydrogen. This combines with a high selectivity to alkanes to produce a final high selectivity to alkanes and a low hydrogen yield.

Recent studies on Pt supported over carbon nanotubes for the aqueous phase reforming of ethylene glycol showed that the activity can be greatly improved by changing the support to a more convenient one [34, 35]. However, increasing the catalyst activity threefold demands increasing the metal concentration seven times. We can see in Table 4 that though the activity level for these novel catalysts is promising, the required metal load is inconveniently high. A considerable reduction in the Pt charge is obtained by adding Re to the carbon support, but the noble metal content remains high [36].

Finally, with a similar support, Kim et al. [37] developed new catalysts with 3 wt.% Pt, modified with Fe in different atomic ratios (Table 4). The catalysts are an improvement in relation to the previously discussed examples. They have a superior conversion capacity and higher H_2 yield. However, the authors did not report the coke content of catalyst after the experiments. This report had very good results though the Pt content (3 wt.%) is higher than ours. It is also important to note that Fe greatly improves the conversion and H_2 yield, in a similar way as Ni and Co did in the present work.

4. Conclusions

At the APR reaction, conditions of this work Pt/Al_2O_3 catalysts prepared by incipient wetness impregnation (IWI) show a hydrogen yield of about 15.2%. The addition of 3% Ni or Co and the use of a different method of metal addition (urea matrix combustion, UMC) produce catalysts that have a higher yield to hydrogen and a higher stability and a lower selectivity to coke and methane. Particularly, the addition of Co by UMC increases the hydrogen yield by 50% in relation to the standard Pt/Al_2O_3 catalyst, while keeping similar hydrogen selectivity and a much lower selectivity to methane. The addition of Ni by UMC increases the hydrogen yield by 75% in comparison to the standard Pt/Al_2O_3 catalyst but increases both the methanation and coking activities.

The simultaneous use of the UMC method and Ni and Co promotion enabled the synthesis of APR catalysts that had a lower metal content than those reported in the literature, but a similar or better activity level.

The decrease of the selectivity to the undesired products such as coke and methane enabled an increase of the conversion by a factor of two and of the hydrogen yield by a factor of 1.5.

A comparison with published results indicates that catalysts of other authors obtained similar or better results than ours, but with the use of higher metal contents or more expensive supports. In some cases, their experiments were carried out with diluted solutions of ethylene glycol or higher reaction temperatures and pressures, conditions that are associated with higher operation costs.

In summary, APR catalyst of low noble metal content and hence of low cost can be obtained by using the UMC method to incorporate Ni or Co promoters to alumina supported Pt catalysts. These catalysts show promising catalytic properties for the improved production of H_2 by APR of biomass related feedstocks.

Conflicts of Interest

There are no conflicts to declare.

Acknowledgments

The authors thank Sasol for the donation of alumina used in this paper. The authors also thank the financial support of CONICET (PIP Grant 2014-560), Universidad Nacional delLitoral (CAI+D Grant 2016-084), and ANPCyT (PICT Grant 2013–3217).

References

[1] G. J. Antos, A. M. Aitani, and J. M. Parera, *Catalytic Naphtha Reforming Science and Technology*, Marcel Dekker, Inc., New York, NY, USA, 1995.

[2] M. W. Twigg, *Catalyst Handbook*, Wolfe Publishing Ltd., London, UK, 1989.

[3] X. Zhang, Q. Zhang, T. Wang, B. Li, Y. Xu, and L. Ma, "Efficient upgrading process for production of low quality fuel from bio-oil," *Fuel*, vol. 179, pp. 312–321, 2016.

[4] X. Zhang, Q. Zhang, T. Wang, L. Ma, Y. Yu, and L. Chen, "Hydrodeoxygenation of lignin-derived phenolic compounds to hydrocarbons over Ni/SiO2–ZrO2 catalysts," *Bioresource Technology*, vol. 134, pp. 73–80, 2013.

[5] X. Zhang, T. Wang, L. Ma, Q. Zhang, X. Huang, and Y. Yu, "Production of cyclohexane from lignin degradation compounds over Ni/ZrO2–SiO2 catalysts," *Applied Energy*, vol. 112, pp. 533–538, 2013.

[6] D. J. Wuebbles and A. K. Jain, "Concerns about climate change and the role of fossil fuel use," *Fuel Processing Technology*, vol. 71, pp. 99–119, 2001.

[7] T. Nozawa, A. Yoshida, S. Hikichi, and S. Naito, "Effects of Re addition upon aqueous phase reforming of ethanol over TiO2 supported Rh and Ir catalysts," *International Journal of Hydrogen Energy*, vol. 40, pp. 4129–4140, 2015.

[8] G. W. Huber and J. A. Dumesic, "An overview of aqueous-phase catalytic processes for production of hydrogen and alkanes in a biorefinery," *Catalysis Today*, vol. 111, pp. 119–132, 2006.

[9] I. Coronado, M. Stekrova, M. Reinikainen, P. Simell, L. Lefferts, and J. Lehtonen, "A review of catalytic aqueous-phase reforming of oxygenated hydrocarbons derived from biorefinery water fractions," *International Journal of Hydrogen Energy*, vol. 41, pp. 11003–11032, 2016.

[10] R. R. Davda, J. W. Shabaker, G. W. Huber, R. D. Cortright, and J. A. Dumesic, "A review of catalytic issues and process conditions for renewable hydrogen and alkanes by aqueous-phase reforming of oxygenated hydrocarbons over supported metal catalysts," *Applied Catalysis B: Environmental*, vol. 56, pp. 171–186, 2005.

[11] J. W. Shabaker, G. W. Huber, and J. A. Dumesic, "Aqueous-phase reforming of oxygenated hydrocarbons over Sn-modified Ni catalysts," *Journal of Catalysis*, vol. 222, pp. 180–191, 2004.

[12] G. W. Huber, J. W. Shabaker, S. T. Evans, and J. A. Dumesic, "Aqueous-phase reforming of ethylene glycol over supported Pt and Pd bimetallic catalysts," *Applied Catalysis B: Environmental*, vol. 62, pp. 226–235, 2006.

[13] A. Martinez and G. Prieto, "Breaking the dispersion-reducibility dependence in oxide-supported cobalt nanoparticles," *Journal of Catalysis*, vol. 245, no. 2, pp. 470–476, 2007.

[14] S. L. González-Cortés, T. Xiao, and M. Green, *Scientific Bases for the Preparation of Heterogeneous Catalysts*, Elsevier, Oxford, UK, 1st edition, 2006.

[15] S. L. González-Cortés and F. E. Imbert, "Fundamentals, properties and applications of solid catalysts prepared by solution combustion synthesis (SCS)," *Applied Catalysis A: General*, vol. 452, pp. 117–131, 2013.

[16] G. Xanthopoulou and G. Vekinis, "Investigation of catalytic oxidation of carbon monoxide over a Cu–Cr-oxide catalyst made by self-propagating high-temperature synthesis," *Applied Catalysis B: Environmental*, vol. 19, pp. 37–44, 1998.

[17] A. Varma, A. S. Mukasyan, A. S. Rogachev, and K. V. Manukyan, "Solution combustion synthesis of nanoscale materials," *Chemical Reviews*, vol. 116, no. 23, pp. 14493–14586, 2016.

[18] M. Boudart, A. Aldag, J. E. Benson, N. A. Dougharty, and C. G. Harkins, "On the specific activity of platinum catalysts," *Journal of Catalysis*, vol. 6, no. 1, pp. 92–99, 1966.

[19] R. R. Davda, J. W. Shabaker, G. W. Huber, R. D. Cortright, and J. A. Dumesic, "Aqueous-phase reforming of ethylene glycol on silica-supported metal catalysts," *Applied Catalysis B: Environmental*, vol. 43, no. 1, pp. 13–26, 2003.

[20] R. D. Cortright, R. R. Davda, and J. A. Dumesic, "Hydrogen from catalytic reforming of biomass-derived hydrocarbons in liquid water," *Nature*, vol. 418, no. 6901, pp. 964–967, 2002.

[21] E. Ko, E. Park, K. W. Seo, H. C. Lee, D. Lee, and S. Kim, "Pt–Ni/γ-Al$_2$O$_3$ catalyst for the preferential CO oxidation in the hydrogen stream," *Catalysis Letters*, vol. 110, pp. 275–279, 2006.

[22] Joint Committee on Powder Diffraction Standards (JCPDS) file No. 86–1410.

[23] K. Persson, A. Ersson, S. Colussi, A. Trovarelli, and S. G. Järas, "Catalytic combustion of methane over bimetallic Pd–Pt catalysts: the influence of support materials," *Applied Catalysis B: Environmental*, vol. 66, no. 3-4, pp. 175–185, 2006.

[24] N. Luo, K. Ouyang, F. Cao, and T. Xiao, "Hydrogen generation from liquid reforming of glycerin over Ni–Co bimetallic catalyst," *Biomass Bioenergy*, vol. 34, pp. 489–495, 2010.

[25] M. C. Rangel, L. S. Carvalho, P. Reyes, J. M. Parera, and N. S. Fígoli, "*n*-octane reforming over alumina-supported Pt, Pt–Sn and Pt–W catalysts," *Catalysis Letters*, vol. 64, pp. 171–178, 2000.

[26] C. Hwang and C. Yeh, "Platinum-oxide species formed by oxidation of platinum crystallites supported on alumina," *Journal of Molecular Catalysis A: Chemical*, vol. 112, no. 2, pp. 295–302, 1996.

[27] L. F. Liotta, G. Pantaleo, A. Macaluso, G. Di Carlo, and G. Deganello, "CoO$_x$ catalysts supported on alumina and alumina-baria: influence of the support on the cobalt species and their activity in NO reduction by C3H6 in lean conditions," *Applied Catalysis A: General*, vol. 245, no. 1, pp. 167–177, 2003.

[28] S. Rane, Ø. Borg, J. Yang, E. Rytter, and A. Holmen, "Effect of alumina phases on hydrocarbon selectivity in Fischer–Tropsch synthesis," *Applied Catalysis A: General*, vol. 388, pp. 160–167, 2010.

[29] D. Nabaho, J. W. Niemantsverdriet, M. Claeys, and E. van Steen, "Hydrogen spillover in the Fischer–Tropsch synthesis: an analysis of platinum as a promoter for cobalt–alumina catalysts," *Catalysis Today*, vol. 261, pp. 17–27, 2016.

[30] M. El Doukkali, A. Iriondo, P. L. Arias et al., "A comparison of sol–gel and impregnated Pt or/and Ni based γ-alumina catalysts for bioglycerol aqueous phase reforming," *Applied Catalysis B: Environmental*, vol. 125, pp. 516–529, 2012.

[31] J. Li, W. P. Tian, X. Wang, and L. Shi, "Nickel and nickel-platinum as active and selective catalyst for the maleic anhydride hydrogenation to succinic anhydride," *Chemical Engineering Journal*, vol. 175, pp. 417–422, 2011.

[32] W. Nabgan, T. Amran, T. Abdullah et al., "Hydrogen production from catalytic steam reforming of phenol with bimetallic nickel-cobalt catalyst on various supports," *Applied Catalysis A: General*, vol. 527, pp. 161–170, 2016.

[33] D. J. M. de Vlieger, B. L. Mojet, L. Lefferts, and K. Seshan, "Aqueous Phase Reforming of ethylene glycol – Role of intermediates in catalyst performance," *Journal of Catalysis*, vol. 292, pp. 239–245, 2012.

[34] H. D. Kim, H. J. Park, T. W. Kim et al., "Hydrogen production through the aqueous phase reforming of ethylene glycol over supported Pt-based bimetallic catalysts," *International Journal of Hydrogen Energy*, vol. 37, pp. 8310–8317, 2012.

[35] H. D. Kim, T. W. Kim, H. J. Park et al., "Hydrogen production via the aqueous phase reforming of ethylene glycol over platinum-supported ordered mesoporous carbon catalysts: effect of structure and framework-configuration," *International Journal of Hydrogen Energy*, vol. 37, pp. 12187–2197, 2012.

[36] H. D. Kim, H. J. Park, T. W. Kim et al., "The effect of support and reaction conditions on aqueous phase reforming of polyol over supported Pt-Re bimetallic catalysts," *Catalysis Today*, vol. 185, pp. 73–80, 2012.

[37] M. C. Kim, T. W. Kim, H. J. Kim, C. U. Kim, and J. W. Bae, "Aqueous phase reforming of polyols for hydrogen production using supported PtFe bimetallic catalysts," *Renewable Energy*, vol. 95, pp. 396–403, 2016.

[38] T. W. Kim, H. D. Kim, K. E. Jeong et al., "Catalytic production of hydrogen through aqueous-phase reforming over platinum/ordered mesoporous carbon catalysts," *Green Chemistry*, vol. 13, pp. 1718–1728, 2011.

2

Converting a Microwave Oven into a Plasma Reactor

Victor J. Law and **Denis P. Dowling**

School of Mechanical and Materials Engineering, University College Dublin, Belfield D04 V1W8, Dublin 4, Ireland

Correspondence should be addressed to Victor J. Law; viclaw66@gmail.com

Academic Editor: Michael Harris

This paper reviews the use of domestic microwave ovens as plasma reactors for applications ranging from surface cleaning to pyrolysis and chemical synthesis. This review traces the developments from initial reports in the 1980s to today's converted ovens that are used in proof-of-principle manufacture of carbon nanostructures and batch cleaning of ion implant ceramics. Information sources include the US and Korean patent office, peer-reviewed papers, and web references. It is shown that the microwave oven plasma can induce rapid heterogeneous reaction (solid to gas and liquid to gas/solid) plus the much slower plasma-induced solid state reaction (metal oxide to metal nitride). A particular focus of this review is the passive and active nature of wire aerial electrodes, igniters, and thermal/chemical plasma catalyst in the generation of atmospheric plasma. In addition to the development of the microwave oven plasma, a further aspect evaluated is the development of methodologies for calibrating the plasma reactors with respect to microwave leakage, calorimetry, surface temperature, DUV-UV content, and plasma ion densities.

1. Introduction

Since the 1990s, tabletop domestic microwave ovens have been converted into plasma reactors and used for a wide range of manufacturing applications. The common feature in these reactors is that they contain a multimode resonant cavity (MRC) which is illuminated through one sidewall of the cavity, using a rectangular transverse electric (TE_{10}) waveguide with an interior waveguide aspect ratio of $2:1$ that houses a packaged cavity magnetron operating in the 2.45 GHz range. Using this configuration, no further impedance-matching apparatus is used between the magnetron and MRC.

As these types of microwave oven plasma reactors exploit dielectric heating and plasma chemistry, it is worth noting that dielectric heating of organic materials has a long and established history ranging from medical therapeutic use (short-wave diathermy) in the 1900s [1] and demonstrations of food cooking at the 1933 Chicago World's Fair [2] to the first microwave cooking of foodstuff, with patent application being filed in 1945 [3], followed by the first commercial microwave cooker built and sold by Raytheon in 1947 and Amana in 1967 [2, 4]. These ovens were of limited commercial success due to their bulkiness and cost, but commercial success came later when the cost-effective, packaged cavity magnetron became available [5, 6]. Although a combination of microwave heating and chemical reactions were reported in the early 1980s, no large-scale oven production was done until rapid synthesis of organic compounds in microwave ovens was performed in 1986 [7, 8]. More recently (2017), carbothermic reduction of zinc oxide and zinc ferrites has also been reported [9]. Once the first conversion of a microwave oven into a plasma reactor was reported in 1978 [10], plasma-induced synthesis of inorganic compounds became available [11–13], followed by plasma modification of polymer surfaces [14]. Interest in the conversion of microwave oven for plasma processing has also been reported for plasma pyrolysis paper [15, 16] and in-liquid plasma decomposition to produce hydrogen gas and carbon films [17–21]. More recently, initial studies of marine diesel exhaust gas abatement within a converted microwave oven have been reported [22]; however, little gas-line, or reactor conversion, details were given.

The success of the packaged cavity magnetron and the rectangular TE_{10} waveguide as found in the standard domestic oven has lead to their reuse in more advanced microwave plasma systems that are employed for microwave chemical deposition of diamond-like films [23], in the semiconductor industry [24] and in microwave plasma systems that are designed for the dissociation of hydrogen from water [25]. Plasma reactors based on the microwave oven have also been built for plasma cleaning of contaminated ion implant ceramics [26, 27] and used for plasma removal of photoresistant substances [28]. In 2009, the US Patent US. 2009/0012223 A1 describing a cylindrical cavity driven by a magnetron that generated atmospheric plasma for the fast food industry was also published [29].

Outside the peer-reviewed journals, microwave oven experiments as performed in schools have been reported which range from using plasma balls to explore eggs and the creation of soup sculptures [30]. Semiamateur studies on the subject of microwave oven plasma reactors have also been written. One particular article by Hideaki Page in the Summer/Fall issue of the Bell Jar provides a useful discussion on the practical problems encountered in converting domestic microwave ovens into plasma reactors operating at subatmospheric pressure [31]. Two of the problems encountered are as follows: (1) finding a suitable location for cutting into the thin (typically, 0.75 to 1 mm) metal sheet walls of the MRC without causing the metal to bend and buckle and (2) achieving sufficient vacuum in the jam jars or inverted bowls for the plasma to strike. Video postings on https://www.youtube.com/ also provide graphic information on domestic microwave oven plasma cleaning experiments [32]. Most of the other postings indicate that you would not want to do them yourself at home. Indeed, Stanley [33] goes as far as to exemplify many YouTube postings as "wacky and downright dangerous." For completeness, five such postings are given here [34–38].

The aim of this paper is to review the technology of microwave oven plasma reactors, the plasma chemical engineering, and the process measurements used. Within the works reviewed here, plasma processes have been reported quoting different pressure values and pressure units; therefore, to ease comparison between the processes, the original pressure values along with the equivalent SI unit of pressure (Pascal) are presented. This review paper is constructed as follows: Section 2 presents the technology used in microwave oven conversion. Section 3 looks at a purpose-built microwave plasma reactor that is based on the microwave oven. Section 4 describes the measurements that are used to calibrate the microwave MRC in terms of microwave leakage, calorimetric, surface temperature, near-field E-probe, and plasma ion density measurement. Section 5 provides a look at the cavity magnetron control drive circuit, and finally, Section 6 provides a conclusion to this review.

2. Microwave Oven Conversions

2.1. Converted Microwave Oven Plasma Reactors. By way of introduction, it is useful to list the 10 claims in Ribner's 1989 patent [10] that relates to the oven conversion process (Figure 1(a)). In brief, the claims are as follows: (1) positioning a vacuum chamber within the MRC with embodiments for admitting gas into the vacuum chamber and

through the cavity and for extracting gas by-products from the vacuum chamber; (2) relating to claim 1, where a means of regulating the gas to generate uniform plasma in the vacuum chamber; (3) a moving antenna for a means of generating a time average of uniform plasma; (4) a rotating antenna for a means of generating a time average of uniform plasma; (5) a means of reducing microwave leakage around each feedthrough; (6) a means of water cooling substrates within the vacuum chamber during plasma etching without thermal damage to the substrate during the plasma etch process; (7) relating to claim 6, where the water tubes have a heat transfer relationship with the vacuum chamber with a means of microwave leakage prevention; (8) a means of controlling microwave power; (9) relating to claim 8, a potentiometer in series with the primary transfer side of the magnetron transformer for controlling the maximum power in the oven; (10) relating to claim 9, when plasma etching of organics from substrate; and finally, (11) relating to claim 10, the use of water cooling of the substrate.

Beyond Ribner's patent, some studies [11–22] show that the microwave oven plasma reactor can be used for a multitude of processes and in many levels of oven reconstruction. The following sections describe the changes required to conventional domestic microwave ovens that range from minimal to major.

2.1.1. The Use of Replaceable Reaction Vessels. An example is the rapid synthesis of phase pure K_3C_{60} [11] and alkali-metal fullerides [12] in replaceable reaction vessels. Only minor changes to the conventional oven are required such as the provision of providing supports for positioning the reaction vessel at the node or antinode of the microwave field, nor the need for a rotating table or a moving (or rotating) antenna as the objective of the plasma process is to focus the microwave energy onto the sample (Figure 1(b)). In this case, the samples were prepared in an argon-filled Pyrex vessel and then positioned using fire bricks at the node or antinode of the microwave field. The plasma process time is however limited due to the fixed amount of residual gas in the reaction vessel.

2.1.2. The Use of Replaceable Desiccators. Ginn and Steinbock [14] have reported oxygen plasma cleaning of poly(dimethylsiloxane) surfaces within a replaceable desiccator that incorporated a steel electrode to promote plasma ignition (Figure 1(c)). The samples are prepared outside of the microwave oven and then placed in the desiccator which is purged with oxygen for 2 minutes and then evacuated to a pressure of about 10^{-3} Torr (0.133 Pascal). When placed in the oven and the microwave power (1100 W) is turned on, the steel wire electrode generates a spark to initiate the oxygen plasma. Here again, the plasma process time is limited, but the use of a steel electrode is found to promote the reaction to end. The subjection of the wire electrode is discussed further in Section 2.1.7.

2.1.3. Pumping through the Wall. In 2010, Singh and Jarvis reported the generation of carbon nanostructures from within

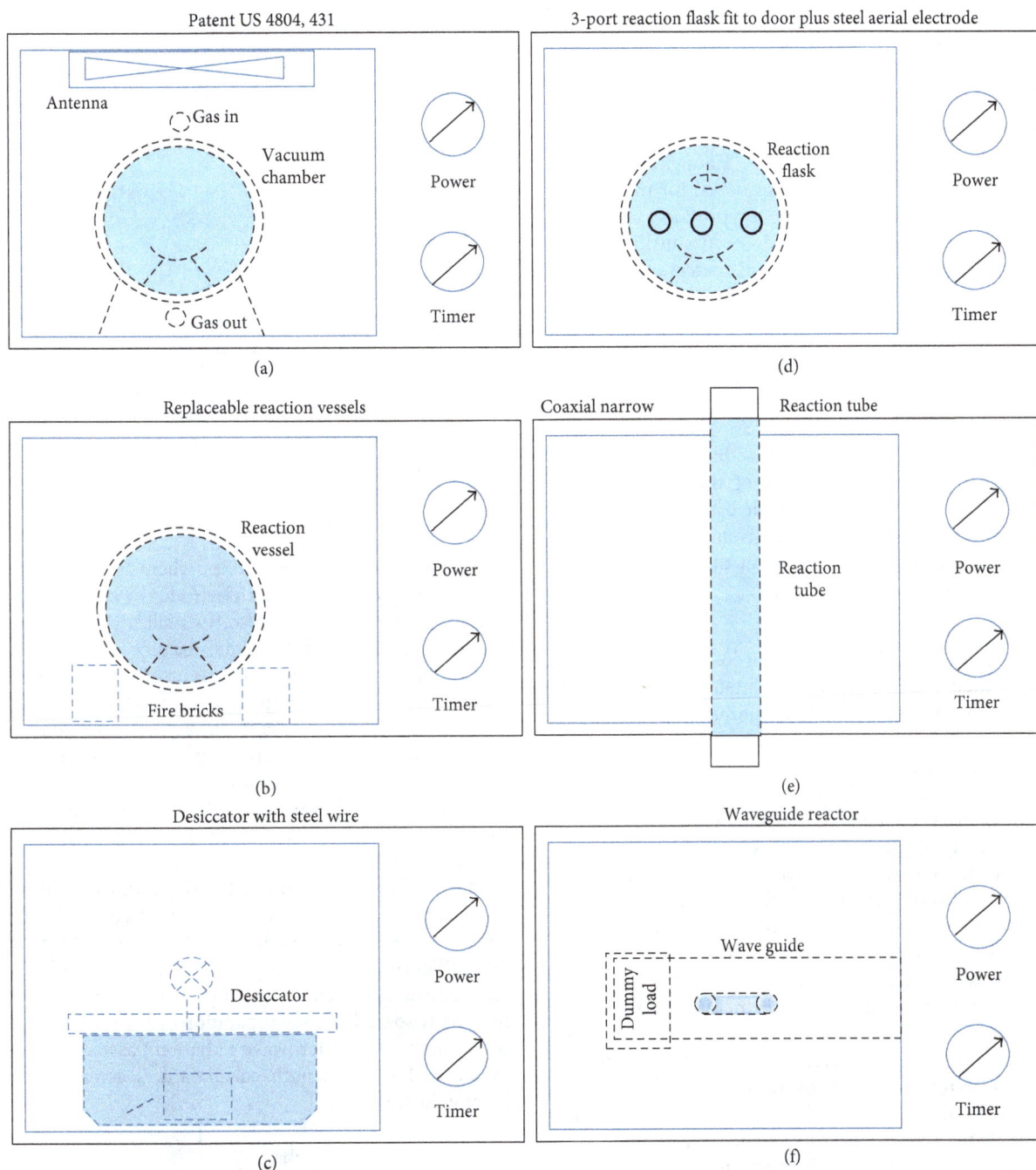

FIGURE 1: Front view schematic of converted microwave ovens. Each oven is scaled to the US 4,804,431 patent oven, and for clarity, the auxiliary gas lines and vacuum systems outside the ovens are not shown. (a) Patent US 4,804, 431. (b) Replaceable reaction vessels. (c) Desiccator with a steel wire. (d) Three-port reaction flask fitted to door plus steel aerial electrode. (e) Coaxial narrow reaction tube. (f) Waveguide reactor.

a continuously pumped 3-port reaction flask (made from borosilicate glass and 1000 ml volume) that was held within the microwave oven [17]. To support the vessel and facilitate access to it, the oven door was replaced with an aluminum plate of the same size that has three apertures, one for each flask port. With the flask supported, the flask was evacuated from the outside using one port, while the other two ports are used for the carrier gas and the selected hydrocarbon precursor gases (either ethanol, xylene, or toluene). To enhance

the reaction, a 2 mm diameter aerial electrode made from Nilo K® (Ni 29%, Fe 53%, and Co 17%) was mounted on a stainless steel base within the reaction flask (Figure 1(d)). As no vacuum pressure or microwave power was reported, it must be assumed that the flask was subatmospheric, and the microwave power was at a maximum (1000 W). Nevertheless, using this approach, no other modifications to the oven were needed. Two variants of this approach which preserve the door access are found in the work of Page [31] who drilled through the bottom of the cavity and Tallaire who drilled through the side of the cavity. In the latter case, Tallaire's YouTube posting provides an example of plasma cleaning of a microscope glass side [32].

2.1.4. Coaxial Narrow Tube Reactor.

Khongkrapan et al. have reported a converted microwave oven for pyrolysis of paper to produce gaseous waste by-products at 800 W [15, 16]. In their reactor, the process occurs inside a cylindrical quartz tube (internal/external diameters of 27/30 mm and length of 250 mm) that coaxially passes vertically through the MRC. Air or argon is used as the precursor gas at a nominal atmospheric pressure (101.3 k·Pa) with the gas flowing from bottom to top of the MRC. The shredded paper (5 g) is suspended in the center of the tube (Figure 1(e)). In [16], Khongkrapan et al. state that an igniter was placed within the tube to generate the plasma, but no direct details were given. Upon further reading of their reference list (reference 17 in their paper), a simple cartoon showing the igniter positioned within the tube is given again without text explanation. The subject of igniters in the form of a metal antenna is discussed in Section 2.1.7.

2.1.5. Internal Waveguide.

In 2004, Brooks and Douthwaite presented their internal waveguide fitted to an 800 W domestic microwave oven for plasma-induced processing of powered metal oxides (Ga_2O_3, TiO_2, and V_2O_5) into binary metal nitrides formed in ammonia (NH_3) plasma [13]. In this design, a slot is cut at the rear of the MRC to allow a 20 mm internal diameter U-shaped tube containing the solid-state sample within an alumina boat to be positioned within the microwave field (Figure 1(f)). Outside of the MRC, one end of the U-shaped tube is fitted to a vacuum pump and the other end is fitted to the carrier and process gases. To prevent microwave leakage at the rear of the oven, extensive gasket and Faraday shielding were arranged. An internal waveguide is then fitted to the MRC iris in such a way to focus the microwave energy in the vicinity of the sample. Furthermore, to prevent reflected power damaging the cavity magnetron and overheating the waveguide, a water-cooled dummy load is fitted to the waveguide output aperture. With these extensive oven conversions, the plasma region may be considered to be operating in the coherent mode rather than in the multimode. Typically, plasma parameters used to convert the metal oxides to nitrides are an NH_3 gas flow rate of 113 cm³·min⁻¹, pressure of 20 mbar (2000 Pascal), and microwave power of 900 W for a plasma exposure time of 2.5 to 6 hours.

2.1.6. Liquid Plasma Vessels.

Microwave in-liquid plasma decomposition of n-dodecane (molecular formula: $C_{12}H_{26}(I)$) to simultaneously produce hydrogen gas and carbide in the hydrocarbon liquid has been achieved using a converted microwave oven at a reported microwave power level of 500 to 750 W [18–20]. A typical representation of these reactors is shown in Figure 2. The reaction is performed in a closed volume Pyrex reaction vessel containing 500 ml of n-dodecane

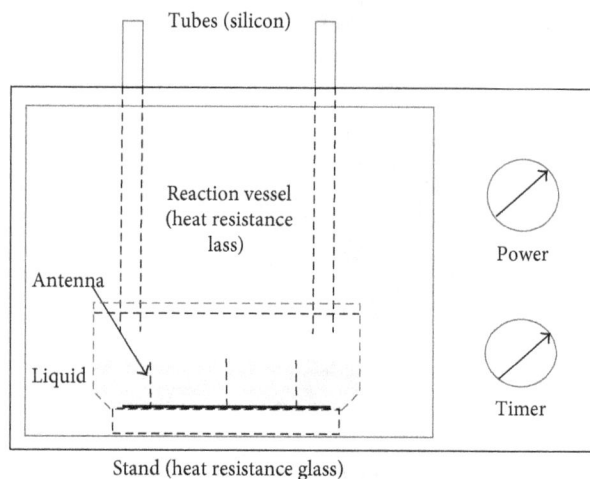

FIGURE 2: A typical front view schematic of a converted microwave oven for liquid processing. For clarity, the auxiliary gas lines outside the ovens are not shown.

liquid with one or more electrodes, where the electrode(s) can be either single-tip steel wire electrodes or copper U-shaped dual-tip aerial electrodes. Also, two silicon/PTFE tubes are inserted from the top of the cavity: one tube is used for sending the carrying gas (argon) as the precursor gas and the second tube is used to collect the spent argon and the by-product gas at a working pressure close to atmospheric pressure.

To understand the purpose of these electrodes, the reaction efficiency of both types of electrodes is examined as a function of the geometry and the number of electrodes in the context of their electromagnetic design and heterogeneous reaction kinetics.

First, consider the single-tip electrodes [18–20]. These metal electrodes have a dimensional length of $L = 21$ mm and a diameter of 1.5 mm, and they are fixed vertically in a single array (Figure 3(a)) or in a multiple array (Figure 3(b)) with 1 electrode in the center and up to 6 electrodes circumferentially spaced at a gap separation of $\lambda_m/4$, where λ_m is the wavelength of the microwave radiation passing through the medium. The wavelength calculation is given in the following equation:

$$\lambda_m \sim \frac{C}{f} \cdot \frac{1}{\sqrt{\varepsilon_r}}. \tag{1}$$

The approximate expression in (1) is used as the operation frequency of the free running cavity magnetron is frequency pulled by changing SWR conditions in the rectangular TE_{10} waveguide in which the magnetron is mounted. All other symbols have their normal meaning: C is the speed of light (2.99792×10^8 m·s⁻¹), f is the magnetron operating frequency (2.45 GHz), and ε_r is the medium in which the radiation is passing through. Thus, for the liquid n-dodecane ($\varepsilon_r = 1.78$ to 2), λ_m approximates to 8.85 cm and $\lambda_m/4$ approximates to 2.2 cm.

Based on the works [18–20] and the work of Pongsopon et al. [21], it is generally considered that the electrodes have three well-defined roles: to confine the plasma to the immediate proximity of the electrode(s) tip, to function as

(a)

(b)

(c)

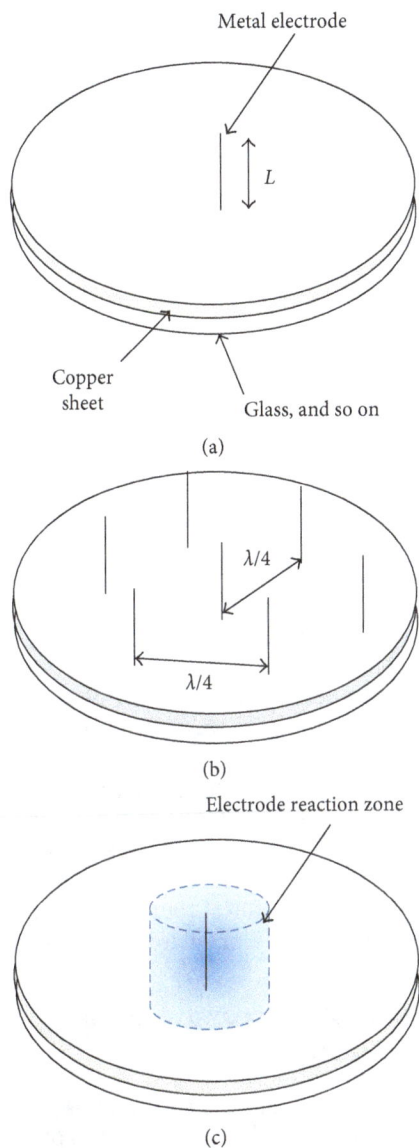

FIGURE 3: Typical single-tip electrode arrangement (a); multiple electrode arrangements (b); electrode reaction zone (c).

a catalytic source for plasma heterogeneous reaction, and in the case of manufacturing carbon nanomaterials, to provide a substrate on which the carbon material can grow. In the first of these roles, increasing the number of electrodes from 1 to 6 has revealed that the efficiency of plasma decomposition of n-dodecane does increase, but beyond 6-7 electrodes, the reaction efficiency becomes rate-limited. This may be due to electromagnetic power loss by the resonant structure of the electrodes [20] or simply that the addition of more than 7 electrodes and their associated surrounding reaction zones (cylindrical volume around each electrode; Figure 3(c)) within a fixed closed volume simply produces a loading effect within the heterogeneous reaction [39]. That is to say, as the percentage of the combined electrode reaction zones approaches the total fixed volume, the amount of fresh reactant flowing to the electrode reaction zone becomes reduced. Therefore, mass transport in and out of

each electrode reaction zone, rather than plasma decomposition, may become the rate-limiting step. To clarify these observations, further investigation is needed.

For the dual-tipped aerial electrode, Toyota et al. [20] have shown that the U-shaped aerial electrodes have distinct optimum lengths of $L \sim 2\lambda_{m}$, $3\lambda_{m}/2$, λ_{m}, and $\lambda_{m}/2$. They also show that the use of the approximation sign in (1) is justified by experimentally determining the $\lambda/2$ FHHW length of the U-shaped dual-tip aerial electrode to be 4.4 to 4.7 cm for n-dodecane.

2.1.7. Igniter. The description of construction and use of wire aerial electrodes for plasma ignition is now used as an aid to outline the construction of the plasma igniter [16] and the drawing in [40] (Figure 4). Assuming that the drawing in [40] may be scaled, the plasma igniter may be constructed in two ways: Firstly, the igniter may be constructed using two wire electrodes opposing each other and bent at 45° so that their tips are aligned with the gas flow, and the fixing location is formed using an insulating ring. The second and more practical arrangement is that the igniter is preformed from a 30 mm diameter × 0.5 mm thick steel steel disc, and a plurality of electrodes are punched from the central portion of the disc and bent to a 45° angle. For the purpose of this second option, the construction of a 4-electrode igniter is exemplified using the 27/30 mm internal/external diameter glass tube in [16] as a reference tube (Figure 1(e)). A schematic of the manufacturing stages of the igniter is given in Figure 4, where it is shown that the first stage is to punch out the form of the igniter, the second stage is to bend the electrodes, and the third stage is to align the igniter to the glass tube. Using this method of construction, the lip of the preform can self-align to enable the 4 aerial electrodes to suit the plasma ignition criteria as described in Section 2.1.5.

2.1.8. Production of Plasmoids (Fireballs). The production of plasmoids, sometimes called fireballs or ball-lightning, within domestic microwave ovens has been posted on YouTube postings [34–38]. Perhaps, the simplest way of producing a fireball without modification to the microwave oven is to place a partially sliced grape (that has its two halves connected via a thin piece of skin) in the microwave oven and then turn on the microwave power for 3–10 seconds. The YouTube posting [34] shows that arc-like plasmoids are generated at the thin skin bridge that connects the two grape halves, with the discharge emission continuing until either the power is turned off or the grapes have shriveled up. This action may be understood by considering that the two freshly cut grape halves have a characteristic dimension of 1.5 to 2 cm and are partially filled with a conducting electrolyte, the combination of which creates an organic conducting dipole antenna not unlike the metal antennas discussed in Sections 2.1.5 and 2.1.6. Given this understanding, it is reasonable to assume that as the free electrons are pushed back and forth through the narrow thin skin bridge of the grape, heat is generated due to the

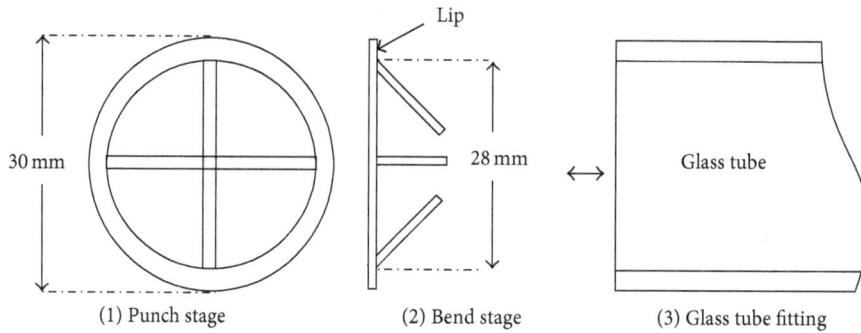

FIGURE 4: Manufacturing stages of the disc igniter that is suitable for a narrow glass tube reactor.

resistance and burns away the skin. In addition, the movement of electrons through the grape electrolyte induces a rapid increase in temperature, causing vaporization of the electrolyte to a cloud of electrons and ions, thus forming the localized plasmoid. The plasmoid continues to be sustained as long as the free electrons are available from the diminishing volume of the grape electrolyte.

Moving away from the organic source for generating plasmoids, a lighted safety match, supported by a wine cork, covered with a glass jar, and placed within the center of the MRC, can also be used [35]. Upon turning on the microwave power, a plasma discharge is generated that rises to the top of the jar, thus forming a buoyant plasmoid. Warren [36] used a similar approach, but this time using a glass jar supported by three wine corks and a lighted cigarette placed in the gap provided by the corks. In this work and the previous example, the plasmoids are maintained when the thermal source is extingusihed. It is only when the microwave power is turned-off does the plasmoid become extinguished. Plasmoids can also be generated within electric light bulbs and fluorescence tubes as shown in [37]: this example also appears to be the basis for the near-field E-probe (Section 4.4).

A more dangerous approach to generating plasmoids is demonstrated in [38] where a cavity magnetron connected to a food tin can is used to lunch microwaves at a domestic light bulb to produce a plasmoid within the bulb. From this experiment, it would appear that the electric filament acts as the initiating electrode.

Before finishing this section, it is worth noting that the cylindrical plasma reactor produced for the fast food industry [29] employed a patented passive plasma catalyst in the form of an electrode to ignite the atmospheric plasma [41], where the passive plasma catalyst can include any object capable of inducing plasma by deforming the local electric field. On the other hand, the patent states that an active plasma catalyst produces particles or a high-energy wave packet capable of transferring a sufficient amount of energy to a gaseous atom (or molecule) to remove at least one electron from the gaseous atom (or molecule) in the presence of electromagnetic radiation. Given these two definitions, it is reasonable to assume that the safety match flame [35], cigarette [36] and grape [34] can be classed as an active plasma catalyst and the metal electrode as a passive plasma catalyst.

2.1.9. Plasmoid Food Cooking. The Korean patents [42, 43] and conference paper [44] report on a form of tuning within the TE_{10} waveguide that fall outside the scope of this review, but they are listed for three reasons: Firstly, the phenomena of plasmoids extend the cooking range of the domestic microwave oven from one of dielectric heating of food stuff to one that provides surface browning and imparting texture and flavor that is similar to the traditional flame-cooking process. Secondly, Jerby et al. [44] have noted that plasmoids produced in this way require wire antenna electrode to ignite the plasmoid and therefore may contain nanoparticles, which might be harmful for the food quality and even make it inedible. Thirdly, the additional use of plasma discharge that generates ozone and ions for the removal of odor-producing materials from the cooking chamber [45] does provide one possible technical route forward in the future development of the domestic microwave oven.

3. Purpose-Built Microwave Oven Plasma Reactor

This section describes the methodology used in the construction of a purpose-built microwave oven plasma reactor. Of particular importance in this regard is the MRC series of plasma reactors that were built in the mid-1990s at Cambridge Fluid Systems Ltd (England, UK). The design concept behind these plasma reactors was to build a simple reliable and cost-effective table-top plasma reactor that could be sold to research laboratories and low-volume productions units. Their main use was for surface engineering enhancement in the microelectronic, semiconductor sector and the manufacture of bodyshell of Formula One racing cars.

The design of the plasma reactor is similar to microwave ovens, where the cavity magnetron antenna is located within a TE_{10} waveguide that is used to illuminate the MRC through a single iris. The cutoff frequency $(f_c)_{m,n}$ of the TE_{10} waveguide is calculated using the following equation:

$$(f_c)_{m,n} = \frac{C}{2}\sqrt{\left(\frac{m}{a}\right)^2 + \left(\frac{n}{b}\right)^2},\qquad(2)$$

where c is the speed of the light and a and b are the internal dimensions (width and height) of the waveguide; in this case, 80 and 38 mm are used, respectively, which equate a cutoff frequency of 1.875 GHz.

With the cavity magnetron antenna positioned 26 mm from the end of the waveguide, the frequency and bandwidth of the magnetron are allowed to be free running. Thus, the noncoherent reflected power passing through the iris travels back to the magnetron thus altering the SWR of the coherent wave within the TE_{10} waveguide, resulting in varying the output power of the magnetron.

The MRC reactor design differs from the domestic microwave oven plasma reactor in the following ways (also cf. Figure 1 with Figure 5):

(i) The chassis, MRC, and waveguide are constructed as one welded component using 1.4 mm thick mild steel sheet. Before each of the three components is welded together, they have all the necessary holes punched and clasp nuts fixed. Once welded, the structure is nickel plated to produce a metal structure that is robust with sufficient stiffness to support all the additional components (the front and rear stainless flanges, gas lines, DC power supply pressure gauge, etc.). Using this construction approach, the MRC has a theoretical maximum unloaded Q-factor (Q_u) in the TE mode that is dependent on the ratio of stored energy in the cavity (V_c) to the energy loss to the cavity walls ($\delta \times A_c$):

$$Q_u = \frac{2V_c}{\delta A_c}, \qquad (3)$$

where δ is the electrical skin depth at the cavity wall per cycle and A_c is the cavity wall area.

For this reactor, the main cavity has an approximately Q_u of 20,000 at a resonant frequency of 2.45 GHz.

(ii) A cylindrical Pyrex glass chamber (190 mm diameter, 300 mm length, and 5 mm wall thickness: producing a volume of 3 liters) is located within the multimode cavity with its longitudinal axis perpendicular to the microwave iris and with the front and rear of the chamber housed within metal flanges that form the part of the multimode cavity wall. The rear flange contains welded vacuum and pressure gauge ports, and the front flanges contain the access door. This design maximizes the chamber volume and removes all fragile glass fittings, plastic tube connectors, and feedthrough microwave leakage gaskets.

(iii) The gas lines are fitted within the chassis and to the side of the MRC, thereby enabling the process gasses to be injected through multiple equally spaced radial ports in the front flange thus reducing the possibility of precursor gas being preionized prior to chamber entry and maximizing uniform gas flow and plasma uniformity along process chamber longitudinal axis.

3.1. Plasma Cleaning of Ion Implant Ceramic Insulators.

Ion implantation is one of the key processes in the high volume (220 wafers per hour) manufacture of silicon

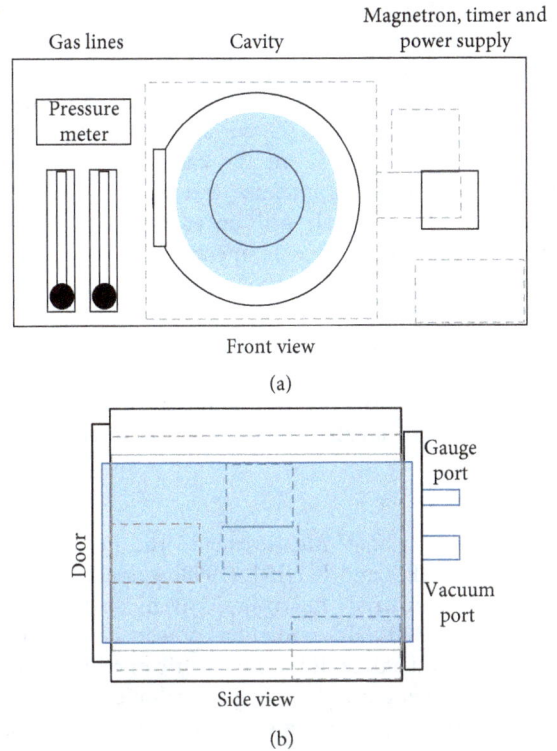

FIGURE 5: Typical front and side view schematic of the MRC plasma reactor chassis and cavity. Photograph of the microwave MRC-200 plasma reactor.

semiconductor devices. These ion implant machines, however, cost between \$1.8M and \$3M. These machines are also highly maintenance-intensive systems with high capital cost; therefore, availability and cost of ownership are major factors to be considered. Many of the parts changed during regular maintenance, and ion source changes are ceramic insulators. In this section, an overview of the plasma cleaning of ion implant ceramic is described: for full details of the process, see [26, 27]. The plasma cleaning process has been performed in the MRC series of plasma reactor using a gas mixture of 5–10% O_2 in CF, with an admixture of 50% by flow of argon. The argon admix is used to stabilize the microwave plasma by moderating the electron energy distribution and to provide a uniform excited species throughout the plasma volume. The plasma etch chemistry at the surface of the ceramic may be considered to proceed by the following representative heterogeneous reaction:

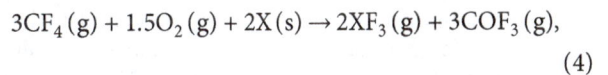

$$3CF_4\,(g) + 1.5O_2\,(g) + 2X\,(s) \rightarrow 2XF_3\,(g) + 3COF_3\,(g),$$
$$(4)$$

in which the addition of O_2 scavenges carbon from the CF_4 through the formation of COF_x species to enhance the steady-state concentration of F atoms in the plasma volume. The element X in reaction (4) represents the group V element (As, P, and Sb) on the ceramic surface, and the XF_3 are the etch products. Given sufficient microwave power, the etch rates of these products are therefore controlled by the production of F atoms (O_2 balance), product volatility, and ceramic microscopic surface area.

For the MRC-100 reactor, typical plasma process parameters obtained were 104 W and 10 mbar with an etch time of 45 minutes, where the surface temperature of the ceramics reached 80 ± 5 K. In the case of the MRC-200 reactor, the process parameters were as follows: 200 W and 10 mbar (1000 Pascal) with an etch time of 20 to 25 minutes, where the surface temperature of the ceramics reached 125 ± 5 K.

4. Microwave Cavity Calibration

This section describes a range of different techniques which are used for the evaluation of the microwave efficiency as well as leakage.

4.1. Microwave Leakage Measurement. The European Directive 2004/40/EC and ICNIRP (1998) guidelines recommend that industrial microwave ovens have surface microwave (3–300 GHz) radiation power density levels >5 mW·cm^{-2} and 5 times less for domestic microwave ovens intended for general use. Applying a quadratic law based on plane wave theory, an operator standing at a distance of 20 cm from an industrial oven, the operator will receive the maximum allowable power density level of 3 mW·cm^{-2}. For domestic oven environment, this equates to a power density level of 0.3 mW·cm^{-2}. In the context of converted ovens intended for plasma, many feedthrough and apertures in the MRC require considerable care in design and construction to prevent microwave leakage.

4.2. Calorimetric Magnetron Power Calibration. The magnetron power entering the MRC may be calibrated using the water open-dish load method, see, for example, British Standard 7509:1995 and IEC 1307:1994. Hence, given the heat capacity of water is 4.184 J/(g·K), the calculated applied power (P) within the cavity may be obtained by placing a known mass of water (m) within the cavity and heating it for a short period of time (t), making sure the water does not boil. With the knowledge of the measured water temperature change (ΔT = final temperature – initial temperature), the microwave power entering into the cavity is calibrated for a given power setting using the following equation (see also [40]):

$$P = \frac{mC\Delta T}{t}. \tag{5}$$

The calibration however must be considered to be an upper value for plasma processing as its dielectric volume will be different from the water calibration. (Note: when the process chamber volume or the geometric shape does not allow the open dish method to be used, the alternative flow method can be used, as outlined in [46]).

Given (4), the microwave power density (W·cm^{-3}) of a system may be calculated by dividing through by the process chamber volume. The following open-dish load method for the MRC 100 is given here as an illustrative

example: the MRC-100 and MRC-200 reactors have a calculated magnetron applied power of 104 W, which equates to a power density of 0.116 W·cm^{-3}. For the MRC-200 reactor, the calorimetric measurements produce magnetron applied power values of 200 W (0.022 W·cm^{-3}) and 450 W (0.05 W·cm^{-3}).

4.3. Surface Temperature Measurement. Knowledge of the surface temperature of materials immersed in the plasma is useful in understanding the heterogeneous plasma-surface interaction. This is of particular importance when local dielectric heating has the potential of thermal runaway, because most materials increase their dielectric loss with temperature [12]. Two simple means of estimating local surface temperature of materials that are immersed in microwave plasma have been used. For surface temperatures below 180 K, liquid crystal temperature-sensitive (20 ± 5 K to 180 ± 5 K) strips attached to the plasma immersed surface can be used [27]. For higher temperatures, salts of known melting point (KCl = 1043 K and NaCl = 1074 K) sealed in silica capillaries have been used [13].

4.4. Near-Field Plasma E-Probe Measurement. Attempts to strike a plasma outside the ignition pressure limits of 0.1 to 20 mbar (10 to 2000 Pascal) result in the microwave radiation being stored "per cycle" in the empty-cavity mode; under these conditions, the rate of energy loss to the cavity wall can substantially heat the MRC structure, and in the extreme case, the microwave radiation leakage can become a health risk. Additionally, if the MRC is loaded with materials (semiconductor wafers and low dielectric strength material), these can be electrically or mechanically damaged. It therefore becomes necessary to have an automatic power cutoff device to prevent microwave leakage and damage to both the reactor and loaded materials. The near-field plasma *E*-probe as described by Law [47] is one such device that helps monitor such events. In this circuit, a neon discharge lamp, a photodiode, and a reference voltage are connected as shown in Figure 6, with one leg of the neon being used protruding into the cavity to act as a near-field *E*-probe. In the original circuit design, a strip chart is used to record the voltage-time-series data, but with today's analogue-to-digital converters and software (such as LabView) the plasma ignition state-space and plasma state-space may be monitored with trigger levels set to give a binary Go/No control output [24, 47].

4.5. Bébésonde Electrostatic Probe Measurement. For plasma processes, it is generally considered that the ion flux arriving and leaving the surface determines the plasma process. However, the use of electrostatic probe techniques to determine the ion density and temperature of plasma driven by modulated power source in the presence of sputtered insulating material is problematic. Such is the case of plasma containing CF_4. This section describes a probe technique that is tolerant of drive modulation and sputtered insulating materials. The following measurements were performed at

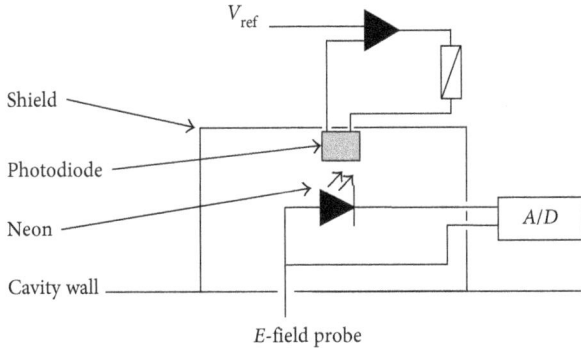

FIGURE 6: Schematic of the E-probe circuit, showing a neon discharge lamp, a photodiode, and an analogue-to-digital converter.

the Oxford Research Unit, Open University, on the MRC-100 reactor. The probe used is an RF-biased ion flux probe (known colloquially as "Bébésone, or BBs"). For full details of the probe and measurement, see [48].

For good probe measurements, visual observations of the plasma volume as a whole and around the probe itself are usually required. For the MRC reactors, it was found necessary to replace the standard front flange door with a flange incorporating a probe port and inspection port, covered by an open mesh. Given this modification, argon plasma within the MRC-100 visually exhibits little structure with a slight brightening close to the dielectric annulus boundary. Nevertheless, there is no evidence of microwave structure in the optical emission, which indicates that energy is rapidly homogenized in the electron population.

Following these visual baseline inspections, the Bébésonde was used to determine the ion flux in the low power and high setting with argon gas flow varying from a maximum ($5 \, l \cdot min^{-1}$) to a minimum for which plasma could be sustained ($2 \, l \cdot min^{-1}$). For the MRC-200 reactor, argon ion densities are measured to be in the range of $2 \times 10^{11} \cdot cm^{-3}$ at 50 mbar ($5 \, k \cdot Pascal$) to $3 \times 10^{12} \cdot cm^{-3}$ at 1 mbar (100 Pascal), and the electron temperature is in the range of 1 to 1.5 eV for an input power of $0.022 \, W \cdot cm^{-3}$: the higher plasma density at low pressure corresponds to a longer mean free path and increased electron power transfer.

4.6. Ultraviolet Fluorescence Microwave Probe. For many processing plasmas, photon energy ($E = hc/\lambda$) varies from 10 eV to 1 eV with preferential intensity at discrete spectral wavelengths. The spectral characteristic is determined by the nature of gas excitation and relaxation processes, both of which are sensitive to the local electron temperature and gas excitation cross sections. Gas composition and mode of plasma production therefore has a major impact on UV production and plasma chemistry.

The ultraviolet fluorescence probe utilizes activated rare-earth salts: Y_2SiO_5:Ce (absorption < 200 nm) and Zn_2SiO_4:Mn and Y_2O_3:Eu^{3+} (absorption < 300 nm). For full details of these probes, see [49]. At these energies, the host lattice (H) undergoes electronic excitation via the production of

electron-hole pair separation (6), followed by the lowest energetic deactivation pathway fluorescence at wavelengths longer than the initial radiation. In general, the fluorescence spectrum is composed of a narrow band with the precise wavelength determined by the intimate relation between the activator (Ce, Mn, and Eu^{3+}) and the host lattice and by the irradiating photon radiation:

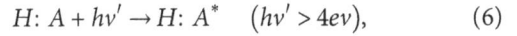

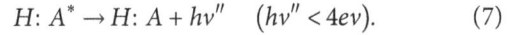

$$H: A + h\nu' \rightarrow H: A^* \quad (h\nu' > 4 ev), \quad (6)$$

$$H: A^* \rightarrow H: A + h\nu'' \quad (h\nu'' < 4 ev). \quad (7)$$

The most simple form of the probes comprises a synthetic DUV grade fused silica capsule (12 mm diameter × 20 mm) containing one of three activated salts at a nominal reduced pressure of 10 mbar. The fused silica has transmittance $T = 0.5$ at 170 nm. When placed in the plasma volume, the probe collects 4π steradians of incoming DUV photon radiation. The emitted fluorescence is viewed through an optical crown glass viewport ($T = 0.5$ at 380 nm). Due to the dielectric capsule, photoluminescence rather than electroluminescence is considered to be the prime mechanism of fluorescence in these probes. The dielectric also acts as a wavelength discriminator and provides the upper working limit of the probe to be 1100 K.

Using these knowledge, the salts Zn_2SiO_4:Mn and Y_2O_3:Eu^{3+} (placed in their own capsule) integrates DUV plasma, while Y_2SiO_5:Ce when placed directly in the plasma volume integrates VUV (<200 nm) plasma, thus their fluorescence appearing in the green, red, and blue, respectively.

5. Magnetron Oven Control Circuits

Microwave ovens generally employed one of two types of magnetron drive circuits. For microwave ovens with low-power outputs (typically <500 W), the output power is achieved by pulse-width modulation of a single incident power to the magnetron with the cooking time set between 0 and 30 minutes. Between the 500 W and 1100 W rating, continuous application of the microwave power is used where the power level is set by the magnetron drive capacitor value:

$$P = \frac{1}{2}(CV^2)f. \quad (8)$$

The choice of either of these two drive circuits can have an impact on the magnetron control circuit ability to carry out the selected plasma process. This choice is exemplified by comparing the short and low power requirements of rapid plasma syntheses of organic compounds [7, 8], cleaning of glass slides and polymers [14] to that of the high power and long processing times of solid-state metal oxides processing [13] and the plasma cleaning of ion implant ceramic insulators [26, 27].

5.1. Cavity Magnetron Capacitor-Controlled Drive Circuit. In this section, the choice of the capacitor controlled drive circuit is considered along with its safety control circuit that is used in the MRC-100/200 plasma reactors. A schematic of

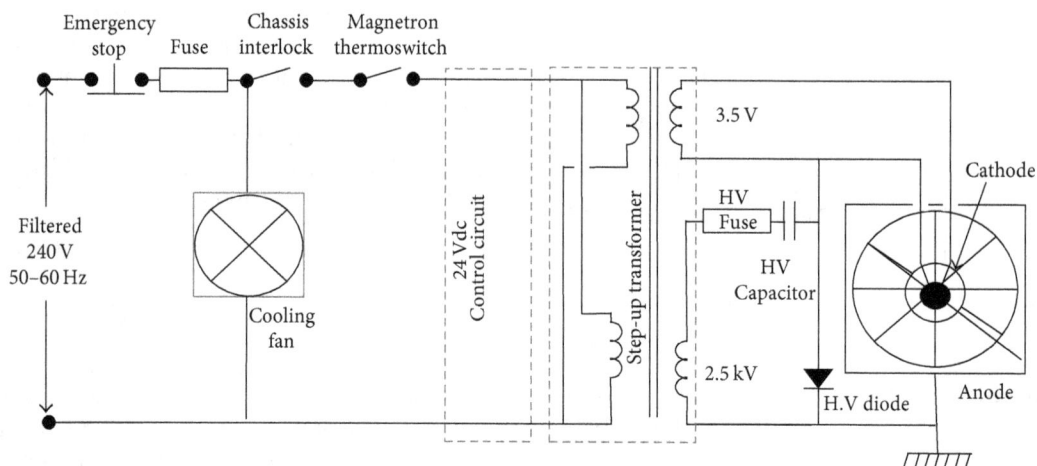

FIGURE 7: Cavity magnetron capacitor-controlled dive circuit. The 24 V DC control circuit represented by the dashed box is discussed in Section 5.2.

the capacitor controlled drive circuit is shown in Figure 7. A similar cavity magnetron capacitor controlled drive circuit is reported in [23]. Chaichumporn et al. have also reported further refinement of the magnetron anode voltage (3.3 to 6.6 kV) in [40].

The transformer comprises two winding circuits: The first provides a winding ratio to produce 240 V at 50–60 Hz to approximately 3.5 V at 50–60 Hz for the cathode heater, and the second provides the step-up voltage (240 V at 50–60 Hz to 2-3 kV at 50–60 Hz). The HV capacitor (0.6 to 1.5 μF) and the HV diode are used to bias the cathode negative with respect to the anode block, which contains the magnetron's cavity structure. Using this arrangement, the capacitor value and diode determines the DC power dissipated in the magnetron cavity structure. For safety reasons, both passive and active control components are incorporated on either side of the setup transformer. These include the following: the circuit is fused on both sides of the transformer plus an emergency stop button, a chassis interlock, a magnetron thermoswitch (135°C), and a chasse cooling fan. In addition, 24 V DC circuit is used to provide remote passive and active control of the drive circuit (see Section 5.2).

5.2. 24 V DC Control Circuit. In purpose-built microwave oven plasma reactors, the auxiliary equipment (vacuum pumps, vacuum valves pressure gauge, gas lines, purge lines, process timer, and microwave power) is synchronized to plasma process in a safe manner. The role of the 24 V DC control circuit is to synchronize the auxiliary components to the plasma process and shut down the system in the case of an emergency failure: this is especially important when using flammable, corrosive, and toxic precursor gases and by-products. The circuit is designed so that all chasse controls are isolated from the cavity magnetron capacitor-controlled drive circuit using relays and solenoids. It should be noted that when converting a domestic microwave oven into a plasma reactor, the synchronization has to be built from scratch.

6. Conclusions

This work has reviewed the conversation of the domestic microwave oven into a source for cleaning as well as chemical reactions. The conversion of domestic systems into plasma reactor is described, as is the construction of purpose-built microwave oven plasma reactors. Calibration of the MRC has been discussed along with identifying the two types of magnetron drive circuitry used. The professional and armature use has also been presented, in the latter case, mainly limited to kitchen top experiments. The proof of principle and small batch processes established in these plasma reactors range from plasma cleaning of glass and polymer surface and removal of toxic metals from ceramic surfaces to the manufacture of carbon nanostructures and the pyrolysis of paper to produce gaseous waste by-products. In all cases, the power source is the package cavity magnetron operating below 1100 W output power and pressures ranging from a few 0.1 s Pascal to a nominal atmospheric pressure (101.3 Pascal).

At or close to atmospheric pressure single or multiple wire antenna electrodes that have a physical length approximating to 1/4 or 1/2 of the microwave length in which that it is immersed have been found to play a catalytic role in instigating plasma production, and in the case of manufacturing carbon nanomaterials provide a substrate on which carbon material can grow. With regard to reaction rate, increasing with the number (1 to 6) of electrodes beyond which the reaction becomes rate-limited, a geometrical loading effect around the wire antenna is proposed in this paper. Whether this or electromagnetic power loss is responsible for the effect, further work is required. In addition, this work has reconstructed a preformed disc aerial electrode (igniter) suitable for a narrow tube reactor [16, 40] (Figure 4).

The safety match flame, lighted cigarette, sliced grape, and metal antenna electrode have been observed to have a catalyst role in the production of plasma and plasmoids. To distinguish between the metal antenna and the thermal-chemical based catalyst, it has been put forward that metal antennas may be classified as a passive catalyst as they only supply a surface that generates free electrons, whilst the safety match flame, lighted cigarette, and sliced grape may be classified as an active catalyst as they supply energy in the form of heat and free electrons from an electrolyte.

Finally, this review also has highlighted plasmoid food cooking within a MRC and the use of plasma discharge for removing food odor from microwave ovens. Given that food safety issues are addressed, it is reasonable to envisage microwave oven plasma reactors, incorporating both plasmoid cooking of food stuff and plasma deodorization of cooking by-products may be realized in the near future.

Conflicts of Interest

The authors declare that there are no conflicts of interest regarding the publication of this paper.

Acknowledgments

The authors would like to acknowledge the support of SFI through the I-Form Advanced Manufacturing Research Center 16/RC/3872.

References

[1] F. Nagelschmidt, *Diathermy Text Book for Physicians and Students*, Springer-Verlag, Berlin, Heidelberg, 1921.

[2] Davis, *A History of the Microwave Oven*, The IEEE News Source, The IEEE News Source, Piscataway, NJ, USA, https://www.theinstitute.ieee.org/tech-history/technology-history/a-history-of-the-microwave-oven.

[3] P. L. Spencer, "Method of treating foodstuffs," US Patent 2,495,429, 1950.

[4] J. M. Osepchuk, "The history of the microwave oven: a critical review," in *Proceedings of Digest IEEE International Microwave Symposium*, pp. 1397–1400, Boston, MA, USA, June 2009.

[5] J. R. Mims, "Microwave magnetron," US Patent 3,739,225, 1973.

[6] T. Koinuma, "Magnetron," US Patent 3,809,590, 1974.

[7] R. N. Gedye, F. Smith, and K. C. Westaway, "The rapid synthesis of organic compounds in microwave ovens," *Canadian Journal of Chemistry*, vol. 6, no. 1, pp. 17–26, 1988.

[8] R. N. Gedye, W. Rank, and K. C. Westaway, "The rapid synthesis of organic compounds in microwave ovens. II," *Canadian Journal of Chemistry*, vol. 69, no. 4, pp. 706–711, 1991.

[9] M. Omran, T. Fabritius, E.-P. Heikkinen, and G. Chen, "Dielectric properties and carbothermic reduction of zinc oxide and zinc ferrite by microwave heating," *Royal Society Open Science*, vol. 4, no. 9, p. 170710, 2017.

[10] A. Ribner, "Microwave plasma etching machine and method of etching," US Patent 4,804,431, 1989.

[11] R. E. Douthwaite, M. L. H. Green, and M. J. Rosseinsky, "Rapid synthesis of phase pure K_3C_{60} using a microwave-induced argon plasma," *Journal of the Chemical Society, Chemical Communications*, no. 18, pp. 2027-2028, 1994.

[12] R. E. Douthwaite, M. L. H. Green, and M. J. Rosseinsky, "Rapid synthesis of alkali-metal fullerides using a microwave-induced argon plasma," *Chemistry of Materials*, vol. 8, no. 2, pp. 394–400, 1996.

[13] D. J. Brooks and R. E. Douthwaite, "Microwave-induced plasma reactor based on a domestic microwave oven for bulk solid state chemistry," *Review of Scientific Instruments*, vol. 75, no. 12, pp. 5277–5279, 2004.

[14] B. T. Ginn and O. Steinbock, "Polymer surface modification using microwave-oven-generated plasma," *Langmuir*, vol. 19, no. 19, pp. 8117-8118, 2003.

[15] P. Khongkrapan, N. Tippayawong, and T. Kiatsiriroat, "Thermochemical conversion of waste papers to fuel gas in a microwave plasma reactor," *Journal of Clean Energy Technologies*, vol. 1, no. 2, pp. 80–83, 2013.

[16] P. Khongkrapan, P. Thanompongchart, N. Tippayawong, and T. Kiatsiriroat, "Fuel gas and char from pyrolysis of waste paper in a microwave plasma reactor," *IJEE*, vol. 4, no. 6, pp. 969–974, 2013.

[17] R. Singh and A. L. L. Jarvis, "Microwave plasma-enhanced chemical vapour deposition growth of carbon nano-structures," *South African Journal of Science*, vol. 106, no. 5-6, p. 4, 2010.

[18] S. Nomura, H. Toyota, S. Mukasa, H. Yamashita, T. Maehara, and A. Kawashima, "Production of hydrogen in a conventional microwave oven," *Journal of Applied Physics*, vol. 106, no. 7, p. 073306, 2009.

[19] S. Nomura, H. Yamashita, H. Toyota, S. Mukasa, and Y. Okamura, "Simultaneous production of hydrogen and carbon nanotubes in a conventional microwave oven," in *Proceedings of International Symposium on Plasma Chemistry (ISPC19)*, vol. 65, Bochum, Germany, July 2009.

[20] H. Toyota, S. Nomura, and S. Mukasa, "A practical electrode for microwave plasma processes," *International Journal of Materials Science and Applications*, vol. 2, no. 3, pp. 83–88, 2013.

[21] R. Pongsopon, T. Chim-Oye, and M. Fuangfoong, "Microwave plasma reactor based on microwave oven," in *PIERS Proceedings*, pp. 2723–2726, Guangzhou, China, August 2014.

[22] N. Manivannan, W. Balachandran, R. Beleca, and M. Abbod, "Microwave plasma system design and modelling for marine diesel exhaust gas abatement of NOx and SOx," *International Journal of Environmental Science and Development*, vol. 6, no. 2, pp. 151–154, 2015.

[23] M. C. Savadori, V. P. Mammana, O. G. Martins, and F. T. Degasperi, "Plasma-assisted chemical vapour deposition in a tunable microwave cavity," *Plasma Sources Science and Technology*, vol. 4, no. 3, pp. 489–494, 1995.

[24] V. J. Law and N. Macgearailt, "Visualization of a dual frequency plasma etch process," *Measurement Science and Technology*, vol. 18, no. 3, pp. 645–649, 2007.

[25] Y. H. Jung, S. O. Jang, and H. J. You, "Hydrogen generation from the dissociation of water using microwave plasmas," *Chinese Physics Letters*, vol. 30, no. 6, p. 065204, 2013.

[26] V. J. Law and D. Tait, "Contaminated ceramic plasma cleaning," *European Semiconductor*, vol. 19, no. 9, pp. S38–S41, 1997.

[27] V. J. Law and D. Tait, "Microwave plasma cleaning of ion implant ceramic insulators," *Vacuum*, vol. 49, no. 4, pp. 273–278, 1998.

[28] T. M. Burke, E. H. Linfield, D. A. Ritchie, M. Peper, and J. H. Burroughs, "Hydrogen radical surface cleaning of GaAs for MBE regrowth," *Journal of Crystal Growth*, vol. 175-176, pp. 416–421, 1997.

[29] D. Tasch, D. J. Brosky, S. Conrad, S. Kumar, and D. Kumar, "Microwave plasma cooking," US Patent 2009/0012223 A1, 2009.

[30] H. Stanley, "Microwave experiments at school," *Science in School*, vol. 12, pp. 31–33, 2009.

[31] H. Page, "Microwave oven plasma reactor," *Summer/Fall*, vol. 10, no. 3-4, pp. 11–13, 2001.

[32] A. Tallaire, "Plasma cleaning in modified microwave oven at LSPM (CNRS)," https://www.youtube.com/channel/UCfG3h7mSltjtsKcXH0dFCHQ.

[33] H. Stanley, "Plasma balls: creating the 4th state of matter with microwaves," *Science in School*, vol. 12, pp. 24–29, 2009.

[34] Soxfreak5243, "Grapes making fireballs in the microwave," https://www.youtube.com/watch?v=JrD6yzemDRw.

[35] Stupideaproductions, "Microwave plasma: awesome experiment," https://www.youtube.com/watch?v=G7lfzA7WzVI.

[36] J. P. Warren, "Microwave plasma experiment," https://www.youtube.com/watch?v=CNMjCggFKzM.

[37] W. Sajado, "10 destructive science experiments with microwave. Be really careful when using microwave!, https://www.youtube.com/watch?v=8Yv9o8aFTuk.

[38] Kreosan, "What microwave oven is capable. Generated plasma," https://www.youtube.com/watch?v=RrOw03gIIQQ.

[39] V. J. Law, G. A. C. Jones, N. Patel, and M. Tewordt, "Loading effects in CH_4 and H_2 Morie of GaAs," *Microelectronic Engineering*, vol. 11, no. 1–4, pp. 611–614, 1990.

[40] C. Chaichumporn, P. Ngamsirijit, N. Brkoonklin, K. Eaiprasetsak, and M. Fuangfoong, "Design and construction of 2.45 GHz microwave plasma source at atmospheric pressure," *Procedia Engineering*, vol. 8, pp. 94–100, 2011.

[41] D. Kumar and S. Kumar, "Plasma assisted joining," US Patent 7,309,843 B2, 2007.

[42] K. Y. Gyeong and R. J. Gwan, "Heating device of microwave oven," KR Patent 20010004084, 2001.

[43] S. C. Bo, L. Y. Woo, and S. S. Wom, "Heater apparatus for microwave oven," KR Patent 100766440, 2007.

[44] E. Jerby, Y. Meir, R. Jaffe, and I. Jerby, "Food cooking by microwave-excited plasmoid in air atmosphere," in *Proceedings of 14th International Conference on Microwave and High Frequency Heating*, pp. 17–30, Nottingham, UK, 2013.

[45] W. H. Lee and H. J. Kim, "Cooking device with deodorization," US Patent 2009/0110592 A1, 2009.

[46] N. F. Alekseev, D. D. Malairov, and I. B. Bensen, "Generation of high-power oscillations with a magnetron in the centimeter band," *Proceedings of the IRE*, vol. 32, no. 3, pp. 136–139, 1944.

[47] V. J. Law, "Microwave near-field plasma probe," *Vacuum*, vol. 51, no. 3, pp. 463–468, 1998.

[48] N. J. Brathwaite, J. P. Booth, and G Gunge, "A novel electrostatic probe method for ion flux measurements," *Plasma Sources Science and Technology*, vol. 5, no. 4, pp. 677–684, 1996.

[49] V. J. Law, "Ultraviolet fluorescence microwave plasma probe," *Vacuum*, vol. 49, no. 3, pp. 217–220, 1998.

3

High-Active Metallic-Activated Carbon Catalysts for Selective Hydrogenation

Nicolás Carrara ⓘ,[1] **Carolina Betti,**[1] **Fernando Coloma-Pascual,**[2] **María Cristina Almansa** ⓘ,[2] **Laura Gutierrez,**[1,3] **Cristian Miranda,**[4] **Mónica E. Quiroga** ⓘ,[1,3] and **Cecilia R. Lederhos** ⓘ[1]

[1]*Instituto de Investigaciones en Catálisis y Petroquímica (INCAPE) (FIQ-UNL, CONICET), Colectora Ruta Nac., No. 168 Km 0, Pje El Pozo, 3000 Santa Fe, Argentina*
[2]*Servicios Técnicos de Investigación, Facultad de Ciencias, Universidad de Alicante, Apartado 99, 03080 Alicante, Spain*
[3]*Facultad de Ingeniería Química, Universidad Nacional del Litoral, Santiago del Estero 2654, 3000 Santa Fe, Argentina*
[4]*Laboratorio de Investigación en Catálisis y Procesos (LICAP), Universidad del Valle, Ciudad Universitaria Meléndez, Calle 13 # 100-00, Cali, Colombia*

Correspondence should be addressed to Cecilia R. Lederhos; clederhos@fiq.unl.edu.ar

Academic Editor: Raghunath V. Chaudhari

A series of low-loaded metallic-activated carbon catalysts were evaluated during the selective hydrogenation of a medium-chain alkyne under mild conditions. The catalysts and support were characterized by ICP, hydrogen chemisorption, Raman spectroscopy, temperature-programmed desorption (TPD), temperature-programmed reduction (TPR), X-ray diffraction (XRD), Fourier transform infrared spectroscopy (FTIR micro-ATR), transmission electronic microscopy (TEM), and X-ray photoelectronic spectroscopy (XPS). When studying the effect of the metallic phase, the catalysts were active and selective to the alkene synthesis. NiCl/C was the most active and selective catalytic system. Besides, when the precursor salt was evaluated, PdN/C was more active and selective than PdCl/C. Meanwhile, alkyne is present in the reaction media, and geometrical and electronic effects favor alkene desorption and so avoid their overhydrogenation to the alkane. Under mild conditions, nickel catalysts are considerably more active and selective than the Lindlar catalyst.

1. Introduction

Olefins are of great academic and industrial interest. They are very important raw material for the synthesis of biologically active compounds, margarine, and lubricants, as well as in plastic industry [1]. Also, olefins present many applications in fine chemistry. The purification of olefins is a very important step in manufacturing polymers with a 0.35% content of impurities, as alkyne compounds, leading to a polymer with undesirable properties. There are several ways of purification of olefins stream, and the most common is solvent extraction, which is a nonecological friendly procedure. Another alternative for obtaining olefins is via the selective hydrogenation of alkynes. This catalytic process is the major challenge because they reduce the production costs and let obtaining good yields and high selectivities.

Different noble metals as Pd, Pt, Ru, and Rh anchored on inorganic supports are highly active and selective for carbon-carbon multiple bond hydrogenations [2–7]. During the partial hydrogenation of alkynes to alkenes, or dienes to olefins, many authors found that supported palladium catalysts presented the highest catalytic activities [8, 9]. The most used catalyst for alkyne partial hydrogenation is the commercial *Lindlar* catalyst, that contains palladium poisoned with lead (Pd/CaCO$_3$, 5 wt.% Pd modified with Pb (OAc)$_2$) [10]. The cost of this noble metal has increased greatly in the last decade, and the world tendency is to develop cheaper catalysts by lowering the Pd content or by using cheaper metals.

In previous papers [11, 12], we have studied the partial hydrogenation reaction of a medium-chain alkyne (1-heptyne) to obtain the corresponding olefin using monometallic catalysts using alumina and carbonaceous supports. In these papers, the influences of different supports, reaction, and reduction temperature and the metallic phase on the activity and the selectivity were evaluated.

Activated carbonaceous surfaces have been studied during the last decades [13, 14] in different reactions showing that they are very complex. Carbonaceous supports are widely used because of their inertness, stability, high specific surface, low cost, and ability to easily recover the metallic phase after ending their useful life. Besides, during the synthesis of the catalysts, on the surface of activated carbon, different oxygenated and nitrogenized functional groups are found, and both can be easily modified when the material is physically or chemically pretreated [15, 16].

The objectives of this work were to synthesize low-loaded Ru, Pd, and Ni catalysts using an activated carbon as support and evaluate their activities and selectivities during 1-heptyne selective hydrogenation reaction under mild conditions. The effect of the type of metal and the precursor salt on the activity and selectivity to the desired product (1-heptene) is also evaluated. The Lindlar catalyst was used as a reference.

2. Experimental

2.1. Catalyst Preparation. CNR-115 pelletized activated carbon provided by NORIT (S_{BET}: 1503 m$^2 \cdot$g^{-1}, micropore volume: 0.738 mL$\cdot$g^{-1}, and total pore volume: 1.010 mL$\cdot$g^{-1}) was used as a support. Chloride acidic solutions of Ru, Pd, and Ni (purity >99.98%) at pH = 1 with HCl were used to prepare PdCl/C, RuCl/C, and NiCl/C catalysts by the incipient wetness technique. On the contrary, PdN/C was prepared from Pd(NO$_3$)$_2$ acidic solution (purity >99.98%) at pH = 1 with HNO$_3$. Solutions have the necessary concentration to obtain 0.4 wt.% of each metal (M: Pd, Ni, and Ru) on the final catalyst. After impregnation, all samples were dried overnight at 373 K and reduced under a hydrogen stream, flow rate of 25 mL$\cdot$min^{-1}, at 393 K for 1 h before each catalytic evaluation.

2.2. Catalyst Characterization. Metal loadings of the catalysts were obtained by the ICP technique in PerkinElmer Optima 2100DV equipment, previous digesting the samples in a *Start D* Microwave Digestion system (Milestone).

Dispersion values were obtained by hydrogen chemisorption in Micromeritics Accusorb 2100e equipment after reducing each sample at 393 K for 1 h under 5% v/v H$_2$/Ar, and an atomic ratio H : M equal to 1 was assumed for all of the calculations as suggested by the other authors [17–19]. The particle diameter of the catalysts was calculated from the chemisorption results following the procedure reported by Paryjczac et al. [20, 21]. The adopted metallic surface area, σ value, was calculated from the average of the densities of the (100), (110), and (111) crystal planes [22, 23]; the values for Pd, Ru, and Ni were 1.27$\cdot$10^{19}, 1.63$\cdot$10^{19}, and 1.54$\cdot$10^{19} atoms$\cdot$m^{-2}, respectively.

Temperature-programmed desorption (TPD) allowed to determine the amount and type of oxygen-containing groups on the activated carbon surface. 10 mg of the sample were treated in a He stream at 100 mL$\cdot$min^{-1} from 293 to 1273 K at 20 K$\cdot$min^{-1}. In the experiments, the effect of the thrust is corrected using a "white" curve. CO and CO$_2$ desorptions were measured in a simultaneous TG-DTA (Mettler Toledo model TGA/SDTA851e/LF/1600) coupled to a quadrupole mass spectrometer (Pfeiffer Vacuum model Thermostar GSD301T).

The FTIR micro-ATR technique was carried in a Shimadzu FTIR-8400 spectrophotometer which is based on the internal reflectance phenomenon. The FTIR micro-ATR technique is the most sensitive technique for surface analysis [24].

The Raman spectra were recorded using a LabRam spectrometer (Horiba–Jobin–Yvon) coupled to an Olympus confocal microscope (a 100x objective lens was used for simultaneous illumination and collection) and equipped with a CCD detector cooled to about 200 K using the Peltier effect. The excitation wavelength was 532 nm in all cases (Spectra-Physics diode pump solid state laser). The laser power was set at 30 mW.

X-ray diffraction (XRD) measurements of powdered samples were obtained using a Shimadzu XD-D1 instrument with CuK$_\alpha$ radiation ($\lambda = 1.5405$ Å) in the $21 < 2\theta < 49°$ at 0.25°$\cdot$min^{-1} scan speed.

The electronic state of superficial species and their atomic ratios were determined by X-ray photoelectron spectroscopy (XPS), following Pd and Ru 3d$_{5/2}$, Ni and Cl 2p$_{3/2}$, N, C, and O 1s peaks' binding energy (BE) for catalysts and support. Measurements were acquired in VG-Microtech Multilab equipment, with MgK$_\alpha$ ($h\nu$: 1253.6 eV) radiation and a pass energy of 50 eV. The analysis pressure during data acquisition was kept at 5.10^{-7} Pa. Samples were treated in situ in the presence of a H$_2$ stream following the same pretreatment conditions for each catalyst. A careful deconvolution of the spectra was made, and the areas of the peaks were estimated by calculating the integral of each peak after subtracting a Shirley background and fitting the experimental peak to a combination of Lorentzian/Gaussian lines of 30–70% proportions. The reference binding energy (BE) was C 1s peak at 285.0 eV. Determinations of the superficial atomic ratios were made by comparing the areas under the peaks after background subtraction and corrections due to differences in escape depth and in photoionization cross sections [25].

The reducibility of the surface species in all the catalysts was determined by temperature-programmed reduction (TPR) which was carried out in an Ohkura TP 2002 S instrument equipped with a thermal conductivity detector. Samples were pretreated at 373 K for 30 min under an argon stream, cooled up to room temperature under an argon flow, and then, the temperature was increased up to 1223 K at 10 K$\cdot$min^{-1} in a 5% hydrogen/argon mixture stream.

Transmission electronic microscopy (TEM) photographs were obtained using an electronic microscope JEOL JEM-2011 at 200 kV. The samples were reduced for 1 h at 393 K and dispersed in distilled water in order to obtain TEM images which are only used for comparative purposes.

2.3. Catalytic Evaluation. The reaction test accessed was 1-heptyne selective hydrogenation, carried out at 303 K, 150 kPa, and 800 rpm during 180 min. The possibility of

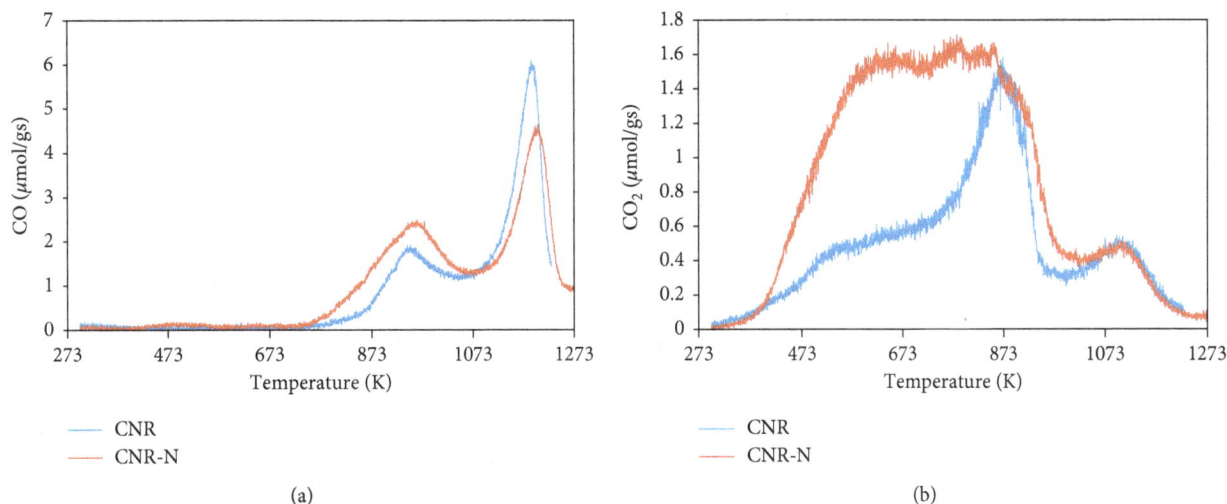

FIGURE 1: TPD spectra: (a) CO and (b) CO_2 desorption profiles as a function of temperature for CNR and CNR-N samples.

external and internal diffusional limitations during the catalytic tests was discarded using a previously described procedure [26]. Experiments were performed in a stainless steel stirred tank reactor equipped with a magnetically driven stirrer with two blades in counterrotation. The inner wall of the reactor was completely coated with PTFE in order to neglect the catalytic action of the steel of the reactor found by other authors [27]. Samples of 0.75 g of the catalyst were used in a total volume of 75 mL of 5% (v/v) solution of 1-heptyne (Fluka, Cat. no. 51950) in toluene (Merck, Cat. no. TX0735-44).

Blank tests were obtained using CNR and CNR-N activated carbons, and no conversion of 1-heptyne was observed. The commercial *Lindlar* catalyst (Aldrich Cat. no. 20,573-7) was used as a reference, maintaining constant S : M (1-heptyne/metal molar ratio). Reactant and products were analyzed by gas chromatography using a flame ionization detector and a HP INNOWax polyethyleneglycol capillary column.

3. Results and Discussion

3.1. Catalysts and Support Characterization. A way to know the number and nature of oxygen functional groups present on the carbon surface is by the temperature-programmed desorption (TPD) technique. It is well known [15, 16] that the treatment of the catalysts with a strong oxidant acid, such as nitric acid, modifies the oxygen amount and the kind of superficial groups in the carbon surface. For this purpose, a sample of the carbon support (CNR) was treated with HNO_3 (CNR-N), and both were analyzed and compared by TPD. This allows studying the effect of nitric acid, which was used during the catalyst preparation step, on the CNR surface.

In Figures 1(a) and 1(b), the concentration profiles of CO and CO_2 as a function of temperature, respectively, for CNR and CNR-N samples are shown. During the TPD experiments, the samples analyzed showed highly different behaviors, mainly during the CO_2 desorption (Figure 1(b)) due to the decomposition of carboxyl and lactone groups. It

TABLE 1: Quantitative results of the TPD-MS analysis for carbon samples.

	CO (μmol/g)	CO_2 (μmol/g)	O (μmol/g)
CNR	2349	1394	5136
CNR-N	2621	2442	7504

can be seen that these surface groups had highly increased in the carbon sample treated with the oxidizing acid (HNO_3).

The quantitative results from TPD-MS analysis for the CNR and CNR-N supports are presented in Table 1. The amount of surface oxygen groups of carbonaceous supports, estimated by CO, CO_2, and O_2 evolved in TPD experiments (Table 1), shows that the oxidizing treatment with HNO_3 produces a large amount of surface oxygenated complexes, with 175% higher amount of carboxylic-type groups that decompose to give CO_2 under the heat treatment [28]. This enhancement is also observed in the TPD profile in Figure 1(b).

The FTIR micro-ART spectra are given in Figure 2(a) for the supports CNR and CNR-N and in Figure 2(b), for the catalysts PdCl/C, PdN/C, NiCl/C, and RuCl/C. On the one hand, in Figure 2(a), it can be seen that the CNR carbonaceous support did not show new bands in the support treated with nitric acid (CNR-N). The bands at 665, 1583, and 3346 cm^{-1} correspond to ring deformation and C-O stretching vibrations in phenolic groups, but in general, the lack of parallel behavior between the growth of bands in 1000–1350 cm^{-1} is attributed to primarily C-O stretching of ether groups, possibly furans, to C-OH species [29], and to O-H or C-N stretching [30].

On the other hand, in Figure 2(b), for all the catalysts, PdN/C, PdCl/C, NiCl/C, and RuCl/C, the following peaks can be seen in general: (i) a broad band centered at 3444 cm^{-1} which was assigned mainly to O-H stretching, associated with the water absorbed on the surface of the material and hydroxylated groups; (ii) a NH_2 stretching band located at 3600–3100 cm^{-1}; and (iii) a C-H stretching band between 3000 and 2800 cm^{-1} [31]. In the 1350–1000 cm^{-1} range, the growth of bands for the PdN/C catalyst can be

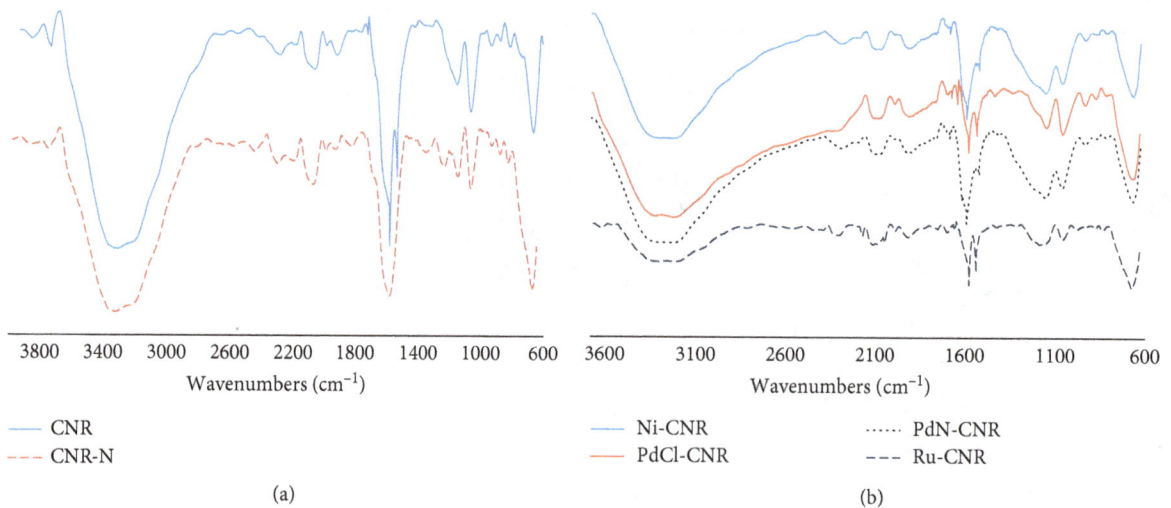

FIGURE 2: FTIR micro-ART spectra for the supports (a) CNR and CNR-N and for the catalysts (b) PdCl/C, PdN/C, NiCl/C, and RuCl/C.

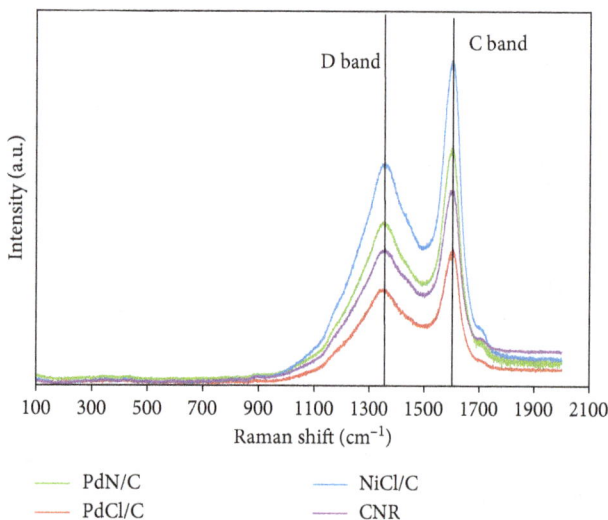

FIGURE 3: Raman spectroscopy for PdN/C, PdCl/C, and NiCl/C catalysts.

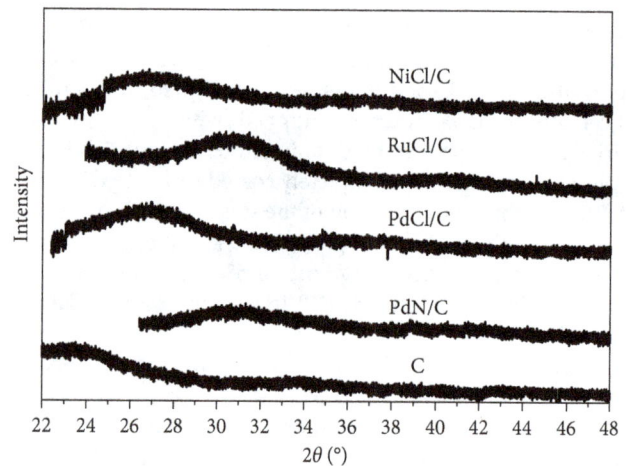

FIGURE 4: DRX diffractograms of PdN/C, PdCl/C, RuCl/C, and NiCl/C catalysts treated in H_2 at 393 K and the CNR support.

attributed mainly to C-N stretching [32] because of the support treatment with nitric acid during the synthesis of the catalyst.

Figure 3 shows the Raman spectra of the catalysts and carbon support in the range of 100–2000 cm^{-1}. The profile of all catalysts analyzed is very similar to the CNR support spectrum. Two intense bands at ca. 1600 cm^{-1} and ca. 1350 cm^{-1} are observed, which are the typical signals of the graphite spectrum. The first signal is known as the *G band* associated with the graphitic order, and the second is named the *D band* and is attributed to structural defects [32].

PdN/C, PdCl/C, RuCl/C, NiCl/C, and the CNR support diffractograms are shown in Figure 4. In this figure, it can be seen that there are no defined signals for the metallic samples or to the support, which could indicate the absence of the crystalline phases in all samples, or the crystals were very small (<100 Å). The absence of the crystalline phase is a consequence of the low-metallic loadings of the samples, 0.4 wt.%, low below the detection limit of the XRD equipment.

Figure 5 shows temperature-programmed reduction profiles of PdN/C, PdCl/C, RuCl/C, and NiCl/C catalysts. In this figure, the TPR profile of CNR carbon is also presented for a comparative way. As can be seen in Figure 5, the support and the catalysts profiles present a broad peak with a maximum at ca. 870 K, which can be attributed to the gasification of the carbonaceous support, the decomposition of chloride or nitrogen species, the reduction of superficial oxygenated groups of the support, or due to the possible presence of impurities (5-6%), producing CH_4, CO, and/or CO_2 [33–35]. de Miguel et al. [36], for carbonaceous supports, found these types of consumption at very high temperatures, between 923 and 1223 K, associated with decomposition of superficial oxygenated groups (C[O]) that can liberate CO in the presence of a reductive stream. In Figure 5, it can be seen that, at temperatures higher than 700 K, all metallic catalysts samples present very similar profile due to the reduction of surface groups of the carbonaceous support.

The palladium catalysts (PdN/C and PdCl/C) presented at high temperatures a much wider peak profiles with respect

TABLE 2: Metal loading, dispersion, particle diameter, XPS, initial reaction rate, and activity (TOF) results for the catalysts.

Catalysts	Metal loading (%)	D (%)	d (nm)	XPS		$r°$ (mol·g_M^{-1}·s^{-1})	TOF (s^{-1})
				M $3d_{5/2}$ BE (eV)	Cl/M (at/at)		
PdN/C	0.39	48	2.4	335.2	—	0.077	17.0
PdCl/C	0.42	62	1.9	335.2	1.56	0.050	8.6
RuCl/C	0.41	20	5.3	282.1	2.96	0.056	28.5
NiCl/C	0.41	4	25.2	857.3	2.05	0.026	38.9
Lindlar	5	2.5	46.4	337.2	—	0.003	13.5

to the support, as can be observed in Figure 5, with a maximum at ca. 875 K. No peaks are present at low reduction temperature, and this is because the palladium can be reduced by carbon at low temperature, even at room temperature as was observed by other authors [37]. In Figure 5, it can be seen that the peak above 700 K is more intense for PdN/C and NiCl/C catalysts. Besides, the TPR profile of the NiCl/C sample shows a small peak at 610 K attributed to the reduction of NiO to Ni° species, which indicates a weak metal-support interaction [38–40]. Lastly, for the ruthenium catalyst (RuCl/C), the TPR profile presents a signal with maximum at 610 K, related to the reduction of ruthenium, $Ru^{3+} \rightarrow Ru^{2+} \rightarrow Ru°$, assigned to the reduction of the ruthenium chloride species [16, 35].

As an early conclusion, at the reduction temperature used during the preparation step of the catalysts (393 K, marked in Figure 5), the palladium metals on PdCl/C and PdN/C could be mostly as Pd° species on the catalyst surface; meanwhile, nickel and ruthenium are present as Ni^{2+} and Ru^{3+} species.

Table 2 summarizes all the catalysts under study, the results of ICP, dispersion and particle diameter by hydrogen chemisorption, XPS, initial reaction rate, and activity TOF values. For a comparative study, the *Lindlar* catalyst was also included in Table 2. The metal loadings of the synthetized catalysts are ca. 0.4 wt.% in the final catalyst, while a concentration of 5 wt.% of Pd was corroborated for commercial *Lindlar*. Dispersion values show the following order PdCl/C > PdN/C > RuCl/C, while NiCl/C and *Lindlar* catalysts present extremely low D values (lower than 5%). Higher dispersions are favored by low loadings of the metallic precursor and by low pretreatment conditions used during the preparation of the catalysts [41]; besides, high surface areas and large pore volume of the carbonaceous materials usually favor very high dispersion of an active phase. All these factors avoid the sintering of the active phase. In our case, the highest dispersion of the Pd catalyst prepared from chlorinated precursor can be due to the presence of complex oxychlorinated species formed during the pretreatment step. The presence of these species improves the metal dispersion due to the greater interaction between these oxychlorinated species and the support as compared to the weaker one displayed by palladium(II) oxide (PdO) [42–44]. Furthermore, the PdN/C catalyst presents a dispersion of ca. 50% attributed to the high concentration of carboxylic-type groups on the surface, as observed by TPD-MS analysis, that exert an electronic and geometrical effect over the nitrate precursor used. On the contrary, due to the high surface area, ruthenium particles would also be highly dispersed on these carbon supports

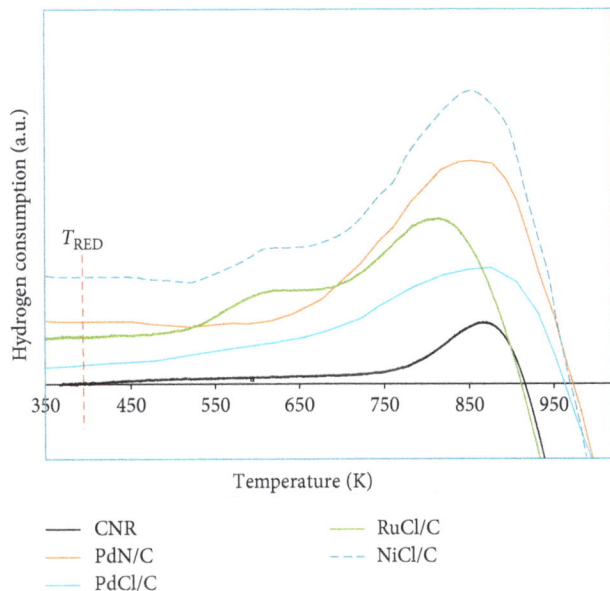

FIGURE 5: TPR profiles of PdN/C, PdCl/C, RuCl/C, and NiCl/C catalysts and the CNR support.

[45]. However, there is no simple relationship between the value of the surface area of the activated carbon and the metal dispersion of the catalysts supported on activated carbon [46]. Some authors [47–51] reported the inhibition and drawbacks due to the use of $RuCl_3$ as a precursor salt because the chloride from this salt is a poison for CO and H_2 chemisorptions and may be partitioned between the metal and the support. In our case, the high degree of sintering of the ruthenium metal particles could be due to the fact that, during the pretreatment process, no oxychlorinated species interacting with the support are formed, as noted earlier while studying ruthenium catalysts of different metal contents over the alumina support [12, 21]. The low dispersion of nickel is a characteristic when this metal is used to prepare inorganic or carbonaceous catalysts [52, 53].

From the H_2 chemisorption results, the metal particle size (d) was determined on assuming a spherical particle in accordance with Badano et al. [21]. The particle size order is PdCl/C < PdN/C < RuCl/C ≪ NiCl/C ≪ *Lindlar*. The smallest particle size of PdCl/C catalysts is due to the higher surface interaction between $Pd^{\delta+}O_xCl_y$ and oxygenated groups of the carbonaceous support, as observed earlier when alumina was used as a support [21]. The higher particle size of NiCl/C could be related with the higher structural defects observed by the Raman spectra in Figure 3.

(a)

(b)

(c)

(d)

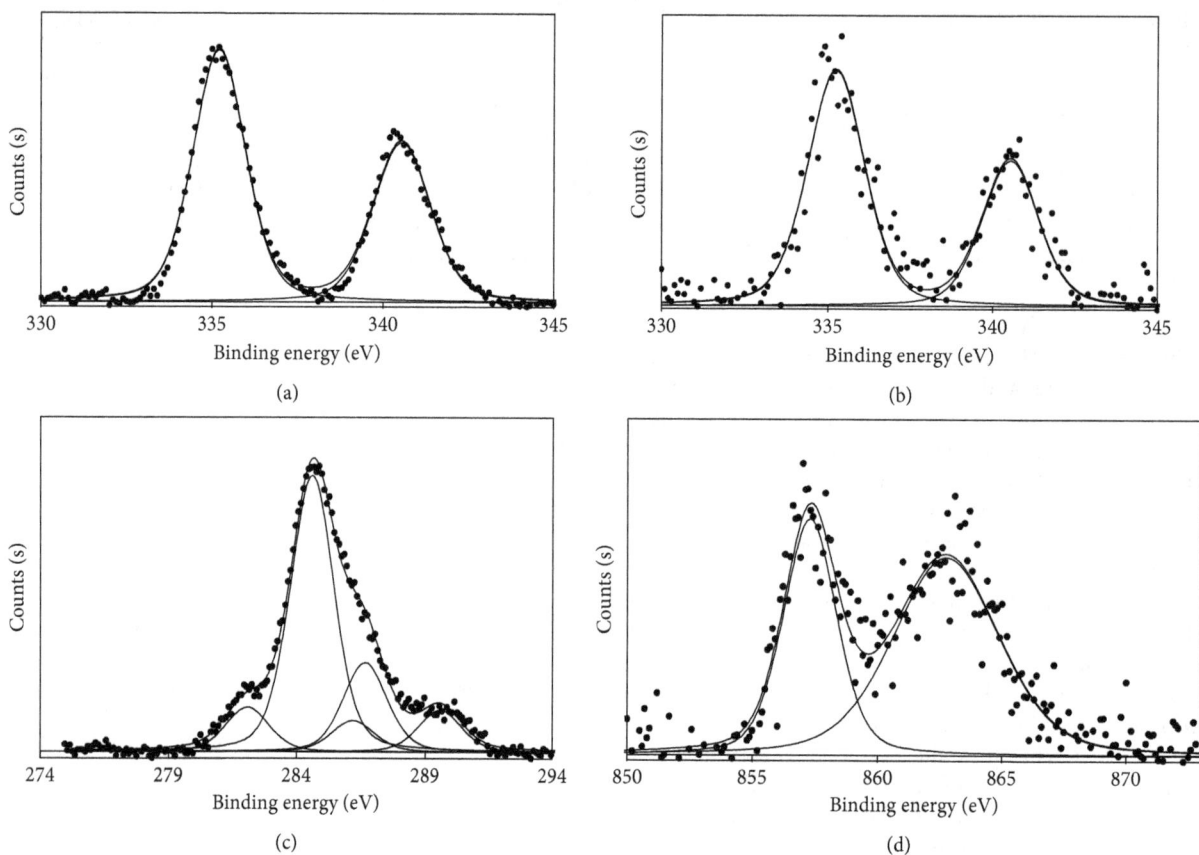

FIGURE 6: XPS spectra of BE Pd 3d for (a) PdN/C, (b) PdCl/C, (c) RuCl/C, and (d) NiCl/C catalysts treated in H_2 at 393 K.

Figure 6 shows the XPS spectra of the Pd $3d_{5/2}$ band for PdN/C and PdCl/C, Ru $3d_{5/2}$ band for RuCl/C, and Ni $2p_{3/2}$ band for the NiCl/C catalyst. Besides, the XPS spectra of the catalysts prepared from chlorine precursors (PdCl/C, RuCl/C, and NiCl/C) show a Cl $2p_{3/2}$ peak at ca. 199.4 eV that corresponds to surface chloride species [54] that were not completely eliminated after reduction because of the high adsorption capacity of the carbonaceous material used. The maximum binding energy (BE) of Pd and Ru $3d_{5/2}$, Ni $2p_{3/2}$ peaks, and Cl/M atomic ratios are also listed in Table 2.

XPS results in Figure 6 show that both palladium catalysts present a Pd $3d_{5/2}$ peak at 335.2 eV, which can be attributed to Pd$^{\delta+}$ ($\delta \approx 0$) [16, 54]. Figure 6 also shows the spectra of palladium catalysts and the $3d_{3/2}$ peak doublet shifted 5.2 eV with respect to the $3d_{5/2}$ peak. No nitrogen was detected by XPS on the Pd catalyst prepared from the nitrate precursor. On the contrary, for the PdCl/C sample, the Cl/Pd atomic ratio was 1.56, confirming that chloride is present even after the reduction treatment. This could be attributed to the high absorption capacity of the carbon support. The chlorine content does not affect the palladium surface as it is totally reduced on the PdCl/C catalyst.

In the case of ruthenium, the BE reference is difficult because the Ru 3d peaks appear at the same region than the C 1s peak. There are also discrepancies in the BE reported in the literature for ruthenium compounds. The Ru $3d_{5/2}$ signal for the RuCl/C catalyst was observed at 282.1 eV corresponding to ruthenium electrodeficient species as RuCl$_3$

[54]. For this catalyst, after the temperature pretreatments, the superficial atomic ratio Cl/Ru was 2.96.

According to the bibliography [54, 55], the Ni $2p_{3/2}$ signal appears at 852.5 eV for Ni° species. For the NiCl/C catalyst, the two signals were observed in the spectrum of Figure 6 at 857.3 and 862.7 eV. The first peak could be assigned to the electrodeficient nickel species (Ni^{n+}, with $n \rightarrow 2$), probably corresponding to nickel interacting with the support [54–56], while the second peak corresponds to the characteristic shake-up satellite structure of Ni (II) [55, 57, 58]. This catalyst also presents a high Cl/Ni atomic ratio (2.05) as a consequence of the adsorption capacity or the carbonaceous support.

For the Lindlar catalyst, in Table 2, the main peaks at 335.2 eV (69%, at/at) and 336.9 eV (31%, at/at) are also registered, assigned to Pd$^{\delta+}$ and electrodeficient Pd^{n+} species (with $\delta \rightarrow 0$ and $n \rightarrow 2$), respectively [53]. On the contrary, on the Lindlar catalyst, also two peaks of the Pb $4f_{7/2}$ spectrum at 136.8 eV (20%, at/at) and 138.6 (80%, at/at) were attributed to Pb and Pb(OAc)$_2$, respectively [53].

In Figure 7, TEM micrographs of the carbonaceous catalysts are shown. The metal particle size for the PdCl/C catalyst is between 1 and 4 nm, presenting a high concentration of 2 nm Pd particles on the carbon surface. For PdN/C, the main palladium particle size is between 4 and 8 nm. The particle sizes of Ru on RuCl/C are mainly between 1 and 5 nm, although several bigger particles were observed on the micrographs. Besides, the NiCl/C catalyst presented the highest particle sizes between 25 and 40 nm.

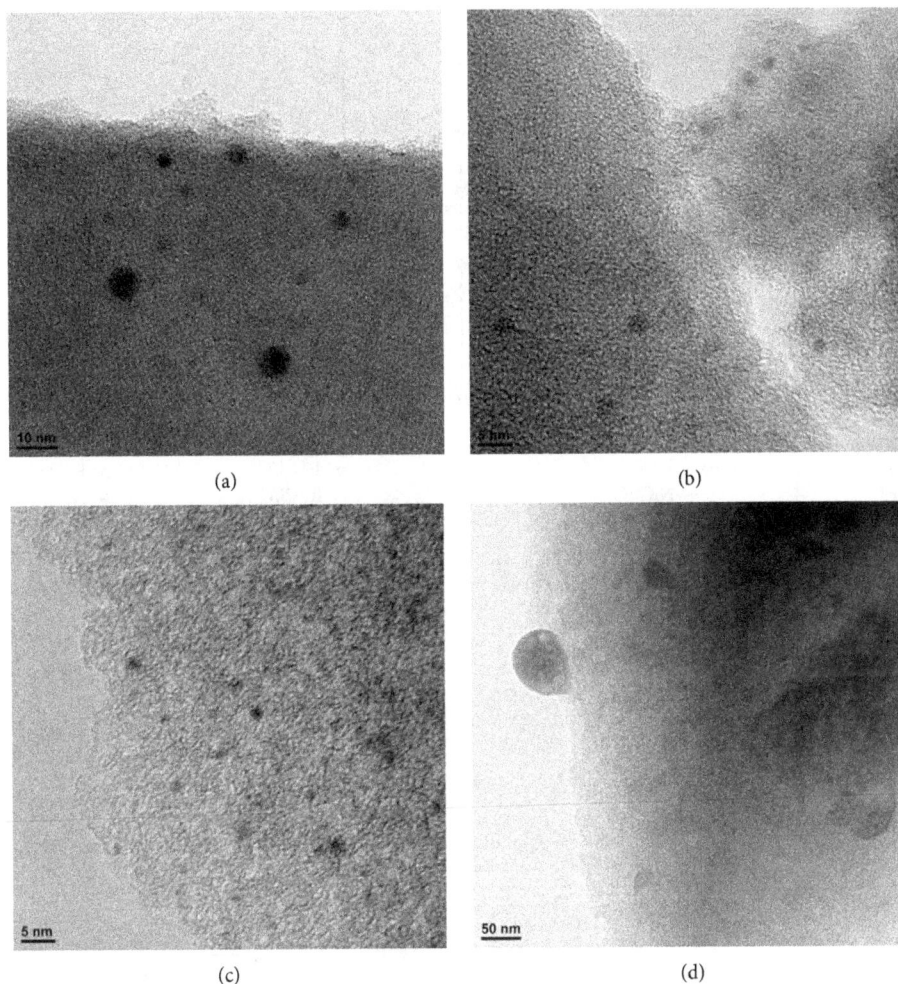

(a)

(b)

(c)

(d)

FIGURE 7: TEM Images for (a) PdN/C, (b) PdCl/C, (c) RuCl/C, and (d) NiCl/C catalysts.

3.2. Catalytic Tests.

The catalytic performance during the selective hydrogenation of 1-heptyne for PdN/C, PdCl/C, RuCl/C, and NiCl/C catalysts is shown in Figure 8. 1-heptyne total conversion (%) and selectivity to 1-heptene (%) as a function of time (min) were plotted in this figure for all the studied catalysts. All the catalysts were active during the hydrogenation of 1-heptyne and presented high selectivity to 1-heptene, the desired product.

Besides, in Figure 9, 1-heptyne total conversion (%) and selectivity to 1-heptene (%) as a function of time (min) for the commercial Lindlar catalyst used as a reference is shown.

According to Figures 8 and 9 the order of selectivities to the corresponding alkene was:

$$PdCl/C < PdN/C \cong RuCl/C < Lindlar < NiCl/C \qquad (1)$$

In order to compare the performance of the catalytic systems, the initial reaction rate of hydrogenation of 1-heptyne (r_A^0) was estimated for all the catalysts using the following formula:

$$r_A^0 = \frac{V \cdot C_A^0}{W_M} \cdot \left(\frac{\partial X_A}{\partial t}\right)_{t=0}, \qquad (2)$$

where r_A^0 is the initial 1-heptyne reaction rate (mol·g_M^{-1}·s^{-1}), $(\partial X_A/\partial t)_{t=0}$ is the tangent value of the 1-heptyne total conversion versus time curve at $t = 0$, C_A^0 is the initial concentration of 1-heptyne (mol·L^{-1}), W_M is the mass of the metal (Pd, Ru, or Ni) in the catalyst (g), V is the reaction volume (L), and t is the reaction time (s).

From these initial reaction rates of 1-heptyne hydrogenation, TOF values were calculated for all the catalysts and are summarized in Table 2. For hydrogenation reactions, TOF values between 10^{-2} and 10^2 indicate a structure-insensitive system, not influenced by particle size [59, 60], so the differences in activities and selectivities between the catalysts may be associated with geometrical and/or geometrical effects. The order of activities of low-loaded catalysts is

$$PdCl/C < Lindlar < PdN/C < RuCl/C < NiCl/C \qquad (3)$$

The effect of the precursor salt was studied by comparing PdCl/C and PdN/C catalysts. The results obtained for these catalysts show differences in both activity and selectivity. For the catalyst synthetized from palladium nitrate as the precursor salt, the obtained TOF was 17 s^{-1} versus the activity of palladium chloride precursor salt 8.6 s^{-1}. That is, each active site of the PdN/C catalyst is almost two times more active than that of PdCl/C. As observed by XPS and RTP, both catalysts have Pd$^{\delta+}$ (with $\delta \cong 0$) as the active site, so no

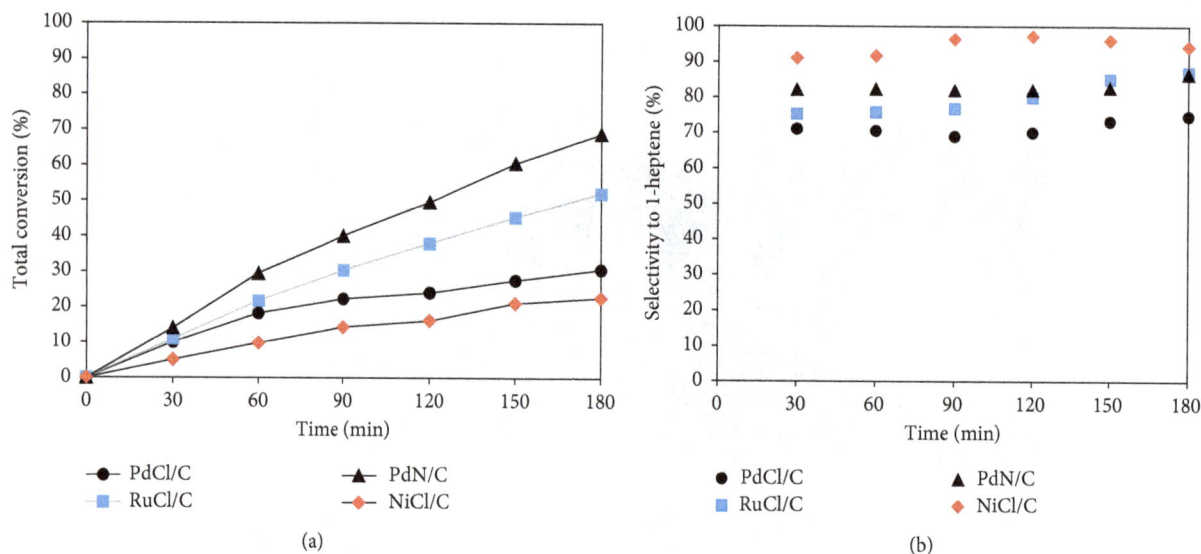

FIGURE 8: Total conversion (a) and selectivity to 1-heptene (b) versus time (min) for PdN/C, PdCl/C, RuCl/C, and NiCl/C catalysts pretreated in H_2 at 393 K.

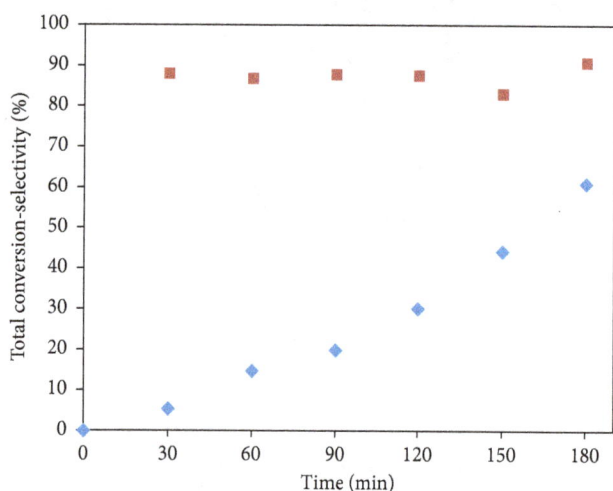

FIGURE 9: Total conversion (◆) and selectivity to 1-heptene (■) versus time (min) for the *Lindlar* catalyst (reproduced from Lederhos et al. [52].

electronic effects are generated by the metallic phase; besides as observed by Raman, quite higher structural effects were found on PdN/C than on the PdCl/C sample. When comparing the catalytic behavior of the different Pd precursors, higher selectivities to 1-heptene were obtained when palladium nitrate was used, ca. 86% at 180 min, while for PdCl/C selectivities, ca. 70% was obtained.

In Figure 10, the main superficial groups present on carbonaceous supports are summarized. Between them carboxylic groups by heterolytic rupture of the O-H bond, could favor the adsorption of the cationic metallic precursors M^{n+}. In our case, the higher distribution of carboxylic and lactone groups on the surface of PdN/C could favor the adsorption of Pd^{2+}, and its reduction to $Pd°$ (as detected by XPS and TPR) during the pretreatment step: for 1 h at 393 K under a hydrogen stream.

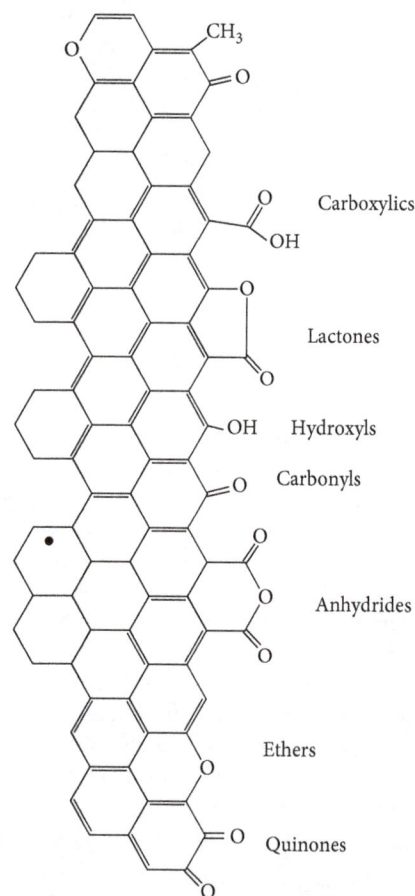

FIGURE 10: Schematic representation of the main superficial groups found on a carbonaceous supports.

Also, the high concentration of carboxylic groups favors the high dispersion of the active sites and the small particle sizes. So, the observed differences in activity between PdN/C and PdCl/C could be originated, at least in part, by *geometrical effects* due to higher structural defects (as noted in D-bands on the Raman spectra) and *electronic effects* originated by the high concentration of carboxylic superficial groups on PdN/C (the favorable effect) and high remnants of chloride on PdCl/C catalysts (the negative effect). Besides, differences in selectivity between the PdN/C and PdCl/C catalysts could be caused on the one hand to the electronic and geometrical effects that favors the formation of 1-heptene and on the other hand to higher amount of carbonaceous superficial groups that favors the desorption of the alkene, avoiding its over-hydrogenation, while alkyne is present in the reaction media.

When analyzing the effect of the metal type used to synthesize the carbonaceous catalysts, PdCl/C, RuCl/C, and NiCl/C, the highest activity and selectivity were achieved when the supported metal is nickel, a very low cost precursor salt. The high activity and selectivity of NiCl/C could be related to electronic or geometrical effects of active sites, which would control the dissociative adsorption of hydrogen being this the controlling step [61], and/or to the aromatic superficial groups of the support that would preferentially adsorb remaining 1-heptyne desorbing 1-heptene from the reaction media, so avoiding its total hydrogenation to the alkane.

4. Conclusions

The effect of metallic salt and the kind of metal was studied using a carbonaceous support. A series of low-loaded catalysts supported on an activated carbon (ca. 0.4 wt.%) were synthetized by the incipient wetness technique. Nanoparticles of palladium, ruthenium, and nickel were deposited using $Pd(NO_3)_2$, $PdCl_2$, $RuCl_3$, and $NiCl_2$ aqueous acidic solutions as precursor salts. The catalysts were pretreated under moderate conditions.

PdN/C, PdCl/, RuCl/C, and NiCl/C were characterized by ICP, TPD, FTIR micro-ATR, Raman, XRD, TPR, hydrogen chemisorption, XPS, and TEM.

All the catalytic systems were assessed during the selective hydrogenation of 1-heptyne, a medium-chain alkyne under moderate operational conditions: 303 K, 150 kPa, and 800 rpm during 180 min. The catalysts were active and selective to 1-heptene. The order of activity and selectivity found was NiCl/C > RuCl/C > PdCl/C. Geometrical and electronic effects over the NiCl/C catalyst favors alkyne adsorption and 1-heptene desorption in the reaction media, avoiding alkyne total hydrogenation to the alkane. Besides, the PdN/C catalyst is more active and selective than PdCl/C due to geometrical and electronic effects, and the first ones are caused by structural defects (observed by Raman), while the seconds ones are originated by the high loading of carboxylic superficial on PdN/C and the chloride remnants on PdCl/C catalysts (as observed by TPD-MS and XPS, resp.).

Conflicts of Interest

The authors declare that they have no conflicts of interest.

Acknowledgments

The financial assistance of UNL (CAI+D), CONICET (PIP 457), and ANPCyT (PICT-2013-2021 and PICT-2016-1453) is greatly acknowledged.

References

[1] A. Molnár, A. Sárkány, and M. Varga, "Hydrogenation of carbon–carbon multiple bonds: chemo-, regio- and stereo-selectivity," *Journal of Molecular Catalysis A: Chemical*, vol. 173, no. 1-2, pp. 185–221, 2001.

[2] D. Teschner, E. Vass, M. Hävecker et al., "Alkyne hydrogenation over Pd catalysts: a new paradigm," *Journal of Catalysis*, vol. 242, no. 1, pp. 26–37, 2006.

[3] P. Kačer, M. Kuzma, and L. Červený, "The molecular structure effects in hydrogenation of cycloalkylsubstituted alkynes and alkenes on platinum and palladium catalysts," *Applied Catalysis A: General*, vol. 259, no. 2, pp. 179–183, 2004.

[4] B. M. Choudary, M. Lakshmi Kantam, N. Mahender Reddy et al., "Hydrogenation of acetylenics by Pd-exchanged mesoporous materials," *Applied Catalysis A: General*, vol. 181, no. 1, pp. 139–144, 1999.

[5] J. Lei, L. B. Su, K. Zeng et al., "Recent advances of catalytic processes on the transformation of alkynes into functional compounds," *Chemical Engineering Science*, vol. 171, pp. 404–425, 2017.

[6] C. Oger, L. Balas, T. Durand, and J.-M. Galano, "Are alkyne reductions chemo-, regio-, and stereoselective enough to provide pure (Z)-olefins in polyfunctionalized bioactive molecules?," *Chemical Reviews*, vol. 113, no. 3, pp. 1313–1350, 2013.

[7] G. A. Attard, J. A. Bennett, I. Mikheenko et al., "Semi-hydrogenation of alkynes at single crystal, nanoparticle and biogenic nanoparticle surfaces: the role of defects in Lindlar-type catalysts and the origin of their selectivity," *Faraday Discussions*, vol. 162, pp. 57–75, 2013.

[8] S. Nishimura, *Handbook of Heterogeneous Catalytic Hydrogenation for Organic Synthesis*, Wiley, New York, NY, USA, 2001.

[9] B. Chen, U. Dingerdissen, J. G. E. Krauter et al., "New developments in hydrogenation catalysis particularly in synthesis of fine and intermediate chemicals," *Applied Catalysis A: General*, vol. 280, no. 1, pp. 17–46, 2005.

[10] H. Lindlar and R. Dubuis, "Palladium catalyst for partial reduction of acetylenes," *Organic Syntheses*, vol. 46, pp. 89–92, 1966.

[11] C. Lederhos, P. C. L'Argentière, and N. S. Fígoli, "1-heptyne selective hydrogenation over Pd supported catalysts," *Industrial & Engineering Chemistry Research*, vol. 44, no. 6, pp. 1752–1766, 2005.

[12] C. Lederhos, P. C. L'Argentière, F. Coloma-Pascual, and N. S. Fígoli, "A study about the effect of the temperature of hydrogen treatment on the properties of Ru/Al2O3 and Ru/C and their catalytic behavior during 1-heptyne semi-hydrogenation," *Catalysis Letters*, vol. 110, no. 1-2, pp. 23–28, 2006.

[13] E. Auer, A. Freund, J. Pietsch, and T. Tacke, "Carbons as supports for industrial precious metal catalysts," *Applied Catalysis A: General*, vol. 173, no. 2, pp. 259–271, 1998.

[14] F. Rodríguez-Reinoso, "The role of carbon materials in heterogeneous catalysis," *Carbon*, vol. 36, no. 3, pp. 159–175, 1998.

[15] J. L. Figueiredo, M. F. R. Pereira, M. M. A. Freitas, and J. J. M. Órfão, "Modification of the surface chemistry of activated carbons," *Carbon*, vol. 37, no. 9, pp. 1379–1389, 1999.

[16] C. R. Lederhos, J. M. Badano, N. Carrara et al., "Metal and precursor effect during 1-heptyne selective hydrogenation using an activated carbon as support," *The Scientific World Journal*, vol. 2013, Article ID 528453, 9 pages, 2013.

[17] J. Okal, M. Zawadzki, L. Kępiński, L. Krajczyk, and W. Tylus, "The use of hydrogen chemisorption for the determination of Ru dispersion in Ru/γ-alumina catalysts," *Applied Catalysis A: General*, vol. 319, pp. 202–209, 2007.

[18] C.-B. Wang, H.-K. Lin, and C.-M. Ho, "Effects of the addition of titania on the thermal characterization of alumina-supported palladium," *Journal of Molecular Catalysis A: Chemical*, vol. 180, no. 1-2, pp. 285–291, 2002.

[19] F. Pompeo, N. N. Nichio, M. M. V. M. Souza, D. V. Cesar, O. A. Ferretti, and M. Schmal, "Study of Ni and Pt catalysts supported on α-Al$_2$O$_3$ and ZrO$_2$ applied in methane reforming with CO$_2$," *Applied Catalysis A: General*, vol. 316, no. 2, pp. 175–183, 2007.

[20] T. Paryjczac and J. A. Szymura, "Electron microscopic and chemisorption comparison studies on the metal dispersion of Pd, Rh, and Ir supported catalysts," *Zeitschrift für Anorganische und Allgemeine Chemie*, vol. 449, no. 1, pp. 105–114, 1979.

[21] J. M. Badano, M. Quiroga, C. Betti, C. Vera, S. Canavese, and F. Coloma-Pascual, "Resistance to sulfur and oxygenated compounds of supported Pd, Pt, Rh, Ru catalysts," *Catalysis Letters*, vol. 137, no. 1-2, pp. 35–44, 2010.

[22] J. J. F. Scholten, A. P. Pijpers, and A. M. L. Hustings, "Surface characterization of supported and nonsupported hydrogenation catalysts," *Catalysis Reviews*, vol. 27, no. 1, pp. 151–206, 1985.

[23] R. J. Matyi, L. H. Schwartz, and J. B. Butt, "Particle size, particle size distribution, and related measurements of supported metal catalysts," *Catalysis Reviews*, vol. 29, no. 1, pp. 41–99, 1987.

[24] S. Shin, J. Jang, S.-H. Yoon, and I. Mochida, "A study on the effect of heat treatment on functional groups of pitch based activated carbon fiber using FTIR," *Carbon*, vol. 35, no. 12, pp. 1739–1743, 1997.

[25] R. Borade, A. Sayari, A. Adnot, and S. Kaliaguine, "Characterization of acidity in ZSM-5 zeolites: an X-ray photoelectron and IR spectroscopy study," *Journal of Physical Chemistry*, vol. 94, no. 15, pp. 5989–5994, 1990.

[26] D. Liprandi, M. Quiroga, E. Cagnola, and P. L'Argentière, "A new more sulfur-resistant rhodium complex as an alternative to the traditional Wilkinson's catalyst," *Industrial & Engineering Chemistry Research*, vol. 41, no. 19, pp. 4906–4910, 2002.

[27] S. Hu and Y. Chen, "Partial hydrogenation of benzene: a review," *Chinese Journal of Chemical Engineering*, vol. 29, pp. 387–396, 1998.

[28] J. A. Díaz-Auñón, M. C. Román-Martínez, C. Salinas-Martínez de Lecea et al., "[PdCl$_2$(NH$_2$(CH$_2$)12CH$_3$)$_2$] supported on an active carbon: effect of the carbon properties on the catalytic activity of cyclohexene hydrogenation," *Journal of Molecular Catalysis A: Chemical*, vol. 153, no. 1-2, pp. 243–256, 2000.

[29] B. J. Meldrum and C. H. Rochester, "In situ infrared study of the surface oxidation of activated carbon in oxygen and carbon dioxide," *Journal of the Chemical Society, Faraday Transactions*, vol. 86, no. 5, pp. 861–865, 1990.

[30] R. M. Silverstein, G. Clayton Basler, and T. C. Morril, *Spectrometric Identification of Organic Compounds* Chapter III, Wiley, Wiley, New York, NY, USA, 5th edition, 1991.

[31] C. J. Pouchert, *The Aldrich Library of Infrared Spectra*, Wiley, New York, NY, USA, 3rd edition, 1981.

[32] R. Haritha Gangupomu, M. L. Sattler, and D. Ramirez, "Comparative study of carbon nanotubes and granular activated carbon: physicochemical properties and adsorption capacities," *Journal of Hazardous Materials*, vol. 302, pp. 362–374, 2016.

[33] L. Oliviero, J. Barbier Jr., D. Duprez, A. Guerrero-Ruiz, B. Bachiller-Baeza, and I. Rodríguez-Ramos, "Catalytic wet air oxidation of phenol and acrylic acid over Ru/C and Ru–CeO$_2$/C catalysts," *Applied Catalysis B: Environmental*, vol. 25, no. 4, pp. 267–275, 2000.

[34] J. Krishna Murthy, S. Chandra Shekar, V. Siva Kumar, and K. S. Rama Rao, "Highly selective zirconium oxychloride modified Pd/C catalyst in the hydrodechlorination of dichlorodifluoromethane to difluoromethane," *Catalysis Communications*, vol. 3, no. 4, pp. 145–149, 2002.

[35] F. Pinna, M. Signoretto, G. Strukul, A. Benedetti, M. Malentacchi, and N. Pernicone, "Ruthenium as a dispersing agent in carbon-supported palladium," *Journal of Catalysis*, vol. 155, no. 1, pp. 166–169, 1995.

[36] S. R. de Miguel, O. A. Scelza, M. C. Román-Martínez, C. Salinas-Martínez de Lecea, D. Cazorla-Amorós, and A. Linares-Solano, "States of Pt in Pt/C catalyst precursors after impregnation, drying and reduction steps," *Applied Catalysis A: General*, vol. 170, no. 1, pp. 93–103, 1998.

[37] S. R. de Miguel, J. I. Vilella, E. L. Jablonski, O. A Scelza, C. Salinas-Martinez de Lecea, and A. Linares-Solano, "Preparation of Pt catalysts supported on activated carbon felts (ACF)," *Applied Catalysis A: General*, vol. 232, no. 1-2, pp. 237–246, 2002.

[38] F. Cardenas-Lizana, S. Gómez-Quero, and M. A. Keane, "Clean production of chloroanilines by selective gas phase hydrogenation over supported Ni catalysts," *Applied Catalysis A: General*, vol. 334, no. 1-2, pp. 199–206, 2008.

[39] P. Kim, H. Kim, J. B. Joo, W. Kim, I. K. Song, and J. Yi, "Effect of nickel precursor on the catalytic performance of Ni/Al$_2$O$_3$ catalysts in the hydrodechlorination of 1,1,2-trichloroethane," *Journal of Molecular Catalysis A: Chemical*, vol. 256, no. 1-2, pp. 178–183, 2006.

[40] G. Li, L. Hu, and J. M. Hill, "Comparison of reducibility and stability of alumina-supported Ni catalysts prepared by impregnation and co-precipitation," *Applied Catalysis A: General*, vol. 301, no. 1, pp. 16–24, 2006.

[41] K. V. R. Chary, D. Naresh, V. Vishwanathan, M. Sadakave, and W. Ueda, "Vapour phase hydrogenation of phenol over Pd/C catalysts: a relationship between dispersion, metal area and hydrogenation activity," *Catalysis Communications*, vol. 8, no. 3, pp. 471–477, 2007.

[42] A. B. Gaspar, G. R. dos Santos, R. de Souza Costa, and M. A. P. da Silva, "Hydrogenation of synthetic PYGAS—effects of zirconia on Pd/Al$_2$O$_3$," *Catalysis Today*, vol. 133–135, pp. 400–405, 2008.

[43] D. O. Simone, T. Kennelly, N. L. Brungard, and R. J. Farrauto, "Reversible poisoning of palladium catalysts for methane oxidation," *Applied Catalysis*, vol. 70, no. 1, pp. 87–100, 1991.

[44] A. B. Gaspar and L. C. Dieguez, "Dispersion stability and methylcyclopentane hydrogenolysis in Pd/Al$_2$O$_3$ catalysts," *Applied Catalysis A: General*, vol. 201, no. 2, pp. 241–251, 2000.

[45] L. Li, Z. H. Zhu, Z. F. Yan, G. Q. Lu, and L. Rintoul, "Catalytic ammonia decomposition over Ru/carbon catalysts: the importance of the structure of carbon support," *Applied Catalysis A: General*, vol. 320, pp. 166–172, 2007.

[46] A. Guerrero-Ruiz, P. Badenes, and I. Rodríguez-Ramos, "Study of some factors affecting the Ru and Pt dispersions over high surface area graphite-supported catalysts," *Applied Catalysis A: General*, vol. 173, no. 2, pp. 313–321, 1998.

[47] M. Nurunnabi, K. Murata, K. Okabe, M. Inaba, and I. Takahara, "Effect of Mn addition on activity and resistance to catalyst deactivation for Fischer–Tropsch synthesis over Ru/Al$_2$O$_3$ and Ru/SiO$_2$ catalysts," *Catalysis Communications*, vol. 8, no. 10, pp. 1531–1537, 2007.

[48] V. Ragaini, R. Carli, C. L. Bianchi, D. Lorenzetti, and G. Vergani, "Fischer—Tropsch synthesis on alumina-supported ruthenium catalysts I. Influence of K and Cl modifiers," *Applied Catalysis A: General*, vol. 139, no. 1-2, pp. 17–29, 1996.

[49] K. Lu and B. Tatarchuk, "Activated chemisorption of hydrogen on supported ruthenium I. Influence of adsorbed chlorine on accurate surface area measurements," *Journal of Catalysis*, vol. 106, no. 1, pp. 166–175, 1987.

[50] T. Narita, H. Miura, M. Ohira et al., "The effect of reduction temperature on the chemisorptive properties of Ru/Al$_2$O$_3$: effect of chlorine," *Applied Catalysis*, vol. 32, pp. 185–190, 1987.

[51] T. Narita, H. Miura, K. Sugiyama, T. Matsuda, and R. D. Gourales, "The effect of reduction temperature on the chemisorptive properties of Ru/SiO$_2$: effect of chlorine," *Journal of Catalysis*, vol. 103, no. 2, pp. 492–495, 1987.

[52] C. R. Lederhos, J. M. Badano, M. E. Quiroga, F. Coloma-Pascual, and P. C. L'Argentière, "Influence of Ni addition to a low-loaded palladium catalyst on the selective hydrogenation of 1-heptyne," *Química Nova*, vol. 33, no. 4, pp. 816–820, 2010.

[53] M. J. Maccarrone, C. R. Lederhos, G. Torres et al., "Partial hydrogenation of 3-hexyne over low-loaded palladium mono and bimetallic catalysts," *Applied Catalysis A: General*, vol. 441-442, pp. 90–98, 2012.

[54] *NIST X-ray Photoelectron Spectroscopy Database-NIST Standard Reference Database 20 Version 4.1*, National Institute of Standards and Technology, Gaithersburg, MD, USA, December 2017, https://srdata.nist.gov/xps/.

[55] M. M. Telkar, J. M. Nadjeri, C. V. Rode, and R. V. Chaudhari, "Role of a co-metal in bimetallic Ni–Pt catalyst for hydrogenation of m-dinitrobenzene to m-phenylenediamine," *Applied Catalysis A: General*, vol. 295, no. 1, pp. 23–30, 2005.

[56] J. Juan-Juan, M. C. Roman-Martinez, and M. J. Illan-Gomez, "Catalytic activity and characterization of Ni/Al$_2$O$_3$ and NiK/ Al$_2$O$_3$ catalysts for CO$_2$ methane reforming," *Applied Catalysis A: General*, vol. 264, no. 2, pp. 169–174, 2004.

[57] B. W. Hoffer, A. D. van Langeveld, J. P. Janssens, R. L. C. Bonné, C. M. Lok, and J. A. Moulijn, "Stability of highly dispersed Ni/AlO catalysts: effects of pretreatment," *Journal of Catalysis*, vol. 192, no. 2, pp. 432–440, 2000.

[58] E. Heracleous, A. F. Lee, K. Wilson, and A. A. Lemonidou, "Investigation of Ni-based alumina-supported catalysts for the oxidative dehydrogenation of ethane to ethylene: structural characterization and reactivity studies," *Journal of Catalysis*, vol. 231, no. 1, pp. 159–171, 2005.

[59] M. Boudart, "Catalysis by supported metals," *Advances in Catalysis*, vol. 20, pp. 153–166, 1969.

[60] J. A. Pajares, P. Reyes, L. A. Oro, and R. Sariego, "Hydrogenation of 1-hexene by rhodium catalysts," *Journal of Molecular Catalysis*, vol. 11, no. 2-3, pp. 181–192, 1981.

[61] M. J. Maccarrone, G. Torres, C. Lederhos et al., "Kinetic study of the partial hydrogenation of 1-heptyne over Ni and Pd supported on alumina," in *Hydrogenation*, I. Karamé, Ed., Chapter 7, pp. 159–184, InTech, Rijeka, Croatia, 2012.

4

Catalytic Activity of a Composition based on Strontium Bismuthate and Bismuth Carbonate at the Exposure to the Light of the Visible Range

K. S. Makarevich(ID)**, A. V. Zaitsev**(ID)**, O. I. Kaminsky**(ID)**, E. A. Kirichenko**(ID)**, and I. A. Astapov**(ID)

Institute of Materials Khabarovsk Scientific Center, Far Eastern Branch of the Russian Academy of Sciences, Khabarovsk, Russia

Correspondence should be addressed to O. I. Kaminsky; kamin_div0@mail.ru

Academic Editor: Gianluca Di Profio

This paper presents the results pertaining to studying the properties of the photocatalytically active composition of strontium bismuthate $SrBi_{2.70}O_{5.05}$, $SrBi_{2.90}O_{5.35}$, and $SrBi_{3.25}O_{5.88}$ and bismuth carbonate $(BiO)_2CO_3$ in molar ratios 1/0.67, 1/0.56, and 1/0.37, respectively. These compositions are obtained through pyrolytic synthesis from organic precursor complexes of strontium and bismuth with sorbitol. It has been established that the synthesised powder materials absorb the light of the visible range up to 500 nm owing to the presence of a narrow-gap semiconductor strontium bismuthate. The presence of a wide-band semiconductor $(BiO)_2CO_3$ ensures an effective separation of electron-hole pairs. The diffuse reflection spectra (DRS) of compositions differ from the analogous spectra of a mechanical mixture of these semiconductor phases with the same composition, which allows one to assume the heterostructural structure of the semiconductor system. The same heterostructure is confirmed by the results of mapping. Catalytic particles that are synthesised at a temperature of 500°C containing 27 mass% $(BiO)_2CO_3$ (corresponding to $SrBi_{2.9}O_{5.35}/(BiO)_2CO_3 = 1/0.56$) have the greatest activity with respect to the photodecomposition of methylene blue (MB). The possibility of controlling the optical properties and photocatalytic activity of the composition is depicted due to the joint formation of the strontium bismuthate phase and the bismuth carbonate phase.

1. Introduction

Organic compounds are often the primary toxic components of wastewater from various origins. Photocatalysis became a promising technology for their purification after the publication of one of the pioneering works in this regard, which was written by Frank and Bard in 1977 [1]. On the other hand, studies concerning photocatalytically active semiconductors began to develop intensively after the publication of the well-known work of Fujishima and Honda [2], which lead to the publication of a large series of articles by various authors dedicated to obtaining hydrogen from water using light energy. Both the aforementioned processes are based on the phenomenon of photolysis of water on the surface of a semiconductor under the action of light. As a result, strong oxidising agents, such as atomic oxygen, free hydroxide radicals, hydrogen peroxide, to name a few, are produced with the capacity of ensuring deep oxidation of organic compounds. The undoubted advantage of this method concerning organic removal is the lack of regular costs for additional chemical reagents. Consequently, the method of photo-catalytic purification has found wide application, from the final stage of the treatment of drinking water at water intake stations to the sewage treatment conducted in industries such as oil refining, paint and varnish, and textile.

The search for new catalytically active systems, capable of photolysis of water under the influence of sunlight, has been conducted for the purpose of creating compositions based on narrow-gap semiconductors having an absorption region in the visible spectral range [3, 4]. In particular, they include semiconductors based on transition metal oxides. However, photocatalysts with this composition have a significant disadvantage that is associated with the toxicity of the heavy metals included in them (e.g., Cd and Pb) [5–7], which makes it impossible to utilise such substances for water purification. It is known that bismuth is an exception in terms of being a toxicological hazard among heavy metals. Its toxicity is so minute that bismuth compounds have been found, for instance, to be widely used in pharmacology as antacid agents for oral administration [8, 9]. This is owing to the fact that mobile cationic forms of bismuth are not formed in alkaline, neutral, and weakly acidic media. Thus, during the purification of water, its saturation with regard to bismuth is excluded. Among bismuth compounds, high catalytic activity comparable to anatase shows the presence of carbonate $(BiO)_2CO_3$. However, the region of its intrinsic absorption lies in the UV range of the spectrum. To sensitise $(BiO)_2CO_3$ to visible light, it is doped with nitrogen [10, 11]. Other bismuth compounds, such as the narrow-band semiconductor $CaBi_6O_{10}$, do not require the use of additives, since they are characterised by intrinsic absorption in the visible spectral range [12, 13]. However, they are less effective with regard to the photodecomposition of organic compounds. One way to increase the reactivity of narrow-band semiconductors is to create heterostructural composites with wide-gap semiconductors.

As a rule, two-phase compositions are synthesised in two stages: first, the main component is obtained, after which a second component is formed on its surface. In this method of production, it is difficult to achieve a uniform mutual distribution of semiconductors in the concerned volume. Therefore, in our study, we employed the previously developed pyrolytic synthesis method based on bismuth complexes with sorbitol [14], which simultaneously produces a composition of two semiconductors. As a wide-gap semiconductor, $(BiO)_2CO_3$ was formed in the composition. Moreover, the absorption of visible light was achieved owing to the introduction of strontium bismuthate, which occupies a central position in the solid solutions presented in the $SrO-Bi_2O_3$ diagram [15]. The purpose of this study was to create and study the properties of $SrBi_xO_y/(BiO)_2CO_3$ compounds of various compositions, depending on the synthesis conditions. Furthermore, it purposes to reveal the phase composition of the catalyst, providing the greatest photocatalytic activity of the composition during the decomposition of the organic dye.

2. Experiment

$SrBi_4O_7$ and several compositions based on it were obtained through a modified method of pyrolytic synthesis. Previously, we employed this method to obtain $CaBi_6O_{10}$. Strontium nitrate $Sr(NO_3)_2$ (ACROS), bismuth nitrate pentahydrate $Bi(NO_3)_3\cdot5H_2O$ (ACROS), and sorbitol $C_6H_{14}O_6$ served as precursors for the synthesis of composites based on $SrBi_xO_y$.

The amount of sorbitol in the precursor mixture's composition increased by 1.5 times in comparison to [16], which allowed the amount of CO_2 and $(BiO)_2CO_3$, together with $SrBi_xO_y$, to increase. The mixture of $Sr(NO_3)_2$,

$Bi(NO_3)_3\cdot5H_2O$, and sorbitol $(Sr:Bi=1:4$ molar ratio) was grinded in mortar. The grinded mixture spontaneously formed a transparent viscous solution due to the hydrated water of nitrates. Pyrolysis of the precursor mixture was conducted at 180°C in the air. Subsequently, the pyrolyzed mixture was heated to temperatures of 300, 400, 500, 600, and 700°C and was held isothermally for 24 hours. The obtained $SrBi_xO_y/(BiO)_2CO_3$ compositions have been designated as SBC-X, where X signifies the synthesis temperature. The concentration of $(BiO)_2CO_3$ in the obtained samples was measured by dissolving the sample in 10% HCl solution. Subsequently, the amount of CO_2 released was estimated by making use of a $Ba(OH)_2$ solution. In addition, Brunauer–Emmett–Teller (BET) specific surface area of the samples was measured via N_2 adsorption at 77 K by using Sorbi 4.1. analyser. The crystal in a phase of the samples was identified through X-ray diffraction by utilising a DRON-7 diffractometer with $CuK\alpha$ ($\lambda = 1.5406$ Å) as the radiation source, which operated at 40 kV and 40 mA. Moreover, the UV and visible diffuse reflectance spectra were conducted on a MDR-41 spectrophotometer system equipped with an integrating sphere attachment, with $BaSO_4$ as the background. Surface morphology and distribution of particles were studied by LEO 1430VP SEM, by employing an accelerating voltage of 15 kV.

Furthermore, photocatalytic experiments were performed using an automated photometric unit [17]. Photocatalytic activities of the samples SBC-X were evaluated through the degradation of methylene blue solution (1.2 mg$\cdot$L^{-1}, 350 mL) under different light irradiations discharge lamp Aqua Arc Osram, Sylvania (250 W, continuous spectrum $\lambda = 380–850$ nm). Mass of photocatalyst sample is 200 mg. The spectrum of the lamp was similar to the spectral distribution of the radiation intensity with regard to sunlight. To cut out the UV range, a filter ($\lambda > 380$ nm) was employed. During the irradiation, the solution was stirred using a IKA RO10 magnetic stirrer. The temperature of the photocatalytic system was maintained at 25°C by utilising the water-cooling thermostat (Julabo F25-ED).

3. Results and Discussion

The study of the process of phase formation in the system under investigation upon its heating yielded the following results. The composition at 300–400°C consists of three phases. In the temperature range, the composition at 500–600°C has a two-phase composition. At 700°C, the composition becomes single-phase. This can be seen from XRD (Figure 1). On the XRD patterns of these samples, there are pronounced reflections belonging to α-Bi_2O_3 and $(BiO)_2CO_3$. In addition, there are peaks corresponding to a solid solution of the $SrO-Bi_2O_3$ system with stoichiometry similar to $SrBi_3O_{5.5}$ (ICDD-45-609). Furthermore, analysis of the XRD patterns of sample SBC-400 revealed that the concentration of Bi_2O_3 is reduced in this regard, which indicates the introduction of Bi_2O_3 into a solid solution of strontium bismuthate. In this sample, $(BiO)_2CO_3$ has not decomposed yet. The intensity of its diffraction peaks does not change in comparison to SBC-300. On the XRD pattern of sample SBC-500, an increase in the intensity of the reflex

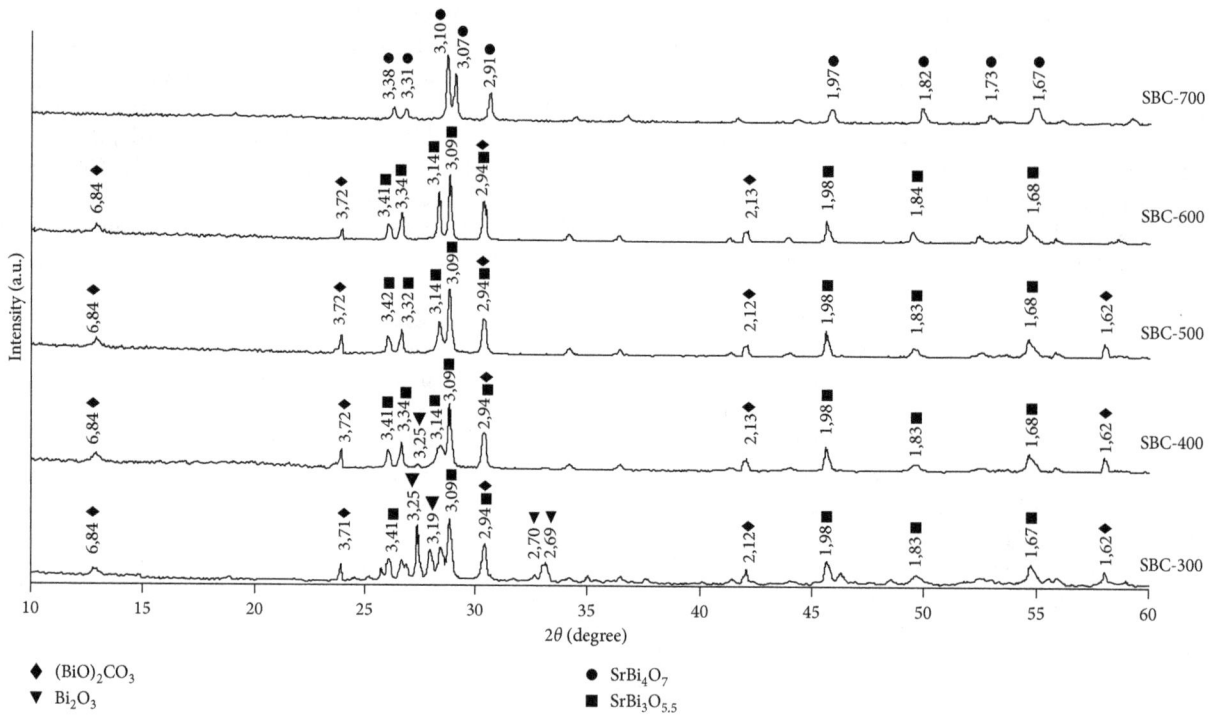

FIGURE 1: XRD patterns of samples obtained at different synthesis temperatures.

3.14 Å is observed. This demonstrates that the solid solution of strontium bismuthate continues to be formed through the incorporation of unreacted bismuth oxide. At temperatures above 600°C, $(BiO)_2CO_3$ is partially decomposed, and the intensity of its reflexes decreases. The bismuth oxide formed during the decomposition of $(BiO)_2CO_3$ enters a solid solution based on strontium bismuthate. This process is characterised by an increase in the intensity of the reflex 3.14 Å. At a temperature of 700°C, the process concerning the formation of a solid solution of strontium bismuthate ends. This is confirmed by the displacement of all the peaks in the region of large angles and through a change in the ratio of the two primary phase reflexes (3.11 and 3.07 Å) of strontium bismuthate. Consequently, the diffractogram assumes the form that is typical of a solid solution of strontium bismuthate with stoichiometry close to $SrBi_4O_7$. Based on the results of XRD, the strontium carbonate phase was not found.

Chemical analysis revealed that the concentration of carbonates in the samples obtained at 300°C to 400°C is approximately the same and equal to 29–32 wt%. The content $(BiO)_2CO_3$ in the sample SBC-500 is about 27% and 18% in the SBC-600 composition.

The results of the mapping show that there are areas containing predominantly strontium and having insignificant quantity of carbon. Comparing with XRD data, these areas were identifiable, like the phase of strontium bismuthate. The remaining surface of the sample contains bismuth, oxygen, and carbon, but so does insignificant quantity of strontium and is identifiable as bismuth carbonate. As can be seen from Figure 2, the bismuth carbonate phase has high roughness, while the bismuth-strontium phase is represented by a smooth surface. The formation of such a structure is possible due to the epitaxial growth in

the process of joint synthesis of both semiconductors that make up the composition. The synthesized compositions are represented by two main particle fractions, where the smaller one ($(BiO)_2CO_3$) has a nanometer size range, and the large ones have a strontium-bismuthate phase with particle sizes of about 1.5–3.5 μm that form agglomerates. The average size of the agglomerates is in the range of 20–50 μm.

By conducting a comparative analysis of XRD data and chemical analysis, it can be assumed that the obtained compositions have the following content (Table 1). As it is commonly known [15], the phase diagram of $SrO-Bi_2O_3$ rather has a wide range of solid solutions. According to refined data [18], a solid solution is formed under the condition that the Sr : Bi atomic ratio is in the range from 1 : 2.5 to 1 : 7.5, meaning that it corresponds to compounds with stoichiometry from $SrBi_{2.5}O_{4.75}$ to $SrBi_{7.5}O_{12.25}$. According to XRD data, the strontium bismuthate formed in the compositions is most close to with Sr : Bi 1 : 3 stoichiometry $SrBi_3O_{5.5}$ (ICDD-45-609) and, consequently, is a representative of these series of solid solutions. The reflexes characteristic of this strontium bismuthate were identified for the samples SBC-400, SBC-500, and SBC-500. However, a more accurate identification of the stoichiometry of strontium bismuthate obtained from XRD data is not possible, since the samples contain $(BiO)_2CO_3$, whose main reflexes are superposition or close to the reflections of strontium bismuthate from the above series of solid solutions. In order to identify more accurately the stoichiometry of the formed bismuth phases, we additionally carried out a chemical analysis that made it possible to determine the content of $(BiO)_2CO_3$ in the samples. Further, a simple calculation was carried out, based on the following facts: firstly, the total atomic Sr : Bi

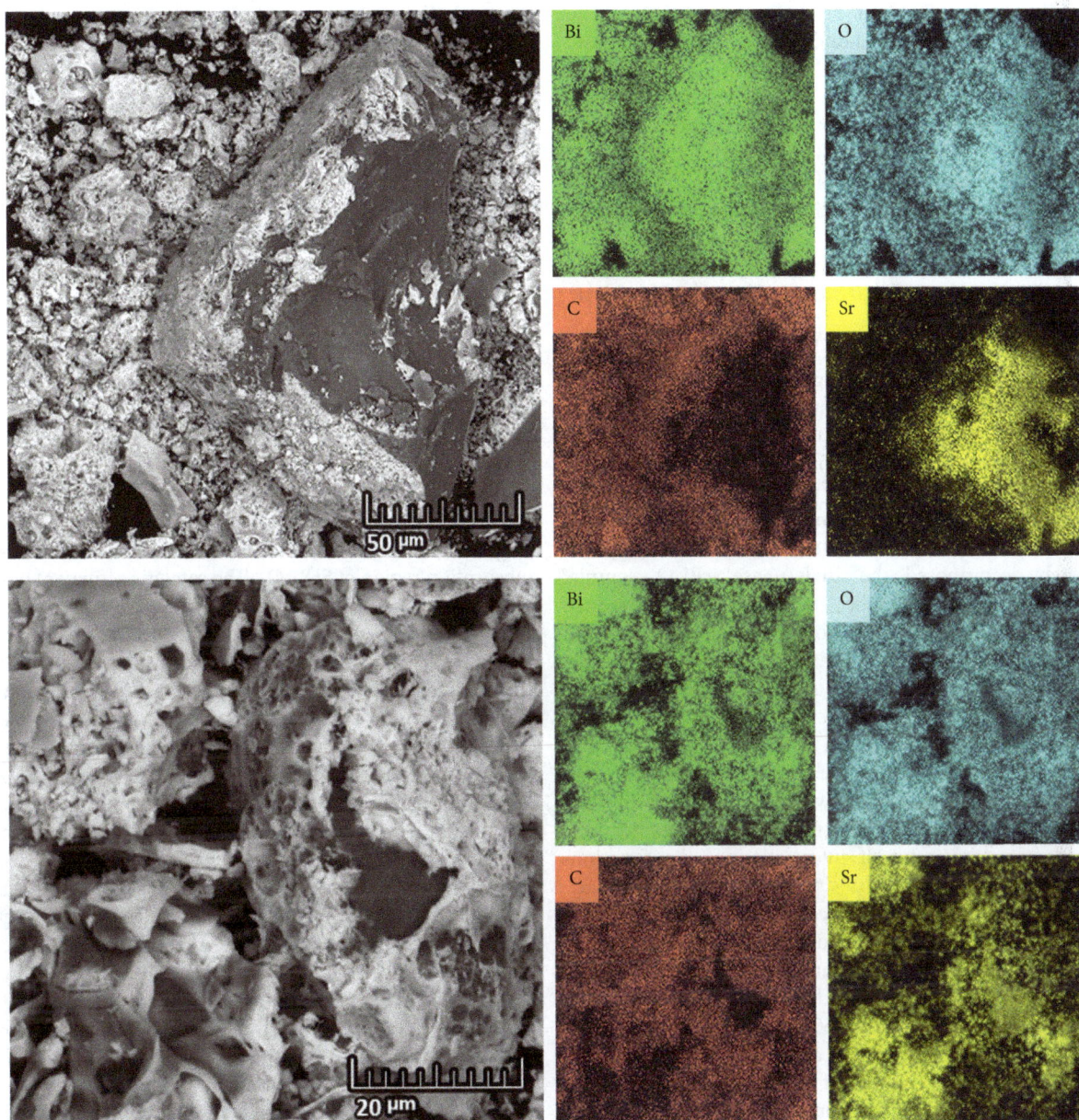

FIGURE 2: SEM images of composed SBC-500 and results of mapping.

TABLE 1: The content of semiconductor phases in the obtained compositions.

Photocatalysts	Total atomic ratio Sr : Bi in the compositions	Weight content $(BiO)_2CO_3$ (%)	Photocatalysts $SrBi_xO_y/(BiO)_2CO_3$	
			Element stoichiometry in phases	Molar ratio of phases in the compositions
SBC-400	1 : 4	32	$SrBi_{2.70}O_{5.05}/(BiO)_2CO_3$	1/0.67
SBC-500	1 : 4	27	$SrBi_{2.90}O_{5.35}/(BiO)_2CO_3$	1/0.56
SBC-600	1 : 4	18	$SrBi_{3.25}O_{5.88}/(BiO)_2CO_3$	1/0.37
$SrBi_4O_7$	1 : 4	0	$SrBi_4O_7$	1/0

ratio in all the obtained compositions was set equal to 1 : 4 during the synthesis, secondly, the samples of SBC-400, SBC-500, and SBC-600 are two-phase, and thirdly, the content of $(BiO)_2CO_3$ is accurately known from the chemical analysis. Therefore, by knowing the share of bismuth that went to the formation $(BiO)_2CO_3$, it is possible to calculate the share of bismuth that went to the formation of strontium bismuthate, i.e., to clarify the stoichiometry of Sr : Bi in the solid solution formed (Table 1). The data obtained in this way are in good agreement with the results of XRD. The adjusted Sr : Bi ratio is actually close to 1 : 3 and is 1/2.7, 1/2.9, and 1/3.25 for the samples of SBC-400, SBC-500, and SBC-600, respectively. The composition SBC-300 (based on XRD data) is three-phase, since it contains, in addition to the two phases described above, also Bi_2O_3. Accordingly, the use of the above sequence of calculations in relation to it will not be correct. For this reason, the sample SBC-300 is not included in Table 1. A more detailed study of it was not the purpose of this paper, since the compositions formed by bismuth oxide and alkaline earth metal bismuth were previously investigated by many authors, for example, in [10–12]; in addition, the sample SBC-300 did not show a high catalytic activity.

The composition SBC-400 has the greatest specific surface area of $3.5\ m^2/g$, according to BET analysis (Table 2). Increasing the synthesis temperature from 400°C to 600°C leads to the fact that the specific surface area of the samples is reduced to $3.0\ m^2/g$. This is probably due to the gradual decomposition of the carbonate phase $(BiO)_2CO_3$ and the coarsening of the particles belonging to the formed solid solution of strontium bismuthate. When the carbonate is completely decomposed, the single-phase $SrBi_4O_7$ sample obtained at 700°C has a specific surface area of $2.4\ m^2/g$. This is comparable to the specific area of pure α-Bi_2O_3 ($2.2\ m^2/g$), which is obtained in the same manner. Hence, the presence of a carbonate phase contributes to the formation of a more developed surface in two-phase compositions.

Based on the DRS for each synthesised composition, $SrBi_4O_7$, and $(BiO)_2CO_3$, the Kubelka–Munk function $F(R) = 0.5(1-R)^2/(R-1)$ was calculated, where R represents the reflection coefficient. The widths of the band gap are determined from the point of intersection of the function graph's linear section $(F(R)\ hv)^{1/2}$ with the axis of abscissae: photon energy hv (Figure 3). According to the concerned estimates, the width of the forbidden band of $SrBi_4O_7$ and $(BiO)_2CO_3$ was $E_g = 2.50\ eV$ and $3.38\ eV$, respectively. Moreover, analysis of the dependencies $(F(R)\ hv)^{1/2}$ for the compositions SBC-500 and SBC-600 showed that they can be divided into two linear Sections I and II (Figure 3). They are a superposition of the graphs of two semiconductors that constitute the composition. Section I corresponds to bismuth carbonate, whereas section II corresponds to strontium bismuthate. As the synthesis temperature increases, section I decreases, and section II, respectively, increases due to the increase in the proportion of strontium bismuthate in the solid solution. Comparing the analogous curves concerning a mechanical mixture of 73% $SrBi_4O_7$ + 27% $(BiO)_2CO_3$, one can note that there is a marked difference between them. Investigation of optical properties has shown that the spectra of DRS compositions do not signify a simple superposition of the semiconductor phases that form them and differ from the spectra of a mechanical mixture with the same composition. Their distinctive feature is the characteristic absorption region in the range of 200–270 nm, which is absent on the individual phases' spectra of the composition's constituents and on the spectrum of a simple mechanical mixture identical to the composition of the relevant composition.

Studying the photocatalytic activity of synthesised materials demonstrated that their use makes it possible for one to increase the decomposition rate of methylene blue upon

TABLE 2: The catalytic activity of the resulting compositions and single-phase photocatalysts.

Sample	Specific surface area, S (m²/g)	Reaction rate constant, $k \cdot 10^{-3}$ (min⁻¹)	Specific reaction rate constant, $k_s \cdot 10^{-4}$ (g·min⁻¹·m⁻²)
α-Bi_2O_3	2.1	4.0	1.90
SBC-300	3.2	9.0	2.81
SBC-400	3.5	1.3	3.71
SBC-500	3.2	1.5	4.69
SBC-600	3.0	1.3	4.33
$SrBi_4O_7$	2.4	8.0	3.33

irradiation with visible light. Kinetic dependences of the change in the concentration of organic dye, in the presence of the synthesised compositions SBC 300–700, $SrBi_4O_7$, and α-Bi_2O_3, have been depicted in Figure 4. The single-phase samples of $SrBi_4O_7$ and α-Bi_2O_3 possess the lowest catalytic activity. Moreover, the activity of $SrBi_4O_7$ is moderately higher than that of bismuth oxide. This is owing to the wider absorption region of $SrBi_4O_7$. The catalytic activity of biphasic compositions in all cases was observed to be higher. We attribute this result to the cocatalytic effect of the interaction of the $SrBi_xO_y/(BiO)_2CO_3$ phases forming the composition. Its occurrence is possible due to the heterostructural nature of the structure of the composition. This is confirmed by the data of the SEM mapping of the images given earlier. The possibility of epitaxial growth is also indicated by similar X-ray structural parameters of the obtained strontium bismuthate and bismuth carbonate. To test this assumption, a mechanical mixture was prepared from the separately obtained $SrBi_4O_7$ and $(BiO)_2CO_3$ phases and an identical SBC-500 sample, exhibiting the greatest catalytic activity. The catalytic activity of the prepared mixture (Figure 4) showed an intermediate value between the activities of its constituent phases and is inferior to the activity of any of the obtained compositions, including SBC-500. Therefore, the joint-phase formation during synthesis is a prerequisite for the efficient operation of the photocatalytic composition.

The sample synthesised at 500°C exhibits greatest catalytic activity. This is probably related to the formation of the required amount of strontium bismuthate, which sensitises the system to visible light. On the contrary, $(BiO)_2CO_3$ is retained in sufficient amount (27%) for the efficient functioning of the catalyst. When its quantity is reduced to 18%, the efficiency of the SBC-600 catalyst decreases. In sample SBS-400 (32%-$(BiO)_2CO_3$), there is not enough strontium bismuthate, and hence the catalyst is not capable of absorbing the energy of visible light. The reaction rate constant (k) was determined by determining the parameters of the linear regression equations for the $(C_t/C_0) = f(t)$ dependencies. The obtained values of k have been presented in Table 2. It is known that the samples under study have different specific surface areas. For a correct comparison of the catalytic activity, the rate constant $k_s \cdot 10^{-4}$, normalised to the specific surface, was calculated (Table 2). With an increase in the synthesis temperature of the samples, their catalytic activity increases and becomes greatest for the SBC-500 composition ($k_s = 4.69$). A further increase in temperature at first leads to a decrease in the k_s constant (4.33)

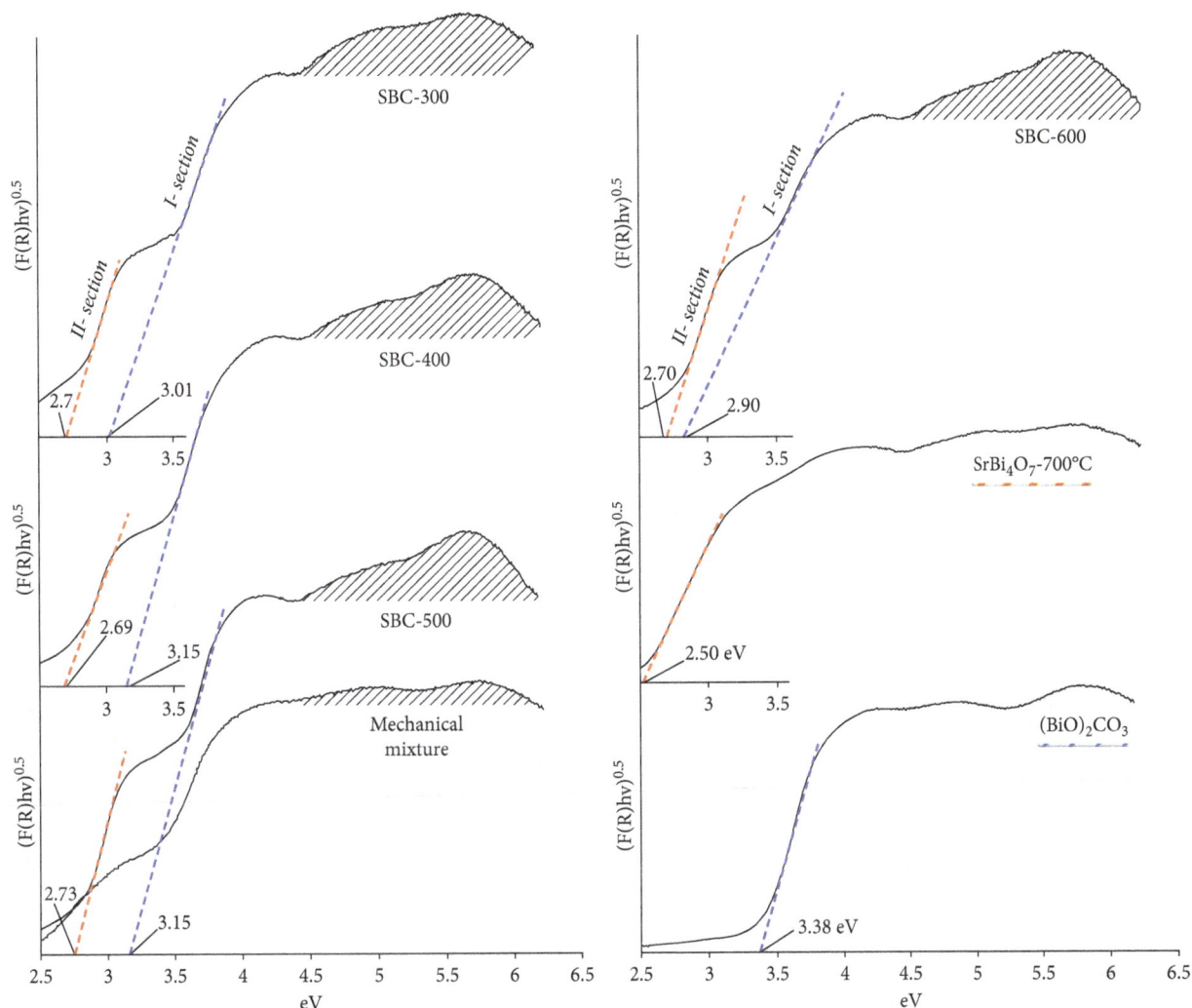

FIGURE 3: Dependences $(F(R)h\nu)^{\frac{1}{2}}$ of single-phase samples of $SrBi_4O_7$ and $(BiO)_2CO_3$ and the compositions SBC-300, SBC-400, SBC-500, and SBC-600 and their mechanical mixture of 73% $SrBi_4O_7$ + 27% $(BiO)_2CO_3$ from photon energy.

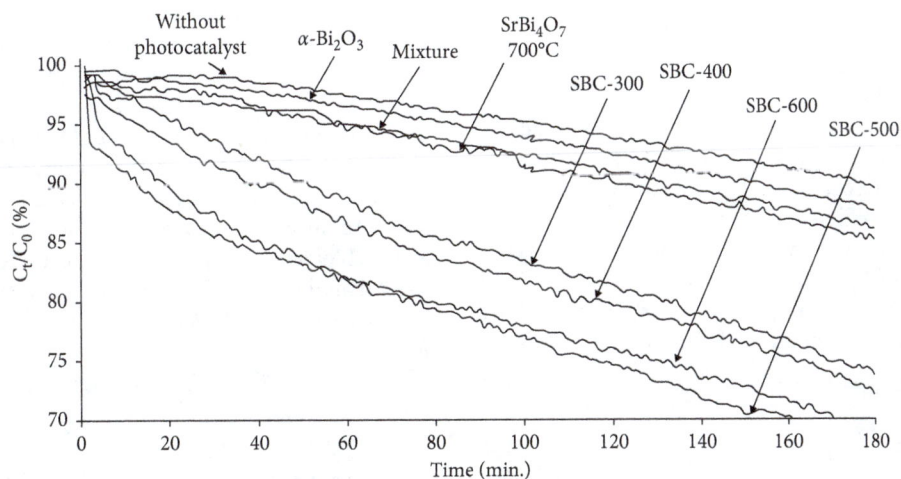

FIGURE 4: Photocatalytic decomposition curves of MB under irradiation with visible light in the presence of SBC-300, SBC-400, SBC-500, and SBC-600 and single-phase samples of $SrBi_4O_7$, and α-Bi_2O_3 and a mechanical mixture of 73% $SrBi_4O_7$ + 27% $(BiO)_2CO_3$.

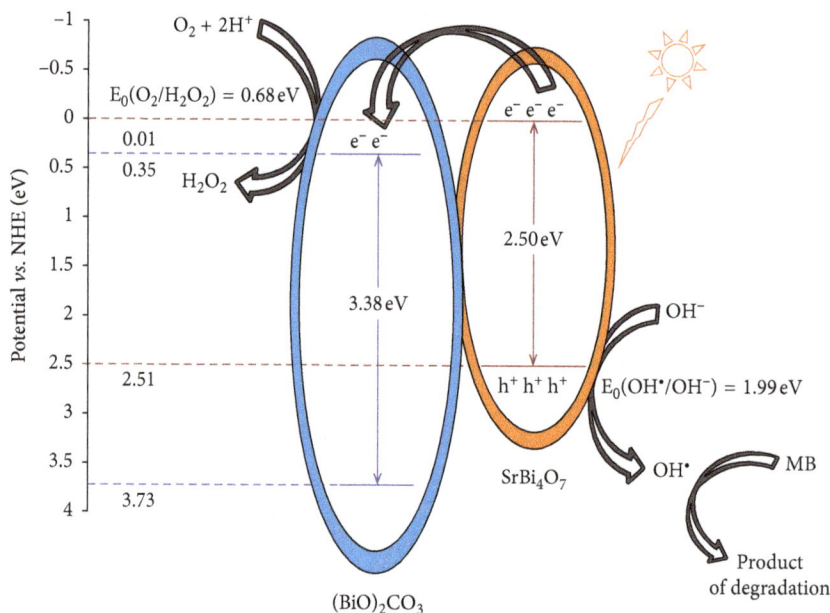

FIGURE 5: The possible degradation mechanism of MB over the prepared composites.

for the SBC-600 sample. Subsequently, it exhibits a significant drop in k_s (3.33) at a temperature of 700°C, when the composition becomes single-phase. A comparative analysis of the obtained data k_s shows that the high catalytic activity of the compositions in comparison with monophases and a mechanical mixture is primarily due to the heterostructural phase content, and not only to the developed surface.

4. Discussion

The results of this study obtained can be explained by considering the cocatalytic effect of the different phases in the composition of $SrBi_xO_y/(BiO)_2CO_3$. The formation of such a structure is possible owing to the epitaxial growth in the process of joint synthesis of both semiconductors constituting the composition. Moreover, the possibility of epitaxial growth is also indicated by similar X-ray structural parameters of the obtained strontium bismuthate and bismuth carbonate. The catalytic activity of the mixture of 73% $SrBi_4O_7$ + 27% $(BiO)_2CO_3$ (Figure 4) demonstrated an intermediate value between the activities of its constituent phases, which are inferior to the activity of any of the resulting compositions, including SBC-500. Consequently, the formation of heterostructural phases in the synthesis process increases the efficiency of the photocatalytic composition. Furthermore, values pertaining to the boundaries of the valence band (VB) and the conduction band (CB) of $SrBi_4O_7$ and $(BiO)_2CO_3$ compounds were calculated by employing the following equations [19]:

$$E_{VB} = XE - E^e + 0.5E_g, \qquad (1)$$

$$E_{CB} = XE - E^e - 0.5E_g, \qquad (2)$$

where XE signifies the absolute electronegativity of the semiconductor, which is calculated as the geometric mean of

the absolute electronegativity of the atoms forming it (the XE values for $SrBi_4O_7$ and $(BiO)_2CO_3$ are 5.76 eV and 6.54 eV, resp.); E^e represents the absolute potential of the standard hydrogen electrode (4.5 eV); E_{VB} denotes the VB edge potential (the VB potential); E_{CB} signifies the CB edge potential (the CB value); E_g represents the energy of a semiconductor's band gap, which is experimentally obtained from an analysis of the spectra of DRS. According to the calculations, the upper boundary of the VB zone and the lower CB boundary for $SrBi_4O_7$ are 2.51 eV and 0.01 eV, respectively, whereas for $(BiO)_2CO_3$, it was 3.73 and 0.35 eV. The band structure of the obtained samples can be depicted in the form of the circuit as shown in Figure 5.

Exposure to visible light results in the generation of electron-hole pairs only in the narrowband semiconductor $SrBi_4O_7$, since $(BiO)_2CO_3$ does not possess the ability to utilise the energy of the spectrum's visible part (Figure 5). The energies of CB and VB $SrBi_4O_7$ are higher than that of $(BiO)_2CO_3$. Moreover, the transition of photogenerated electrons from the conduction band of $SrBi_4O_7$ to the conduction band $(BiO)_2CO_3$ is energetically favorable. Consequently, electrons accumulate in the conduction band of a wide-gap semiconductor, whereas holes remain in the valence band of $SrBi_4O_7$, which leads to a spatial separation of the charge carriers. Thus, the lifetime of charge carriers is significantly increased, which contributes to formation of a more efficient flow concerning photolysis of water. The value of the CB potential of both semiconductors (0.01 and 0.35 eV) is lower than the O_2/H_2O_2 potential, which ensures the reduction of dissolved oxygen to hydrogen peroxide:

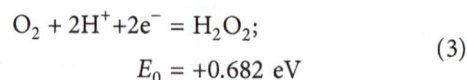

$$O_2 + 2H^+ + 2e^- = H_2O_2;$$
$$E_0 = +0.682 \text{ eV} \qquad (3)$$

Subsequently, peroxide is able to recombine with hydroxide radicals, which are active agents for the destruction of organic compounds. In addition, the formation of hydroxide radicals also occurs in a narrow-gap semiconductor, since the value of VB for $SrBi_4O_7$ (2.51 eV) is more positive in comparison to the oxidation potential of hydroxide ions in a photohole:

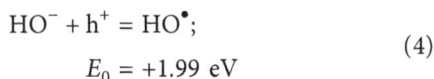

$$HO^- + h^+ = HO^\bullet;$$
$$E_0 = +1.99 \text{ eV} \tag{4}$$

Photoholes in VB $SrBi_4O_7$, in turn, can directly oxidise MB when absorbing dye molecules on the semiconductor surface.

5. Conclusions

(i) Through the method pertaining to oxidative pyrolysis of complexes of the corresponding metals with sorbitol, it is depicted that it is possible to obtain a composition based on strontium bismuthate and bismuth carbonate. The photocatalytic activity of such a composition under the action of visible light is higher than that of a mechanical mixture of the same composition, the α-Bi_2O_3 and $SrBi_4O_7$ phases. The greatest catalytic activity is possessed by compositions that are obtained at 500 to 600°C containing 73–82 mass.%, $SrBi_xO_y$, and 27–18 mass.%, $(BiO)_2CO_3$; it corresponds of strontium bismuthate $SrBi_{2.9}O_{5.35}$, $SrBi_{3.25}O_{5.88}$, and bismuth carbonate $(BiO)_2CO_3$ in molar ratios 1/0.56; 1/0.37, respectively.

(ii) Investigation of optical properties has shown that the spectra of DRS compositions do not signify a simple superposition of the semiconductor phases that form them and differ from the spectra of a mechanical mixture with the same composition. Their distinctive feature is the characteristic absorption region in the range of 200–270 nm, which is absent on the individual phases' spectra of the composition's constituents and on the spectrum of a simple mechanical mixture identical to the composition of the relevant composition.

(iii) Analysis of DRS and the results of mapping suggest heterostructural nature of the compositions. The formation of such a structure is possible due to epitaxial growth, bismuth carbonate on strontium bismuthate during the synthesis of $SrBi_xO_y/(BiO)_2CO_3$.

(iv) The mechanism concerning the photocatalytic activity of the semiconductor phases forming studied the composition is explained on the basis of the band structure's analysis.

(v) The results of the work present the prospects for creating environmentally safe and efficient photocatalysts based on bismuth compounds that are sensitive to visible light irradiation

Conflicts of Interest

The authors declare that they have no conflicts of interest.

References

[1] S. N. Frank and A. J. Bard, "Semiconductor electrodes. Photoassisted oxidations and photoelectrosynthesis at polycrystalline titanium dioxide electrodes," *Journal of the American Chemical Society*, vol. 99, no. 14, pp. 4667–4675, 1977.

[2] A. Fujishima and K. Honda, "Electrochemical photolysis of water at a semiconductor electrode," *Nature*, vol. 238, no. 5358, pp. 37–45, 1972.

[3] W. Mekprasart and W. Pecharapa, "Synthesis and characterization of nitrogen-doped TiO_2 and its photo-catalytic activity enhancement under visible light," in *Proceedings of the Eco-Energy and Materials Science and Engineering Symposium*, vol. 9, pp. 509–514, 2011.

[4] R. Ahmadkhani and A. Habibi-Yangjeh, "Facile ultrasonic-assisted preparation of Fe_3O_4/Ag_3VO_4 nanocomposites as magnetically recoverable visible-light-driven photocatalysts with considerable activity," *Journal of the Iranian Chemical Society*, vol. 14, no. 4, pp. 863–872, 2017.

[5] H. Zhu, B. Yang, J. Xu et al., "Construction of Z-scheme type CdS-Au-TiO_2 hollow nanorod arrays with enhanced photocatalytic activity," *Applied Catalysis B*, vol. 90, no. 3-4, pp. 463–469, 2009.

[6] Q. Gu, H. Zhuang, J. Long et al., "Enhanced Hydrogen production over C-doped CdO photocatalyst in Na_2S/Na_2SO_3 solution under visible light irradiation," *International Journal of Photoenergy*, vol. 2012, Article ID 857345, 7 pages, 2012.

[7] S. Peng, A. Ran, Z. Wu, and Y. Li, "Enhanced photocatalytic hydrogen evolution under visible light over $Cd_x Zn_{1-x} S$ solid solution by ruthenium doping," *Reaction Kinetics, Mechanisms and Catalysis*, vol. 107, no. 1, pp. 105–113, 2012.

[8] Y. M. Yuxin and Y. I. Mixajlov, *Chemistry of Bismuth Compounds and Materials*, Iizdatelstvo SO RAN, Novosibirsk, Russia, 2001.

[9] A. Slikkerveer and F. de Wolff, "Toxicity of bismuth and its compounds," *Toxicology Methods*, no. 2, pp. 439–454, 1996.

[10] C. Wang, Z. Zhao, B. Luo, M. Fu, and F. Dong, "Tuning the morphology structure and photocatalytic activity of nitrogen-doped $(BiO)_2CO_3$ by the hydrothermal temperature, corporation," *Journal of Nanomaterials*, vol. 2014, Article ID 192797, 10 pages, 2014.

[11] X. Dong, W. Zhang, W. Cui et al., "Pt Quantum pots deposited on N-doped $(BiO)_2CO_3$: enhanced visible light photocatalytic no removal and reaction pathway," *Catalysis Science & Technology*, vol. 7, no. 6, pp. 1324–1332, 2017.

[12] Y. J. Wang, Y. M. He, T. T. Li, J. Cai, M. F. Luo, and L. H. Zhao, "Photocatalytic degradation of methylene blue on $CaBi_6O_{10}/Bi_2O_3$ composites under visible-light," *Chemical Engineering Journal*, vol. 189–190, pp. 473–481, 2012.

[13] P. Kanlaya, N. Wetchakun, W. Kangwansupamonkon, K. Ounnunkad, B. Inceesungvorn, and S. Phanichphant, "Photocatalytic mineralization of organic acids over visible-light-driven $Au/BiVO_4$ photocatalyst," *International Journal of Photoenergy*, vol. 2013, Article ID 943256, 7 pages, 2013.

[14] A. V. Shtareva, D. S. Starev, K. S. Makarevich, and M. B. Pereginyak, "Sposob polucheniya fotokatalizatora na osnove vismutata shhelochnozemelnogo metalla i sposob ochistki vody ot organicheskix zagryaznitelej fotokatalizatorom," "A method for producing a photocatalyst based on an

alkaline earth metal bismuthate and a method for purifying water from organic contaminants with a photocatalyst, Patent RUS, no. 2595343, 2014.

[15] R. S. Roth, C. J. Rawn, B. P. Burton, and F. Beech, "Beech phase equilibria and crystal chemistry in portions of the system SrO-CaO-Bi$_2$O$_3$-CuO, part II—the system SrO-Bi$_2$O$_3$-CuO," *Journal of Research of the National Institute of Standards and Technology*, vol. 95, no. 3, pp. 291–335, 1990.

[16] Y. V. Karyakin and I. I. Angelov, "Pure chemicals," *M: Ximiya*, p. 408, 1974.

[17] A. V. Zajtsev, O. I. Kaminsky, K. S. Makarevich, and S. A. Pjachin, "Avtomatizirovannyj kompleks dlya issledovaniya sorbcionnoj i fotokataliticheskoj aktivnosti s obedinennoj reakcionnoj i izmeritelnoj chastyu," (The automated complex for examination of getter and photocatalytic activity with the incorporated reactionary and measuring part), Sbornik nauchnyx soobshhenij (DVGUPS), no. 22, pp. 57–63, 2017.

[18] K. T. Jacoba and K. P. Jayadevan, "System Bi–Sr–O: synergistic measurements of thermodynamic properties using oxide and fluoride solid electrolytes," *Journal of Materials Research*, vol. 13, no. 7, pp. 1905–1918, 1998.

[19] A. H. Nethercot, "Prediction of fermi energies and photoelectric thresholds based on electronegativity concepts," *Physical Review Letters*, vol. 33, no. 18, pp. 1088–1091, 1974.

5

Mass Transfer during Osmotic Dehydration of Tunisian Pomegranate Seeds and Effect of Blanching Pretreatment

Basma Khoualdia, Samia Ben-Ali ⓘ, and Ahmed Hannachi

University of Gabes, National School of Engineers of Gabes, Laboratory of Research Process Engineering and Industrial Systems, LR11ES54, St. Omar Ibn El Khattab 6029, Gabes, Tunisia

Correspondence should be addressed to Samia Ben-Ali; benali.samia@gmail.com

Academic Editor: Bhaskar Kulkarni

In this work, the osmotic dehydration (OD) of Tunisian pomegranate seeds of "El Gabsi" variety was investigated. To optimize the process operating conditions, the effect of temperature, hypertonic solution solid content, and stirring speed was studied. The best conditions resulting in the higher water loss and the minimum of fruit damages found are 40°C, 50°Bx, and 440 rpm. In these conditions, the effect of blanching pretreatments on the solute and water transfer kinetics during the OD was investigated. The blanching pretreatments were carried out using two methods: blanching in a boiling water bath and in a microwave oven. The mass diffusion kinetic depends on time, temperature, hypertonic solution solid content, stirring speeds, and pretreatment process. Peleg's model showed a good fit to the experimental data. By applying blanching pretreatments, the water and solute effective diffusivities passed from the order of 10^{-9} to the order of 10^{-8}, and the OD equilibrium time was significantly reduced.

1. Introduction

Pomegranate (*Punica granatum* L.) is one of the oldest known edible fruits belonging to the family Punicaceae [1]. This fruit is frequently grown in arid and semiarid areas, especially in parts of Asia, North Africa, around the Mediterranean, and in the Middle East [2, 3]. Tunisia is one of the main producers in the Mediterranean region, and its production is mainly located in Gabes in southeastern Tunisia. The El-Gabsi variety represents about 35% of the annual Tunisian pomegranate production [4].

The pomegranate fruit is an important source of beneficial bioactive and nutritive compounds for human beings. It is rich in organic acids, minerals (such as potassium), vitamins (C, A, and K) [2, 5], and phenolic compounds such as hydrolysable tannins, condensed tannins, and phenolic acids [1,6–9]. Particularly, the edible pomegranate part contains sugars, organic acids, anthocyanins, minerals, proteins, unsaturated fatty acids, polysaccharides, and vitamins [10, 11].

The rich composition of this fruit gives it hypolipidemic, antioxidant, antiviral, antineoplastic, anticancer, antibacterial, antidiabetic, antidiarrheal, helminthic, vascular, and digestive protection properties. Due to its immense potential for health benefits, pomegranate has achieved the title of "superfood" [12], and it has been used in traditional medicine over centuries [6].

Thus, the conservation of the edible part of pomegranate is of great interest. The traditional conservation methods affect seriously the fruit's quality. OD reduces the product water activity by maintaining its sensory and nutritional characteristics. Water activity reduction slows down deteriorative reactions and increases microbial stability, thus prolonging the fruit shelf life [13]. OD allows partial water removal from cellular tissue by immersion in a concentrated aqueous solution [14]. The driving force for water removal is the difference in osmotic pressure between the fruit and the hypertonic solution [15]. The water removal from the product is always accompanied by a simultaneous counter diffusion of solutes from the osmotic solution into the tissue [16]. The cell membrane, the exchange surface between the osmotic solution and the product, exerts high resistances to mass transfer and slows the OD rate [17]. For this reason, the mass transfer can be supported by combining the DO with pulsed vacuum [18, 19], pulsed electric field [20–22], ultrasound [14, 23], freezing [24, 25], centrifugation [20, 26], and blanching [13, 17].

The benefits of such pretreatment processes have been widely reported in the literature. Corrêa et al. [27] affirmed that the application of a vacuum pulse on tomato slices is strongly recommended for reducing NaCl incorporation in osmotic processes with ternary solutions. Also, the centrifugal force combined with the OD of carambola slices gives higher water loss and less solid gain [28]. However, pretreating kiwi slices with ultrasound more than 10 min causes the formation of microchannels through the membranes which improve the mass exchange by OD [14]. Similarly, freezing pretreatment was used with several fruits to improve the OD kinetics. It causes the formation of cracks on the cell membrane which facilitate both solute and water transfer. Some of frozen pretreated products are as follows: pomegranate seeds [25], pumpkin [17], tomatoes [24], and mango [29]. The pulsed electric field was also used as a pretreatment or simultaneously with OD to accelerate the mass transfer. Using this pretreatment, the cell membrane seems not to be affected. It was applied on apple slices [30], on carrot tissue [20], and on bell peppers [22].

Blanching is an alternative pretreatment for OD. It is a heat treatment process that can be carried out either by the immersion of the samples in a hot solution bath [17, 27], by a hot steam exposure, or by ohmic heating in a microwave [13] for a short time (a few minutes) in a temperature range of 85°C to 100°C [4]. This pretreatment increases the cell membrane permeability and removes the gas occluded in plant tissues [28]. Thereafter, the mass transfer becomes faster during the further processing [13]. Thus, the main objective of the present study is to accelerate the mass transfer during the OD of "El Gabsi" pomegranate seeds in a sucrose solution. So, blanching was used as a pretreatment, and its effects on the solute and water transfer were investigated.

2. Materials and Methods

2.1. Sample and Solution Preparation.
Fresh pomegranate fruits (*Punica granatum* L.) at full ripeness, from the same region with homogeneous size were bought at a local market in Gabes, south of Tunisia. The fruits were cleaned with wet paper, wiped very well with blotting paper, and then stored at 5°C until use.

Sucrose solutions of 30°Bx, 40°Bx, 50°Bx, and 60°Bx, were used in the OD. It was prepared by dissolving analytical grade D(+)-sucrose crystals supplied by Carlo Erba Reagents Laboratory, France, in distilled water with a solute/water ratio of 1/2. The solid content of the solution was verified before using.

2.2. Blanching Pretreatment.
The pomegranate fruit was hand peeled, and the blanching pretreatments prior to OD were performed in two different ways:

(i) Microwave blanching: the pomegranate seeds were put in a sucrose solution (50°Bx) with a sample/solution ratio of 1/4. The mixture was put in a microwave at 600 W for one minute.

(ii) Boiling water bath blanching: the pomegranate seeds were immersed in a water bath at 92°C during one minute and were wiped very well with blotting paper.

2.3. Osmotic Dehydration.
Blanched and unblanched pomegranate seed samples of "El Gabsi" variety were submitted to OD. OD was carried out in batch mode in a sucrose solution with a magnetic stirring of 440 rpm. The sample to solution ratio was 1/4. During OD, samples were withdrawn at regular intervals (30 min), quickly rinsed with distilled water to remove osmotic solution from the surface, drained over by absorbent paper to eliminate excess of water, weighed, and oven-dried as described in Section 2.4 to determine their dry matter. The OD kinetics was monitored for seven hours.

2.4. Parameter Measurement.
The dry matter of samples was determined by stove-drying at 105°C until reaching a stable weight.

The solid content was directly measured using a refractometer model Sopelem 3127. The zero of the refractometer was adjusted using distilled water.

2.5. Theoretical Consideration.
Equations (1)–(4) were used to calculate the weight reduction, the solid gain, the water loss, and the dehydration rate, respectively:

$$WR = \frac{(w_0 - w_t)}{w_0} * 100, \tag{1}$$

$$WL = WR + SG, \tag{2}$$

$$SG = \frac{(s_t - s_0)}{w_0} * 100, \tag{3}$$

$$DR = \frac{WL}{t}, \tag{4}$$

where w is the pomegranate seeds' weight; s is the pomegranate seeds' dry matter; and t is the OD duration (min). The indexes 0 and t refer to initial or after an osmotic treatment for a period of time t, respectively.

Peleg's model was used to describe the mass transfer kinetics. It is presented by the following equation:

$$Y_t = Y_0 \pm \frac{t}{k_1 + k_2 t}, \tag{5}$$

where Y is the water loss or the solid gain; k_1 and k_2 are the Peleg parameters; and t is the OD duration. The indexes 0 and t refer to initial or after an osmotic treatment for a period of time t, respectively. The "±" symbol is a minus for the water loss and a plus sign for the solid gain.

For a long processing time, Equation (5) gives the Equation (6) leading to the equilibrium values of water losses and solid gains. These values will be used to determine the effective diffusivities using Fick's second law.

$$Y_{eq} = Y_0 \pm \frac{1}{k_2}, \tag{6}$$

where Y_{eq} is the water loss or the solid gain at equilibrium (g/g of dry matter).

The differential form of Fick's law is given by the following equation:

$$\varphi_i = -D_{eff,i}\, S\, \frac{\partial C_i}{\partial x}, \qquad (7)$$

where φ_i is the material flow through the surface S during the unit of time; S is the outer surface of a seed; C is the solute concentration; $D_{eff,i}$ is the effective diffusivities; and x is the distance in a direction normal to the seeds' surface. The index i is relative to the diffusing species, $i = w$ for water and $i = s$ for solute.

By considering that seeds are homogeneous spheres, the initial solid and water contents are uniform throughout the volume of the material, the equilibrium solid and the water contents are uniform throughout the surface, the diffusion coefficient is constant, and the resistance to the external mass transfer is negligible compared to the internal resistance, the solution of Fick's equation is given by

$$W_i = \frac{Y - Y_{eq}}{Y_0 - Y_{eq}} = \sum_1^n \frac{6}{(n\pi)^2} \exp\!\left(-(n\pi)^2 D_{eff,i}\, \frac{t}{R^2}\right), \qquad (8)$$

where $D_{eff,i}$ is the effective diffusivity of water or solute; n is the number of series terms; R is the equivalent radius of sphere; t is the time; W_i is the dimensionless amount of water losses or solid gains, respectively; and i is an index indicating water ($i = w$) or solute ($i = s$).

3. Results and Discussion

3.1. Effect of Processing Time on the Mass Transfer during OD.
OD incorporates a double product transformation in its drying process: there is water removal as well as the solute incorporation, resulting in the overall product weight reduction. Figure 1 shows the evolution of the water loss and solid gain during the OD of pomegranate seeds at $T = 40°C$, $C = 50°Bx$, $N = 440$ rpm, and sample/solution weight ratio equal to 1:4. Both dehydration and sugar impregnation depend on processing time. Indeed, kinetics is faster at the beginning where the potential between the osmotic solution and the pomegranate seed separated by the cell membrane is greater (phase 1), and then it slows down gradually (phase 2) until achieving a nearly constant value, beyond 300 minutes (phase 3: the near equilibrium) [24, 31, 32]. The quasiequilibrium is caused by the progressive decrease in the driving force and the formation of a solute layer on the cell membrane surface which can prevent the passage of molecules of water and solute [24, 33]. On the other hand, there is a quantitative difference between the water loss and the solid gain occurring during OD reflecting the cell membrane selectivity. Indeed, given its small size, the passage of the water molecule across the cell membrane is favored over that of sucrose which in most will be trapped in the pores.

3.2. Effect of Hypertonic Solution Solid Content on the Mass Transfer during OD. Figure 2 gives the water loss and solid

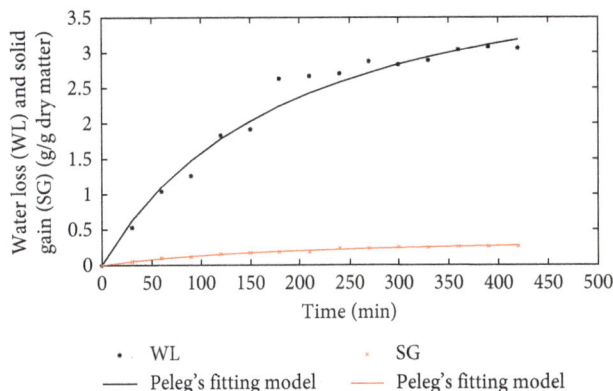

FIGURE 1: Effect of processing time and membrane selectivity on the mass transfer during OD of pomegranate seeds: ($T = 40°C$, $C = 50°Bx$, $N = 440$ rpm, and sample/solution weight ratio = 1:4).

gain evolution during the pomegranate seeds' OD as a function of osmotic solution's soluble solid content (Brix). Experiments were carried out at 20°C and 330 rpm. The equilibrium was not reached for all the concentrations given the low temperature and stirring speed. This does not preclude the clarity of the effect of the osmotic solution's initial soluble solid content on water loss and solid gain.

On the first hand, an increase in the osmotic solution concentration results in an increase in water loss and solid gain. On the other hand, the slopes of mass transfer curves as a function of time increase with concentration, which reflects the improvement in transfer kinetics. Indeed, by increasing the concentration, the difference in the water load between the osmotic solution and the product increases, and subsequently the transfer becomes faster and more important. When moving from 30°Bx to 40°Bx and to 50°Bx, a remarkable increase in water loss and solid gain was noted. For the 50°Bx and 60°Bx, the water loss curves are almost confounded while the solid gain continues to increase. This can be explained by the formation of a layer of sugar on the surface of the pomegranate seed, which forms a barrier to water transfer and leaves the sugar exposed to the pores. Finally, we can conclude that the concentration has a significant effect on mass transfer during OD of pomegranate seeds, but very high concentrations are not helpful to not prevent the water transfer while favoring that of sugar. 50°Bx was chosen as the optimal concentration for the OD of pomegranate seeds.

3.3. Effect of Temperature on the Mass Transfer during OD.
OD is achievable even at low temperatures, which avoids the adverse effects of heat on the food. After seven hours of OD of pomegranate seed at room temperature (20°C), the water loss was about 41% accompanied with 3% of solid gain. A reasonable increase of temperature leads to an increase in water loss and solid gain as well as an acceleration of mass transfer kinetics, as shown in Figure 3. Indeed, the increase of temperature causes the cell membrane softening and the opening of its pores which allows a more intense and faster mass transfer. By increasing the temperature from 20°C to 30°C, the water loss was almost not affected, but the solid

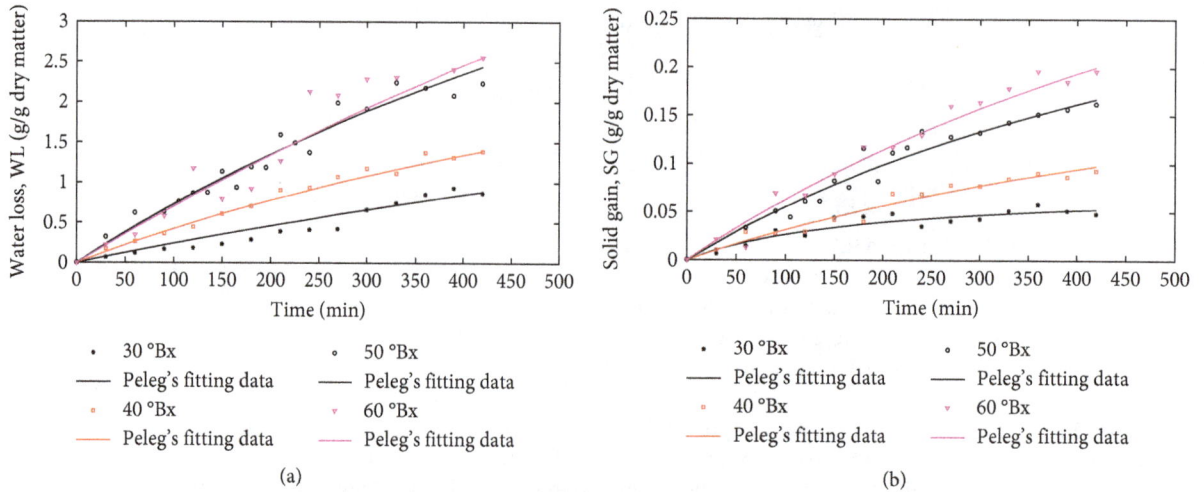

FIGURE 2: Effect of osmotic solution solid content on the mass transfer during OD of pomegranate seeds: (a) water loss and (b) solid gain (T = 30°C, N = 330 rpm, and sample/solution weight ratio = 1 : 4).

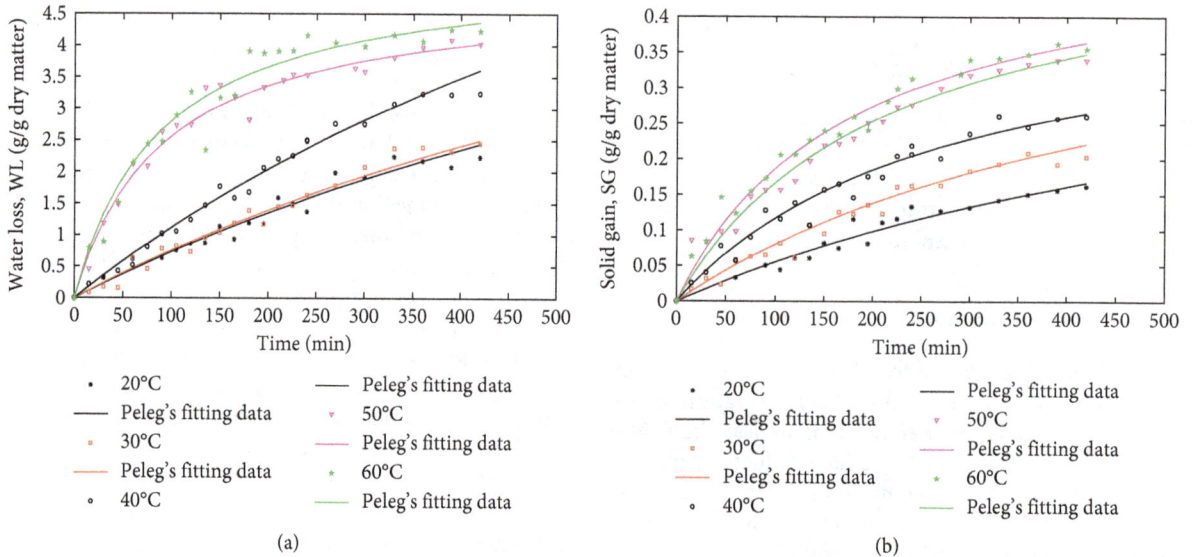

FIGURE 3: Effect of temperature on mass transfer during the OD of pomegranate seeds: (a) water loss and (b) solid gain (C = 50°Bx, N = 330 rpm, and sample/solution weight ratio = 1 : 4).

gain increased. So, the cellular structure was affected, but the sugar impregnation took precedence over dehydration. Moving from 30°C to 40°C, both water loss and solid gain increased. By going to higher temperatures, there is a very significant acceleration of mass transfer kinetics with close to perfect pseudo first-order curves. From the point of view of mass transfer, fast kinetics and clear equilibrium bearing accompanied with a significant water removal are desirable. But, at these two temperatures, the osmotic solution becomes pink indicating the hard affection of product texture and the loss of cell membrane semipermeability. The appearance of this color reflects the transfer of pomegranate seeds' own solutes, in particular the anthocyanin which is responsible for its pink coloration, to the osmotic solution and the assignment of the product quality. It has been previously reported that such high temperatures caused

irreversible damage and a selectivity loss of cell membrane [34, 35]. Similarly, Khan defined 50°C as a reasonable limit temperature for vegetables and fruit to the deterioration of flavor, texture, and heat-sensitive compounds of product [36]. Indeed, enzymatic browning and deterioration of fruit flavor begins at a temperature of 49°C. In recap, a moderate temperature increases to improve the water and solute transfer during OD. But, relatively high temperatures cause undesirable irreversible effects on aliment texture. So, in the case of pomegranate seeds, 40°C was chosen to reach the maximum water removal without damaging the cell membrane and losing the fruit quality.

3.4. Effect of Stirring Speed on OD Kinetics. Magnetic or mechanical agitation is very useful in similar cases. It makes it possible to ensure the mixture homogeneity and the temperature uniformity, and it reduces the mass transfer resistance of the cell membrane [37]. Figure 4 shows the effect of stirring speed on water loss and solid gain during pomegranate OD at 50°Bx, 40°C, and a product/solution ratio of ¼. At 330 rpm, the water loss and solid gain evolution are almost linear: this low stirring speed did not sufficiently reduce the cell membrane resistance to mass transfer. By going to 440 rpm, it is noted that the effect of the stirring speed is located mainly at the beginning of OD process without actually affecting the final transferred solute and water. The resistance to mass transfer was reduced, and the water loss and solid gain curves are getting closer to pseudo first-order shape. Similar results were found by [36]. By using a stirring speed of 550 rpm, the water loss and solid gain increased significantly, and the osmotic solution changed the color to pink. Indeed, the high stirring speed causes the damage of seeds and produces some cracks on its surface which explain the osmotic solution's new color caused by the transfer of some own solutes' pomegranate seeds and the quantitatively larger transfer of water and sugar through the damaged cell membrane.

Finally, 50°Bx, 40°C, and 440 rpm were chosen as the optimal OD operating conditions allowing to higher water loss and less product damage. The optimal conditions were used to unfold the effect of blanching on the pomegranate seeds' OD kinetics.

3.5. Effect of Blanching Pretreatment on OD Kinetics. The difference of the OD from one cellular tissue to another depends on the cellular tissue characteristics, the solute nature, the operating conditions, and the pretreatment. These conditions can accelerate or slow down the mass transfer. In this work, the blanching pretreatment affects the pomegranate seeds' texture and of course the mass transfer during the following treatment. Figure 5 shows the water loss and solid gain evolution during the OD of pomegranate seeds in sucrose solution for blanched and raw samples and the corresponding kinetic Peleg's model fitting. During the first and the second phase of OD, remarkable differences in both water loss and solid gain were noted between the untreated sample and the blanched samples. But, at the end of process, at the near-equilibrium, water loss and solid gain in the case of blanched samples tend to reach almost the same amounts such as the control sample. This behavior can probably be explained by the reversibility of the cell membrane permeabilization [38, 39]: The blanching pretreatment opens the cell membrane pores but does not destroy the cellular texture. The cellular tissue returns to its initial state after a period. Therefore, the OD near-equilibrium for the pretreated samples has the same level as the control sample.

Peleg's parameters (k_1 and k_2) were determined for water loss and solid gain using equation (5), as shown in Table 1. All the correlation coefficient values demonstrate the goodness of fit.

The Tunisian pomegranate seeds have already a high sugar content. Wherefore, we need to eliminate water with a minimum of sucrose impregnation. Therefore, the final water loss to solid gain ratio values were calculated for raw and blanched seeds and reported in Table 1. They were 8.729, 11.432, and 12.764 for control sample, microwave-blanched seeds, and boiling-water-blanched seeds, respectively. The pomegranate seeds blanched in a boiling water bath gave the higher WL/SG. It corresponds to the best treatment.

3.6. Effect of Blanching Pretreatment on the Dehydration Rate of Pomegranate Seeds. Figure 6 reports the dehydration rate as a function of time for both blanched and raw samples. For economic reasons, the OD must be stopped when reaching the half of the maximum of the dehydration rate [35]. The maximum dehydration rates of pomegranate seeds during the OD were in the order of 0.0176 g/min, 0.054 g/min, and 0.056 g/min for unblanched samples, microwave-blanched samples, and boiling-water-blanched samples, respectively. The maximum dehydration rates for the two blanched samples were slightly different. In particular, the time necessary to osmodehydrate unblanched pomegranate seeds was 300 min. Nevertheless, it was about 85 min for the two blanched samples. In fact, a processing time reduction of 72% was achieved by applying blanching as a pretreatment of the OD.

The reduction of OD's processing time was mostly due to cellular membrane softening caused by the blanching treatment, which facilitated the water and solute transfer as previously described in Section 3.5.

3.7. Solute and Water Effective Diffusivities Prediction. The solution of Fick's diffusion law, given by equation (8) was used to evaluate effective diffusion coefficients. For long periods of dehydration, the sum in equation (8) can be limited to the first term of the series [40].

Table 2 shows the effective diffusivities values for water and solute calculated using Fick's model, which also presented a good fit to experimental data, showing an average correlation coefficients (R^2) close to unity. The water and solute effective diffusivities of the control sample were the lowest one. They were, respectively, 8.9×10^{-9} and 5.2×10^{-9}. Indeed, the cell membrane was intact and had the highest firmness. Therefore, its permeability for solute and water was lower. These values are comparable to those found by Herman-Lara et al. [41] working on the OD of radish in NaCl solutions. Bchir et al. [42] working on the OD of pomegranate seeds in sucrose solution found lower values of effective diffusivities, in the order of 10^{-12}. This comparison should however take into account the experimental conditions, the different estimation methods employed, and the varieties of pomegranate fruit studied.

The diffusion coefficients increased by applying the blanching pretreatment using the two methods. The moisture diffusivities increased from 8.9×10^{-9} to 2.8×10^{-8} for the microwave blanching and to 1.8×10^{-8} for the boiling water bath blanching. Similarly, solute diffusivities increased from 5.2×10^{-9} to 1.8×10^{-8} for the microwave blanching and to 4.1×10^{-8} for the boiling water bath blanching. This behavior was probably due to cellular tissue

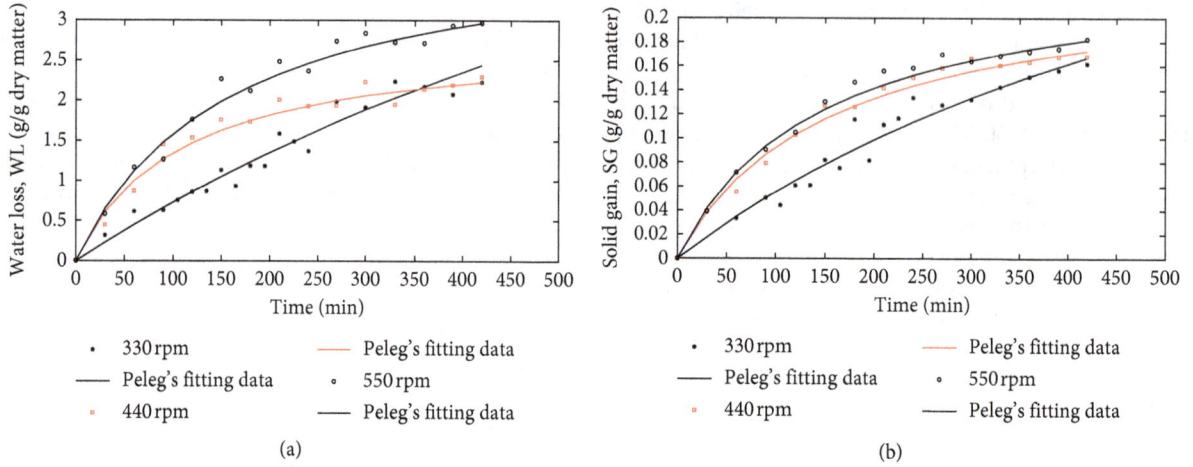

FIGURE 4: Effect of stirring speed on mass transfer during the OD of pomegranate seeds: (a) water loss and (b) solid gain ($C = 50°\text{Bx}$, $T = 40°\text{C}$, and sample/solution weight ratio = 1 : 4).

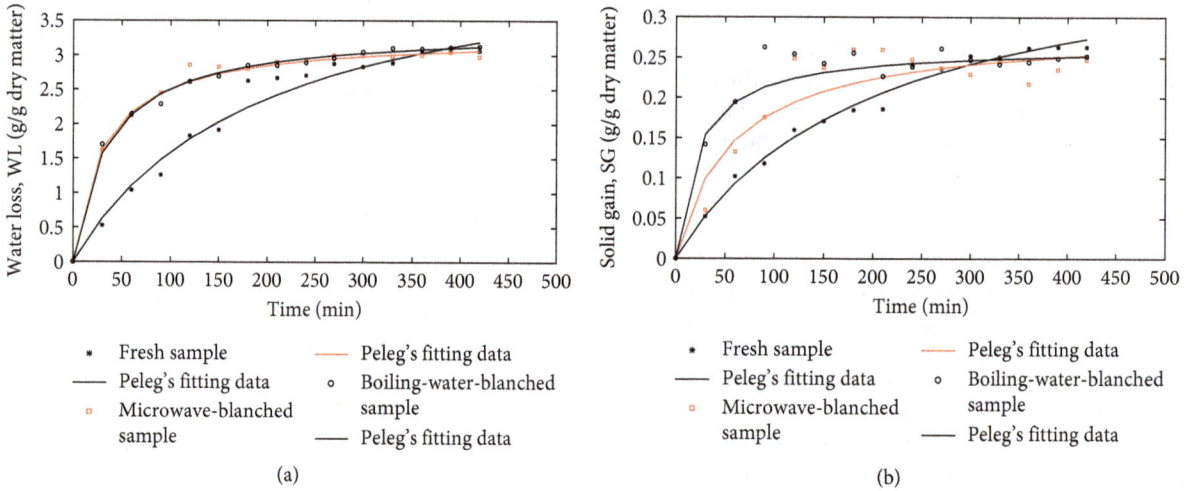

FIGURE 5: Effect of blanching pretreatment on the (a) water loss and (b) solid gain during OD of pomegranate seeds ($C = 50°\text{Bx}$, $T = 40°\text{C}$, $N = 440$ rpm, and sample/solution weight ratio = 1 : 4).

TABLE 1: Peleg's parameters.

Sample		Water loss (WL)			Solid gain (SG)			$\text{WL}_{eq}/\text{SG}_{eq}$
		k_1	k_2 (min)	R^2	k_1	k_2 (min)	R^2	
Temperature	20°C	126.006	0.110	0.977	1580.3	2.2	0.976	—
	30°C	120.700	0.112	0.989	1033.6	2.1	0.985	—
	40°C	81.044	0.084	0.992	620.892	2.299	0.983	—
	50°C	18.560	0.204	0.988	406.691	1.904	0.989	—
	60°C	16.693	0.189	0.978	339.998	1.936	0.991	—
Concentration	30°Bx	400.000	0.1781	0.969	2458.9	13.1	0.941	—
	40°Bx	214.811	0.2065	0.994	2815.2	3.5	0.980	—
	50°Bx	126.006	0.1102	0.978	1580.3	2.2	0.976	—
	60°Bx	135.010	0.0688	0.965	1423.9	1.6	0.988	—
Stirring speed	330 rpm	126.006	0.110	0.978	1580.1	2.2	0.976	—
	440 rpm	38.699	0.357	0.987	654.785	4.238	0.994	—
	550 rpm	38.706	0.245	0.990	593.496	4.094	0.995	—
Fresh seeds		19	0.284	0.96	496	2.48	0.99	8.729
Microwave-blanched seeds		9	0.305	0.99	99	3.47	0.94	11.432
Boiling-water-blanched seeds		10	0.296	0.99	82	3.78	0.97	12.764

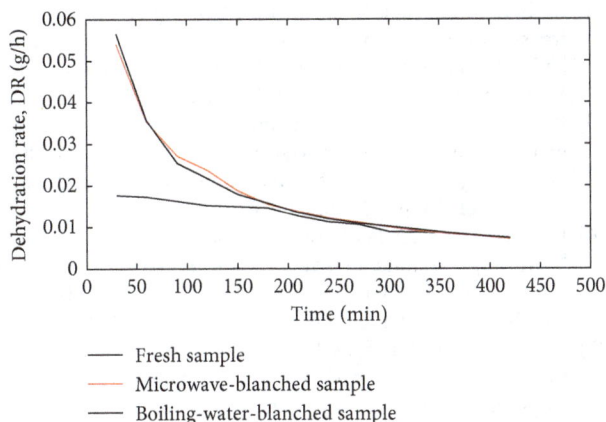

FIGURE 6: Effect of blanching pretreatment on the dehydration rate of pomegranate seeds ($C = 50°$Bx, $T = 40°$C, $N = 440$ rpm, and sample/solution weight ratio = 1 : 4).

TABLE 2: Solute and water effective diffusivities during OD.

Sample	Water loss (WL)		Solid gain (SG)	
	D_{effw} m²·s⁻¹	R^2	D_{effs} m²·s⁻¹	R^2
Fresh seeds	8.885×10^{-9}	0.988	5.157×10^{-9}	0.981
Microwave-blanched seeds	2.846×10^{-8}	0.987	1.764×10^{-8}	0.939
Boiling-water-blanched seeds	1.797×10^{-8}	0.961	4.072×10^{-8}	0.975

softening caused by the blanching treatment, which facilitated the water and solute transfer [39]. Different authors reported similar results for various vegetables and fruits: for pumpkin by Kowalska et al. [17], for leek slices by Doymaz [43], and so on.

For samples blanched in the sucrose solution in the microwave, the solute effective diffusivity is lower than those for the sample blanched in boiling water. Indeed, the blanching in the hypertonic solution causes a solute impregnation inside the sample during the pretreatment stage. Therefore, when going to the dehydration stage, pomegranate seeds were already enriched in sucrose and their pores were partially occupied by solute, which makes the sugar diffusion during the OD more difficult [39].

4. Conclusions

In this study, the optimal operating conditions for the OD of Tunisian pomegranate seeds were assessed. The best conditions giving the higher water losses and the best fruit quality were found to be 50°Bx, 40°C, and 440 rpm. At these optimum values of temperature, hypertonic solution solid content, and stirring speeds, the effect of blanching on water and solute transfer during the OD was investigated. Blanching has been applied in two techniques: microwave blanching and blanching in a boiling water bath. Pretreatment such as blanching improves the OD kinetics and reduces the dehydration processing time. The effective moisture and solute diffusivities are in the order of 10^{-9} for

the raw samples and in the order of 10^{-8} for the blanched samples. In terms of processing duration and WL/SG ratio, the best combination is by coupling the OD to the blanching in boiling water bath.

Conflicts of Interest

The authors declare that there are no conflicts of interest regarding the publication of this paper.

Acknowledgments

This work was financially supported by the Ministry of the Higher Education and Scientific Research of Tunisia.

References

[1] R. R. Mphahlele, O. A. Fawole, L. M. Mokwena, and U. L. Opara, "Effect of extraction method on chemical, volatile composition andantioxidant properties of pomegranate juice," *South African Journal of Botany*, vol. 103, pp. 135–144, 2016.

[2] M. E. Pena-Estévez, F. Artés-Hernández, F. Artés et al., "Quality changes of pomegranate arils throughout shelf life affected by deficit irrigation and pre-processing storage," *Food Chemistry*, vol. 103, pp. 302–311, 2016.

[3] I. Hmid, D. Elothmani, H. Hanine, A. Oukabli, and E. Mehinagic, "Comparative study of phenolic compounds and their antioxidant attributes of eighteen pomegranate (*Punica granatum* L.) cultivars grown in Morocco," *Arabian Journal of Chemistry*, vol. 10, pp. S2675–S2684, 2013.

[4] B. Bchir, S. Besbes, J. M. Giet, H. Attia, and C. Blecker, "Synthèse des connaissances sur la déshydratation osmotique," *Biotechnologie Agronomie Société et Environnement*, vol. 15, pp. 129–142, 2010.

[5] M. Sanchez-Rubio, A. Taboada-Rodríguez, R. Cava-Roda, A. Lopez-Gomez, and F. Marín-Iniesta, "Combined use of thermo-ultrasound and cinnamon leaf essential oil to inactivate *Saccharomyces cerevisiae* in natural orange and pomegranate juices," *Journal of LWT-Food Science and Technology*, vol. 73, pp. 140–146, 2016.

[6] F. Khosravi, M. H. Fathi Nasri, H. Farhangfar, and J. Modaresi, "Nutritive value and polyphenol content of pomegranate seed pulp ensiled with different tannin-inactivating agents," *Journal of Animal Feed Science and Technology*, vol. 207, pp. 262–266, 2015.

[7] R. R. Mphahlele, M. A. Stander, O. A. Fawole, and U. L. Opara, "Effect of fruit maturity and growing location on the postharvestcontents of flavonoids, phenolic acids, vitamin C and antioxidantactivity of pomegranate juice (cv. Wonderful)," *Journal of Scientia Horticulturae*, vol. 179, pp. 36–45, 2014.

[8] E. Sentandreu, M. Cerdan-Calero, and J. M. Sendra, "Phenolic profile characterization of pomegranate (*Punica granatum*) juice by high-performance liquid chromatography with diode array detection coupled to an electrospray ion trap mass

analyzer," *Journal of Food Composition and Analysis*, vol. 30, no. 1, pp. 32–40, 2013.

[9] M. Kazemi, R. Karim, H. Mirhosseini, and A. Abdul Hamid, "Optimization of pulsed ultrasound-assisted technique for extraction of phenolics from pomegranate peel of Malas variety: punicalagin and hydroxybenzoic acids," *Journal of Food Chemistry*, vol. 206, pp. 156–166, 2016.

[10] P. J. Szychowski, M. J. Frutos, F. Burlao, A. J. Paerez-Laopez, A. A. Carbonell-Barrachina, and F. Hernandez, "Instrumental and sensory texture attributes of pomegranate arils and seeds as affected by cultivar," *LWT-Food Science and Technology Journal*, vol. 60, no. 2, pp. 656–663, 2015.

[11] S. A. Al-Maiman and A. Dilshad, "Changes in physical and chemical properties during pomegranate (*Punica granatum* L.) fruit maturation," *Food Chemistry*, vol. 76, no. 4, pp. 437–441, 2002.

[12] J. A. Teixeira da Silva, T. S. Rana, D. Narzary, N. Verma, D. T. Meshram, and S. A. Ranade, "Pomegranate biology and biotechnology: a review," *Journal of Scientia Horticulturae*, vol. 160, pp. 85–107, 2013.

[13] J. Moreno, A. Chiralt, I. Escriche, and J. A. Serra, "Effect of blanching/osmotic dehydration combined methods on quality and stability of minimally processed strawberries," *Food Research International*, vol. 33, no. 7, pp. 609–616, 2000.

[14] M. Nowacka, U. Tylewicz, L. Laghi, M. Dalla Rosa, and D. Witrowa-Rajchert, "Effect of ultrasound treatment on the water state in kiwifruit during osmotic dehydration," *Food Chemistry Journal*, vol. 144, pp. 18–25, 2014.

[15] M. A. da Conceicao Silva, Z. E. da Silva, V. C. Mariani, and S. Darche, "Mass transfer during the osmotic dehydration of West Indian cherry," *LWT-Food Science and Technology*, vol. 45, no. 2, pp. 246–252, 2012.

[16] J. P. Maran, V. Sivakumar, K. Thirugnanasambandham, and R. Sridhar, "Artificial neural network and response surface methodology modeling in mass transfer parameters predictions during osmotic dehydration of *Carica papaya* L," *Alexandria Engineering Journal*, vol. 52, no. 3, pp. 507–516, 2013.

[17] H. Kowalska, A. Lenart, and D. Leszczyk, "The effect of blanching and freezing on osmotic dehydration of pumpkin," *Journal of Food Engineering*, vol. 86, no. 1, pp. 30–38, 2008.

[18] P. Fito, "Modelling of vacuum osmotic dehydration of food," *Journal of Food Engineering*, vol. 22, no. 1-4, pp. 313–328, 1994.

[19] H. Allali, L. Marchal, and E. Vorobiev, "Effects of vacuum impregnation and ohmic heating with citric acid on the behavior of osmotic dehydration and structural changes of apple fruit," *Biosystems Engineering*, vol. 106, no. 1, pp. 6–13, 2010.

[20] E. Amami, A. Fersi, E. Vorobiev, and N. Kechaou, "Osmotic dehydration of carrot tissue enhanced by pulsed electric field, salt and centrifugal force," *Journal of Food Engineering*, vol. 83, no. 4, pp. 605–613, 2007.

[21] E. Amami, E. Vorobiev, and N. Kechaou, "Modelling of mass transfer during osmotic dehydration of apple tissue pre-treated by pulsed electric field," *LWT-Food Science and Technology*, vol. 39, no. 9, pp. 1014–1021, 2006.

[22] B. I. O. Ade-Omowaye, P. Talens, A. Angersbach, and D. Knorr, "Kinetics of osmotic dehydration of red bell peppers as influenced by pulsed electric field pretreatment," *Food Research International*, vol. 36, no. 5, pp. 475–483, 2003.

[23] F. A. N. Fernandes, F. E. Linhares Jr., and S. Rodrigues, "Ultrasound as pre-treatment for drying of pineapple," *Ultrasonics Sonochemistry*, vol. 15, no. 6, pp. 1049–1054, 2008.

[24] M. G. Athanasia and N. L. Harris, "Modeling of mass and heat transfer during combined processes of osmotic dehydration and freezing (Osmo-Dehydro-Freezing)," *Chemical Engineering Science*, vol. 82, pp. 52–61, 2012.

[25] B. Bchir, S. Besbes, H. Attia, and C. Blecker, "Osmotic dehydration of pomegranate seeds (*Punica granatum* L.): effect of freezing pre-treatment," *Journal of Food Process Engineering*, vol. 30, no. 3, pp. 335–354, 2010.

[26] E. Amami, A. Fersi, L. Khezami, E. Vorobiev, and N. Kechaou, "Centrifugal osmotic dehydration and rehydration of carrot tissue pre-treated by pulsed electric field," *LWT-Food Science and Technology*, vol. 40, no. 7, pp. 1156–1166, 2007.

[27] J. L. G. Corrêa, D. B. Ernesto, and K. S. de Mendonça, "Pulsed vacuum osmotic dehydration of tomatoes: sodium incorporation reduction and kinetics modeling," *LWT-Food Science and Technology*, vol. 71, pp. 17–24, 2016.

[28] N. Barman and L. S. Badwaik, "Effect of ultrasound and centrifugal force on carambola (*Averrhoa carambola* L.) slices during osmotic dehydration," *Ultrasonics Sonochemistry*, vol. 34, pp. 37–44, 2017.

[29] J. Floury, A. Le Bail, and Q. T. Pham, "A three-dimensional numerical simulation of the osmotic dehydration of mango and effect of freezing on the mass transfer rates," *Journal of Food Engineering*, vol. 85, no. 1, pp. 1–11, 2008.

[30] R. Simpson, C. Ramírez, V. Birchmeier et al., "Diffusion mechanisms during the osmotic dehydration of Granny Smith apples subjected to a moderate electric field," *Journal of Food Engineering*, vol. 166, pp. 204–211, 2015.

[31] C. C. Ferrari and M. D. Hubinger, "Evaluation of the mechanical properties and diffusion coefficients of osmodehydrated melon cubes," *International Journal of Food Science and Technology*, vol. 43, pp. 2065–2074, 2008.

[32] N. M. Misljenovic, G. B. Koprivica, L. R. Jevric, and L. J. B. Levic, "Mass transfer kinetics during osmotic dehydration of carrot cubes in sugar beet molasses," *Romanian Biotechnological Letters*, vol. 16, pp. 6790–6799, 2011.

[33] P. P. Sutar and D. K. Gupta, "Mathematical modeling of mass transfer in osmotic dehydration of onion slices," *Journal of Food Engineering*, vol. 78, no. 1, pp. 90–97, 2007.

[34] C. F. Cristhiane and D. H. Miriam, "Evaluation of the mechanical properties and diffusion coefficients of osmodehydrated melon cubes," *International Journal of Food Science and Technology*, vol. 43, no. 11, pp. 2065–2074, 2008.

[35] M. R. Khoyi and H. Javad, "Osmotic dehydration kinetics of apricot using sucrose solution," *Journal of Food Engineering*, vol. 78, no. 4, pp. 1355–1360, 2007.

[36] M. R. Khan, "Osmotic dehydration technique for fruits preservation-a review," *Pakistan Journal of Food Sciences*, vol. 22, pp. 71–85, 2012.

[37] S. Bahadur, S. P. Parmjit, N. Vikas, and F. K. John, "Optimisation of osmotic dehydration process of carrot cubes in mixtures of sucrose and sodium chloride solutions," *Food Chemistry*, vol. 123, no. 3, pp. 590–600, 2010.

[38] H. Vega-Mercado, O. Martin-Belloso, B. Qin et al., "Non thermal food preservation: pulsed electric fields," *Trends in Food Science and Technology*, vol. 81, no. 5, pp. 151–157, 1997.

[39] C. Severini, A. Baiano, T. De Pilli, B. F. Carbone, and A. Derossi, "Combined treatments of blanching and dehydration: study on potato cubes," *Journal of Food Engineering*, vol. 68, no. 3, pp. 289–296, 2005.

[40] A. Aghfir, S. Akkad, M. Rhazi, C. S. E. Kane, and M. Kouhila, "Détermination du coefficient de diffusion et de l'énergie d'activation de la menthe lors d'un séchage conductif en

régime continu," *Revue des Energies Renouvelables*, vol. 11, pp. 385–394, 2008.

[41] E. Herman-Lara, C. E. Martinez-Sanchez, H. Pacheco-Angulo, R. Carmona-Garcia, H. Ruiz-Espinosab, and I. I. Ruiz-Lopez, "Mass transfer modeling of equilibrium and dynamic periods during osmotic dehydration of radish in NaCl solutions," *Food and Bioproducts Processing*, vol. 91, no. 3, pp. 346–354, 2012.

[42] B. Bchir, S. Besbes, H. Attia, and C. Blecker, "Osmotic dehydration of pomegranate seeds: mass transfer kinetics and differential scanning calorimetry characterization," *International Journal of Food Science and Technology*, vol. 44, no. 11, pp. 2208–2217, 2009.

[43] İ. Doymaz, "Influence of blanching and slice thickness on drying characteristics of leek slices," *Chemical Engineering and Processing: Process Intensification*, vol. 47, no. 1, pp. 41–47, 2008.

6

Reactive Extrusion of Polyethylene Terephthalate Waste and Investigation of its Thermal and Mechanical Properties after Treatment

Mahmoud A. Mohsin,[1] Tahir Abdulrehman,[2] and Yousef Haik[2]

[1]*Department of Chemistry, University of Sharjah, P.O. Box 27272, Sharjah, UAE*
[2]*College of Science and Engineering, Hamad Bin Khalifa University, P.O. Box 34110, Doha, Qatar*

Correspondence should be addressed to Mahmoud A. Mohsin; mmohsin@sharjah.ac.ae

Academic Editor: Jose C. Merchuk

This study investigates treating polyethylene terephthalate (PET) waste water bottles with different mass of ethylene glycol (EG) using reactive extrusion technique at a temperature of 260°C. The study puts emphases on evaluating the thermal, mechanical, and chemical characteristics of the treated polyethylene terephthalate. The properties of the treated PET from the extruder were analyzed using FT-IR, TGA, DSC, and nanoindentation. The melt flow indexes (MFI) of both treated and untreated PET were also measured and compared. Thermal properties such as melting temperature (T_m) for treating PET showed an inversely proportional behavior with the EG concentrations. The FT-IR analysis was used to investigate the formation of new linkages like hydrogen bonds between PET and EG due to the hydroxyl and carbonyl groups. Nanoindentation results revealed that both the mechanical characteristics, elastic modulus and hardness, decrease with increasing EG concentration. On the other hand, the melt flow index of treated PET exhibited an increase with increasing EG concentration in the PET matrix.

1. Introduction

Synthetic polymers have undergone remarkable growth in terms of diversity, quality, and production volume in recent years [1]. The main difference between the two classes of polymers, namely, natural and synthetic polymers, is that the former are biologically or environmentally degradable, which makes them less durable and have high production cost, whereas the latter last considerably longer in all applications, being almost nonbiodegradable under normal circumstances. The nonbiodegradability of synthetic polymers has triggered a major unease for environmentalists in terms of polymer waste management [2]. In recent years, recycling of synthetic polymers has caught the attention of many research groups, for two major motives: to minimize the ever growing volume of polymer waste and to create value-added items from low cost resources by transforming synthetic polymeric waste into beneficial and useful items having similar properties to virgin materials [3–6].

PET is considered as a multipurpose polymer used for the manufacture of products which differ extensively in their properties and thus differ in their use. Owing to its extensive use over the years, PET is considered as one of the most important polymers in the present world [7, 8]. PET is semicrystalline, transparent, thermoplastic polyester. It has good tensile strength, chemical resistance, and suitable thermal stability. Huge quantities of PET are used in the production of food packaging materials and water bottles [9, 10]. With the enormous development and usage of PET, a large fraction of PET gets added to the waste system on a yearly basis. The demand for nonbiodegradable PET in various applications has been increasing at an alarming rate [11, 12]. Although PET is not hazardous, many factors contribute to the accumulation of PET in the environment and this necessitates serious attention to recycling as the accumulation has become a global environmental issue [2, 13, 14]. The recycling of PET through chemical methods [10, 15–18], especially depolymerization or lysis of PET using solvents and ionic liquids, has gained considerable interest as it can help in the synthesis of different kinds of end products. Depolymerization of PET can be achieved by hydrolysis using water [19] or lysis using alcohol [20] (methanolysis using methanol) [21, 22] or amines [23] or acids [24]. The most established or commercialized processes for the depolymerization of PET are glycolysis, hydrolysis, and methanolysis [5]. The depolymerization of PET can also be classified based on the type of catalyst or ionic liquids or supercritical conditions used in the process [25].

Glycolysis is the process where glycols like ethylene glycol (EG) are used in the lysis of PET [26] and glycolysis also has applications in the formulation of resins comprising unsaturated polyesters. Ethylene glycol is used in the reduction of metal salts to metals [27, 28] and is a precursor to polyesters, in addition to its various other applications [27]. The rate of PET glycolysis is found to be of second order in terms of EG concentration [29]. From the glycolysis reaction, the monomer bis(2-hydroxyethyl) terephthalate (BHET) is the main product and the rate of reaction is dependent on various factors which include temperature, pressure, ethylene glycol concentration, and the catalyst involved. For a recycling process to be desirable, it should be effective in terms of energy consumption and expenses related to processing. Glycolysis is usually carried out in boiling ethylene glycol [30] in the presence of catalysts like zinc acetate [31] or lead acetate [16]. The use of such catalysts in large scale may cause pollution due to their toxic nature [5]. The properties of PET can be modulated by treating it with suitable reagents or dry processes for use in various applications [32, 33]. The wettability or hydrophilic property of PET was enhanced using two dry processes, namely, atmospheric pressure plasma and ultraviolet excimer light [32]. Reactive extrusion was used to modify the rheological and mechanical properties of PET by treating with acrylic epoxy resin as the chain extender [33]. Composites prepared from melt blending of PET and nanoclay were found to have enhanced Young's modulus compared to PET and also prevented the migration of terephthalic acid into yoghurt drinks which were maintained in PET-nanoclay composites [34]. Antimicrobial PET fabric was synthesized from PET-MgO composites that were produced by the melting, mixing, and extrusion of PET and magnesium oxide nanoparticles [35]. Composites of PET with microencapsulated carbon microspheres (MCMSs) were prepared by melt blending and these composites were found to have good flame retarding and mechanical properties [36, 37]. Hence, the properties of PET can be enhanced by melting blending or extrusion with certain characteristic materials.

Coprocessing is another method of using plastic wastes as raw material or as a source of energy or to replace both natural mineral resources and fossil fuels such as petroleum, coal, and gas in energy intensive industries such as cement, steel, and power generation [38–40].

In this work, PET was treated with 0, 0.4, 0.6, 1, 1.4, and 2 grams of EG using a single screw extruder at the temperature of 260°C. The effect of different mass of EG on PET degradation was investigated. The chemical, thermal, and mechanical properties of the treated PET were investigated.

2. Experimental

2.1. Materials and Methods. The collected PET waste bottles were cleaned, dried, and shredded into small pieces. Ethylene glycol (99.5% (G.C) minimum assay, Panreac, Spain) was used for the treatment of PET. A single screw extruder (HAAKE POLYLAB QC, Thermo Scientific, Germany) was used for the reactive processing of PET with different mass of EG. Electrical Grinder (Geepas Industrial Co. Ltd., China) was used to mill the extruded PET and convert it into powder for further analysis and melt flow index measurements.

2.2. Treated PET Sample Preparation. The preparation procedure was divided into two parts. In the first part, 20 g sample of the shredded PET waste was weighed. Then it was placed in the hopper of the extruder, after the extruder barrel was heated to the temperature of 260°C. The PET was maintained in the barrel at 260°C with very low screw speed for 15 minutes to achieve a complete melting of the PET. The screw speed of the extruder was adjusted to 10 rpm in order to obtain a homogenous spread of melted PET in the extruder barrel. Then the speed of the screw was increased to 50 rpm to extrude the entire PET from the extruder. In the second part of the experiment, a mixture of PET and EG was prepared. A 20 g sample of PET was placed in a flask, and then 0.4 g of EG was added. The mixture was placed inside the extruder at 260°C for 15 min at 10 rpm in order to obtain a homogenous mixture. The speed of the screw was increased to 50 rpm to extrude the treated PET from the extruder. The same procedure was repeated to analyze the effect of 0.6 g, 1 g, 1.4 g, and 2 g of EG on the PET. Each treated PET sample was kept at room temperature in sealed plastic bags.

2.3. Fourier Transform Infrared Spectroscopy (FT-IR). Fourier transform infrared spectrometer (MAGNA-IR 560 Spectrometer, Thermo Nicolet Corporation) was used to identify the functional groups of the extruded PET treated with different mass of EG. The extruded PET was grinded to a fine powder and a few milligrams of the sample was grinded with dry KBr to a very fine powder. This powder was then compressed into a thin disc. First, the FT-IR instrument was calibrated by scanning the background. Then, the KBr disc was placed in the sample holder which was inserted into the FT-IR spectrometer for analysis. The infrared transmission spectral results of the samples were attained at room temperature in the wavenumber range 4000–400 cm^{-1} using 32 scans with a resolution of 2 cm^{-1}.

2.4. Thermogravimetric Analysis (TGA). The TGA analysis of the treated PET was carried out utilizing a Thermogravimetric Analyzer (Q50, TA Instruments). Samples of ~15 mg weight were taken in the sample pan. Then loaded sample pan was placed in the balance suspended inside the furnace tube, which was then heated from room temperature to 600°C at the rate of 10°Cmin^{-1} in N_2 atmosphere. The obtained TGA measurements were evaluated with the software, TA Universal Analysis 2000 (TA Instruments).

2.5. Differential Scanning Calorimetry (DSC). The DSC analysis of the treated PET was performed with a Differential Scanning Calorimeter (Q200, TA Instruments). The treated PET sample (~5 mg) was placed in an aluminum T_{zero} sample pan. Then the sample was sealed with aluminum T_{zero} lid using a pressing machine. The sealed pan was placed in the DSC device, where the reference cell was an empty sealed aluminum pan. All samples were subjected to heating from ambient conditions to 400°C at a rate of 10°Cmin^{-1} in nitrogen atmosphere. The DSC graphs obtained were evaluated with the software, TA Universal Analysis 2000 (TA Instruments). The melting temperature (T_m) was determined from the DSC thermogram as the pinnacle temperature value of the endothermic outcome.

2.6. Nanoindentation. The mechanical properties of the treated PET were obtained using nanoindentation, (NanoTest, Micro Materials, UK), using a Berkovich diamond indenter. The indentations were made by applying a maximum load of 10 mN at a constant rate of displacement of 0.05 mNs^{-1}. On attaining the maximum load, the indenter was removed from the surface of the sample at the same rate of 0.05 mNs^{-1} till the indenter was totally withdrawn from the sample. The purpose of maintaining a uniform displacement rate for indentation of the samples is to prevent hardening effects due to strain on the measurements. A total of five indents were made on each sample with a distance of 50 μm between the indentations to prevent interactions.

The load-depth data from the indentations was used to determine the hardness (H) and reduced moduli (E) of the samples.

The deformation associated with both elastic and plastic behavior occurs during the indentation of the specimen, but as a result of unloading of the indenter only the elastic portion is recovered.

The hardness (H) value as a result of indentation can be defined as in

$$H = \frac{P_{\max}}{A} = \frac{P_{\max}}{24.5h_c^2}. \tag{1}$$

The load value determined at the maximum level or depth of penetration (h) during an indentation cycle is represented by $P_{\max}$, whereas "A" represents the projected contact area and h_c denotes the depth of contact during the indentation. The value of h_c can be calculated by means of

$$h_c = \frac{h - 0.75P_{\max}}{S}. \tag{2}$$

S is derived from the early or initial region of the unloading curve and denotes the slope (dp/dh) when $h = h_{\max}$. The value 0.75 in (2) is a constant and is mainly dependent on the geometry of the indenter.

The initial unloading contact stiffness (S) can be used to determine the elastic modulus of the sample. The equation for the relationship between contact stiffness, reduced modulus, and contact area is represented by

$$S = 2\beta E_r \left(\frac{A}{\pi}\right)^{1/2}. \tag{3}$$

The constant β is dependent on the indenter geometry. For a Berkovich indenter, β has a value of 1.034. The reduced elastic modulus is represented by E_r. The reduced modulus takes into consideration the elastic deformation of not only the specimen but also the indenter. The reduced modulus, E_r, is evaluated by determining the contact stiffness (S) and contact area (A) from the load-displacement graph obtained by the indentation process. The elastic modulus (E_s) of the sample is calculated as shown in

$$E_s = \left(1 - v_s^2\right) \left[\frac{1}{E_r} - \frac{\left(1 - v_i^2\right)}{E_i}\right]^{-1}. \tag{4}$$

Poisson's ratios of the specimen and indenter are represented by v_s and v_i, respectively. Poisson's ratio for the indenter has the value of 0.07. E_i represents the modulus of the Berkovich diamond indenter which has a value of 1141 GPa. Poisson's ratio v_s for the specimen was estimated to be 0.35 which is the value for semicrystalline polymeric materials. This value was used for all calculations.

2.7. Melt Flow Index (MFI). The melt flow index for the treated PET was performed using XRL-400A/B/C series, Material Flow Velocity Machine (Chengde Jingmi Testing Machine Co., Ltd., China). The melt flow index apparatus consists of two testing modes: displacement testing mode and time testing mode. The displacement testing mode measures the flow index as the distance covered by a certain tested sample, expressed as grams/centimeter. Alternatively, the time testing mode measures the time needed to cut a certain amount from the tested sample, expressed as gram/10 minutes. The latter mode was selected for measuring the melt flow index. The treated and untreated PET samples were grinded and converted into powder using a grinder. Then the samples were weighed and place into the MFI instrument for testing. The parameters (cutting time, temperature, and load) were adjusted by performing several tests on the samples in order to optimize the testing parameters. The samples were heated to 256°C and kept at this temperature for 10 minutes. Then the load of 2160 grams was applied to samples to force the melted samples out of the nozzle to measure the MFI in terms of g/10 min. To determine the temperature at which the treated and untreated PET samples start flowing through the nozzle of the MFI instrument, the samples were heated to 220°C and maintained at this temperature for 10 minutes with the nozzle closed. After 10 minutes, the load was applied, the nozzle was opened with the temperature increasing at a constant rate, and the temperature at which the sample started flowing out of the nozzle was noted for each sample.

3. Results and Discussion

FT-IR defines the existence of specific chemical or functional groups in the polymeric material. The combined FT-IR spectra for EG, PET, and treated PET with different mass (grams) of EG are shown in Figure 1.

The treated PET with different EG mass demonstrated the presence of new hydroxyl groups (O–H) that appeared

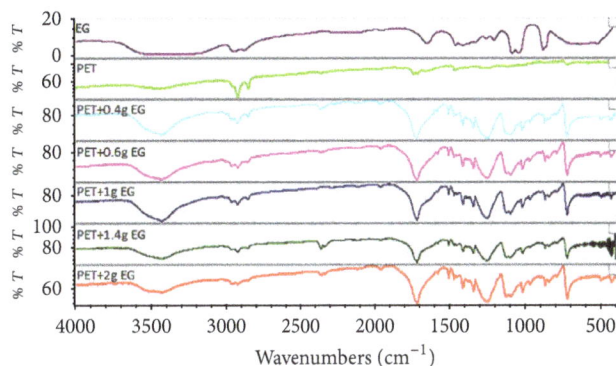

FIGURE 1: FT-IR spectra of the untreated and treated PET with different mass (grams) of ethylene glycol EG.

at wavenumber between 3500 cm^{-1} and 3400 cm^{-1} as shown in the FT-IR spectra which may contribute to the formation of inter- and intramolecular hydrogen bonds in the EG treated PET polymer. The hydroxyl groups in the treated PET polymer are contributed by EG during the interaction between the two components in the extruder. On the other hand, the nontreated PET polymer does show the presence of hydroxyl groups but with reduced intensity than the treated PET. Moreover, at wavenumber 2950 cm^{-1}, an alkyl group (C–H stretch) appeared in the PET polymer, whereas this group was reflected with less intensity in the FT-IR spectra for the treated PET with EG, and its intensity decreases gradually with increasing EG concentration and was found to be the least for the PET treated with 2 g of EG. Furthermore, a carbonyl group (C=O stretch) appeared in the treated PET with EG at a wavenumber of 1715 cm^{-1}, while this group is also reflected in the FT-IR spectra for the untreated PET polymer but with a lesser intensity as shown in Figure 1. Besides, an ether group (C–O–C stretch (diaryl)) is obvious for the treated PET with different EG concentrations at an approximate wavenumber of 1250 cm^{-1}, whereas it is demonstrated at a very low intensity for the untreated PET polymer. Another band that is clearly shown in the FT-IR spectra for the treated PET with EG is reflected at a wavenumber of about 1100 cm^{-1}, while it is not shown for the untreated PET polymer. This wavenumber indicates the availability of C–C stretch in the treated PET with different EG concentrations but not in PET polymer. In addition, the FT-IR clearly shows a band at the wavenumber of about 720 cm^{-1} for the treated PET with various EG concentrations, whereas it is demonstrated at a lesser intensity for the PET polymer. This peak indicates the existence of CH$_2$ bending in treated PET with EG to a higher extent than in the untreated PET polymer. The variation in the intensity of the peaks or the presence of new peaks for the FT-IR spectra of the EG treated PET when compared to untreated PET could be due to the presence of EG in the polymeric matrix or due to the transesterification or condensation reactions of PET with EG or the formation of monomers like BHET.

Thermogravimetric analysis (TGA) was carried out to assess the thermal stability, the stages of degradation, and the moisture percentage in the sample. The TGA for the

untreated PET and treated PET with 0.4, 0.6, 1, 1.4, and 2 grams of EG was carried out. The combined results of the TGA analysis for all the samples are shown in Figure 2(a). TGA readings, including initial decomposition temperature (T_{dec}), percent weight loss ($W\%$), and percentage of moisture are summarized in Table 1.

The thermal degradation of polymers is a complex process and consists of initiation, propagation, and termination stages [40]. Initial decomposition temperature of the PET treated with EG as well as neat PET polymer was found to be at about 374°C. The weight loss at 100°C which corresponds to the moisture content was found to be very low for all the PET samples. From the TGA test, the moisture content in the PET polymer was shown to be 0.1%. On the other hand, the moisture content in the treated PET with 0.4 g of EG was 0%. This value increased gradually with the increase in the EG concentration as shown in Table 1. The weight loss percent at the decomposition temperature was investigated for all the samples. The weight loss for the EG treated PET increased with increase in the EG mass; that is, the PET treated with 0.4 g of EG showed 2.59% weight loss, whereas this value rose to 5.24% for PET with 2 g of EG. On the contrary, the value of the weight loss percent for the neat PET polymer was found to be 0.73, which is less compared to the PET samples treated with EG. From Figure 2(b), in the derivative thermogravimetric (DTG) analysis, the peak temperature values were found to be 437, 434, 436, 444, 448, and 450°C for the PET treated with 0, 0.4, 0.6, 1, 1.4, and 2 grams of EG, respectively.

The organic macromolecules or the low-molecular weight organic molecules that constitute the polymer matrix can withstand a particular temperature range and this forms the basis for the thermal degradation of polymers. The thermal stability of the polymers relies on the intrinsic features of the specimen and also the particular interface or interactions among the different macromolecular or molecular components of the polymeric matrix [41]. When certain low-molecular weight materials called plasticizers are added to the polymer matrix, it results in the modification of the 3-dimensional organization of the polymeric structure. Plasticizers also reduce the intermolecular forces of attraction, thereby increasing the mobility and volume of the polymeric matrix. This phenomenon is achieved by the formation of linkages in the form of hydrogen (H) bonds between the hydroxyl groups and the polar functional groups present in the polymer and plasticizer [42].

From the TGA curves for the untreated PET and PET treated with different EG mass, it is observed that all the specimens are relatively stable in the temperature range of 25–350°C. In addition, the percentage of moisture content in all the samples measured at about 100°C is almost negligible. Besides, it is clear from the graph that the TGA curves for all the samples indicate one-step degradation, where the degradation temperature (T_d) is within the range of 370–410°C. The weight loss for all the samples at 374°C is shown in Table 1. The onset thermal degradation point for all treated PET starts at 374°C. Beyond the temperature of 374°C, the PET/EG polymer becomes significantly degraded. Moreover, as the mass of EG increases in the PET sample, the

(a)

(b)

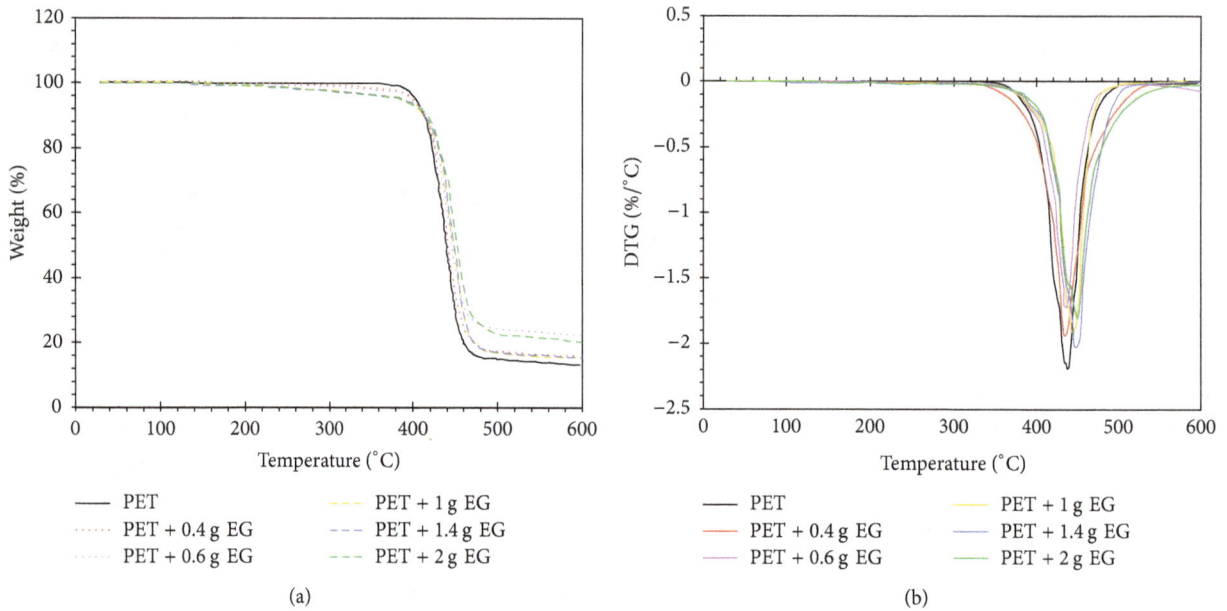

FIGURE 2: (a) TGA curves and (b) DTG analysis for the untreated and EG treated PET.

TABLE 1: Total weight loss of all samples.

Sample	PET + 0 g EG	PET + 0.4 g EG	PET + 0.6 g EG	PET + 1 g EG	PET + 1.4 g EG	PET + 2 g EG
Decomposition temperature (°C)	374.64	374.71	374.71	374.71	374.71	374.71
% weight loss	0.73	2.59	2.81	4.37	4.89	5.42
% moisture content @ 100°C	0.1	0	0.02	0.12	0.18	0.12

weight loss increases at the temperature of 374°C, where the weight loss percentage increased from 2.59 for PET treated with 0.4 g of EG to 5.42 for PET treated with 2 g of EG. This could be due to the volatile nature of the plasticizer at higher temperature.

Differential scanning calorimetry analysis (DSC) was carried out for all the samples to determine the melting points (T_m) of untreated PET and the treated PET with 0.4 g, 0.6 g, 1 g, 1.4 g, and 2 g of EG. From the TGA curves, the degradation of the PET samples begins at 374°C, beyond which there is enhanced weight loss and degradation and hence the DSC thermograms were concluded at 400°C. The combined DSC thermograms of the PET/EG for various EG masses are shown in Figure 3. In addition, a summary of the DSC analysis results, mainly the melting point (T_m), is shown in Table 2.

The main physical property investigated by using DSC test was the melting point of all the samples. As shown in Table 2, the melting point of PET polymer was found to be 243.49°C. It is clear that the melting point of the treated PET with EG decreased with increase in the EG mass.

From the DSC curve, it can be seen that the relaxation in the thermal region of 230–260°C is triggered by the softening of the PET crystalline domains due to melting. With increase in the content of EG in the PET matrix, a shifting trend in the

DSC curve was observed as shown in Figure 3(a). This shift noticeably shows a decrease in the melting point of the treated PET with increase in EG mass, where the melting point of the treated PET decreased from 250.08°C to 242.33°C with the increase in EG mass from 0.4 g to 2 g, as shown in Table 2. The DSC thermograms for the entire temperature range used in the experiment are shown in Figure 3(b). This decrease in the melting temperature might signify that the systematic compact organization of the PET molecules was reduced with the introduction of EG into the polymeric matrix. Also, it can be inferred that EG enhances the mobility of PET segments and reduces PET crystalline regions.

The mechanical properties of PET as well as the PET treated with different EG mass were investigated by nanoindentation. The maximum load applied for each nanoindentation on a sample was 10.03 mN. A total of five indentations were made on each sample. Figures 4(a)–4(f) show the loading-depth curves of PET treated with 0 g, 0.4 g, 0.6 g, 1 g, 1.4 g, and 2 g of EG, respectively. The main mechanical properties examined using the nanoindentation test were the maximum depth covered by the indenter through the sample, the hardness, and the reduced modulus of the tested samples. These results are listed in Table 3 for all samples.

It is clear from the table that the maximum depth reached by the indenter in the PET polymer without EG was found

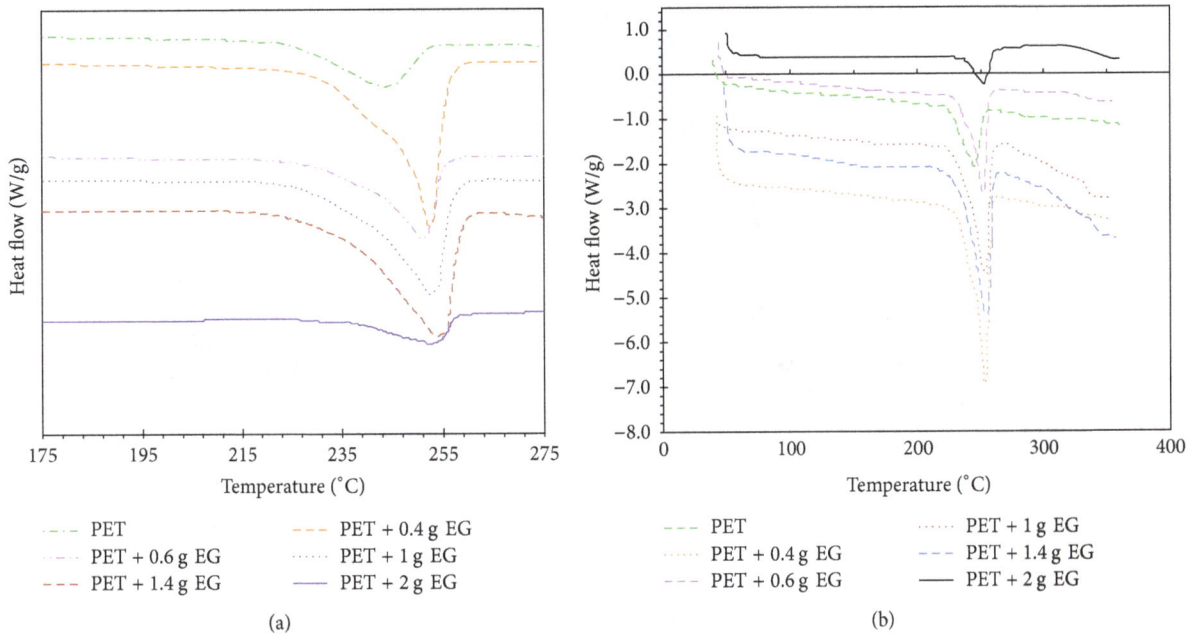

(a)

(b)

FIGURE 3: DSC curves for untreated and EG treated PET. (a) DSC for temperature range between 175 and 275°C, and (b) DSC for the entire temperature range used in the experiment.

TABLE 2: Melting point values for all the nontreated PET and EG treated PET samples.

Sample	PET	PET + 0.4 g EG	PET + 0.6 g EG	PET + 1 g EG	PET + 1.4 g EG	PET + 2 g EG
Melting temperature (°C)	243.49	250.08	247.37	246.03	245.83	242.33

to be 1043.03 nm. On the other hand, the PET treated with EG exhibited an increase in the maximum depth attained by the indenter with increase in the EG concentration used for treatment. The maximum depth attained was found to be 1169.08 nm for PET treated with 0.4 g of ethylene glycol and 1546.05 nm for PET treated with 2 g of EG. Moreover, the hardness and reduced modulus of the PET treated with ethylene glycol showed lower values when compared to the untreated PET polymer. Furthermore, the hardness of the treated polymer showed a decreasing trend with increase in the mass of EG when compared to untreated PET polymer. The nanoindentation test showed that the hardness of PET with 0.4 g EG was 0.41722 GPa, while this value decreased to 0.23520 GPa for PET treated with 2 g of EG. Similar to the hardness trend, the reduced modulus also showed a decreasing trend with increase in the concentration of EG. The maximum value for the reduced modulus (E_r) was recorded to be 8.13187 GPa for the PET without EG treatment. On the contrary, the lowest E_r value was found to be 5.23870 GPa for the PET treated with 2 g of EG. Similarly, the elastic modulus for PET without EG treatment was found to be 7.185 GPa and the lowest elastic modulus was obtained for the PET treated with 2 g of EG, which was found to be 4.618 GPa.

Plasticizers are also known to influence the mechanical properties of polymers. With increase in the plasticizer content in the polymeric matrix, the tensile strength of the polymer decreases whereas the elongation period to break is increased [39]. As the EG mass in the PET was increased, the hardness as well as the reduced modulus decreased. On the other hand, the maximum depth attained by the indenter in the sample increased with the increase in EG concentration in PET. The intramolecular forces in PET were decreased by the presence of EG. Subsequently, the presence of plasticizer will also result in the reduction in the melting point of the treated PET as indicated by DSC. Low-molecular weight small molecules of EG may get implanted between the PET chains or matrix, thereby increasing the free volume and spacing, allowing the polymeric chains to slide past each other at lower temperatures.

The melt flow index of PET and PET treated with EG was measured. The results are summarized in Table 4. From the results, it can be seen that the flow index for PET increases with increase in the mass of EG added during treatment. The flow index was 0.891 g/10 min for untreated PET, whereas the flow index was 13.627 and 12.17 g/10 min for PET treated with 1.4 and 2 grams of EG, respectively.

To determine the temperature at which the samples start flowing out of nozzle of the MFI instrument, the samples were heated to 220°C with nozzles closed followed by opening the nozzle, applying the load, and increasing the temperature at a constant rate. With increase in the mass of EG used

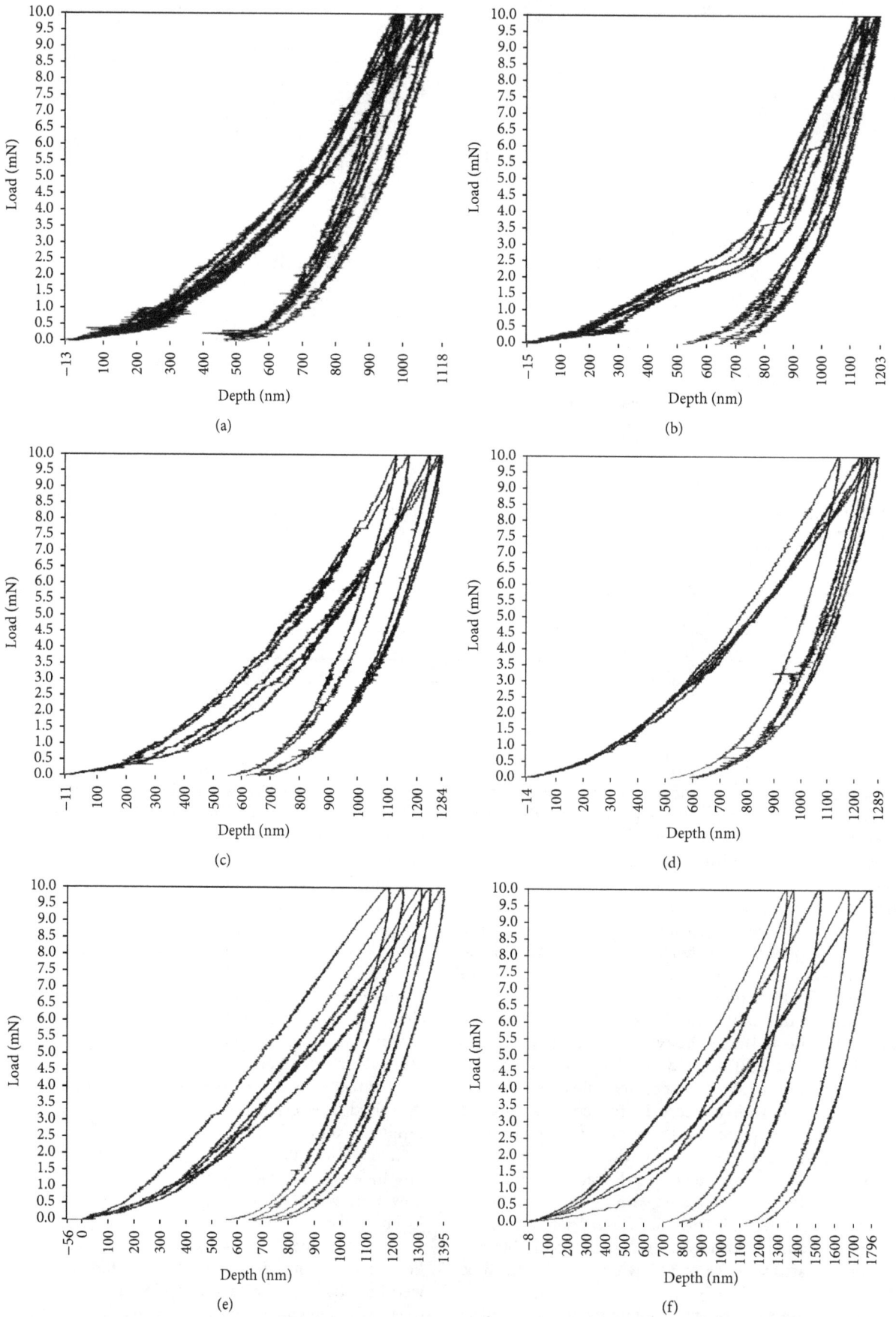

FIGURE 4: Nanoindentation graphs of PET treated with 0, 0.4, 0.6, 1, 1.4, and 2 grams of EG, respectively.

TABLE 3: Nanoindentation results of all samples.

Sample	PET	PET + 0.4 g EG	PET + 0.6 g EG	PET + 1 g EG	PET + 1.4 g EG	PET + 2 g EG
Maximum depth (nm)	1043.03	1169.08	1221.94	1235.42	1296.2	1546.05
Hardness (GPa)	0.57571	0.41722	0.39487	0.38957	0.35241	0.23520
Reduced modulus (GPa)	8.13187	7.70008	6.66154	6.35526	5.92548	5.23871
Elastic modulus (GPa)	7.185	6.802	5.879	5.607	5.226	4.618

TABLE 4: Flow index results for untreated and EG treated PET samples.

Sample	PET	PET + 0.4 g EG	PET + 0.6 g EG	PET + 1 g EG	PET + 1.4 g EG	PET + 2 g EG
Flow index (g/10 min)	0.891	2.501	7.329	7.762	13.627	12.170

TABLE 5: Melt flow start temperature for the untreated and EG treated PET samples.

Sample	PET	PET + 0.4 g EG	PET + 0.6 g EG	PET + 1 g EG	PET + 1.4 g EG	PET + 2 g EG
Flow start temperature (°C)	256	244.5	240	236	233	230

for treating PET, the temperature at which the samples start flowing through the nozzle was decreased. The results of the flow start temperature for the treated and untreated PET samples with EG are given in Table 5.

The flow index of the treated PET has clearly shown the effect of EG concentration in the PET polymer. The melt flow index increases with increase in the EG mass but the flow start temperature decreases with increase in EG concentration. Consequently, the presence of EG resulted in less viscosity of the PET samples at lower temperatures and hence ethylene glycol plays the role of a plasticizer in the PET samples.

4. Conclusion

In this work, PET was placed inside an extruder at 260°C with different EG mass, allowed to react, followed by extruding the samples. The chemical, thermal, and mechanical properties as well as the flow index of the treated PET were tested using FT-IR, TGA, DSC, nanoindentation, and melt flow index. The FT-IR spectra showed that addition of EG resulted in the occurrence of new hydroxyl groups in the treated PET. The intensities of specific peaks were intensified indicating that EG had interacted with PET polymeric matrix. Furthermore, the percent weight loss at the onset of thermal degradation increased with increase in EG mass used for PET treatment as indicated by the thermogravimetric analysis. Moreover, from the DSC thermograms, it can be observed that the melting point of the treated PET decreased with increase in the mass of EG. Also, the mechanical properties were affected by the presence of EG in the PET matrix. The results of the melt flow index analysis indicated that, with increase in the concentration of EG in the PET polymer, the flow index increased and consequently caused the sample to become less viscous. It can be inferred that EG interacts with PET and plays the role of a plasticizer filling the gaps between the polymer chains. This also explains the decrease in hardness

and reduced modulus with increase in EG concentration in the PET polymeric matrix. The reactive extrusion of polymers can be utilized to modulate their mechanical and thermal properties. Also, further research can be conducted by the utilization of suitable catalysts or ionic liquids to modify the properties of extruded PET.

Conflicts of Interest

The authors declare that they have no conflicts of interest.

References

[1] M. Peplow, "The plastics revolution: how chemists are pushing polymers to new limits," *Nature*, vol. 536, no. 7616, pp. 266–268, 2016.

[2] S. R. Shukla and A. M. Harad, "Glycolysis of polyethylene terephthalate waste fibers," *Journal of Applied Polymer Science*, vol. 97, no. 2, pp. 513–517, 2005.

[3] A. Stoski, M. F. Viante, C. S. Nunes, E. C. Muniz, M. L. Felsner, and C. A. P. Almeida, "Oligomer production through glycolysis of poly(ethylene terephthalate): effects of temperature and water content on reaction extent," *Polymer International*, vol. 65, no. 9, pp. 1024–1030, 2016.

[4] A. M. Al-Sabagh, F. Z. Yehia, A.-M. M. F. Eissa et al., "Glycolysis of poly(ethylene terephthalate) catalyzed by the Lewis base ionic liquid [Bmim][OAc]," *Industrial & Engineering Chemistry Research*, vol. 53, no. 48, pp. 18443–18451, 2014.

[5] S. R. Shukla, V. Palekar, and N. Pingale, "Zeolite catalyzed glycolysis of polyethylene terephthalate bottle waste," *Journal of Applied Polymer Science*, vol. 110, no. 1, pp. 501–506, 2008.

[6] R. V. Shah, V. S. Borude, and S. R. Shukla, "Recycling of PET waste using 3-amino-1-propanol by conventional or microwave irradiation and synthesis of bis-oxazin there from," *Journal of Applied Polymer Science*, vol. 127, no. 1, pp. 323–328, 2013.

[7] M. Khoonkari, A. H. Haghighi, Y. Sefidbakht, K. Shekoohi, and A. Ghaderian, "Chemical Recycling of PET Wastes with

Different Catalysts," *International Journal of Polymer Science*, vol. 2015, Article ID 124524, pp. 1–11, 2015.

[8] S. R. Shukla and A. M. Harad, "Aminolysis of polyethylene terephthalate waste," *Polymer Degradation and Stability*, vol. 91, no. 8, pp. 1850–1854, 2006.

[9] F. Chen, Q. Zhou, R. Bu, F. Yang, and W. Li, "Kinetics of poly(ethylene terephthalate) fiber glycolysis in ethylene glycol," *Fibers and Polymers*, vol. 16, no. 6, pp. 1213–1219, 2015.

[10] N. George and T. Kurian, "Recent developments in the chemical recycling of postconsumer poly(ethylene terephthalate) Waste," *Industrial & Engineering Chemistry Research*, vol. 53, no. 37, pp. 14185–14198, 2014.

[11] Q. F. Yue, L. F. Xiao, M. L. Zhang, and X. F. Bai, "The glycolysis of poly(ethylene terephthalate) waste: Lewis acidic ionic liquids as high efficient catalysts," *Polymer*, vol. 5, no. 4, pp. 1258–1271, 2013.

[12] S. H. Park and S. H. Kim, "Poly (ethylene terephthalate) recycling for high value added textiles," *Fashion and Textiles*, vol. 1, no. 1, 2014.

[13] A. Al-Sabagh, F. Yehia, G. Eshaq, A. Rabie, and A. ElMetwally, "Greener routes for recycling of polyethylene terephthalate," *Egyptian Journal of Petroleum*, vol. 25, no. 1, pp. 53–64, 2016.

[14] A. Ghaderian, A. H. Haghighi, F. A. Taromi, Z. Abdeen, A. Boroomand, and S. M.-R. Taheri, "Characterization of rigid polyurethane foam prepared from recycling of PET waste," *Periodica Polytechnica Chemical Engineering*, vol. 59, no. 4, pp. 296–305, 2015.

[15] A. Aguado, L. Martínez, L. Becerra et al., "Chemical depolymerisation of PET complex waste: Hydrolysis vs. glycolysis," *Journal of Material Cycles and Waste Management*, vol. 16, no. 2, pp. 201–210, 2014.

[16] L. Bartolome, M. Imran, B. Gyoo, W. A, and D. Hyun, "Recent Developments in the Chemical Recycling of PET," in *Material Recycling - Trends and Perspectives*, pp. 10–5772, InTech, 2012.

[17] A. Koç, "Studying the utilization of plastic waste by chemical recycling method," *Open Journal of Applied Sciences*, vol. 3, no. 7, pp. 413–420, 2013.

[18] N. S. Todorov, M. F. Radenkov, and D. D. Todorova, "Utilization of crude glycerol and waste polyethylene terephthalate for production of alkyd resins," *Journal of Chemical Technology and Metallurgy*, vol. 50, no. 3, pp. 240–248, 2015.

[19] L. Zhang, J. Gao, J. Zou, and F. Yi, "Hydrolysis of poly(ethylene terephthalate) waste bottles in the presence of dual functional phase transfer catalysts," *Journal of Applied Polymer Science*, vol. 130, no. 4, pp. 2790–2795, 2013.

[20] J. Chen, J. Lv, Y. Ji et al., "Alcoholysis of PET to produce dioctyl terephthalate by isooctyl alcohol with ionic liquid as cosolvent," *Polymer Degradation and Stability*, vol. 107, pp. 178–183, 2014.

[21] H. Kurokawa, M.-A. Ohshima, K. Sugiyama, and H. Miura, "Methanolysis of polyethylene terephthalate (PET) in the presence of aluminium tiisopropoxide catalyst to form dimethyl terephthalate and ethylene glycol," *Polymer Degradation and Stability*, vol. 79, no. 3, pp. 529–533, 2003.

[22] F. Liu, X. Cui, Z. Li, and X. Ge, in *Proceedings of the 5th ISFR*, pp. 272–276, Chengdu, China, 2009.

[23] C. N. Hoang and Y. H. Dang, "Aminolysis of poly(ethylene terephthalate) waste with ethylenediamine and characterization of a,u-diamine products," *Polymer Degradation and Stability*, vol. 98, no. 3, pp. 697–708, 2013.

[24] N. E. Kamber, Y. Tsujii, K. Keets et al., "The Depolymerization of Poly(ethylene terephthalate) (PET) Using N-Heterocyclic

[25] M. Imran, B.-K. Kim, M. Han, B. G. Cho, and D. H. Kim, "Sub- and supercritical glycolysis of polyethylene terephthalate (PET) into the monomer bis(2-hydroxyethyl) terephthalate (BHET)," *Polymer Degradation and Stability*, vol. 95, no. 9, pp. 1685–1693, 2010.

[26] H. Wang, Y. Liu, Z. Li, X. Zhang, S. Zhang, and Y. Zhang, "Glycolysis of poly(ethylene terephthalate) catalyzed by ionic liquids," *European Polymer Journal*, vol. 45, no. 5, pp. 1535–1544, 2009.

[27] H. Yue, Y. Zhao, X. Ma, and J. Gong, "Ethylene glycol: properties, synthesis, and applications," *Chemical Society Reviews*, vol. 41, no. 11, p. 4218, 2012.

[28] S. K. Kanojiya, G. Shukla, S. Sharma et al., "Hydrogenation of Styrene Oxide to 2-Phenylethanol over Nanocrystalline Ni Prepared by Ethylene Glycol Reduction Method," *International Journal of Chemical Engineering*, vol. 2014, Article ID 406939, 2014.

[29] A. N. A. S. A. A. Syariffuddeen, "Glycolysis of Poly (Ethylene Terephthalate) (PET) waste under conventional convection-conductive glycolysis," *International Journal of Engineering Research and Technology*, 2012.

[30] V. Sharma, P. Shrivastava, and D. D. Agarwal, "Degradation of PET-bottles to monohydroxyethyl terephthalate (MHT) using ethylene glycol and hydrotalcite," *Journal of Polymer Research*, vol. 22, no. 12, article no. 241, pp. 1–10, 2015.

[31] K. Ertas and G. Güçlü, "Alkyd resins synthesized from glycolysis products of waste PET," *Polymer—Plastics Technology and Engineering*, vol. 44, no. 5, pp. 783–794, 2005.

[32] K. Gotoh, A. Yasukawa, and Y. Kobayashi, "Wettability characteristics of poly(ethylene terephthalate) films treated by atmospheric pressure plasma and ultraviolet excimer light," *Polymer Journal*, vol. 43, no. 6, pp. 545–551, 2011.

[33] S. Makkam and W. Harnnarongchai, "Rheological and mechanical properties of recycled PET modified BY reactive extrusion," in *Proceedings of the Eco-Energy and Materials Science and Engineering, EMSES 2014*, pp. 547–553, Thailand, December 2013.

[34] N. Dardmeh, A. Khosrowshahi, H. Almasi, and M. Zandi, "Study on Effect of the Polyethylene Terephthalate/Nanoclay Nanocomposite Film on the Migration of Terephthalic Acid into the Yoghurt Drinks Simulant," *Journal of Food Process Engineering*, vol. 40, no. 1, Article ID e12324, 2017.

[35] Y. Zhu, Y. Wang, L. Sha, and J. Zhao, "Preparation of antimicrobial fabric using magnesium-based PET masterbatch," *Applied Surface Science*, vol. 425, pp. 1101–1110, 2017.

[36] M. Niu, X. Wang, Y. Yang et al., "The structure of microencapsulated carbon microspheres and its flame retardancy in poly(ethylene terephthalate)," *Progress in Organic Coatings*, vol. 95, pp. 79–84, 2016.

[37] A. Al-Mulla, "Enthalpy-entropy compensation in polyester degradation reactions," *International Journal of Chemical Engineering*, Article ID 782346, 2012.

[38] W. D. Q. Lamas, J. C. F. Palau, and J. R. D. Camargo, "Waste materials co-processing in cement industry: Ecological efficiency of waste reuse," *Renewable & Sustainable Energy Reviews*, vol. 19, pp. 200–207, 2013.

[39] R. Baidya, S. K. Ghosh, and U. V. Parlikar, "Co-processing of Industrial Waste in Cement Kiln – A Robust System for Material and Energy Recovery," *Procedia Environmental Sciences*, vol. 31, pp. 309–317, 2016.

[40] F. Welle, "Twenty years of PET bottle to bottle recycling—an overview," *Resources, Conservation & Recycling*, vol. 55, no. 11, pp. 865–875, 2011.

[41] M. Mohsin, A. Hossin, and Y. Haik, "Thermal and mechanical properties of poly(vinyl alcohol) plasticized with glycerol," *Journal of Applied Polymer Science*, vol. 122, no. 5, pp. 3102–3109, 2011.

[42] M. Mohsin, A. Hossin, and Y. Haik, "Thermomechanical properties of poly(vinyl alcohol) plasticized with varying ratios of sorbitol," *Materials Science and Engineering: A Structural Materials: Properties, Microstructure and Processing*, vol. 528, no. 3, pp. 925–930, 2011.

7

Application of Carboxymethyl Chitosan-Benzaldehyde as Anticorrosion Agent on Steel

Handoko Darmokoesoemo ⓘD,[1] Suyanto Suyanto,[1] Leo Satya Anggara,[1] Andrew Nosakhare Amenaghawon,[2] and Heri Septya Kusuma ⓘD[3]

[1]Department of Chemistry, Faculty of Science and Technology, Airlangga University, Surabaya 60115, Indonesia
[2]Department of Chemical Engineering, Faculty of Engineering, University of Benin, PMB 1154, Ugbowo, Benin City, Edo State, Nigeria
[3]Department of Chemical Engineering, Faculty of Industrial Technology, Institut Teknologi Sepuluh Nopember, Surabaya 60111, Indonesia

Correspondence should be addressed to Handoko Darmokoesoemo; handoko.darmokoesoemo@gmail.com
and Heri Septya Kusuma; heriseptyakusuma@gmail.com

Academic Editor: Sébastien Déon

Corrosion is one of the problems that is often found in daily life especially in petroleum and gas industry. Carboxymethyl chitosan-(CMC-) benzaldehyde was synthesized as corrosion inhibitor for steel. Corrosion rate was determined by potentiostatic polarization method in HCl 1 M. Dripping and coating, two different treatment, were used to drop and coat steel by CMC-benzaldehyde. The results showed that CMC-benzaldehyde could inhibit the corrosion rate of steel with concentration of 1 g, 3 g, 5 g, and 7 g in 60 mL of solvent. Coating steel with CMC-benzaldehyde with concentration of 7 g/60 mL of solvent and starch of 0.1 g/mL showed the highest efficiency to inhibit corrosion rate of steel. These treatments give corrosion efficiency of 99.8%.

1. Introduction

Corrosion is one of the most common problems found in everyday life, especially in the oil and gas processing industry. Corrosion cannot be prevented or stopped but its rate of destruction can be controlled. Corrosion is the degradation (destruction or degradation of quality) of metal properties through electrochemical reactions that are natural and take place by themselves due to chemical phenomena with the environment. The factors that cause corrosion include air pollution levels, temperature, humidity, and the presence of chemicals that are corrosive [1].

Corrosion-induced impacts can be direct and indirect impacts. Direct impacts include damage to equipment, machinery, and building structures. Indirect impacts result from cessation of production activities due to the replacement of equipment damaged by corrosion. Indirect costs incurred are generally larger than direct costs [2]. Therefore, various attempts are made to inhibit corrosion.

In the oil drilling and processing industry, corrosion becomes an inseparable problem. Uncontrolled corrosion rate can result in a malfunction of the equipment used, especially on offshore oil drilling platforms with high saline environments.

Corrosion that occurs in the metal causes a loss that is not small financially. The economic factor becomes a very important motivation for many current studies to be able to overcome and inhibit the rate of corrosion. According to a recent study, the losses suffered by industry and government in the United States amount to approximately 276 billion US dollar or about 3.1 percent of Gross Domestic Product (GDP). Studies on corrosion in Australia, England, Japan, and other countries have also been conducted. Nearly every country spends about 3-4 percent of Gross Domestic Product (GDP) to overcome corrosion. Total losses due to corrosion can be avoided around 25–30 percent if corrosion prevention can be done effectively [3].

There are various efforts of metal protection to overcome corrosion, among others; surface coating and cathode protection systems require a high cost because of the selection of inhibitors used. The effort has also not been effective in overcoming corrosion. In this research, corrosion inhibitor from organic material as anticorrosion is used. This inhibitor is chosen because it has a high affinity on the metal with high efficiency and being friendly to the environment. The corrosion inhibitor used is, namely, carboxymethyl chitosan (CMC) which is substituted with benzaldehyde.

Carboxymethyl chitosan (CMC) is a derivative of chitosan derived from chitin isolated from terrestrial invertebrates, marine invertebrates, and fungi which are numerous in nature. In invertebrates, chitin serves as an exoskeleton composite matrix, whereas in fungi it functions as a cell wall shaper. Chitosan is a soluble solid in acetic acid and easily degraded, but the application of chitosan is limited because it is not water soluble [4].

The use of CMC in this study is preferred over chitosan itself. This is because CMC has an important characteristic that is water soluble, high gel forming capacity, low toxicity, and good biocompatibility so that the application will be wider [5]. In addition, CMC is widely used because it is amphiprotic, which contains the -COOH and $-NH_2$ groups in its molecules that have many free electron pairs.

In this study, the $-NH_2$ group present in CMC (acting as base) is substituted with benzaldehyde to form carboxymethyl chitosan-benzaldehyde (CMC-benzaldehyde) compound. These compounds are expected to be used as an effective and effective anticorrosion in inhibiting the corrosion rate on steel. CMC-benzaldehyde is used as a corrosion inhibitor due to the presence of heteroatoms (O and N), the phi bond formed between CMC and benzaldehyde, and the number of free electrons that can support the inhibitor in chemisorption with the metal by coordination [6].

Therefore, in this study the inhibition of corrosion rate was carried out on the steel in the medium of HCl 1 M using the CMC-benzaldehyde inhibitor. The method used to determine the corrosion rate is the measurement of the intensity of corrosion current on steel with potentiostatic polarization.

2. Materials and Methods

2.1. Materials and Tools. The materials used in this study were chitosan from Good Manufacturing Practice, glacial acetic acid, NaOH, chloroacetic acid, isopropanol, ethanol 99.8%, distilled water, double-distilled water, benzaldehyde, commercial steel, filter paper, cassava, and HCl 37%, while the tool used in this research is Fourier Transform Infrared Spectrometer (FTIR) SHIMADZU, Potentiostat PGSTAT302N corrosion test instrument coupled with computer and Autolab NOVA software and SEM-EDX Carl Zeiss EVO MA 10.

2.2. Synthesis of CMC and CMC-Benzaldehyde

2.2.1. Synthesis of CMC. A total of 10 g of chitosan was dissolved in 400 mL of acetic acid 2%. The soluble chitosan was added to 13.5 g of NaOH and was reacted over the water

bath at 50°C for 1 hour. The solution was refluxed and added to 15 g of chloroacetic acid which has been dissolved in 20 mL of isopropanol. The mixture was reacted for 4 hours at 50°C. The treated mixture was filtered off with a Buchner funnel and washed with ethanol 70%. The obtained precipitate is dried at room temperature. The result is a CMC which is then characterized using FTIR [8].

2.2.2. Synthesis of CMC-Benzaldehyde. A total of 20 g of CMC was dissolved in 400 mL of distilled water. The solution was reacted with benzaldehyde-ethanol with a ratio of 1:1 (5 mL of benzaldehyde mixed with 5 mL of ethanol). The temperature used during the reaction takes approximately 50–60°C. The reaction was carried out for 5 hours and then filtrate and the precipitate was filtered using a Buchner funnel and washed with ethanol 70%. The precipitate obtained is the result of the synthesis of CMC-benzaldehyde. CMC-benzaldehyde obtained was characterized using FTIR [9].

2.3. Preparation of Starch. A total of 1 kg of cassava is cleaned and peeled. Then the cassava is smoothed using a grinder. Finely ground cassava is then soaked in distilled water for 24 hours. The filtrate and the precipitate are separated by decantation. The precipitate obtained is dried to obtain a starch made from cassava. A total of 5 g of starch was dissolved in 50 mL of hot aquades by stirring until a homogeneous solution was obtained. This formed starch solution is mixed in CMC-benzaldehyde.

2.4. Treatment for Steel. The steel used in this study was cleaned by sand and cut with length of 20 mm, width of 10 mm, and thick of 1 mm. The chemical composition content of the steel was tested using EDX (Energy Dispersive X-Ray) analysis.

2.5. Testing of Corrosion Rate on Steel in HCl 1 M

2.5.1. Testing of Corrosion Rate on Steel without Addition of CMC-Benzaldehyde. The steel is first immersed in 100 mL of HCl 1 M for 120 hours. After the corrosion process runs during that time, the steel is taken and washed with double-distilled water. Furthermore, the steel is left for a while and then heated in an oven at a temperature of 45–50°C for 10 minutes. Testing of corrosion rates on steels in HCl 1 M without addition of CMC-benzaldehyde was performed by potentiostatic polarization method as shown in Figure 1.

2.5.2. Testing of Corrosion Rate on Steel with the Addition of CMC-Benzaldehyde. Tests of corrosion rates on steels in HCl 1 M were carried out with two treatments, dripping and coating, wherein each treatment was added to CMC-benzaldehyde without starch and with starch. Dripping with the addition of CMC-benzaldehyde was performed by dissolving CMC-benzaldehyde of 1 g, 3 g, 5 g, and 7 g in a mixture of 20 mL of distilled water, 20 mL of acetic acid 2%, and 20 mL of alcohol at a ratio of 1:1:1. Subsequently a 100 mL of HCl 1 M was added to the solution of CMC-benzaldehyde by stirring until homogeneous. The solution is

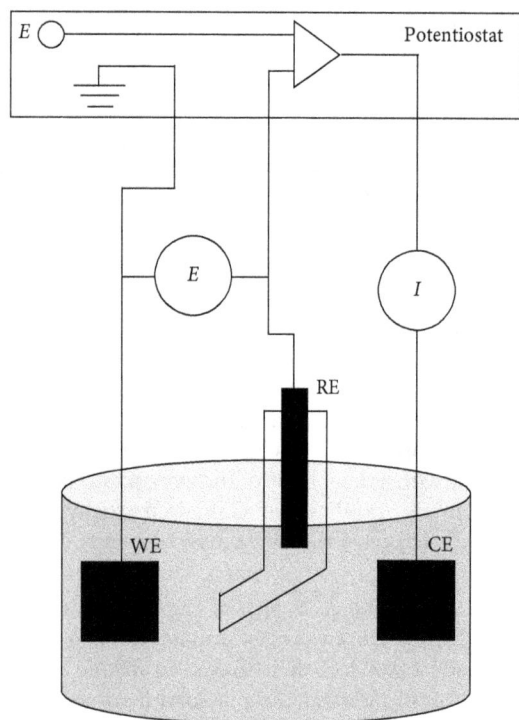

FIGURE 1: The series of equipment for potentiostatic polarization method (working electrode is carbon steel, reference electrode is Hg_2Cl_2, and counter electrode is platinum).

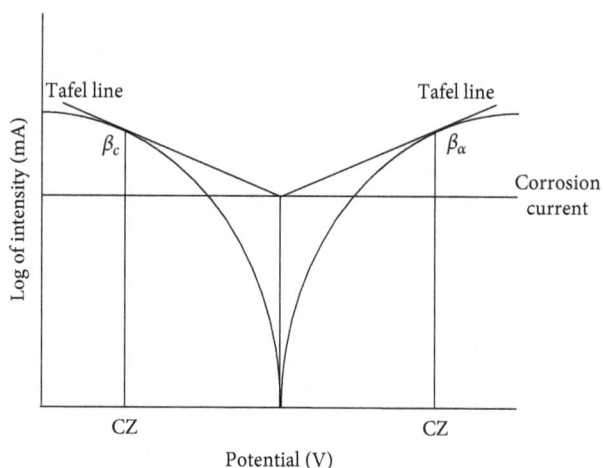

FIGURE 2: The curve between potential against log of current intensity.

closed and left for 72 hours. Next, the steel is immersed in the solution for 120 hours.

Dripping with the addition of CMC-benzaldehyde using starch was performed by dissolving CMC-benzaldehyde of 1 g, 3 g, 5 g, and 7 g in a mixture of 20 mL of distilled water, 20 mL of acetic acid 2%, and 20 mL of alcohol at a ratio of 1 : 1 : 1. Furthermore, the solution of CMC-benzaldehyde was added to starch solution of 1 mL, 3 mL, 5 mL, and 7 mL with a concentration of 0.1 g/mL and 100 mL of HCl 1 M. The solution is closed and left for 72 hours. Next, the steel is immersed in the solution for 120 hours.

Coating with the addition of CMC-benzaldehyde was performed by dissolving CMC-benzaldehyde of 1 g, 3 g, 5 g, and 7 g into a mixture of 20 mL of aquades, 20 mL of acetic acid 2%, and 20 mL of alcohol at a ratio of 1 : 1 : 1 The steel is then immersed in the solution for 72 hours. After a predetermined time, the steel is taken and put into the oven so that CMC-benzaldehyde can adhere to the steel. Furthermore, the steel is immersed in 100 mL of HCl 1 M for 120 hours.

Coating with the addition of CMC-benzaldehyde using starch was carried out by dissolving CMC-benzaldehyde of 1 g, 3 g, 5 g, and 7 g in 20 mL of aquades, 20 mL of acetic acid 2%, and 20 mL of alcohol at a ratio of 1 : 1 : 1. Each solution was added to starch solution of 1 mL, 3 mL, 5 mL, and 7 mL with a concentration of 0.1 g/mL. Furthermore, steel is immersed in the solution for 72 hours. After a predetermined time, the steel is taken and put into oven so that CMC-benzaldehyde and starch can adhere to the steel. Furthermore, the steel is immersed in 100 mL of HCl 1 M for 120 hours.

After the corrosion process takes place with the dripping and coating treatment during that time, the steel is removed and washed with double-distilled water. Furthermore, the steel is left for a while and then heated in an oven at a temperature of 45–50°C for 10 minutes. The corrosion rate test on steel in HCl 1 M with the addition of CMC-benzaldehyde was performed by potentiostatic polarization method as shown in Figure 1.

2.6. Determination of Corrosion Rate. In this study, the determination of corrosion rate was done by determining the intensity of corrosion current with potentiostatic polarization using the correlation between potential with log of current intensity to obtain intensity of corrosion current (I_{corr}) and corrosion rate (v_{corr}) as shown in Figure 2. The relationship between the intensity of corrosion current (I_{corr}) and corrosion rate (v_{corr}) is illustrated by the following equation:

$$v_{corr} = \frac{0.13 \times I_{corr} \times EW}{\rho}, \qquad (1)$$

where v_{corr} is the rate of corrosion (mpy), I_{corr} is the intensity of corrosion current ($\mu A/cm^2$), EW is the equivalent weight (atomic weight/valence) (g), and ρ is the density (g/cm^2).

The smaller corrosion rate of a material indicates the greater resistance of the inhibitor in inhibiting corrosion. Conversely, the greater corrosion rate of a material indicates the smaller resistance of the inhibitor in inhibiting corrosion. The intensity value of corrosion current (I_{corr}) is obtained by performing Tafel analysis with semimanual way by extrapolating the linear part of a plot log of the current intensity with potential at the current meeting of the anode and cathode.

The determination of potential calculation zone (CZ) of the anode and cathode curves affects the slope of cathode curve (β_c) and the slope of anode curve (β_α) which directly

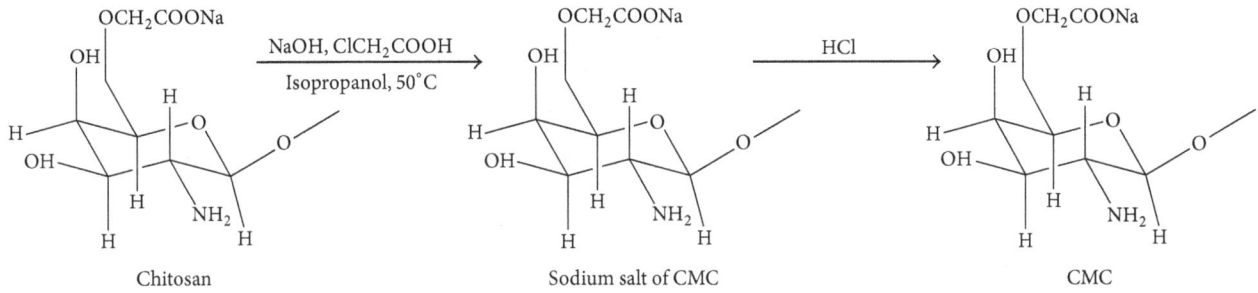

FIGURE 3: Synthesis of CMC from chitosan.

determines the intensity value of corrosion current (I_{corr}). This is shown by the Stern-Geary equation as follows:

$$I_{corr} = \frac{\beta_c \times \beta_a}{2.3 \times (\beta_c + \beta_a) \times A \times R_p},$$ (2)

where I_{corr} is the intensity of the corrosion current ($\mu A/cm^2$), β_c is the slope of cathode curve, β_a is the slope of anode curve, A is the area (cm^2), and R_p is the polarization resistance ($k\Omega/cm^2$). The slope value of the cathode and anode curves for each element or type of metal is not necessarily same, depending on the corresponding valence in the corrosion reaction that occurs.

2.7. Testing Morphology on Steel.

The morphological test on steel aims to find out the steel surface structure as a result of the addition of inhibitors to corrosive media [10]. For morphological testing of steel samples, the steel samples are placed on top of the preparations and then observed by Scanning Electron Microscope (SEM) so that the surface structure and corrosion type on the steel can be seen clearly as a result of the addition of CMC-benzaldehyde.

2.8. Data Analysis.

Data obtained from test results in the form of current and potential using potentiostat PGSTAT302N were processed by Tafel analysis with semimanual way to obtain the intensity of corrosion current (I_{corr}) and corrosion rate (v_{corr}). The Tafel analysis is performed by extrapolating the linear part of a plot log of the current intensity with the potential at the current meeting of the anode and cathode. From the intersection of the line, the intensity of corrosion current (I_{corr}) is obtained that can be converted to obtain the rate of corrosion (v_{corr}) in accordance with (1). In this research, potentiostat PGSTAT302N with Autolab NOVA software is used so that the corrosion rate (v_{corr}) can be obtained together with the intensity of corrosion current (I_{corr}) at the time of Tafel analysis from the relationship between the curve of the potential and the log of current intensity. Data analysis of the effect of variation concentration of CMC-benzaldehyde on the inhibition of corrosion rate on steel is shown in the form of graph and table.

2.9. Inhibition Efficiency of CMC-Benzaldehyde against Corrosion Rate.

Inhibitor efficiency is an inhibitor's ability to inhibit corrosion rate efficiently when compared to without

using inhibitors. To determine the efficiency of inhibitor in reducing and controlling corrosion rate on carbon steel, in this study the efficiency of inhibitor is calculated using the following equation:

$$\text{Inhibition efficiency (\%)} = \frac{V_{k0} - V_{k1}}{V_{k0}} \times 100\%,$$ (3)

where V_{k0} is the corrosion rate without using inhibitor and V_{k1} is the corrosion rate using inhibitor.

3. Results and Discussion

3.1. Synthesis of CMC-Benzaldehyde.

The synthesis of CMC-benzaldehyde is done by reacting CMC with benzaldehyde. In the formation of CMC-benzaldehyde, CMC need to be synthesized by reacting chitosan with NaOH and chloroacetic acid as shown in Figure 3 [7]. CMC formed can dissolve completely in water.

The formation of CMC-benzaldehyde occurs through the reaction mechanism of the formation of an imine. The stage of imine formation which essentially occurs in two stages is addition and elimination. The first stage is the addition of the nucleophilic amine to carbonyl carbon which has a partial positive charge followed by the release of protons from nitrogen and the acquisition of protons in oxygen. The second stage is the protonation of the OH group which can be released as water in an elimination reaction. The reaction mechanism of CMC-benzaldehyde formation is shown in Figure 4.

3.2. Characterization of Chitosan, CMC, and CMC-Benzaldehyde Using Fourier Transform Infrared Spectrometer (FTIR).

Characterization using Fourier Transform Infrared Spectrometer (FTIR) is used to determine the functional group of a compound formed on a particular wave number. The FTIR spectrum of chitosan and CMC according to Zheng et al. [7] is shown in Table 1.

The FTIR spectrum of chitosan used for the synthesis of CMC can be seen in Figure 5. The FTIR spectrum of chitosan shows the following wave numbers: 1033.77–$1083.92 \, cm^{-1}$ (C-O stretch), $1153.35 \, cm^{-1}$ (bridge-O stretch), $1421.44 \, cm^{-1}$ (N-H bending), $2883.38 \, cm^{-1}$ (C-H stretch), and $3440.77 \, cm^{-1}$ (O-H stretch).

The FTIR spectrum of CMC can be seen in Figure 6. The FTIR spectrum of CMC from the synthesis results

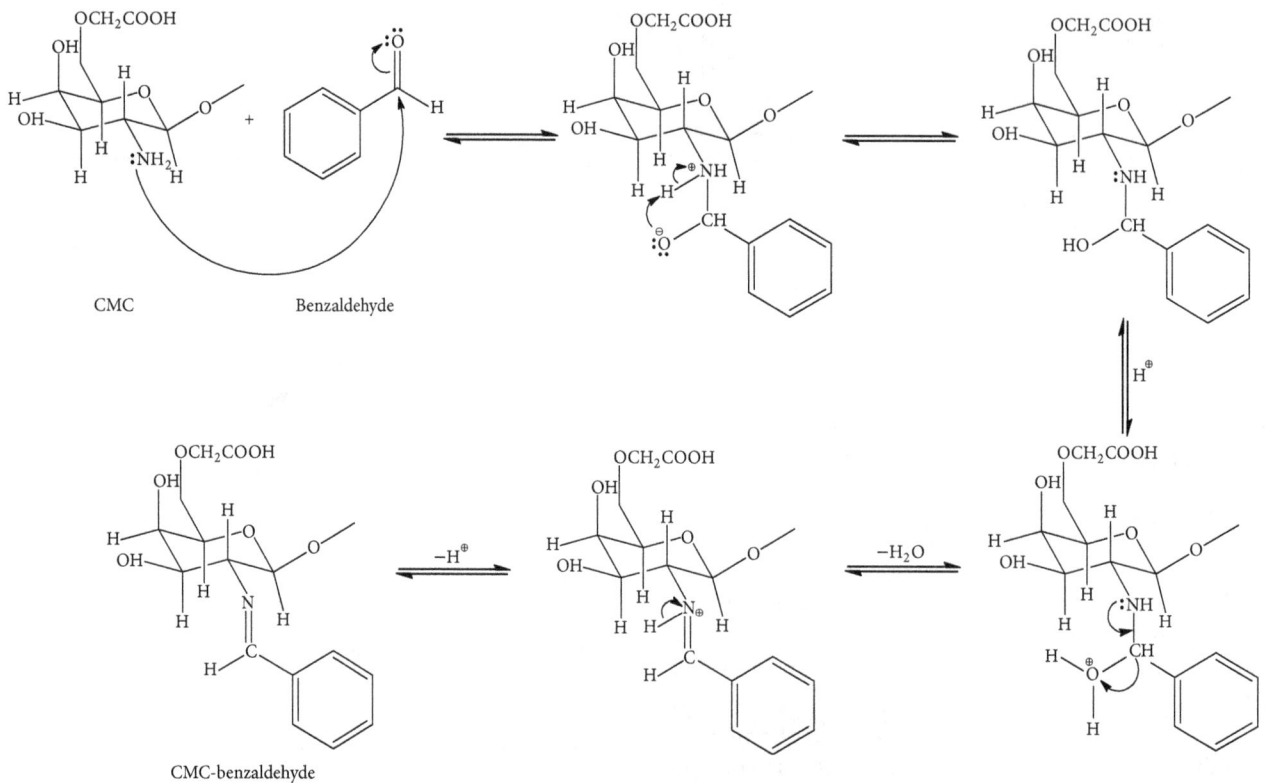

FIGURE 4: Reaction mechanism of CMC-benzaldehyde formation.

TABLE 1: Analysis of functional groups of chitosan and CMC using FTIR.

Functional groups	Wave numbers according to Zheng et al. [7] (cm^{-1})	Wave numbers of chitosan (cm^{-1})	Wave numbers of CMC from the synthesis results (cm^{-1})
C-O stretch	1030–1094	1033.77–1083.92	1033.77–1081.99
Bridge-O stretch	1153	1153.35	1151.42
N-H bending	1556	1421.44	1598.88
C-H stretch	2881	2883.38	2889.17
O-H stretch	3421	3440.77	3434.98
COO$^-$	1407–1598	-	1407.94

shows the following wave numbers: 1033.77–1081.99 cm^{-1} (C-O stretch), 1151.42 cm^{-1} (bridge-O stretch), 1598.88 cm^{-1} (N-H bending), 2889.17 cm^{-1} (C-H stretch), 3434.98 cm^{-1} (O-H stretch), and 1407.94 cm^{-1} (COO$^-$), while the wave number of 1407.94 cm^{-1} shows the existence of a new group formed from the esterification reaction of chitosan into CMC, that is, COO$^-$.

The functional groups formed on chitosan and CMC can be seen in Table 1. The FTIR spectrum of CMC from the synthesis results obtained in this study showed similar results with the FTIR spectrum of CMC according to Zheng et al. [7].

The formation of CMC-benzaldehyde is demonstrated by the presence of new functional group, the imines (C=N) and C=C aromatic, which can be seen at the wave numbers for those groups in the FT-IR spectrum shown in Figure 6. According to Pretsch et al. [11], the wave number

for imine (C=N) is 1645 cm^{-1} and for C=C aromatic is 1600 cm^{-1}. The wave number of CMC-benzaldehyde from the synthesis results is 1641.31 cm^{-1} for imines (C=N) and 1450–1600.81 cm^{-1} for C=C aromatic as shown in Table 2. So based on the characterization using FTIR, it can be said that CMC-benzaldehyde from the synthesis results has formed new functional group which is imine (C=N) and C=C aromatic. The shift of wave numbers in the FTIR spectrum of chitosan, CMC, and CMC-benzaldehyde in this study is shown in Figure 7.

3.3. Mechanism of Corrosion.
Corrosion in the metal is an irreversible oxidation-reduction reaction occurring between the metal and the oxidizing agent in an environment. There are various chemicals in the environment that can accelerate the occurrence of corrosion such as acidic, salt, and alkaline

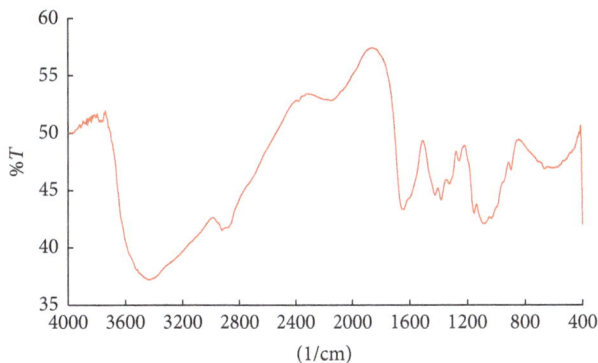

FIGURE 5: FTIR spectrum of chitosan.

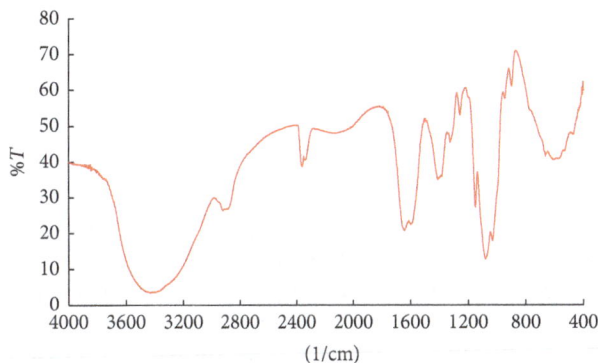

FIGURE 6: FTIR spectrum of CMC.

FIGURE 7: FTIR spectrum of chitosan, CMC, and CMC-benzaldehyde.

TABLE 2: Analysis of new functional groups of CMC-benzaldehyde using FTIR.

Functional groups	Wave numbers according to Pretsch et al. (cm^{-1})	Wave numbers of CMC-benzaldehyde from the synthesis results (cm^{-1})
C=N	1645	1641.31
C=C aromatic	1600	1450–1600.81

substances. The higher concentration of these substances causes the corrosion rate of a metal to become faster.

In this study, HCl 1 M is used as an oxidizing agent on steel. HCl 1 M acts as an acid medium causing corrosion of steel, so that the corrosion rate can become faster in a relatively short time. Reactions that occur in this study can be described as follows:

$$Fe_{(s)} + 2HCl_{(aq)} \longrightarrow FeCl_{2(aq)} + H_{2(g)} \qquad (4)$$

The redox reaction consists of two half-cell reactions. These reactions are

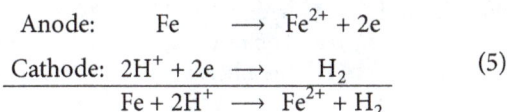

$$\begin{array}{rcl} \text{Anode:} & Fe & \longrightarrow Fe^{2+} + 2e \\ \text{Cathode:} & 2H^+ + 2e \longrightarrow & H_2 \\ \hline & Fe + 2H^+ & \longrightarrow Fe^{2+} + H_2 \end{array} \qquad (5)$$

The half-cell reaction of the anode and cathode shows the exchange of electrons during the redox reaction process.

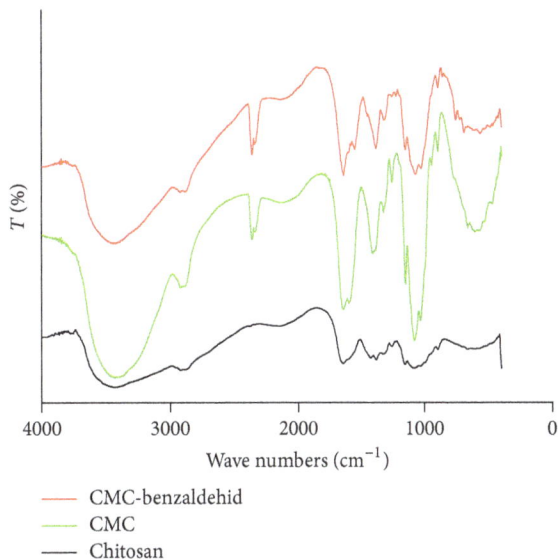

The corrosion mechanism of carbon steel without CMC-benzaldehyde inhibitors provides an opportunity of H^+ from the acid to be captured directly by the electrons that present in the steel so that the corrosion rate becomes fast. Steel without CMC-benzaldehyde inhibitors was used as a negative control in this study. Illustration of corrosion mechanisms on steel without CMC-benzaldehyde inhibitors can be seen in Figure 8(a).

The corrosion mechanism of steel with CMC-benzaldehyde inhibitors gives the chance of H^+ from the acid to be captured by the electrons present in the steel becoming smaller due to the less direct contact that occurs with the addition of CMC-benzaldehyde. Thus, the corrosion rate that occurs becomes slower when compared to the corrosion rate without the addition of CMC-benzaldehyde. Addition of CMC-benzaldehyde cannot cover all parts of steel. This is because CMC-benzaldehyde is a polymer that has an amorphous structure so that the coating on the steel still provides pores that allow the occurrence of contact with H^+. Illustration of corrosion mechanisms on steel with CMC-benzaldehyde inhibitors can be seen in Figure 8(b).

Starch is added to CMC-benzaldehyde in order to cover the pores that are still not covered completely. It aims to reduce the contact between electrons on steel with H^+. Illustration of corrosion mechanisms on steel with CMC-benzaldehyde inhibitors and starch can be seen in Figure 8(c).

3.4. Effect of Addition of CMC-Benzaldehyde on Corrosion Rate. Addition of CMC-benzaldehyde to steel is done by dripping and coating methods and the corrosion rate of the steel is carried out by potentiostatic polarization method. In this study, used variations in concentration on the addition of CMC-benzaldehyde are 1, 3, 5, and 7 g for 60 mL of solvent. In addition, in this study some treatments were added to starch of 0.1 g/mL in CMC-benzaldehyde and some treatments were

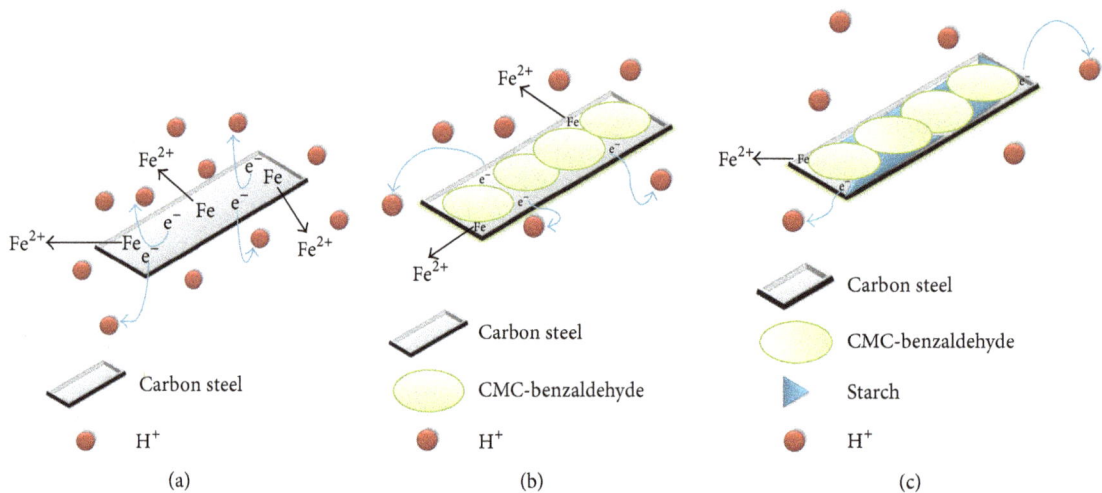

FIGURE 8: Hypothesis of corrosion mechanisms for the following: (a) without the addition of CMC-benzaldehyde inhibitors, (b) with the addition of CMC-benzaldehyde inhibitors, and (c) with the addition of CMC-benzaldehyde and starch inhibitors.

not added to starch. Addition of starch aims to reduce pores that are still not covered with CMC-benzaldehyde inhibitor, so optimal results are expected.

The negative control performed on steel without the addition of CMC-benzaldehyde showed that the corrosive rate caused by HCl 1 M as corrosive medium was 5.89 mm/year with a corrosive current of 506.89 $\mu A/cm^2$.

Dripping with CMC-benzaldehyde showed good results, where the corrosion rate and corrosion current decrease with the concentration of CMC-benzaldehyde that has been given. It also shows that the addition of CMC-benzaldehyde can cause the corrosion rate to be slower than without the addition of CMC-benzaldehyde.

Dripping with CMC-benzaldehyde and starch showed better results than dripping without the addition of starch. In dripping with CMC-benzaldehyde and starch, corrosion rates and corrosion currents decreased dramatically with increasing concentrations of CMC-benzaldehyde and starch that has been given.

And coating with CMC-benzaldehyde also showed decreasing corrosion rate and corrosion current along with increasing concentration of CMC-benzaldehyde that has been given.

Coating with CMC-benzaldehyde and starch also showed a decrease in corrosion rate and corrosion current along with the added concentration of CMC-benzaldehyde that has been given. In coating with CMC-benzaldehyde with a concentration of 7 g for 60 mL of solvent gives very low corrosion rate and corrosion current value compared to other treatments (dripping with CMC-benzaldehyde, dripping with CMC-benzaldehyde and starch, and coating with CMC-benzaldehyde). The values of corrosion rate and corrosion current are 0.0119 mm/year and 1.0203 $\mu A/cm^2$, respectively. The effect of concentration of CMC-benzaldehyde with and without the addition of starch on the dripping and coating methods to the corrosion current and corrosion rate can be seen in Table 3.

3.5. Surface Morphology of Steel.
In this study to determine the type of corrosion that occurs on steel is done using Scanning Electron Microscope (SEM). The morphology of the steel surface as a result of corrosion can be seen clearly when compared with seeing with the naked eye. Based on the SEM results, it can be seen that steel surfaces that have been added using CMC-benzaldehyde with and without starch after potentiostatic polarization testing have resulted in carbon steel having pitting corrosion. The analysis results of surface morphology using SEM for steel after potentiostatic polarization testing can be seen in Figure 9.

3.6. Inhibition Efficiency of CMC-Benzaldehyde against Corrosion Rate.
Based on the calculated data of inhibition efficiency, the highest efficiency value to inhibit the corrosion rate is shown by coating method with CMC-benzaldehyde concentration of 7 g/60 mL of solvent and starch of 0.1 g/mL is 99.8%. This indicates that CMC-benzaldehyde is able to inhibit corrosion rate of 99.8% and there is corrosion rate of 0.2% which can still occur in the steel that has been given treatment by adding CMC-benzaldehyde and starch. Inhibition efficiency from the results of research that has been done by dripping and coating methods using CMC-benzaldehyde with and without starch can be seen in Table 3.

In the research that has been done by Erna et al. [12], the efficiency of corrosion inhibition of CMC for steels in water gives optimum results at pH 5 and CMC concentration of 1 ppm is 77%. While in this study the use of CMC-benzaldehyde gives optimum results at concentration of 7 g/60 mL of solvent and the addition of starch as much as 0.1 g/mL in HCl 1 M with pH 2.53 is 99.8%.

In addition, based on research that has been done by Finsgar and Jackson [13], the efficiency of CMC-benzaldehyde can also be compared with the use of other corrosion inhibitors in the oil and gas industry, as the use of N,N'-ortho-phenylen acetyle acetone imine with concentrations of 50–400 mg/L in HCl 1 M gives inhibitory efficiency of 24.9%–82.6%. However,

TABLE 3: The effect of concentration of CMC-benzaldehyde with and without the addition of starch on the dripping and coating methods to the corrosion current, corrosion rate, and inhibition efficiency.

Treatment	Concentration (g/60 mL of solvent)	Corrosion current (μA/cm^2)	Corrosion rate (mm/year)	Efficiency of inhibitor (%)
Negative control	-	506.89	5.8900	-
Dripping with CMC-benzaldehyde	1	234.97	2.7303	53.65
	3	157.88	1.8346	68.85
	5	142.21	1.6525	71.94
	7	131.13	1.5237	74.13
Dripping with CMC-benzaldehyde and starch	1	404.55	4.7008	20.19
	3	209.64	2.4360	58.64
	5	4.9576	0.0576	99.02
	7	2.2381	0.0260	99.56
Coating with CMC-benzaldehyde	1	388.48	4.5141	23.36
	3	365.21	4.2437	27.95
	5	79.701	0.9261	84.28
	7	14.661	0.1704	97.11
Coating with CMC-benzaldehyde and starch	1	353.53	4.1080	30.25
	3	298.71	3.4710	41.07
	5	2.5404	0.0295	99.50
	7	1.0203	0.0119	99.80

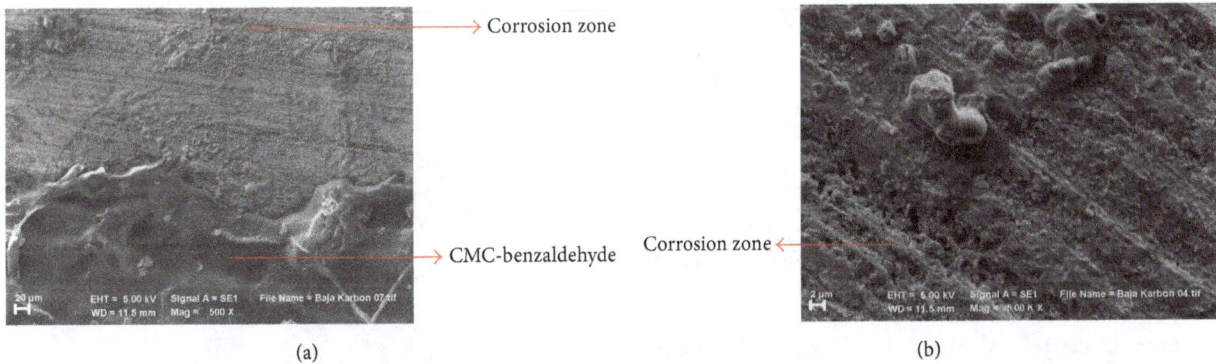

FIGURE 9: The analysis results of surface morphology using SEM for steel surfaces that have been added using (a) CMC-benzaldehyde with starch and (b) CMC-benzaldehyde without starch after potentiostatic polarization (coating method, concentration of 7 g/60 mL).

in this study the efficiency of CMC-benzaldehyde in the steels with concentrations of 1 g/60 mL of solvent to 7 g/60 mL of solvent by dripping and coating methods with and without starch in HCl 1 M gives corrosion efficiency with range of 20.19%–99.8%.

4. Conclusion

Based on the research that has been done and the data obtained, it can be concluded that CMC-benzaldehyde can be used as anticorrosion agent on steel. The effect of concentration on the corrosion rate of steel is that the higher concentration of CMC-benzaldehyde can make the corrosion rate on steel become slower. Conversely, the smaller CMC-benzaldehyde concentration that has been used makes the corrosion rate on steel become faster. The highest efficiency to inhibit corrosion rate on steel is 99.8% which can be obtained

by coating method using CMC-benzaldehyde with concentration of 7 g/60 mL of solvent and starch of 0.1 g/mL. The use of CMC-benzaldehyde inhibitors with concentrations of 1 g/60 mL of solvent to 7 g/60 mL of solvent gives corrosion efficiency with range of 20.19%–99.8%.

Disclosure

The paper was represented as an abstract which is part of undergraduate thesis and can be seen in the link: http://repository.unair.ac.id/28231/1/gdlhub-gdl-s1-2014-anggara-leo-37493-5.-abstr-k.pdf.

Conflicts of Interest

The authors declare no competing financial interests.

References

[1] A. Gunaatmaja, *Pengaruh Waktu Perendaman terhadap Laju Korosi pada Baja Karbon Rendah dengan Penambahan Ekstrak Ubi Ungu sebagai Inhibitor Organik di Lingkungan NaCl 3,5%*, Universitas Indonesia, Java, Indonesia, 2011.

[2] K. R. Trethewey and J. Chamberlain, *Corrosion for Students of Science and Engineering*, Longman Scientific & Technical, New York, NY, USA, 1988.

[3] R. W. Revie and H. H. Uhlig, *Corrosion and Corrosion Control: An Introduction to Corrosion Science and Engineering*, John Wiley & Sons, Hoboken, New Jersey, USA, 4th edition, 2008.

[4] F. R. de Abreu and S. P. Campana-Filho, "Characteristics and properties of carboxymethylchitosan," *Carbohydrate Polymers*, vol. 75, no. 2, pp. 214–221, 2009.

[5] X. Xue, L. Li, and J. He, "The performances of carboxymethyl chitosan in wash-off reactive dyeings," *Carbohydrate Polymers*, vol. 75, no. 2, pp. 203–207, 2009.

[6] F. S. de Souza and A. Spinelli, "Caffeic acid as a green corrosion inhibitor for mild steel," *Corrosion Science*, vol. 51, no. 3, pp. 642–649, 2009.

[7] M. Zheng, B. Han, Y. Yang, and W. Liu, "Synthesis, characterization and biological safety of O-carboxymethyl chitosan used to treat Sarcoma 180 tumor," *Carbohydrate Polymers*, vol. 86, no. 1, pp. 231–238, 2011.

[8] X.-G. Chen and H.-J. Park, "Chemical characteristics of O-carboxymethyl chitosans related to the preparation conditions," *Carbohydrate Polymers*, vol. 53, no. 4, pp. 355–359, 2003.

[9] T. F. Jiao, J. Zhou, L. Gao, Y. Y. Xing, and X. Li, "Synthesis and characterization of chitosan-based schiff base compounds with aromatic substituent groups," *Iranian Polymer Journal*, vol. 20, no. 2, pp. 123–136, 2011.

[10] W. D. Callister, *Material Science and Engineering: An Introduction*, Wiley, New York, NY, USA, 1985.

[11] E. Pretsch, Th. Clerc, J. Seibl, and W. Simon, *Tables of Spectral Data for Structure Determination of Organic Compounds*, Springer-Verlag Berlin Heidelberg, Berlin, Germany, 2nd edition, 1989.

[12] M. Erna, E. Emriadi, A. Alif, S. Arief, and M. J. Noordin, "Sintesis dan aplikasi karboksimetil kitosan sebagai inhibitor korosi pada baja karbon dalam air," *Jurnal Natur Indonesia*, vol. 12, no. 1, pp. 87–92, 2009.

[13] M. Finsgar and J. Jackson, "Application of corrosion inhibitors for steels in acidic media for the oil and gas industry: A review," *Corrosion Science*, vol. 86, pp. 17–41, 2014.

8

Study of the Equilibrium, Kinetics, and Thermodynamics of Boron Removal from Waters with Commercial Magnesium Oxide

Javier Paul Montalvo Andia ⓘ,[1] **Lidia Yokoyama** ⓘ,[2] **and Luiz Alberto Cesar Teixeira** ⓘ[3]

[1]*Instituto de Energía y Medio Ambiente, Universidad Católica San Pablo, Arequipa, Peru*
[2]*Escola de Química, Departamento de Processos Inorgânicos, Universidade Federal do Rio de Janeiro, Rio de Janeiro, RJ, Brazil*
[3]*Departamento de Engenharia Química e de Materiais, Pontifícia Universidade Católica do Rio de Janeiro,*
Rio de Janeiro, RJ, Brazil

Correspondence should be addressed to Lidia Yokoyama; lidia@eq.ufrj.br

Academic Editor: Gianluca Di Profio

In the present work, the equilibrium, thermodynamics, and kinetics of boron removal from aqueous solutions by the adsorption on commercial magnesium oxide powder were studied in a batch reactor. The adsorption efficiency of boron removal increases with temperature from 25°C to 50°C. The experimental results were fitted to the Langmuir, Freundlich, and Dubinin–Radushkevich (DR) adsorption isotherm models. The Freundlich model provided the best fitting, and the maximum monolayer adsorption capacity of MgO was 36.11 mg·g^{-1}. In addition, experimental kinetic data interpretations were attempted for the pseudo-first-order kinetic model and pseudo-second-order kinetic model. The results show that the pseudo-second-order kinetic model provides the best fit. Such result suggests that the adsorption process seems to occur in two stages due to the two straight slopes obtained through the application of the pseudo-first-order kinetic model, which is confirmed by the adjustment of the results to the pseudo-second-order model. The calculated activation energy (E_a) was 45.5 kJ·mol^{-1}, and the values calculated for $\Delta G°$, $\Delta H°$, and $\Delta S°$ were −4.16 kJ·mol^{-1}, 21.7 kJ·mol^{-1}, and 87.3 kJ·mol^{-1}, respectively. These values confirm the spontaneous and endothermic nature of the adsorption process and indicated that the disorder increased at the solid-liquid interface. The results indicate that the controlling step of boron adsorption process on MgO is of a physical nature.

1. Introduction

The exploration and production of petroleum from marine subsoil generate large amounts of liquid effluents, also known as produced water, extracted along with crude oil [1]. Produced water is the water found along with oil in the marine reservoir or the water injected into the reservoir in order to recover petroleum. The produced water flow is low at the early stages of production of the reservoir. Nevertheless, it can reach up to 80% of the crude oil extraction in the final years of exploitation of the well [2].

The produced water composition can vary according to the well and usually contains high salinity, organic and inorganic substances, and levels of dissolved solids over 40%, which makes it toxic to the environment [2, 3].

Boron is an element present in the produced water, and its concentration may vary from 4 mg·L^{-1} to 350 mg·L^{-1} [3, 4]. Although boron is considered a micronutrient essential to the development of microorganisms, plants, microalgae, and animals, this element can be toxic at concentration of 0.3 mg·L^{-1} to sensitive plants, at 2 mg·L^{-1} to semitolerant ones, and at 4 mg·L^{-1} to tolerant ones. Therefore, this is one of the reasons why boron is required to be removed from water and other effluents by environmental protection agencies [5, 6].

The maximum boron concentration recommended by the World Health Organization (WHO) for potable water was 0.5 mg·L^{-1} in 1998. However, this value was modified to 2.4 mg·L^{-1} in 2011. According to standards set by legislation, the limit allowed by the European Union, the UK, and Japan is $1.0 \, mg \cdot L^{-1}$. In South American countries such as Brazil and Peru, the limits for fresh water and wastewater are $0.5 \, mg \cdot L^{-1}$ and $5 \, mg \cdot L^{-1}$, respectively. On the contrary, in the USA, the limit is not subject to federal regulations on this issue. The states of Minnesota, Florida, and California have allowed limits of 0.6, 0.63, and $1 \, mg \cdot L^{-1}$, respectively [7–11].

Consequently, the produced water containing boron requires treatment before being released back into the sea or being employed as a hydrosource of potable water.

There are many possible treatments for the boron removal such as the electrocoagulation, process which achieved efficiencies over 98% from synthetic solutions and real produced waters with initial boron concentration between 10 and $30 \, mg \cdot L^{-1}$ [12]. Other treatments are the adsorption by different types of adsorbents, including activated carbon, fly ash, clay, mesoporous silica, oxides, nanoparticles, biological material, layered double hydroxides, and natural minerals [13–15]. Some additional methods studied at laboratory, for example, electrodialysis [16], phytoremediation [17, 18], and bioelectrochemical systems, presented removal efficiencies up to 90% [19, 20].

However, the most extensive processes employed for boron removal from produced water and seawater, on a larger scale, are reverse osmosis and ion exchange [21, 22].

As far as reverse osmosis is concerned, boron removal can reach values of 98% at pH = 10.5 since, at this pH, the predominant boron specie is a borate ion $B(OH)^{4-}$ that has a negative charge and larger size when compared to the boric acid (H_3BO_3) [2, 22]. In the case of ion exchange, the removal of boron occurs via its adsorption in specific resins, which are usually synthesized by means of macroporous cross-linked polystyrene resins, and functionalized by the N-methyl-D-glucamine (1-amino-1-deoxy-D-glucitol; NMDG) group achieving removal efficiencies up to 99% [23–26].

The most common commercial resins used in ion exchange process are Diaion CRB 02, Purolite S 108, and Amberlite IRA-743 [6, 27–30].

In general, the initial concentration of boron in produced waters treated with resins can vary between 15 and $60 \, mg \cdot L^{-1}$. However, after a certain period of time, this resin saturates, reaching the breakthrough, and needs to be regenerated to be reused [12, 31].

Regeneration step consists of passing an acid solution (usually H_2SO_4) through a saturated resin, resulting in an acid solution with high concentration of boron, usually between 350 and $700 \, mg \cdot L^{-1}$, which must also be treated.

The volume of this effluent containing boron is relevant when it refers by example to oil waterway terminals, such as São Sebastião Waterway Terminal from Petrobras/Brazil, with a capacity of $1,585,345 \, m^3$ [32]. Due to their magnitude, they generate a large quantity of effluent, which is product of the elution of the resins used at wastewater treatment plants of these terminals.

The treatment of this effluent by conventional technologies is not adequate due to the high concentration of boron. Therefore, under this circumstance, technologies such as precipitation or adsorption seem to be viable alternatives.

Through these treatments, boron concentration could be reduced up to 15–$30 \, mg \cdot L^{-1}$ that may allow the mixing of treated effluent with production water and these could be subsequently treated by ion exchange resins or reverse osmosis. The resulting solid residue of precipitation/adsorption process can be disposed in landfills or be used as raw material in the manufacture of glass due to its high content of boron. In this context, magnesium oxide is provided as a good alternative as it is environmentally friendly, cost-effective, and nontoxic. Furthermore, it shows low solubility in water, and it is an effective sorbent for the removal of contaminants such as fluoride and toxic dyes [33, 34].

The aim of the present work was to evaluate the kinetics, thermodynamics, and equilibrium of boron removal from aqueous solutions with high content ($350 \, mg \cdot L^{-1}$) through adsorption in magnesium oxide.

2. Experimental

2.1. Materials and Methods. For each experiment, a synthetic solution of $350 \, mg \cdot L^{-1}$ was prepared by dissolution of boric acid (H_3BO_3) PA in distilled water. The pH of the solution was adjusted with 1 M NaOH and/or 1 M HCl solutions provided by Sigma-Aldrich.

The magnesium oxide used in the adsorption process was supplied by Magnesita SA (MgO-500).

2.2. Characterization of MgO and Boron Concentration Analysis. The sizes of the MgO particles were analyzed by a dual laser liquid dry dispersion particle size analyzer, CILAS 1064L. Boron concentrations in aqueous solutions were determined by optical emission spectrometry analysis with inductively coupled plasma from PerkinElmer, Optima 4300DV. The morphology of the MgO was determined by scanning electron microscopy (MRV).

2.3. Experimental Procedure. The experiments were carried out in a batch reactor with 500 mL of solution. The use of borosilicate glass beaker was to prevent further contamination by dissolution of boron substances from the silicate material (beaker), mainly because the experiments were performed at alkaline pH. The solution was stirred at a speed of 150 rpm, $40 \, g \cdot L^{-1}$ of MgO was added, and the pH was readily adjusted after starting the experiment. The pH of the solution was measured using a pH meter (Bel Engineering W3B) and adjusted with 1 M NaOH and/or 1 M HCl solutions (Sigma-Aldrich).

The study of reaction kinetic was performed by collecting the sample every 5 minutes. The samples were vacuum-filtered through a cellulose nitrate membrane of $0.4 \, \mu m$ pore diameter (Nalgene). After 240 minutes of reaction, the stirring was turned off.

The filtered samples were kept for boron concentration analysis by inductively coupled plasma spectrometer (Optima 4300DV, from PerkinElmer).

Simultaneously, a sludge sample was dried at 70°C and adequately stored for further analysis of the surface morphology by scanning electron microscopy (MRV).

TABLE 1: Chemical composition of MgO-500 (Magnesita SA).

Reagent/composition	Surface area (S_{BET}) (m^2·g^{-1})	MgO (%)	Fe$_2$O$_3$ (%)	Al$_2$O$_3$ (%)	SiO$_2$ (%)	MnO (%)	CaO (%)
MgO-500	31.47	98	0.4	0.1	0.15	0.1	0.9

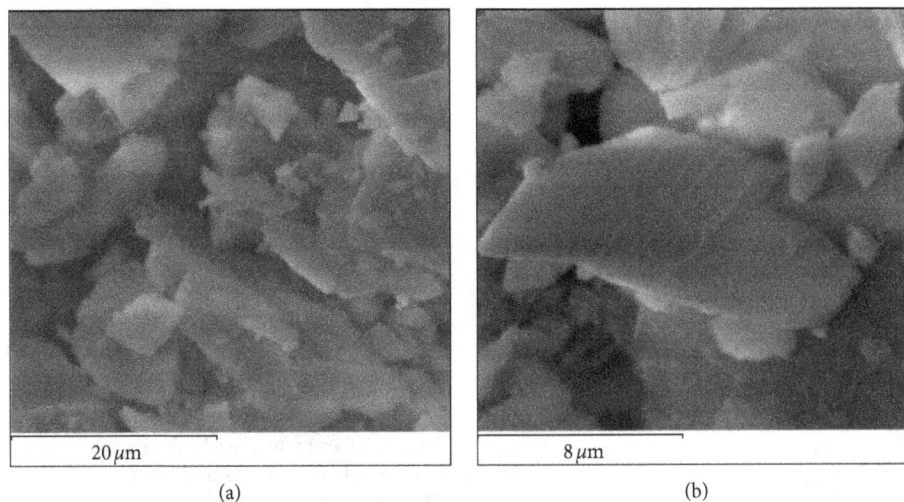

(a)

(b)

FIGURE 1: Scanning electron microscope microphotography of MgO-500 before the adsorption process: (a) 20 μm and (b) 8 μm.

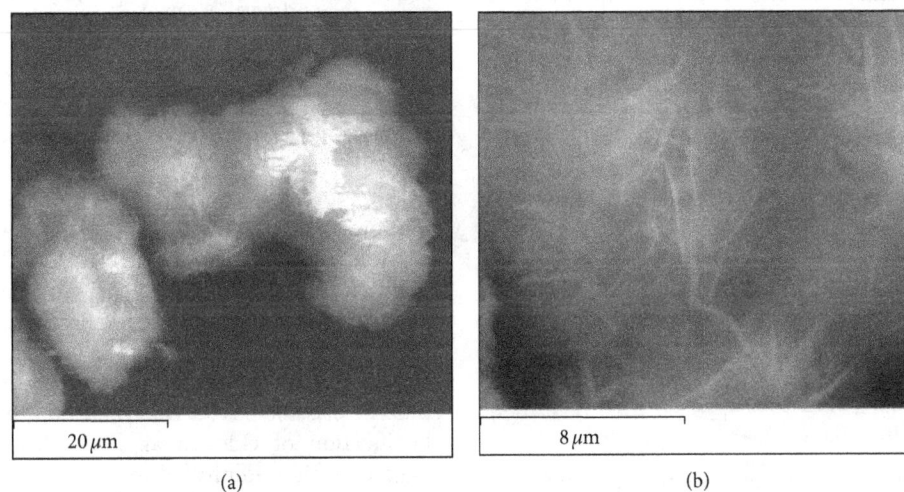

(a)

(b)

FIGURE 2: Scanning electron microscope microphotography of MgO-500 after the adsorption process: (a) 20 μm and (b) 8 μm.

The effect of temperature was evaluated in the process of boron removal at range of 5°C to 50°C, with initial concentration of boron of 350 mg·L^{-1}. For constant temperature maintenance during the experiments, a cooling or heating system was used.

2.3.1. Adsorption Isotherms. The experiments for adsorption isotherm evaluation were carried out in a batch system with MgO concentration range of 8 g·L^{-1} to 64 g·L^{-1} at three different temperatures of 25°C, 40°C, and 50°C, during 240 min of reaction.

The initial concentration of boron was 350 mg·L^{-1}, the stirring speed was 150 rpm, the pH was 10, the temperature was at 25°C, and the volume of the solution was 500 mL. After filtration, the samples were sent to boron concentration analysis.

3. Results and Discussions

3.1. Characterization of MgO. The chemical composition provided by the manufacturer is shown in Table 1.

Figures 1 and 2 show the surface morphology of MgO-500 before and after the adsorption process, respectively. The difference in structure presented in Figure 2 was due to the formation of a layer of H$_3$BO$_3$ on the surface of MgO and the hydration of MgO into Mg(OH)$_2$ after the adsorption process.

3.2. Adsorption Isotherms. The adsorption isotherms describe the equilibrium between the concentration of a material in aqueous phase and its concentration on the surfaces of particle adsorbents. This study employed the Langmuir, Freundlich, and Dubinin–Radushkevich models to describe the equilibrium adsorption.

3.2.1. Dubinin-Radushkevich Isotherm.

Dubinin–Radushkevich (DR) isotherm is a model that considers the adsorption in multilayers and in the heterogeneous surfaces.

DR model is expressed mathematically as follows:

$$\ln C_s = \ln X_m - k \, \epsilon^2, \tag{1}$$

where C_s is the amount of boron adsorbed per MgO $(mg \cdot g^{-1})$, X_m is the maximum adsorption capacity, k is the constant related to sorption energy $(mol^2 \cdot kJ^{-2})$, and ϵ is the Polanyi potential. ϵ and E are expressed by (2) and (3), respectively [35]:

$$\epsilon = RT \ln \left(1 + \frac{1}{C_e} \right), \tag{2}$$

$$E = \sqrt{\frac{1}{2k}}, \tag{3}$$

where R is the gas constant $(J \cdot mol^{-1} \, K^{-1})$ and T is the temperature (K). E $(kJ \cdot mol^{-1})$ is the sorption energy, and the magnitude of its value indicates if the adsorption is of chemical or physical nature.

3.2.2. Langmuir Isotherm.

The Langmuir isotherm is a model that considers monolayer adsorption onto a uniform surface with a finite number of adsorption sites and uniform adsorption energy, and this model is given by the following equation [36]:

$$q_e = \frac{K_L q_{max} C_e}{1 + K_L C_e}. \tag{4}$$

Equation (4) can be linearized as follows:

$$\frac{C_e}{q_e} = \frac{C_e}{q_{max}} + \frac{1}{q_{max} K_L}, \tag{5}$$

where C_e is the equilibrium concentration of boron in solution $(mg \cdot L^{-1})$, q_e is the quantity of boron adsorbed onto the MgO $(mg \cdot g^{-1})$, q_{max} is the maximum monolayer adsorption capacity of MgO $(mg \cdot g^{-1})$, and K_L is the Langmuir adsorption constant related to the energy sorption $(L \cdot mg^{-1})$.

3.2.3. Freundlich Isotherm.

This model is applied to the sorption processes on heterogeneous surfaces and reversible adsorption and admits multilayer adsorption [37].

The Freundlich equation and its linear form can be given as follows:

$$q_e = K_F C_e^{1/n},$$

$$\log q_e = \log K_F + \left(\frac{1}{n} \right) \log C_e, \tag{6}$$

where q_e is the boron concentration adsorbed at equilibrium $(mg \cdot g^{-1})$, C_e is the equilibrium concentration of boron in

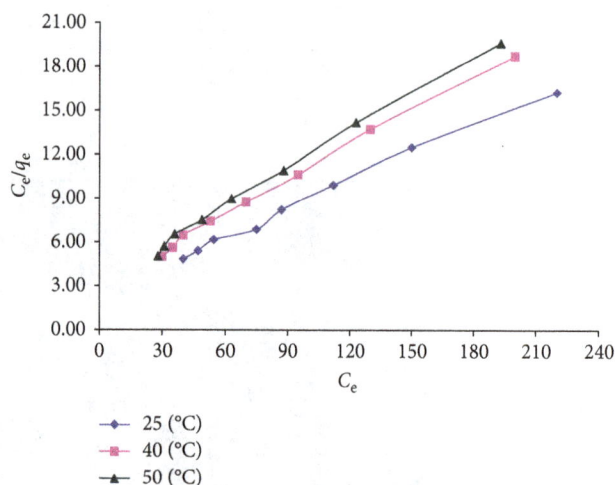

FIGURE 3: Adsorption isotherms of boron MgO at 25°C, 40°C, and 50°C.

TABLE 2: Parameters for the isotherms obtained from the Langmuir, Freundlich, and Dubinin–Radushkevich models at 25°C, 40°C, and 50°C, respectively.

Temperature (°C)	Langmuir			Freundlich			DR
	q_{max} $(mg \cdot g^{-1})$	B $(mg \cdot L^{-1})$	R^2	n	K_F	R^2	R^2
25	34.72	0.004	0.91	1.392	0.33	0.99	0.84
40	35.08	0.005	0.93	1.476	0.51	0.99	0.83
50	36.11	0.006	0.93	1.481	0.57	0.99	0.83

DR: Dubinin–Radushkevich isotherm. Initial boron concentration = 350 mg·L^{-1}, stirring speed = 150 rpm, pH = 10, and t = 240 min.

solution $(mg \cdot L^{-1})$, and K_F $(L \cdot g^{-1})$ and n are the Freundlich sorption isotherm constants, related to the adsorption capacity and the adsorption intensity, respectively.

Figure 3 shows the adsorption isotherms of boron MgO-500 at 25°C, 40°C, and 50°C.

It can be observed that the adsorption isotherm at 25°C could be classified as an H4-type isotherm according to the classification of Giles et al. [38]. This type of isotherm suggests a high affinity between adsorbate-adsorbent, and the subgroup 4 suggests the formation of monolayers of adsorbate on the surface of adsorbent.

On the contrary, for the isotherms at 40°C and 50°C, it is observed that they are similar L-type isotherms according to the classification of Giles et al. [38].

This type of isotherm assumes the existence of an affinity between ion B(OH)$_4^-$ and MgO, and if more sites of the adsorbent are filled, it will be more difficult to fill the empty sites with other solute molecules. This type of isotherm is commonly represented under the following mechanism: (1) molecules are adsorbed in layers and (2) there is competition for the active sites on the adsorbent surface between adsorbate molecules and solvent molecules [39–41].

In the present work, the Langmuir, Freundlich, and Dubinin–Radushkevich models were evaluated, which are the most widely used models to describe an adsorption process.

According to the results shown in Table 2, the model that provides a better fit to the experimental data is the isotherm

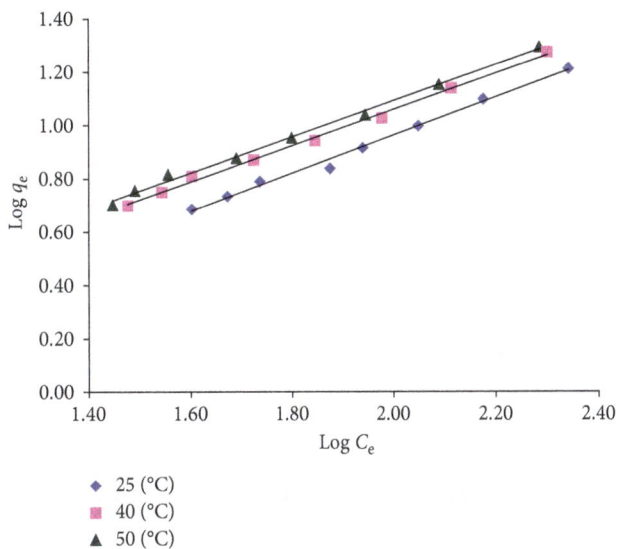

FIGURE 4: Data adjustment of the adsorption isotherms for the Freundlich model at temperatures of 25°C, 40°C, and 50°C. Initial boron concentration = 350 mg·L^{-1}, stirring speed = 150 rpm, pH = 10, and t = 240 min.

TABLE 3: Isotherm constants of the Freundlich isotherm (K_F, n) for the adsorption of boron in different adsorbents.

Adsorbent	K_F	n	Reference
Mining tailing (Woolley Edge)	0.004	1.4	[42]
Pural (76% Al$_2$O$_3$)	0.031	0.6	[41]
Siral (28% SiO$_2$, 72% Al$_2$O$_3$)	0.057	1.3	[41]
Activated alumina	0.440	1.4	[42]
MgO	0.570	1.5	Present work
Al$_2$O$_3$ (72%) and SiO$_2$ (28%)	0.057	1.3	[41]
N-methyl-D-glucamine onto SPC	0.282	2.3	[23]
Tannin gel	0.113	1.4	[44]

It can also be noticed from Table 3 that the adsorption constants (K_F) for boron removal process at different temperatures are higher than those obtained for boron removal processes with activated charcoal, activated alumina, Pural (76% Al$_2$O$_3$), Siral (28% SiO$_2$, 72% Al$_2$O$_3$), tannin gel, and mineral wastes. It is clear that the capacity of boron adsorption by MgO is higher than other adsorbents with exception to N-methyl-D-glucamine onto SPC and neutralized red mud.

of Freundlich at 25°C, 40°C, and 50°C with 99% correlation. Figure 4 shows the data adjusting to the Freundlich isotherm, and Table 3 compares the Freundlich isotherm parameters between MgO used at present work and other adsorbents.

The Freundlich isotherms suggest a heterogeneous adsorbent surface and a reversible adsorption process, which considers the formation of multilayer. However, the adsorption of ion B(OH)$^{4-}$ may involve different mechanisms, such as ion exchange, microprecipitation, complexation/chelation, and electrostatic attraction [40–42].

The Freundlich isotherms were obtained for K_F values of 0.33, 0.51, and 0.57 at temperatures of 25°C, 40°C, and 50°C, respectively. The values of n, which are related to the distribution of ion B(OH)$^{4-}$ linked to the active sites on the adsorbent, were 1.392, 1.476, and 1.481, respectively, for 25°C, 40°C, and 50°C (Table 2).

The value of n is a constant that indicates the strength of adsorption and is also known as a measure of linearity since n is equal to one. In this case, the adsorption process is linear; therefore, the adsorption sites are homogeneous concerning energy, and there would not be any interactions between the adsorbate and the adsorbent. For values of n smaller than the unit, the connection between the adsorbate and the adsorbent is very weak, and consequently, the process is not favorable for the adsorption and its adsorptive capacity decreases.

However, for values of n greater than the unit, the adsorption process is favorable; therefore, the adsorption capacity increases, indicating that the adsorption of ions is favorable under experimental conditions studied [41–43]. As seen in Table 3, all values of n are greater than the unit for the Freundlich isotherm, and the temperature increases from 25°C to 50°C, indicating that the adsorption of boron MgO-500 is favored by the temperature.

3.3. Kinetic Analysis.

The kinetics of boron removal by adsorption onto MgO was studied in the present work. The rate in which boron is removed by adsorption is the most important parameter for the design of a continuous or batch system in an effluent treatment plant. Therefore, it is essential to establish the dependency of time in the adsorption process for different operational conditions.

The kinetics of adsorption depends on the adsorbate-adsorbent interaction. The rate of adsorbate removal determines the residency time required for complete adsorption and could be calculated by kinetic analysis.

3.3.1. Equilibrium Kinetics.

During an adsorption process, the sorbed solute tends to desorb and return to the solution, and vice versa. The process occurs continuously until, at a given time, the adsorption and desorption rates reach an equilibrium state. Hence, in this stage, there will not be any additional adsorption of the pollutant in the solution [45–47].

Tests were conducted to determine the equilibrium of the reaction under the following conditions: 500 mL solution with a concentration of 350 mg·L^{-1} of boron was prepared; then, MgO was added to obtain a concentration of 40 g·L^{-1}, maintaining the pH at 10 and stirring speed of 150 rpm. Samples were collected at different time periods (5, 10, 15, 20, 25, 30, 40, 50, 60, 90, 140, 190, and 240 min).

Figure 5 shows the effect of temperature on boron removal.

The maximum adsorption capacities were 7.2, 7.7, and 8.0 mg·g^{-1} at temperatures of 25°C, 40°C, and 50°C, respectively.

It can also be observed, according to Figure 5, that the rate of the process is greatly affected by an increase in temperature, and at 25°C, 40°C, and 50°C, the equilibrium of reaction was reached around 40 min, 30 min, and 15 min,

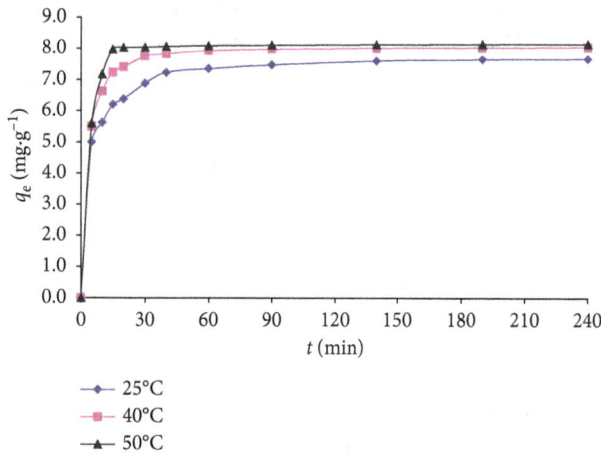

FIGURE 5: Profile of boron adsorption time as a function of temperature. Boron concentration = 350 mg·L^{-1}, pH = 10, MgO concentration = 40 g·L^{-1}, and stirring speed = 150 rpm.

respectively. From this stage, the variations in adsorption capacities were insignificant. Figure 5 also shows that the period of time for reaching equilibrium at temperatures of 40°C and 50°C is relatively short, which could be an advantage for future industrial applications.

3.3.2. Kinetic Models. In 1898, Lagergren presented the kinetic model known as "pseudo-first-order equation" for a liquid-solid system based on the adsorption capacity of the solid.

Later, Ho and Mckay [46] presented a modification of the Lagergren model based on the concentration of the pollutant in the solution, known as "pseudo-second-order equation."

Next, a kinetic analysis based on those models will be presented, which are the most usual and well-established ones for this type of adsorption process.

(1) Pseudo-First-Order Kinetic Model. The Lagergren model was developed for solid-liquid adsorption systems and is based on the adsorption capacity of the adsorbent. Equation (7) shows how this model is usually expressed, being one of the most used models to study the kinetics of adsorption processes:

$$\log(q_m - q_t) = \log(q_m) - \frac{k}{2.303}t. \tag{7}$$

Figures 6 and 7 present the application of the experimental data to the pseudo-first-order model at 25°C, 40°C, and 50°C, respectively.

It is observed in Figure 6 that there is an apparent multilinearity of the kinetic curves, which may lead us to think that the process happens in two stages. Figure 7 shows the good fit of the experimental kinetic data at 25°C in the pseudo-first-order model with $R^2 = 0.96$.

(2) Pseudo-Second-Order Kinetic Model. The pseudo-second-order model is based on the concentration of the

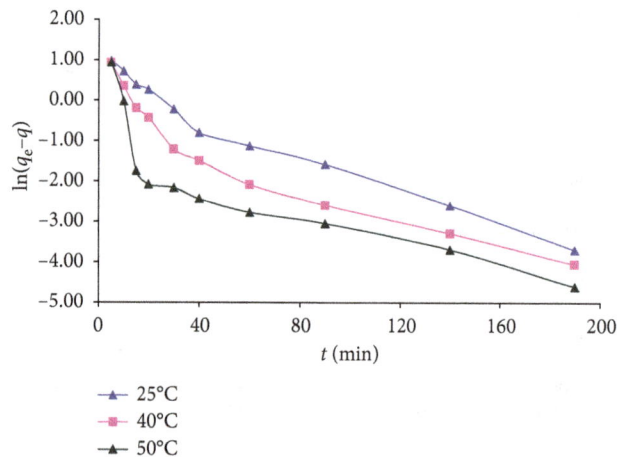

FIGURE 6: Application of the kinetic results to the pseudo-first-order model at temperatures of 25°C, 40°C, and 50°C. Concentration of boron = 350 mg·L^{-1}, pH = 10, concentration of MgO = 40 g·L^{-1}, and stirring speed = 150 rpm.

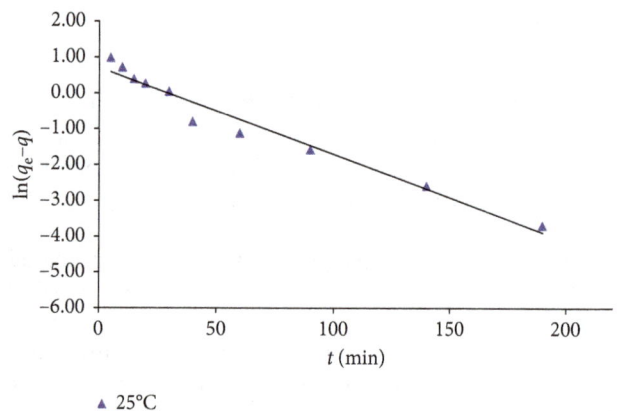

FIGURE 7: Adjustment of the adsorption data to the pseudo-first-order model at 25°C. Concentration of boron = 350 mg·L^{-1}, pH = 10, concentration of MgO = 40 g·L^{-1}, and stirring speed = 150 rpm.

pollutant in the solution. The adsorption rate equation can be written as follows:

$$\frac{t}{q_t} = \frac{1}{kq_e^2} + \frac{1}{q_e}t, \tag{8}$$

where k is the adsorption rate constant (g·mg^{-1}·min^{-1}), q_e is the concentration of boron adsorbed at equilibrium (mg·g^{-1}), and q_t is the concentration of boron on the surface of the adsorbent at time t (mg·g^{-1}). The constants can be determined experimentally by plotting t/q_t versus t.

Figure 8 illustrates the adjustment of the experimental data to the pseudo-second-order kinetic model. Table 4 presents the values of the kinetic parameters obtained from this setting.

It can be observed from the values presented in Table 4 that the pseudo-second-order model is the one with the best fit of the experimental data of the adsorption process, since R^2 values of 0.99 were obtained for this model at the three temperatures studied (25°C, 40°C, and 50°C) compared to

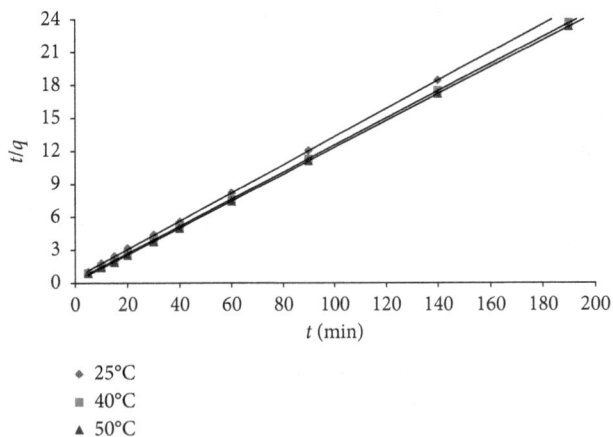

FIGURE 8: Adjustment of the adsorption data to the pseudo-second-order model at 25°C, 40°C, and 50°C. Boron concentration = 350 mg·L^{-1}, pH = 10, concentration of MgO = 40 g·L^{-1}, and stirring speed = 150 rpm.

the R^2 values obtained for the pseudo-first-order model of 0.96, 0.83, and 0.43 at the same temperatures, respectively.

The values of q_e obtained at 25°C, 40°C, and 50°C by the pseudo-second-order model were 7.81 mg·g^{-1}, 8.12 mg·g^{-1}, and 8.17 mg·g^{-1}, respectively, while the experimental values obtained at the same temperatures were 7.23 mg·g^{-1}, 7.75 mg·g^{-1}, and 7.98 mg·g^{-1}, respectively. It can be observed from the comparison of the values of q_e obtained experimentally and by the pseudo-second-order model that the values of q_e are very similar, which confirms that the process of removal of boron by adsorption with MgO follows a pseudo-second-order kinetics.

Studies available in the literature show that many models provide a clarification for the overall adsorption process. However, in many cases, that is not possible when graphs present multilinear feature [48].

In order to understand these cases, it is usual to divide the graph into two or more straight lines and to suggest that the adsorption mechanism is controlled by each straight line. This practice may help to understand the adsorption mechanism to some extent [48].

From Figure 6, which represents the application of the experimental data to a pseudo-first-order kinetic model, it is observed that the adsorption process at the three temperatures seems to occur in two stages due to the two straight slopes obtained from the application of the model, which is confirmed by the adjustment of the results to the pseudo-second-order model that describes a two-step reaction occurring consecutively.

3.3.3. Apparent Activation Energy.

The apparent activation energy of the adsorption process was calculated. The variation in the rate constant according to an increase in temperature can be described by the Arrhenius equation (9):

$$k = k_o \exp^{(-E_A/RT)}, \qquad (9)$$

where k is the sorption rate constant (g·mg^{-1}·min^{-1}), k_o is the independent factor of temperature (g·mg^{-1}·min^{-1}), E_A is

the adsorption activation energy (kJ·mol^{-1}), R is the gas constant (8.314 J·mol^{-1}·K^{-1}), and T is the temperature of the solution (K).

Equation (10) can be expressed in linear form as follows:

$$\ln k = \ln A - \left(\frac{E_A}{R}\right) \cdot \left(\frac{1}{T}\right). \qquad (10)$$

As shown in Figure 8, there is a linear correlation between the rate constant of the pseudo-second-order model and its corresponding absolute temperature, with a correlation coefficient of 0.98. Figure 9 shows the correlated experimental data in the linearized Arrhenius equation. From Figure 9, the relation between k and T can be represented as follows:

$$k = 3.02 \times 10^6 \exp^{(-45.54/8.314T)}. \qquad (11)$$

From (11), we can observe that the frequency factor A is 3.02×10^6 g·mg^{-1}·min^{-1}, and the activation energy is 45.54 kJ·mol^{-1}, which is very slightly outside of the suggested normal range (8–40 kJ·mol^{-1}) for a typical physical adsorption processes [49–52].

3.4. Thermodynamics of the Process.

In order to understand the thermodynamics of the process of boron removal by adsorption with MgO, some thermodynamic parameters were determined. Adsorption tests at different temperatures were performed using a thermostatic bath, under the following conditions: initial boron concentration = 350 mg·L^{-1}, pH = 10, MgO concentration = 40 g·L^{-1}, and stirring speed = 150 rpm.

The temperatures studied were 25°C, 40°C, and 50°C, and the parameters calculated were standard Gibbs free energy variation ($\Delta G°$), standard enthalpy variation ($\Delta H°$), and standard entropy variation ($\Delta S°$).

The standard free energy variation in the adsorption process is related to the equilibrium constant (K_c) and can be calculated according to the following equation:

$$\Delta G° = RT \ln K_c, \qquad (12)$$

where R is the gas constant (8.314 J·mol^{-1}·K^{-1}), T is the absolute temperature, and K_c is the equilibrium constant, which can be estimated from the following equation:

$$K_c = \frac{C_s}{C_e}, \qquad (13)$$

where C_e is the concentration of boron at equilibrium in the solution (mg·L^{-1}) and C_s is the concentration of boron at equilibrium in the adsorbent (mg·L^{-1}).

The variations in standard enthalpy ($\Delta H°$) and standard entropy ($\Delta S°$) can be calculated according to the following Van't Hoff equation:

$$\ln K_c = \frac{\Delta S°}{R} - \frac{\Delta H°}{RT}. \qquad (14)$$

Figure 9 represents the plot of $\ln K_c$ versus $1/T$ from the Van't Hoff equation, where the slope of the line is $-\Delta H°/R$ and the intercept with y-axis is $\Delta S°/R$, from which the

TABLE 4: Kinetic parameters for the adsorption of boron with MgO at 25°C, 40°C, and 50°C.

Temperature	Pseudo-first-order model			Pseudo-second-order model			
	q_e (mg·g^{-1})	k (min^{-1})	R^2	q_e (mg·g^{-1})	k (g·mg min^{-1})	V_o (g·mg min^{-1})	R^2
25°C	5.158	0.056	0.96	7.81	0.033	2.012	0.99
40°C	—	—	0.83	8.12	0.067	4.409	0.99
50°C	—	—	0.46	8.17	0.141	9.381	0.99

Concentration of boron = 350 mg·L^{-1}, pH = 10, concentration of MgO = 40 g·L^{-1}, and stirring speed = 150 rpm.

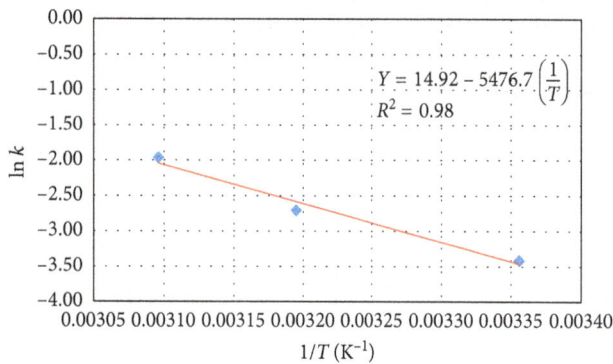

FIGURE 9: Variation of ln (k) versus 1/T.

TABLE 5: Thermodynamic parameters for the adsorption of boron with MgO at 25°C, 40°C, and 50°C.

Temperature	K_c	$\Delta G°$ (kJ·mol-1)	$H°$ (kJ·mol^{-1}·K^{-1})	$\Delta S°$ (kJ·mol^{-1})
25°C	5.36	−4.16	21.75	87.33
40°C	9.61	−5.88	—	—
50°C	10.29	−6.26	—	—

Concentration of boron = 350 mg·L−1, pH = 10, concentration of MgO = 40 g·L−1, stirring speed = 150 rpm, and time = 240 min. The positive value of the standard enthalpy $\Delta H° = 21.75$ kJ·mol^{-1}·K^{-1} for boron adsorption onto MgO shows the endothermic feature.

corresponding thermodynamic parameters were obtained and are presented in Table 5.

From Table 5, it can be observed that when the temperature of the adsorption process increases from 25°C to 40°C and from 40°C to 50°C, the $\Delta G°$ values become increasingly negative varying from −4.16 kJ·mol^{-1} to −5.88 kJ·mol^{-1} and from −5.88 kJ·mol^{-1} to −6.26 kJ·mol^{-1}, respectively. That means the process is spontaneous at all temperatures due to the negative value of $\Delta G°$. Likewise, as the temperature increases, the B(OH)$_4^-$ ion has higher affinity to be adsorbed by MgO.

The $\Delta G°$ for physical sorption is between −20 and 0 kJ·mol^{-1} and is ranged from −80 to −400 kJ·mol^{-1} for chemical sorption [52, 53]. The values of $\Delta G°$ presented in Table 5 indicate that the sorption process is controlled by physical adsorption.

The adsorption enthalpy is also a parameter used to indicate the intensity of the interaction between the adsorbate and the adsorbent. In the phenomenon of physisorption, this parameter has low values (up to about 40 kJ·mol^{-1}) since it is characterized by a low degree of interaction, being the forces involved of the order of magnitude of van der Waals forces. The chemisorption phenomenon is characterized by a high degree of interaction between the adsorbate and the surface of the adsorbent; the enthalpy values, in this case, are around 800 kJ·mol^{-1} [51]. The $\Delta H°$ calculated in this work was 21.75 kJ mol^{-1}·K^{-1}, indicating that the sorption process is controlled by physical adsorption.

Positive value of $\Delta S°$ (87.33 kJ·mol^{-1}) indicates that the degrees of freedom increase at the solid-liquid interface during the adsorption of boron onto magnesium oxide particles.

4. Conclusions

Equilibrium, kinetics, and thermodynamic studies for the adsorption of boron onto magnesium oxide powder were carried out.

The experimental data of adsorption were conducted by the Langmuir, Freundlich, and DR models, wherein the Freundlich isotherm was found to have a better fit for the equilibrium data for adsorption of boron under ambient temperature (25°C) and at higher temperatures (40°C and 50°C), respectively.

Additionally, the investigation of the kinetics of the overall adsorption process was conducted for the pseudo-first-order kinetic and pseudo-second-order kinetic models. Results show that the pseudo-second-order kinetic model generates the best fit to all experimental data. Such result suggests that the adsorption process at these temperatures seems to occur in two stages due to the two straight slopes obtained through the application of the pseudo-first-order kinetic model, which is confirmed by the adjustment of the results to the pseudo-second-order model.

The calculated activation energy (E_a) was 45.54 kJ·mol^{-1}, which is very slightly outside the normal range (8–40 kJ·mol^{-1}) for a typical physical adsorption process.

The values calculated for $\Delta G°$ and $\Delta H°$ were −4161.43 kJ·mol^{-1} and 21.75 kJ·mol^{-1}, respectively. These values confirm the spontaneous and endothermic nature of the process and also suggest a physical adsorption process.

Conflicts of Interest

The authors declare that there are no conflicts of interest regarding the publication of this paper.

Acknowledgments

The funding for this research was provided by the Research and Development Center of Petrobras and the National Council for Scientific and Technological Research of Brazil (CNPq). The authors thank Mariana Lima for assistance during the experimental work.

References

[1] G. Kelvin and M. M. Arvind, "Current perspective on produced water management challenges during hydraulic fracturing for oil and gas recovery," *Environmental Chemistry*, vol. 12, no. 3, pp. 261–266, 2015.

[2] H. E. Ezerie, H. I. Mohamed, and R. B. M. K. Shamsul, "Boron in produced water: challenges and improvements: a comprehensive review," *Journal of Applied Sciences*, vol. 12, no. 5, pp. 402–415, 2012.

[3] F. A. F. Cedric, J. S. Vincent, A. M. Bruce, and M. Farshid, "Determination of boron in produced water using the carminic acid assay," *Talanta*, vol. 150, pp. 240–252, 2016.

[4] H. E. Ezerie, H. I. Mohamed, R. M. K. Shamsul, and Y. Asim, "Boron removal from produced water using electrocoagulation," *Process Safety and Environmental Protection*, vol. 92, no. 6, pp. 509–514, 2014.

[5] P. Howe, "A review of boron effects in the environment," *Biological Trace Element Research*, vol. 66, no. 1–3, pp. 153–165, 1998.

[6] J. Parks and M. Edwards, "Boron in the environment," *Critical Reviews in Environmental Science and Technology*, vol. 35, no. 2, pp. 81–114, 2005.

[7] Brazilian NR, *Conama 357 Resolution*, National Environment Council of Brazil, São Paulo, SP, Brazil, 2017, http://www.mma.gov.br/port/conama/.

[8] MINAM, *Ministry of Environment of Peru*, November 2017, http://sinia.minam.gob.pe/normas.

[9] W. Boyang, G. Xianghai, and B. Peng, "Removal technology of boron dissolved in aqueous solutions–A review," *Colloids and Surfaces A*, vol. 444, p. 338, 2014.

[10] C. T. Onur, V. Jan, and T. Cengiz, "Constructed wetlands for boron removal: a review," *Ecological Engineering*, vol. 64, pp. 350–359, 2014.

[11] WHO, *Guidelines for Drinking-Water Quality*, World Health Organization, Geneva, Switzerland, 4th edition, 2011.

[12] H. E. Ezerie, H. I. Mohamed, R. M. K. Shamsul, and A. Zubair, "Electrochemical removal of boron from produced water and recovery," *Journal of Environmental Chemical Engineering*, vol. 3, no. 3, pp. 1962–1973, 2015.

[13] G. Zhimin, L. Jiafei, B. Peng, and G. Xianghai, "Boron removal from aqueous solution by adsorption–A review," *Desalination*, vol. 383, pp. 29–37, 2016.

[14] K. Tomohito, O. Jumpei, and Y. Toshiaki, "Use of Mg-Al oxide for boron removal from an aqueous solution in rotation: kinetics and equilibrium studies," *Journal of Environmental Management*, vol. 165, pp. 280–285, 2016.

[15] J. Kluczka, T. Korolewicz, M. Zołotajkin, W. Simka, and M. Raczek, "A new adsorbent for boron removal from aqueous solutions," *Environmental Technology*, vol. 34, no. 11, pp. 1369–1376, 2013.

[16] K. Maciej, B. Z. Barbara, D. Piotr, and T. Marian, "The concept of a system for electrodialytic boron removal into alkaline concentrate," *Desalination*, vol. 310, pp. 75–80, 2013.

[17] C. Zhifan, A. T. Alicia, R. A. Savina, X. Junliang, and T. Y. Norman, "Removal of boron from wastewater: evaluation of seven poplar clones for B accumulation and tolerance," *Chemosphere*, vol. 167, pp. 146–154, 2017.

[18] C. T. Onur, T. Cengiz, B. Harun, C. Arzu, and Y. Anıl, "Role of plants and vegetation structure on boron (B) removal process in constructed wetlands," *Ecological Engineering*, vol. 88, pp. 143–152, 2016.

[19] P. Qingyun, M. A. Ibrahim, and H. Zhen, "Enhanced boron removal by electricity generation in a microbial fuel cell," *Desalination*, vol. 398, pp. 165–170, 2016.

[20] P. Qingyun, M. A. Ibrahim, and H. Zhen, "Mathematical modeling based evaluation and simulation of boron removal in bioelectrochemical systems," *Science of the Total Environment*, vol. 570, pp. 1380–1389, 2016.

[21] K. Nalan, K. Pelin, Y. Duygu, Y. Ümran, and Y. Mithat, "Coupling ion exchange with ultrafiltration for boron removal from geothermal water-investigation of process parameters and recycle tests," *Desalination*, vol. 316, pp. 17–22, 2013.

[22] N. Hilal, G. J. Kim, and C. Somerfield, "Boron removal from saline water: a comprehensive review," *Desalination*, vol. 273, no. 1, pp. 23–35, 2011.

[23] X. Li, R. Liu, S. Wua, J. Liu, S. Cai, and D. Chen, "Efficient removal of boron acid by N-methyl-D-glucamine functionalized silica–polyallylamine composites and its adsorption mechanism," *Journal of Colloid and Interface Science*, vol. 361, no. 1, pp. 232–237, 2011.

[24] J. Wolska and M. Bryjak, "Methods for boron removal from aqueous solutions—A review," *Desalination*, vol. 310, pp. 18–24, 2013.

[25] E. Güler, C. Kaya, N. Kabay, and M. Arda, "Boron removal from seawater: State-of-the-art review," *Desalination*, vol. 356, pp. 85–93, 2015.

[26] E. Güler, N. Kabay, M. Yüksel, E. Yavuz, and U. Yüksel, "A comparative study for boron removal from seawater by two types of polyamide thin film composite SWRO membranes," *Desalination*, vol. 273, no. 1, pp. 81–84, 2011.

[27] N. Nadav, "Boron removal from seawater reverse osmosis permeate utilizing selective ion exchange resin," *Desalination*, vol. 124, no. 1–3, pp. 131–135, 1999.

[28] M. Simonnot, C. Castel, M. Nicolai, C. Rosin, M. Sardin, and H. Jauffret, "Boron removal from drinking water with a boron selective resin: is the treatment really selective?," *Water Research*, vol. 34, no. 1, pp. 109–116, 2000.

[29] N. Kabay, M. Bryjak, S. Scholosser, and M. Kitisa, "Adsorption-membrane filtration (AMF) hybrid process for boron removal from seawater: an overview," *Desalination*, vol. 223, no. 1–3, pp. 38–48, 2008.

[30] M. Tagliabue, A. Reverberi, and R. Bagatin, "Boron removal from water: needs, challenges and perspectives," *Journal of Cleaner Production*, vol. 77, pp. 56–64, 2014.

[31] J. Cao, S. Monroe, and D. Hendry, "Environmentally preferred process for efficient boron and hardness removal for produced water reuse," in *Proceedings of SPE Annual Technical Conference and Exhibition*, Houston, TX, USA, 2015.

[32] Petrobras Transporte SA, July 2017, http://www.transpetro.com.br/pt_br/areas-de-negocios/terminais-e-oleodutos/terminais-aquaviarios.html.

[33] Z. Jin, Y. Jia, K. Zhang et al., "Effective removal of fluoride by porous MgO nanoplates and its adsorption mechanism," *Journal of Alloys and Compounds*, vol. 675, pp. 292–300, 2016.

[34] Z. Bai, Y. Zheng, and Z. Zhang, "One-pot synthesis of highly efficient MgO for the removal of Congo red in aqueous solution," *J. Mater. Chem. A*, vol. 5, pp. 6630–6637, 2017.

[35] M. Dubinin and L. Radushkevich, "Equation of the characteristic curve of activated charcoal," *Proceedings of the USSR Academy of Sciences*, vol. 55, pp. 331–333, 1947.

[36] I. Langmuir, "The constitutional and fundamental properties of solids and liquids,," *Journal of the American Chemical Society*, vol. 38, no. 11, pp. 2221–2295, 1916.

[37] H. Freundlich, "Over the adsorption in solution," *Zeitschrift für Physikalische Chemie*, vol. 57, pp. 385–470, 1906.

[38] C. H. Giles, A. P. D´Silva, and A. S. Trivedi, *Surface Area Determination*, Butterworth, London, UK, 1970.

[39] R. Masel, *Principles of Adsorption and Reaction on Solid Surfaces*, Wiley, New York, NY, USA, 2nd edition, 1951.

[40] S. Goldberg, "Reactions of boron with soils," *Plant and Soil*, vol. 193, no. 2, pp. 35–48, 1997.

[41] Y. Seki, S. Seyhan, and M. Yurdakoc, "Removal of boron from aqueous solution by adsorption on Al_2O_3 based materials using full factorial design," *Journal of Hazardous Materials*, vol. 38, pp. 60–66, 2006.

[42] Y. Cengeloglu, "Removal of boron from aqueous solution by using neutralized red mud," *Journal of Hazardous Materials*, vol. 142, pp. 412–417, 2007.

[43] R. Sivaraj and C. Namasivayam, "Orange peel as an adsorbent in the removal of acid violet 17 (acid dye) from aqueous solutions," *Waste Management*, vol. 21, no. 1, pp. 105–110, 2001.

[44] S. Morisada, T. Rin, T. Ogata, KY. Ho, and Y. Nakano, "Adsorption removal of boron in aqueous solutions by amine-modified tannin gel," *Water Research*, vol. 45, no. 13, pp. 4028–4034, 2011.

[45] P. Cheremisinoff, *Water and Water Pollution Handbook*, Wiley, New York, NY, USA, 2nd edition, 2002.

[46] Y. S. Ho and G. McKay, "A comparison of chemisorption kinetic models applied to pollutant removal on various sorbents," *Process Safety and Environmental Protection*, vol. 76, no. 4, pp. 332–340, 1998.

[47] Metcalf and Eddie, *Wastewater Engineering: Treatment, Disposal, and Reuse*, McGraw-Hill, Inc., Boston, MA, USA, 4th edition, 2003.

[48] X. Yang and B. Al-Duri, "Kinetic modeling of liquid-phase adsorption of reactive dyes on activated carbon," *Journal of Colloid and Interface Science*, vol. 287, no. 1, pp. 25–34, 2005.

[49] U. Isah, G. Abdulraheem, S. Bala, S. Muhammad, and M. Abdullahi, "Kinetics, equilibrium and thermodynamics studies of C.I. Reactive Blue 19 dye adsorption on coconut shell based activated carbon," *International Biodeterioration and Biodegradation*, vol. 102, pp. 265–273, 2015.

[50] A. Elekli, G. lgüna, and H. Bozkurtb, "Sorption equilibrium, kinetic, thermodynamic, and desorption studies of Reactive Red 120 on Chara contraria," *Chemical Engineering Journal*, vol. 191, pp. 228–235, 2012.

[51] S. P. D. Monte Blanco, F. B. Scheufele, A. N. Módenes et al., "Kinetic, equilibrium and thermodynamic phenomenological modeling of reactive dye adsorption onto polymeric adsorbent," *Chemical Engineering Journal*, vol. 307, pp. 466–475, 2017.

[52] H. Nollet, M. Roels, P. Lutgen, P. Van der Meeren, and W. Verstraete, "Removal of PCBs from wastewater using fly ash," *Chemosphere*, vol. 53, no. 6, pp. 655–665, 2003.

[53] C. H. Wu, "Adsorption of reactive dye onto carbon nanotubes: Equilibrium, kinetics and thermodynamics," *Journal of Hazardous Materials*, vol. 144, no. 1-2, pp. 93–100, 2007.

9

Theoretical Study of the Effect of Instrument Parameters on the Flow Field of Air-Flow Impacting based Mechanochemical Synthesis

Yang Tao ⓘ**, Jun Lin, Zhao Zhang, Qiuting Guo, Jin Zuo, and Bo Lu**

High Speed Aerodynamics Institute, China Aerodynamic Research and Development Center, Mianyang 621000, China

Correspondence should be addressed to Yang Tao; 50323222@qq.com

Academic Editor: Jose C. Merchuk

The air-flow impacting based mechanochemical synthesis is an alternative strategy to traditional mechanochemical preparations, which has many advantages in terms of reaction temperature, preparation speed, and cleanness. Herein, we theoretically study the effect of instrument parameters, including the axial position of physical target, the diameter difference between nozzle throat and suction pipe, and divergence angles of uniform speed region, on the flow field of the air-flow impacting based mechanochemical synthesis. The optimized parameters have been obtained. Under the optimal conditions, a stable and high-speed air flow is obtained, in which the speed can achieve a Mach number of approximately 2.6. The high-speed air flow is able to easily carry the reacting substances to arrive at the physical target, triggering a chemical reaction. These findings undoubtedly provide a key guideline for further development and application of the air-flow impacting based mechanochemical synthesis.

1. Introduction

The mechanochemical synthesis has received more and more attention due to its unique advantages such as simpleness, economy, and environmental friendliness [1]. It can facilely proceed in the solid state without the use of solvents. A variety of substances spanning from organic compounds to metal-organic frameworks to nanomaterials have been successfully prepared using the mechanochemical synthesis techniques [1–7]. Traditionally, the mechanochemical synthesis is carried out by hand grinding, ball milling, and twin screw extrusion [8–14]. However, these methods suffer from many drawbacks. For instance, these methods will produce a high temperature during the procedures, which are not suitable to prepare thermally sensitive compounds. Additionally, the preparation speed is limited at $kg·h^{-1}$, making them insufficient to realize the large-scale synthesis.

In order to overcome these shortcomings, we recently developed a novel air-flow impacting based mechanochemical synthesis [15]. In this strategy, the air flow can be accelerated to a maximum velocity going up to 300–600 m/s at room temperature (approximately 25°C). When the reacting substances are carried by the air flow with a supersonic speed, they will tempestuously collide with each other or a physical target, accompanying that the mechanical energy effectively initiates the chemical reaction. The gaseous medium is employed instead of physical ball, pestle, and screw, resulting in a more clean and renewable synthesis. More importantly, it has been demonstrated that the air-flow impacting based mechanochemical synthesis is able to achieve large-scale and fast preparation at $kg·min^{-1}$ rates, which is at least sixty times faster than the traditional mechanochemical synthesis methods. Obviously, the performance of air-flow impacting based mechanochemical synthesis is highly dependent on the air-flow field in the instrument. To better understand the preparation process, it is highly desirable to study the air-flow field in the instrument.

Here, we theoretically investigate the effect of aerodynamic design parameters of the instrument on the flow field of the air-flow impacting based mechanochemical synthesis. Several instrument parameters, including the physical target, the diameter difference between nozzle throat inwall and suction pipe of solid reaction particles, and divergence angles of acceleration pipe, have been investigated. Based on the analysis results, we obtain the optimized aerodynamic design parameters. This work provides a guideline for the air-flow impacting based mechanochemical synthesis.

2. Methods

2.1. Design of De Laval Nozzle. De Laval nozzle is employed to produce a high-speed air flow with a supersonic speed. Isentropic flow was assumed in the annular nozzle. The produced Mach number (M_a) is depended on the area ratio of exit surface and throat A_{ex}/A_{thr}, according to the following equation [16, 17]:

$$\frac{A_{ex}}{A_{thr}} = \frac{1}{M_a}\left(\frac{2}{\gamma+1}\left(1+\frac{\gamma-1}{2}M_a^2\right)\right)^{(\gamma+1)/(2(\gamma-1))}, \quad (1)$$

where γ is the ratio of specific heats, which is taken to be constant of 1.4 for the diatomic gas molecules, A_{ex} is the area of the exhaust, and A_{thr} is the area of the throat.

The pressure ratio of the total pressure to the static pressure at the intake side and exhaust side must maintain a high enough value, and the critical pressure ratio can be detected by the following relationships [18–20]:

$$\frac{p_0}{p_{ex}} = \left(\frac{T_0}{T_{ex}}\right)^{\gamma/(\gamma-1)} = \left(1+\frac{\gamma-1}{2}M_a^2\right)^{\gamma/(\gamma-1)}, \quad (2)$$

where p_0 and T_0 are the total pressure and total temperature at intake side and p_{ex} and T_{ex} are the static pressure and static temperature at the exhaust side of the nozzle.

Also, the various parts of the De Laval nozzle, including converge section, diverge section, and uniform speed section, are calculated to obtain the physical appearance such as size, length, and angle through a series of equations as shown in the supporting information (available here). To ensure that the theoretical Mach number can reach 3, the physical sizes of the De Laval nozzle are as follows: converge section (length: 40 mm, diameter: 9.26 mm), throat diameter (3 mm), and diverge section (length: 24.69 mm, diameter: 6.17 mm).

2.2. Numerical Simulation. We employ the commercial CFD code Fluent 6.3 for the numerical simulation based on the above-mentioned De Laval nozzle and variable instrumental parameters as mentioned in our manuscript. The computational analysis uses the Navier–Stokes equations, which can be written in a conservational form as follows [21]:

$$\frac{\partial \rho}{\partial t} + \frac{\partial}{\partial x_j}(\rho u_j) = 0, \quad (3)$$

$$\frac{\partial(\rho u_i)}{\partial t} + \frac{\partial}{\partial x_j}(\rho u_i u_j + \delta_{ij}p) = \frac{\partial \tau_{ij}}{\partial x_j}, \quad (4)$$

$$\frac{\partial(\rho E)}{\partial t} + \frac{\partial}{\partial x_j}(\rho H u_j) = \frac{\partial}{\partial x_j}\left(\tau_{ij}u_i + \lambda\frac{\partial T}{\partial x_j}\right), \quad (5)$$

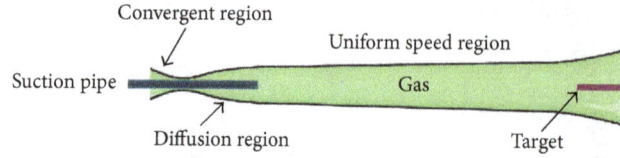

FIGURE 1: Schematic representation of the air-flow impacting based mechanochemical synthesis instrument.

where ρ, p, and T denote flow density, pressure, and temperature, respectively; u_j is the velocity component along the Cartesian coordinate x_j; E and H are the total energy and the total enthalpy per unit volume, respectively; τ_{ij} is the stress tensor; λ is the heat transfer coefficient; and δ_{ij} is the Kronecker delta. The equation of state of ideal gas is introduced to close the system.

The RANS equations with the k-ω SST turbulence model are used for the computational analysis. The SST turbulence model is employed for high-accuracy boundary-layer simulation. It is a hybrid method that couples the standard k-ε and k-ω models in an efficient manner, with the k-ω model used in the near-wall region and the standard k-ε model in the far-field region, blending them together at the interface between the regions. It includes the modeling of transport of shear stress via a modified definition of the turbulent viscosity.

The pressure inlet boundary conditions were used at the intakes of the nozzle and suction pipe. The pressure outlet boundary condition was applied at the exit. The no-slipped wall was used for all of the solid walls. Moreover, all walls were assumed as ideal rigid solid surfaces without deformation during the collision. The overall analysis of the flow field of air-flow impacting based mechanochemical synthesis was carried out on ANSYS 16.0.

3. Results and Discussion

3.1. The Structure of the Air-Flow Impacting Instrument. As shown in Figure 1, the air-flow impacting instrument is consisting of suction pipe, convergent region, diffusion region, uniform speed region, and a physical target. When compressed air is injected into this instrument, the air is accelerated to a subsonic speed in the convergent region. The speed of the air is further increased to the supersonic speed in the diffusion region. The high-speed air can carry the reacting substances to collide with a physical target, triggering a chemical reaction. The speed of the air is a key factor for the chemical preparation, which is mainly dependent on the diameter difference between nozzle throat and suction pipe, the position of the physical target, and the divergence angles.

3.2. Optimization of Aerodynamic Design Parameters for the Flow Field. To describe the speed of the air flow, the Mach number is introduced, which is expressed as $M_a = u/c$, where u is the local flow velocity with respect to the boundaries and c represents the speed of sound in the medium. It can be expressed as $c = \sqrt{\gamma RT}$, where γ is the adiabatic exponent of

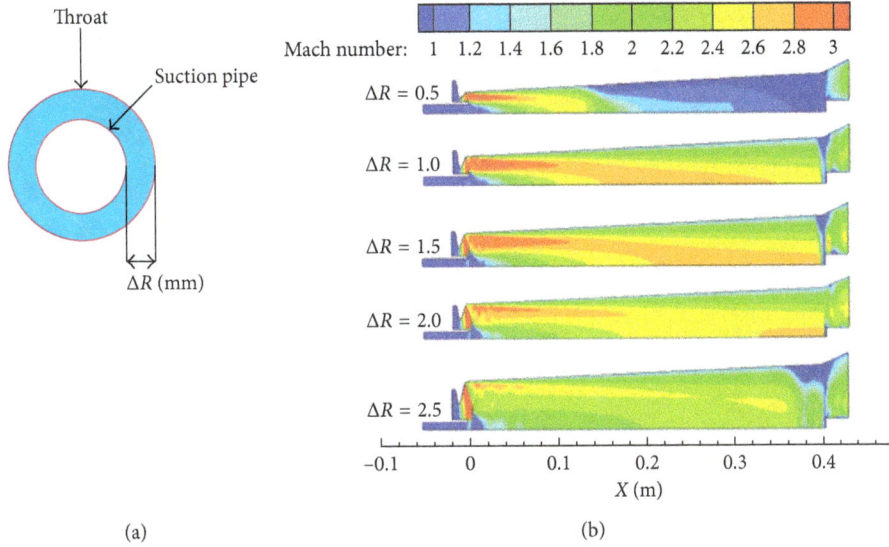

FIGURE 2: (a) The front view of the throat and suction pipe. (b) Effect of the diameter difference between nozzle throat and suction pipe (ΔR) on the flow field of the air flow.

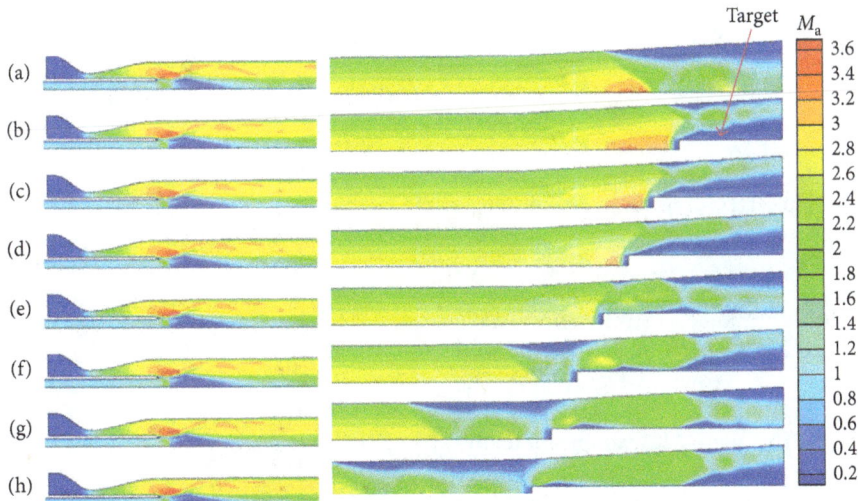

FIGURE 3: Cross-sectional flow field of air flow in the absence (a) and presence of a physical target placed under different positions: (b) 410, (c) 405, (d) 400, (e) 395, (f) 390, (g) 385, and (h) 381 mm.

air, R is the gas constant, and T is the thermodynamic temperature [22]. We can solve the Navier–Stokes equations (5) to obtain the T at different positions, which are further employed for calculating c. The effect of the diameter difference between nozzle throat and suction pipe (ΔR) on the flow field is firstly investigated. As shown in Figure 2, when the value of the ΔR is lower than 0.5 mm, the speed of the compressed air quickly decreases to a sonic speed after spraying. Additionally, if the ΔR is above 2.5 mm, the air flow is apt to form a curved shock wave, which is unstable (Figure 2(b)). Therefore, in order to obtain a stable high-speed air flow, the value of the ΔR should be kept in the range of 1.0-2.0 mm.

Apart from the diameter difference between throat inwall and suction pipe ΔR, the axial position of the physical target apart from the spraying outlet is another important

factor. As displayed in Figure 3(a), it cannot obtain a stable flow field in the absence of the physical target, and the speed of the air flow is relatively low. As the physical target is placed in the instrument, a stable and high-speed air flow can be obtained as shown in Figures 3(b)–3(e). Nevertheless, the physical target is also able to disturb the flow field of the air flow if it is too near the spraying outlet (Figures 3(f)–3(h)). Simultaneously, under this condition, it can result in local congestion, leading to moving the curved shock wave forward. While the distance between the physical target and the spraying outlet is in the range from 395 to 410 mm, a stable and high-speed air flow can be obtained, and the average of M_a is approximately 2.6. Therefore, the optimized axial distance between the physical target and the spraying outlet is ranging from 395 to 410 mm.

(a)

(b)

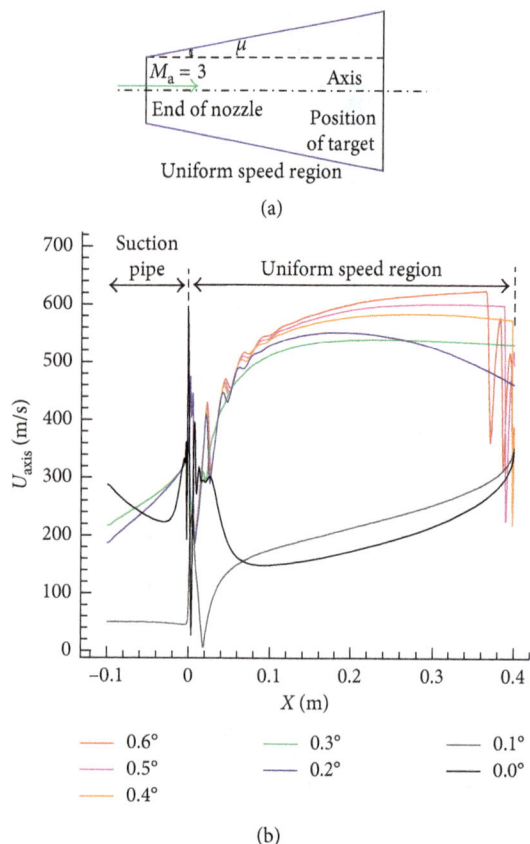

FIGURE 4: (a) The scheme for the divergence angle (μ). (b) The relationship between the axis speed (U_{axis}) and the divergence angle from 0° to 0.6° in the uniform speed region.

In order to avoid the appearance of strong shock wave and keep a high air flow, the divergence angle is introduced in the uniform speed region as shown in Figure 4(a). The effect of the divergence angle on the axial speed (U_{axis}) is examined at a M_a of 3. The high M_a can promote the collisional processes, which effectively initiates the chemical reaction. In the present stage, it is very difficult to achieve the equipment with the M_a number of more than 3 because of the limitations in terms of materials, carrier gas, and pressure. It can be seen that the absence of the divergence angle is able to result in gas congestion and a fast decrease of the U_{axis}, which decreases to the subsonic speed. This result confirms that the introduction of the divergence angle is quite necessary. In addition, with the increasing the divergence angle from 0.1° to 0.3°, the value of the U_{axis} gradually increases and reaches a stable value at 0.3°. However, when the divergence angle is larger than 0.3°, the unstable shock wave appears in the end of the uniform speed region. Accordingly, the optimized divergence angle is 0.3° to obtain a stable and high-speed air flow.

3.3. Distribution of Air-Flow Field.
Under the optimal aerodynamic design parameters, we further investigate the distribution of air-flow field in the air-flow impacting based mechanochemical synthesis instrument. As indicated in

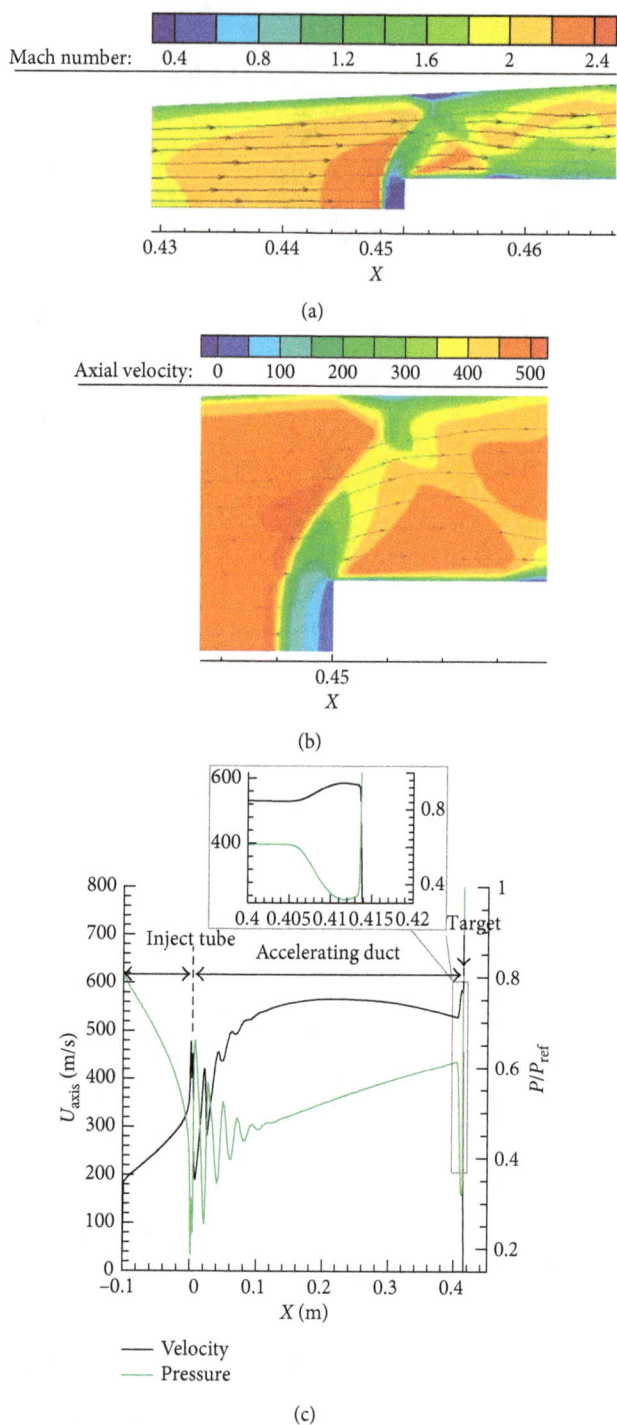

(a)

(b)

(c)

FIGURE 5: (a, b) The distribution of air-flow field at the diffusion region and the physical target. (c) The velocity and pressure versus the position at the axis line, where P and P_{ref} represent the local static pressures and reference pressure with mean value of 1 atm.

Figure 5, the high pressure gas is accelerated in the convergent region, and the gas achieves a relatively high speed. Besides, an extremely relative static pressure ($P/P_{ref} = 0.1$, where P and P_{ref} are the local static pressures and reference pressure with mean value of 1 atm, resp.) is obtained in the exit of the suction pipe. The great pressure difference is in favor of the injection of various reaction substances into the reaction instrument. Furthermore, the low-speed air flow from the suction pipe is carried by the air flow with a high speed in the diffusion region, which crosses over the uniform speed region. Eventually, a high-speed air flow and strong bow shock wave are formed near the physical target. The maximum speed of the air flow is about 550 m/s with a local M_a of 2.6, which is capable of supplying mechanochemical energy for powerful effective chemical preparation.

4. Conclusions

In summary, we investigate the effect of the instrument parameters, such as the physical target, the diameter difference between throat inwall and suction pipe, and the divergence angles, on the flow field of air-flow impacting based mechanochemical synthesis. The results illustrate that these parameters strongly affect the speed of air flow. The average M_a of the air flow can reach 2.6. Additionally, the air flow with a high speed forms a stable flow field, which can carry the reaction substances to collide with the physical target, inducing a chemical reaction. These findings may serve as a guide to the air-flow impacting based mechanochemical synthesis and further broaden the application field of the mechanochemical synthesis.

Conflicts of Interest

The authors declare that there are no conflicts of interest regarding the publication of this paper.

Acknowledgments

The support of this research by the National Natural Science Foundation of China (Grant no. 51327804) is gratefully acknowledged.

References

[1] B. P. Biswal, S. Chandra, S. Kandambeth, B. Lukose, T. Heine, and R. Banerjee, "Mechanochemical synthesis of chemically stable isoreticular covalent organic frameworks," *Journal of the American Chemical Society*, vol. 135, no. 14, pp. 5328–5331, 2013.

[2] S. L. James, C. J. Adams, C. Bolm et al., "Mechanochemistry: opportunities for new and cleaner synthesis," *Chemical Society Reviews*, vol. 41, no. 1, pp. 413–447, 2012.

[3] G. Bharath, R. Madhu, S.-M. Chen, V. Veeramani, D. Mangalaraj, and N. Ponpandian, "Solvent-free mechanochemical synthesis of graphene oxide and Fe_3O_4–reduced graphene oxide nanocomposites for sensitive detection of nitrite," *Journal of Materials Chemistry A*, vol. 3, no. 30, pp. 15529–15539, 2015.

[4] P. A. Julien, K. Užarević, A. D. Katsenis et al., "In situ monitoring and mechanism of the mechanochemical formation of a microporous MOF-74 framework," *Journal of the American Chemical Society*, vol. 138, no. 9, pp. 2929–2932, 2016.

[5] D. Prochowicz, K. Sokołowski, I. Justyniak et al., "A mechanochemical strategy for IRMOF assembly based on predesigned oxo-zinc precursors," *Chemical Communications*, vol. 51, no. 19, pp. 4032–4035, 2015.

[6] G.-W. Wang, "Mechanochemical organic synthesis," *Chemical Society Reviews*, vol. 42, no. 18, pp. 7668–7700, 2013.

[7] A. D. Katsenis, A. Puskaric, V. Strukil et al., "In situ X-ray diffraction monitoring of a mechanochemical reaction reveals a unique topology metal-organic framework," *Nature Communications*, vol. 6, no. 1, p. 6662, 2015.

[8] M. Y. Masoomi, A. Morsali, and P. C. Junk, "Rapid mechanochemical synthesis of two new Cd(II)-based metal-organic frameworks with high removal efficiency of Congo red," *CrystEngComm*, vol. 17, no. 3, pp. 686–692, 2015.

[9] I. Y. Jeon, S. Y. Bae, J. M. Seo, and J. B. Baek, "Scalable production of edge-functionalized graphene nanoplatelets via mechanochemical ball-milling," *Advanced Functional Materials*, vol. 25, no. 45, pp. 6961–6975, 2015.

[10] X. Ma, W. Yuan, S. E. Bell, and S. L. James, "Better understanding of mechanochemical reactions: Raman monitoring reveals surprisingly simple 'pseudo-fluid' model for a ball milling reaction," *Chemical Communications*, vol. 50, no. 13, pp. 1585–1587, 2014.

[11] D. E. Crawford, L. A. Wright, S. L. James, and A. P. Abbott, "Efficient continuous synthesis of high purity deep eutectic solvents by twin screw extrusion," *Chemical Communications*, vol. 52, no. 22, pp. 4215–4218, 2016.

[12] D. E. Crawford, C. K. G. Miskimmin, A. B. Albadarin, G. Walker, and S. L. James, "Organic synthesis by twin screw extrusion (TSE): continuous, scalable and solvent-free," *Green Chemistry*, vol. 19, no. 6, pp. 1507–1518, 2017.

[13] W. Huang, Z. Xie, Y. Deng, and Y. He, "3,3′,5,5′-tetramethylbenzidine-based quadruple-channel visual colorimetric sensor array for highly sensitive discrimination of serum antioxidants," *Sensors and Actuators B: Chemical*, vol. 254, pp. 1057–1060, 2018.

[14] R. Li, H. An, W. Huang, and Y. He, "Molybdenum oxide nanosheets meet ascorbic acid: tunable surface plasmon resonance and visual colorimetric detection at room temperature," *Sensors and Actuators B: Chemical*, vol. 259, pp. 59–63, 2018.

[15] B. Sun, Y. He, R. Peng, S. Chu, and J. Zuo, "Air-flow impacting for continuous, highly efficient, large-scale mechanochemical synthesis: a proof-of-concept study," *ACS Sustainable Chemistry & Engineering*, vol. 4, no. 4, pp. 2122–2128, 2016.

[16] S. He, Y. Li, and R. Z. Wang, "Progress of mathematical modeling on ejectors," *Renewable and Sustainable Energy Reviews*, vol. 13, no. 8, pp. 1760–1780, 2009.

[17] N. H. Aly, A. Karameldin, and M. M. Shamloul, "Modelling and simulation of steam jet ejectors," *Desalination*, vol. 123, no. 1, pp. 1–8, 1999.

[18] Y. A. Çengel and M. A. Boles, *Thermodynamics: An Engineering Approach*, McGraw-Hill, New York, NY, USA, 8th edition, 2015, ISBN 978-0-07-339817-4.

[19] J. R. Partington, *An Advanced Treatise on Physical Chemistry, Fundamental Principles. The Properties of Gases*, Vol. 1, Longmans, Green & Co., London, UK, 1949.

[20] J. Fan, J. Eves, H. M. Thompson et al., "Computational fluid dynamic analysis and design optimization of jet pumps," *Computers & Fluids*, vol. 46, no. 1, pp. 212–217, 2011.

[21] F. R. Menter, "Two-equation eddy-viscosity turbulence models for engineering applications," *AIAA Journal*, vol. 32, no. 8, pp. 1598–1605, 1994.

[22] V. L. Streeter and E. B. Wylie, *Fluid Mechanics*, McGraw-Hill Higher Education, New York, NY, USA, 9th edition, 1988.

10

Evaluation of Separate and Simultaneous Kinetic Parameters for Levulinic Acid and Furfural Production from Pretreated Palm Oil Empty Fruit Bunches

Misri Gozan (iD),[1] Jabosar Ronggur Hamonangan Panjaitan,[1] Dewi Tristantini,[1] Rizal Alamsyah,[2] and Young Je Yoo[3]

[1]*Chemical Engineering Department, Faculty of Engineering, Universitas Indonesia, Depok 16424, Indonesia*
[2]*Centre for Agro-Based Industry, Bogor 16122, Indonesia*
[3]*School of Chemical & Biological Engineering, Seoul National University, Seoul, Republic of Korea*

Correspondence should be addressed to Misri Gozan; misrigozan@gmail.com

Academic Editor: Doraiswami Ramkrishna

Palm oil empty fruit bunches (POEFBs) can be converted into levulinic acid (LA) and furfural, which are among the top building-block chemicals. The purpose of this study was to investigate separate and simultaneous kinetic model parameters for LA and furfural production from POEFBs, which were pretreated by soaking in aqueous ammonia (SAA). The highest LA yield, which was obtained at a reaction temperature of 170°C after 90 min in an acidic solution with a concentration of 1 M, was 52.1 mol%. The highest furfural yield was 27.94 mol%, which was obtained at a reaction temperature of 170°C after 20 min in an acidic solution with a concentration of 0.5 M. SAA pretreatment affected activation energy in glucose degradation reactions and favoured direct conversion of hemicellulose to furfural. The activation energy of LA production ($E_a k_{HMF}$) increases with higher acid catalyst concentration, and the activation energy of furfural production ($E_a k_{XYN}$) decreases with higher acid concentration. These trends in the activation energy occurred in both separate and simultaneous kinetic models. Simultaneous kinetic model is better to calculate kinetic parameters of LA and furfural production than separate kinetic models because the simultaneous kinetic model had a lower sum of square error (SSE) when estimating kinetic parameters.

1. Introduction

The palm oil industry continues to grow in response to increased consumption and demand for palm oil. Thus, the palm oil industry represents a major potential source of biomass. During the processing of palm oil, palm oil empty fruit bunches (POEFBs), which are a type of lignocellulosic biomass, are produced as waste. Lignocellulosic biomass consists of cellulose, hemicellulose, lignin, and minor components, such as ash, proteins, and extractives [1].

The largest component of POEFBs is cellulose. Cellulose chains dissociate to produce cellulose fibrils in wood components [2]. Cellulose is surrounded by hemicellulose and lignin, which are the other major fractions from lignocellulosic biomass [3]. Hemicellulose is a branched amorphous polymer compound, of which the main monomer is xylose [4].

Lignin consists of a complex and amorphous three-dimensional network of phenolic polymers; it acts as a support structure for plant cell walls, making these walls resistant to microbial attack [5].

Pretreatment of lignocellulosic biomass is necessary to facilitate hydrolysis of its cellulose and hemicellulose fractions. The pretreatment process is important because it affects production costs. Due to its low boiling point, ammonia can potentially be used for pretreatment of lignocellulosic biomass by the processing industry, and it can easily be recycled back into the process via ventilation [6]. In addition, ammonia is favoured by many studies, including this one, because it is inexpensive and reduces the formation of by-products [7].

Hydrolysis of lignocellulose to produce sugar monomers or other degradation compounds generally uses acid as a catalyst. Acid hydrolysis of lignocellulosic biomass produces solid and liquid fractions. The solid fraction is rich in cellulose and can be used for bioethanol production. The liquid fraction is rich in hemicellulose, hydrolysate (xylose and arabinose), and fermentation inhibitor compounds [8]. The two fermentation inhibitors produced from acid hydrolysis are levulinic acid (LA) and furfural. LA and furfural are two of the top 12 chemicals, with their potential uses as building blocks for a variety of chemicals and derivatives [9].

LA is a short-chain fatty acid with a ketone carbonyl group and an acidic carboxyl group, enabling it to produce a variety of chemical substances [10]. LA production commences with depolymerisation of the biomass cellulose fraction into oligosaccharides and glucose [11]. Next, six-carbon sugars are hydrolysed to 5-hydroxymethylfurfural (HMF), which is then dehydrated into LA and formic acid [12, 13]. LA has applications in biofuels, in which it is converted into γ-valerolactone, 2-methyltetrahydrofuran, and levulinate esters. Both γ-valerolactone (GVL) and 2-methyltetrahydrofuran can be blended directly with gasoline as alternative fuel for vehicles. Levulinate esters can be used in biodiesel because they have the same properties as fatty acid methyl esters (FAME) [14].

Biofine technology refers to a traditional technology used to produce LA from lignocellulosic biomass [15]. The technology consists of a two-stage process. In the first stage, hexose sugar is converted to HMF using a sulfuric acid catalyst (1–4%) at temperatures of 200–230°C and ambient pressure (20–25 bar) for a few seconds [12]. In the second stage, the product produced in the first stage is hydrolysed into LA at temperatures of around 190–220°C and 10–15 bar pressure for 15–30 min [12].

Furfural derived from lignocellulose has two functional groups: aldehyde and a furan ring system. Furfural can be utilized in various applications, including the production of chemicals and fuels [16]. Furfuryl alcohol is one of the most widely used furfural derivatives. Furfural production generally commences with the initial hydrolysis of the hemicellulose fraction of lignocellulose into pentose, which is then dehydrated into furfural [17]. In this process, in the batch mode, a sulfuric acid catalyst reacts with the biomass at temperatures of 170–185°C to obtain a furfural yield of approximately 40–50%. Examples of commercial furfural production processes are Quaker Oats, Westpro-modified Huaxia Technology, SupraYield, and Vedernikov [15].

Many parameters, such as temperature, acid concentration, and biomass characteristics, influence the rate of LA and furfural reactions. Previous studies have investigated the kinetic reactions of LA and furfural production in various types of biomass using sulfuric acid catalysts [16–26]. Chin et al. [20] performed a kinetic study of POEFB acid hydrolysis, which produced xylose decomposition products and LA. Dussan et al. [21] performed a kinetic study of furfural and LA production from *Miscanthus x giganteus*. These kinetic studies used separate kinetic models; the LA kinetic model was evaluated using a cellulose degradation kinetic model, and the furfural kinetic model was evaluated using a hemicellulose degradation kinetic model. However, LA and furfural kinetic models should be evaluated using

a combined kinetic model because they can be produced simultaneously from biomass in the acid hydrolysis process. Therefore, evaluation of the kinetic model of simultaneous production of furfural and levulinic acid in acid hydrolysis process is important.

There have been no studies evaluating the kinetic parameters of LA and furfural from POEFBs pretreated with ammonia in both separate and simultaneous kinetic models. Therefore, the purpose of this study was to investigate separate and simultaneous kinetic parameters of LA and furfural production using the sulfuric acid hydrolysis process from POEFBs pretreated (soaked) with aqueous ammonia.

2. Materials and Methods

2.1. Biomass and Chemicals. POEFBs were obtained from the palm oil industry PTPN 5 Kertajaya, Banten, Indonesia. The particle size of POEFBs was reduced to 20 mesh, and the POEFBs were washed, dried at temperature 100°C until the moisture content reached 1–5%, and then stored in a plastic bag until further use. Ammonia solution (25%), which was supplied by Merck, Germany, was used as the pretreatment chemical. Sulfuric acid (96.1%) from Mallinckrodt, England, was used as the acid catalyst. Standard analytical grades of glucose, xylose, HMF, furfural, and LA were purchased from Sigma-Aldrich, United States.

2.2. Soaking in Aqueous Ammonia (SAA) Pretreatment. POEFBs were soaked with aqueous ammonia solution (13.13%) for 14 hours at room temperature (27°C) [27]. The solid-to-liquid ratio was 1 : 6. After soaking for 14 hours, the pretreated POEFBs were washed with water to remove ammonia until neutrality and dried until the moisture content reached 1–5%.

2.3. Kinetic Experiments. The kinetic experiments were performed using a pressurized reactor (1 L volume, 16-bar max pressure, and 100 rpm impeller velocity). The reactants were pretreated POEFBs and 0.5 M and 1 M sulfuric acid solution with a mass ratio of 1 : 20. First, pretreated POEFBs and water were added to the reactor. Next, the reactor was tightly sealed and heated. After the reactor reached the desired temperature, a sulfuric acid catalyst was injected into the reactants. The start of the reaction time was when the sulfuric acid catalyst was released into the reactor. Duplicate samples were obtained after 10, 20, 30, 45, 90, and 120 min at 150°C, 160°C, and 170°C (120 min reaction).

2.4. Analytical Methods. Sample compositional analysis of POEFBs before and after the ammonia pretreatment was determined by referring to the method of the National Renewable Energy Laboratory (NREL) [28, 29], and scanning electron microscopy (SEM) of POEFBs was performed to determine the condition of the fibres. The liquid product concentration from kinetic experiments such as glucose, xylose, HMF, furfural, and LA was analysed using high-performance liquid chromatography (HPLC), with an

Aminex HPX-87H ion exclusion column (Bio-Rad, Life Science Group Hercules, CA). The eluent was 0.006 N of H_2SO_4, and the flow rate was 0.6 ml/min. HPLC detector was a refractive index detector, and the temperatures of the detector and column were 60°C.

2.5. Kinetic Modeling. This research used two steps to evaluate kinetic parameters of LA and furfural production which are separate kinetic evaluation and simultaneous kinetic evaluation in the same experimental data. Separate kinetic evaluation assumed that there is no interference between LA and furfural kinetic reaction, so the kinetic parameter calculation of LA and furfural would be done in a separate kinetic model. On the other side, the simultaneous kinetic evaluation calculated LA and furfural kinetic reaction in one kinetic model. The effects of other components in POEFBs such as lignin and ash were negligible from LA and furfural kinetic models, and kinetic parameter calculation since ammonia pretreatment was done before kinetic experiments.

2.6. Separate Kinetic Model of LA and Furfural. In separate kinetic evaluation, LA production was independently evaluated using kinetic model of Girisuta et al. [22], who examined the kinetics of acid hydrolysis of sugarcane bagasse in the production of LA. The reaction model of Girisuta et al. [22] is consecutive reactions from glucan to glucose and humin, then to HMF before finally transforming to LA + FA. The model can be computed with the following kinetic equations:

$$\frac{dC_{glucan}}{dt} = -k_{GLN} \cdot C_{glucan},$$

$$\frac{dC_{glucose}}{dt} = k_{GLN} \cdot C_{glucan} - k_{GLC1} \cdot C_{glucose} - k_{GLC2} \cdot C_{glucose},$$

$$\frac{dC_{HMF}}{dt} = k_{GLC2} \cdot C_{glucose} - k_{HMF} \cdot C_{HMF},$$

$$\frac{dC_{LA(+FA)}}{dt} = k_{HMF} \cdot C_{HMF},$$

$$\frac{dC_{HUM}}{dt} = k_{GLC2} \cdot C_{glucose}.$$

$$(1)$$

As the HPLC analysis of xylose and furfural did not reveal the presence of xylose in most of the samples, xylose was not included in this furfural kinetic model. The furfural kinetic model reaction scheme for the first stage is shown in Figure 1. From Figure 1, we can compute the following kinetic equations:

$$\frac{dC_{xylan}}{dt} = -k_{XLN} \cdot C_{xylan},$$

$$\frac{dC_{furfural}}{dt} = k_{XYN} \cdot C_{xylan} - k_{FUR} \cdot C_{furfural}, \quad (2)$$

$$\frac{dC_{RES}}{dt} = k_{FUR} \cdot C_{furfural}.$$

2.7. Simultaneous Kinetic Model of LA and Furfural. Simultaneous kinetic evaluation would calculate LA and furfural production in one kinetic model. The reaction

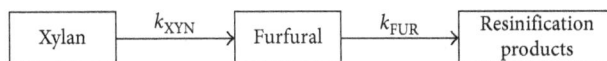

FIGURE 1: Reaction scheme of furfural production.

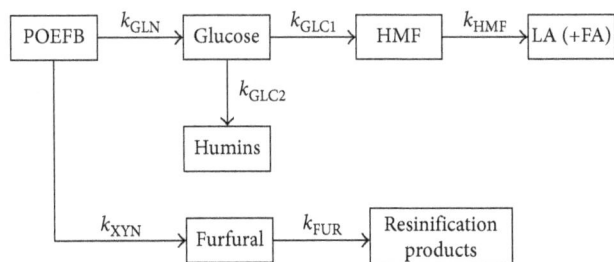

FIGURE 2: Simultaneous reaction scheme of LA and furfural production.

scheme is shown in Figure 2. From Figure 2, we can compute the following kinetic equations:

$$\frac{dC_{POEFB}}{dt} = -k_{GLN} \cdot C_{POEFB} - k_{XLN} \cdot C_{POEFB},$$

$$\frac{dC_{glucose}}{dt} = k_{GLN} \cdot C_{POEFB} - k_{GLC1} \cdot C_{glucose} - k_{GLC2} \cdot C_{glucose},$$

$$\frac{dC_{HMF}}{dt} = k_{GLC2} \cdot C_{glucose} - k_{HMF} \cdot C_{HMF},$$

$$\frac{dC_{LA(+FA)}}{dt} = k_{HMF} \cdot C_{HMF}, \quad (3)$$

$$\frac{dC_{HUM}}{dt} = k_{GLC2} \cdot C_{glucose},$$

$$\frac{dC_{furfural}}{dt} = k_{XYN} \cdot C_{POEFB} - k_{FUR} \cdot C_{furfural},$$

$$\frac{dC_{RES}}{dt} = k_{FUR} \cdot C_{furfural}.$$

2.8. Kinetic Parameter Estimation for Separate Kinetic Model of LA and Furfural. Reaction rate constants were optimized by minimizing errors between the experimental data, such as the concentrations of glucose, xylose, HMF, furfural, and LA at various temperatures and concentrations of acid catalyst with kinetic models. The concentrations were transformed into yields to compensate the large spread in concentrations. The sum of square error (SSE) of LA and furfural model prediction was formulated:

$$\text{SSE} = \sum_x \sum_n \left(Y_{n,predicted} - Y_{n,experiment} \right)^2, \quad (4)$$

where $Y_{n,predicted}$ is the model yield predicted for every sampling time (mol/mol)%, $Y_{n,experiment}$ is the experimental yield for every sampling time (mol/mol)%, and x is the data of glucose, xylose, HMF, furfural, and LA.

The yield of LA from the POEFBs on a molar basis (Y_{LA}) was defined as the ratio of the LA concentration in the acid hydrolysis product (C_{LA}) to the initial C6 sugar concentration in the POEFBs (C_6):

$$Y_{LA} \text{ (mol/mol)} \% = \frac{C_{LA}}{C_6} \times 100\%. \tag{5}$$

The yield of furfural from the POEFBs on a molar basis (Y_F) was defined as the ratio of the furfural in the acid hydrolysis product ($C_{furfural}$) to the initial C5 sugar concentration in the POEFBs (C_5):

$$Y_F \text{ (mol/mol)} \% = \frac{C_{furfural}}{C_5} \times 100\%. \tag{6}$$

Minimization of the sum of square error values was done by fminsearch optimization using MATLAB optimization routine to get LA and furfural reaction rate constants at optimum slope. The value of the reaction rate constants was used to determine the value of the preexponential factor (A) and activation energy (E_a), using the Arrhenius equation. Systematically, the relationship between the reaction rate constant (k) and temperature (T) was expressed by the Arrhenius equation:

$$k = A \exp\left(-\frac{E_a}{RT}\right), \tag{7}$$

where k is the reaction rate constant, A is the preexponential factor, E_a is the activation energy (kJ/mol), R is the gas constant (kJ/mol·K), and T is the temperature (K).

The effect of acid concentration on LA and furfural production can be predicted using acid reaction order equation that in this equation, there is correlation between reaction rate constant and acid concentration, which can be formulated:

$$k = A_x [C]^n, \tag{8}$$

where k is the reaction rate constant, A_x is the preexponential factor from acid reaction order equation, C is the acid concentration (M), and n is the acid reaction order. The calculation of acid reaction order was counted based on (8) from the average value of kinetic experiment temperature.

2.9. Kinetic Parameter Estimation for Simultaneous Kinetic Model of LA and Furfural. The calculation of kinetic parameters for simultaneous kinetic model of LA and furfural was same with the separate kinetic model, but the yield from experimental data was obtained from mass basis.

The yield of LA and furfural from the POEFBs on mass basis was defined as the ratio of the LA and furfural mass concentration in the acid hydrolysis product (m_{LA} and $m_{furfural}$) to the mass of cellulose and hemicellulose in the POEFBs (m_b):

$$Y_{LA} \text{ (gr/gr)} \% = \frac{m_{LA}}{m_b} \times 100\%,$$
$$Y_F \text{ (gr/gr)} \% = \frac{m_{furfural}}{m_b} \times 100\%. \tag{9}$$

3. Results and Discussion

3.1. SAA Pretreatment. The purpose of the pretreatment process was to improve the conversion of cellulose and hemicellulose to LA and furfural. The results of the compositional analysis of different types of biomass with SAA pretreatment are shown in Table 1.

According to Table 1, the composition of the POEFBs before and after the SAA pretreatment did not differ greatly, with just a small reduction in lignin composition (from 22.8% to 21.8%). The composition of cellulose and hemicellulose in the untreated POEFBs increased slightly from 39.3% to 29.8%, as compared to the SAA pretreated POEFBs (42% and 32.0%, resp.). As shown in Table 1, previous studies also reported insignificant changes in lignin composition in untreated and SAA pretreated samples at various temperatures, soaking times, and ammonia concentrations [30–33].

Lignin can affect LA and furfural production. However, according to Daorattanachai et al. [34], lignin can promote the isomerization reaction of glucose to fructose, which can then be dehydrated to HMF using a phosphoric acid catalyst. As is well known, lignin removal from lignocellulose biomass is difficult and costly. Therefore, in this study, lignin was not removed from the biomass.

In this study, based on Zulkiple et al. [27], SSA pretreatment produced more sugar than raw POEFBs after enzyme hydrolysis. Ammonia pretreatment is an alkali pretreatment. The alkali agent saponifies the ester bonds in the xylan backbone, resulting in the production of carboxyl groups and the breakdown of lignin-hemicellulose bonds [6].

SEM images before and after the SAA pretreatment are presented in Figure 3. As shown in Figure 3(a), prior to the SSA pretreatment, the POEFB fibres looked hard and stiff, with a flat, smooth surface structure. After the SAA pretreatment, the lignin-carbohydrate bonds in the POEFB fibres broke down and the surface contained pores, indicating that the surface area of the POEFBs had increased (Figure 3(b)). Thus, SAA pretreatment appears able to break the bonds between lignin and carbohydrate. As reported earlier, ammonia will also increase the accessibility of cellulose because it acts as a swelling agent [35].

3.2. Evaluation of the Separate Kinetic Parameters of LA and Furfural Production

3.2.1. Evaluation of the Kinetic Parameters of LA Production. LA optimization results are presented in Figure 4. According to Figure 4(f), the highest yield of LA in the kinetic experiments was 52.1 mol%; this was obtained at a reaction temperature of 170°C, after a 90 min reaction, using an acid concentration of 1 M. Girisuta et al. [22] and Dussan et al. [21] examined the kinetics of LA production from sugarcane bagasse and a *Miscanthus giganteus* cross, respectively, at a temperature of 150–200°C, with a sulfuric acid catalyst (0.1–0.5 M), and obtained yields of around 60–70% mol. The lower yield found in the present study may be due to the operating temperature, which was quite low (150–170°C),

TABLE 1: Compositional analysis results of different types of biomass with SAA pretreatment.

Reference	[30]		[31]		[32]		[33]		This study	
Biomass	Barley hull		Switchgrass		Manure fibres		Poplar		POEFBs	
t^a	168		8		72		72		14	
T^b	30		40		22		22		27	
C^c (%)	30		15		15		32		13.13	
Composition (%)	Untreated	SAA	Untreated	SAA	Untreated	SAA	Untreated	SAA	Untreated	SAA
α–Cellulose (glucan)	33.6	33.6	37.0	33.2	17.1	18.7	32.7	29.1	39.3	42.0
Hemicellulose (xylan)	30.5	30.0	15.4	17.1	12.7	10.9	16.8	12.0	29.8	32.0
Lignin	19.3	13.4	25.0	22.1	24.8	22.3	34.2	34.4	22.8	21.8
Ash	NA	NA	NA	NA	NA	NA	NA	NA	1.7	1.8

$^a t$: soaking time (hours); $^b T$: soaking temperature (°C); $^c C$: ammonia concentration (%).

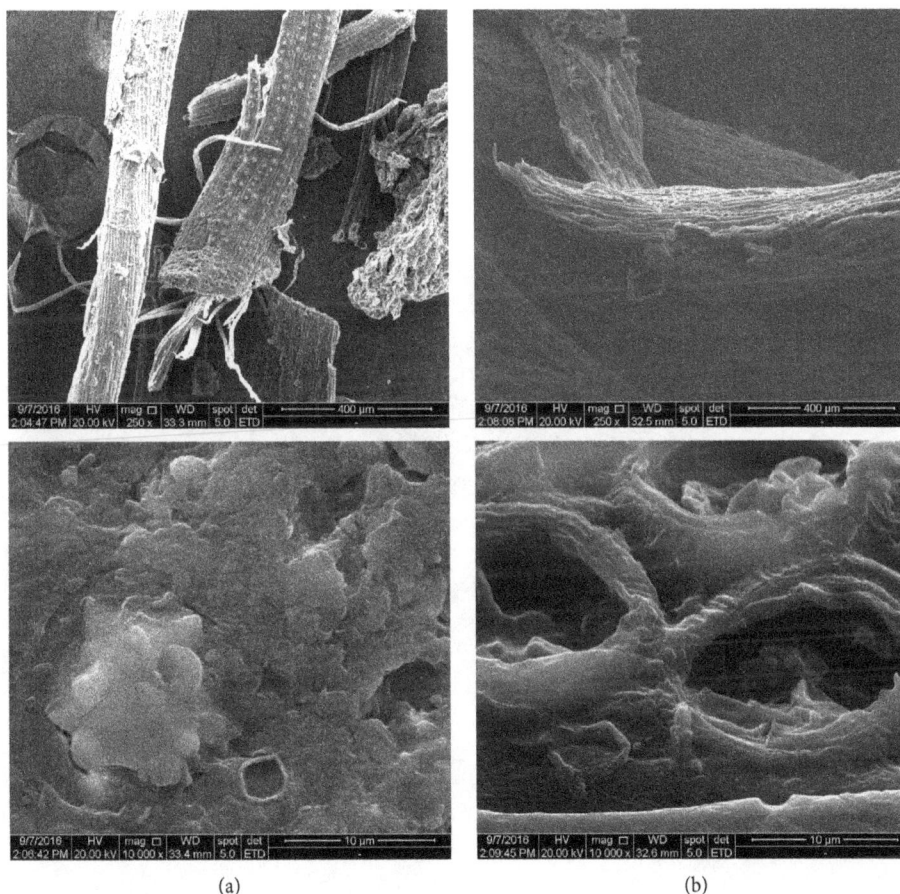

FIGURE 3: SEM images of OPEFB: (a) before the ammonia pretreatment; (b) after the ammonia pretreatment. Images of (a) and (b) have 2 different magnifications: with 400 μm (upper images) and 10 μm (lower images).

although the acid catalyst concentration was high (0.5–1 M) compared to that used by Girisuta et al. [22] and Dussan et al. [21]. This result shows that temperature affects LA production.

Comparison of activation energy values of LA production at different acid concentrations can be seen in Table 2. According to Table 2, activation energy tended to decrease in higher acid concentrations because sulfuric acid functioned as a catalyst to reduce the activation energy for LA production.

The activation energy for glucose formation was lower (132.37 kJ/mol and 108.48 kJ/mol) than the activation energy

for HMF formation (212.40 kJ/mol and 119.49 kJ/mol) in 0.5 and 1 M acid concentration, indicating that the glucose formation reaction was faster than the HMF formation reaction. This proved that higher temperature made the rate of the HMF formation reaction faster.

The HMF formation reaction at 0.5 and 1 M acid concentrations had higher activation energies (212.40 kJ/mol and 119.49 kJ/mol) and acid reaction order (2.00) than the activation energies (188.31 kJ/mol and 62.12 kJ/mol) and acid reaction order (1.66) of humin formation. These indicated that more glucose decomposed into HMF rather than into humins at higher temperatures. Likewise, the HMF

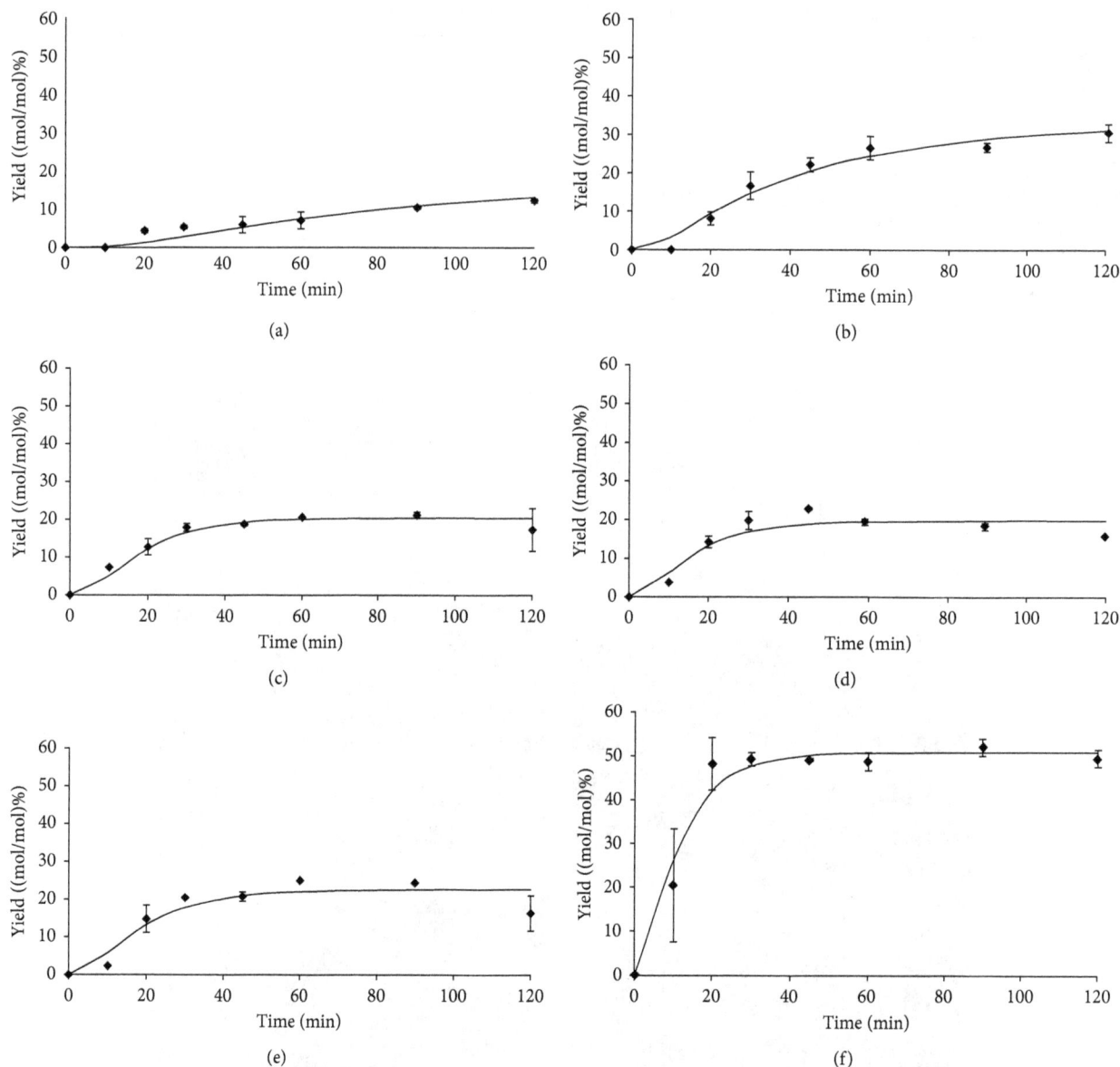

FIGURE 4: LA optimization results between experimental data (◆, LA) and kinetic model (lines): (a) 150°C and 0.5 M; (b) 150°C and 1 M; (c) 160°C and 0.5 M; (d) 160°C and 1 M; (e) 170°C and 0.5 M; and (f) 170°C and 1 M.

formation reaction had the highest activation energy and acid reaction order compared to all other reactions. These indicated that both temperature and acid concentration have greater effects on this reaction than on other reactions.

The effect of acid concentration also had an important role in humin formation, which can be seen from the second highest acid reaction order value after HMF formation. The higher acid concentration triggered the formation of humins. Humin formation is undesirable in the production of LA because this reaction competes with HMF formation when using glucose as the raw material. Therefore, temperature and acid concentration become important factors to prevent humin formation.

The lowest activation energy was recorded for LA formation (42.55 kJ/mol and 56.08 kJ/mol), suggesting that HMF was quickly converted into LA. The HMF concentration

during the kinetic experiments was low due to rapid HMF degradation and LA formation.

The comparison of kinetic parameters in LA production (0.5 M acid concentration) with other research is shown in Table 3. The results from this study were compared with Girisuta et al. [22] and Dussan et al. [21], who examined the

TABLE 2: Kinetic parameters for LA production.

Reaction rate constants	Activation energy E_a (kJ/mol)		Acid reaction order n (–)
	0.5 M	1 M	
k_{GLN}	132.37	108.48	0.78
k_{GLC1}	212.40	119.49	2.00
k_{GLC2}	188.31	62.12	1.66
k_{HMF}	42.55	56.08	1.15

TABLE 3: Comparison of kinetic parameters for LA production.

	Reaction rate constants	[22]	[21]	This study
Activation energy E_a (kJ/mol)	k_{GLN}	144.85	188.9	132.37
	k_{GLC1}	152.14	155.5	212.40
	k_{GLC2}	161.41	186.2	188.31
	k_{HMF}	101.63	121.3	42.55
Acid reaction order n (–)	k_{GLN}	1.57	1.40	0.78
	k_{GLC1}	1.14	1.39	2.00
	k_{GLC2}	1.08	0.90	1.66
	k_{HMF}	1.32	1.95	1.15

kinetics of LA production from sugarcane bagasse and a *Miscanthus giganteus* cross, respectively, at a temperature of 150–200°C with a lower sulfuric acid catalyst (0.1–0.5 M).

The activation energy of glucose formation in the present study (132.37 kJ/mol) was not too different from the studies by Girisuta et al. [22] and Dussan et al. [21], who reported figures of 144.85 kJ/mol and 188.9 kJ/mol, respectively. The differences could be due to the different characteristics of the biomass used.

In the present study, the lowest activation energy for LA formation was the same as that recorded in other research. Girisuta et al. [22] and Dussan et al. [21] reported low activation energies of about 101.63 kJ/mol and 121.3 kJ/mol, respectively. In this study, the activation energy conversion of LA formation was much smaller (42.55 kJ/mol), indicating that LA readily forms at a lower activation energy.

SAA pretreatment was the most likely cause of the higher activation energy in HMF formation (212.40 kJ/mol) as compared to humin formation (188.31 kJ/mol) in this study. In contrast, Girisuta et al. [22] and Dussan et al. [21] reported that the activation energy of HMF formation was lower than the activation energy of humin formation. The ammonia pretreatment likely improved the accessibility of the glucose, thereby resulting in increased production of HMF rather than humins. The results suggested that, at higher temperatures, the reaction pathway will tend to lead HMF formation rather than humins. The humins in this study may be products from cellulose conversions other than HMF.

The acid concentration had a larger effect on LA formation, which is consistent with the data in earlier studies. Girisuta et al. [22] and Dussan et al. [21] reported acid reaction orders of approximately 1.32 and 1.95, respectively, which were relatively similar to the acid reaction order in the present study (1.15). Therefore, higher acid concentration will enhance LA formation.

3.2.2. Evaluation of the Kinetic Parameters of Furfural Production. Furfural optimization results are presented in Figure 5. As shown in Figure 5(e), the largest furfural yield in this study (27.94 mol%) was obtained at a temperature of 170°C, with a 20 min reaction and a 0.5 M acid concentration. The furfural yield in this study was low. This was likely due to the high temperature (170°C), resulting in the formation of degradation products (formic acid and tar), as reported previously by Danon et al. [36]. In addition, the presence of lignin may have affected furfural production. Lamminpaa

et al. [37] showed that lignin can increase the pH of a reactant solution, leading to low conversion of xylose into furfural.

In the present study, xylose was detected only at a temperature of 150°C, after a 60 min reaction time in a 0.5 M acid concentration. Xylose was undetectable at higher temperatures, higher acid concentrations, and longer reaction times, in contrast to the findings of Dussan et al. [21] and Chin et al. [20]. Both studies detected xylose at higher reaction temperatures of about 200°C and 180°C, respectively. The kinetic model by Dussan et al. [21], which incorporated xylose, was not in line with the experimental data. Therefore, xylose was not included in the furfural kinetic model.

The SAA pretreatment explains the differences between xylose production in this study as compared to those observed in the kinetic experiments conducted by Dussan et al. [21] and Chin et al. [20]. Hemicellulose is an amorphous polymer, which is more easily degraded than cellulose. In the presence of ammonia pretreatment, the increased accessibility of hemicellulose means it is likely to be converted into a variety of products. Therefore, during acid hydrolysis, xylose located in hemicellulose amorphous fibres will be directly converted into furfural.

In this study, according to Table 4, the activation energy of furfural formation (76.76 kJ/mol) was higher than the activation energy of furfural decomposition (9.89 kJ/mol) in 0.5 M acid concentration. This means that, at higher temperatures, reactions can produce more furfural. The largest furfural yield in this study (27.94 mol%) was obtained at a temperature of 170°C in 0.5 M acid concentration.

In contrast, the higher acid concentration (1 M) made the activation energy of furfural formation (59.22 kJ/mol) lower than the activation energy of furfural decomposition (77.08 kJ/mol). This means that, at higher temperatures, reactions can degrade furfural, as indicated by the fact that the furfural yield was lower at a temperature of 170°C in 1 M acid concentration in the present study.

3.3. Evaluation of Simultaneous Kinetic Parameters of LA and Furfural Production. The results of the evaluation of simultaneous kinetic parameters of LA and furfural are presented in Figure 6, and the comparison of separate and simultaneous LA and furfural kinetic parameters at different acid concentrations can be seen in Table 5, which shows that the activation energy of LA production was reduced because of the sulfuric acid catalyst. The LA activation energy trends are the same in separate and simultaneous LA production, except for $E_a k_{GLN}$.

As shown in Table 5, the activation energy of glucose formation ($E_a k_{GLN}$) in the simultaneous kinetic model tended to increase when the acid concentration was higher. The higher activation energy of the glucose formation ($E_a k_{GLN}$) indicated that there were complex hydrolysis interaction reactions of pretreated POEFBs when LA and furfural kinetic parameters were calculated simultaneously. Glucose was the main monomer derived from the pretreated POEFB cellulose fraction, which was the highest fraction (42%) in pretreated POEFBs; therefore, glucose was the largest monomer composition in pretreated POEFBs.

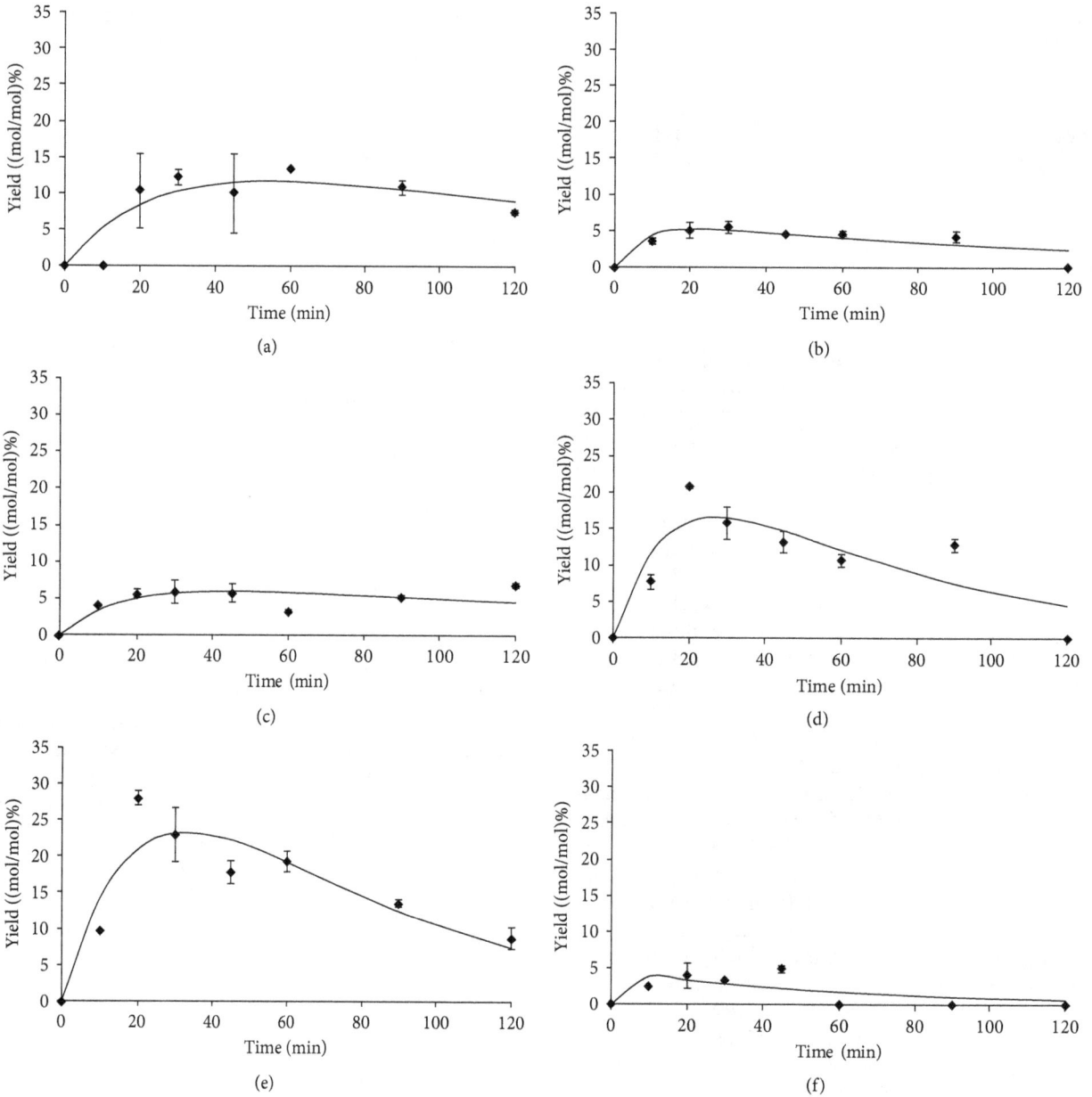

FIGURE 5: Furfural optimization results between experimental data (◆, furfural) and kinetic model (lines): (a) 150°C and 0.5 M; (b) 150°C and 1 M; (c) 160°C and 0.5 M; (d) 160°C and 1 M; (e) 170°C and 0.5 M; and (f) 170°C and 1 M.

The degradation process from cellulose fraction into glucose was a complex reaction in biomass hydrolysis. In contrast, LA was the product of a series of reaction steps in the acid hydrolysis of biomass. LA formation was a complex reaction produced from C6 sugars (glucose and fructose) in lignocellulose biomass [12]. Therefore, the activation energy of glucose formation ($E_a k_{GLN}$) depends on reaction conditions and the interaction between compounds that affect its degradation into other products (HMF and humins).

According to Table 5, the activation energy of furfural formation (171.74 kJ/mol) was higher than the activation energy of furfural decomposition (98.21 kJ/mol) in 0.5 M acid concentration, and the activation energy of furfural formation (88.81 kJ/mol) was lower than the activation energy of furfural

TABLE 4: Kinetic parameters for furfural production.

Reaction rate constants	Activation energy E_a (kJ/mol)		Acid reaction order n (–)
	0.5 M	1 M	
k_{XYN}	76.76	59.22	1.02
k_{FUR}	9.89	77.08	1.64

decomposition (175.14 kJ/mol) in 1 M acid concentration. This means that, at higher temperatures, reactions can produce more furfural in 0.5 acid concentration, but furfural would be degraded in 1 M acid concentration. Therefore, it was shown that the interaction effect on biomass hydrolysis is very low for furfural formation because hemicellulose is

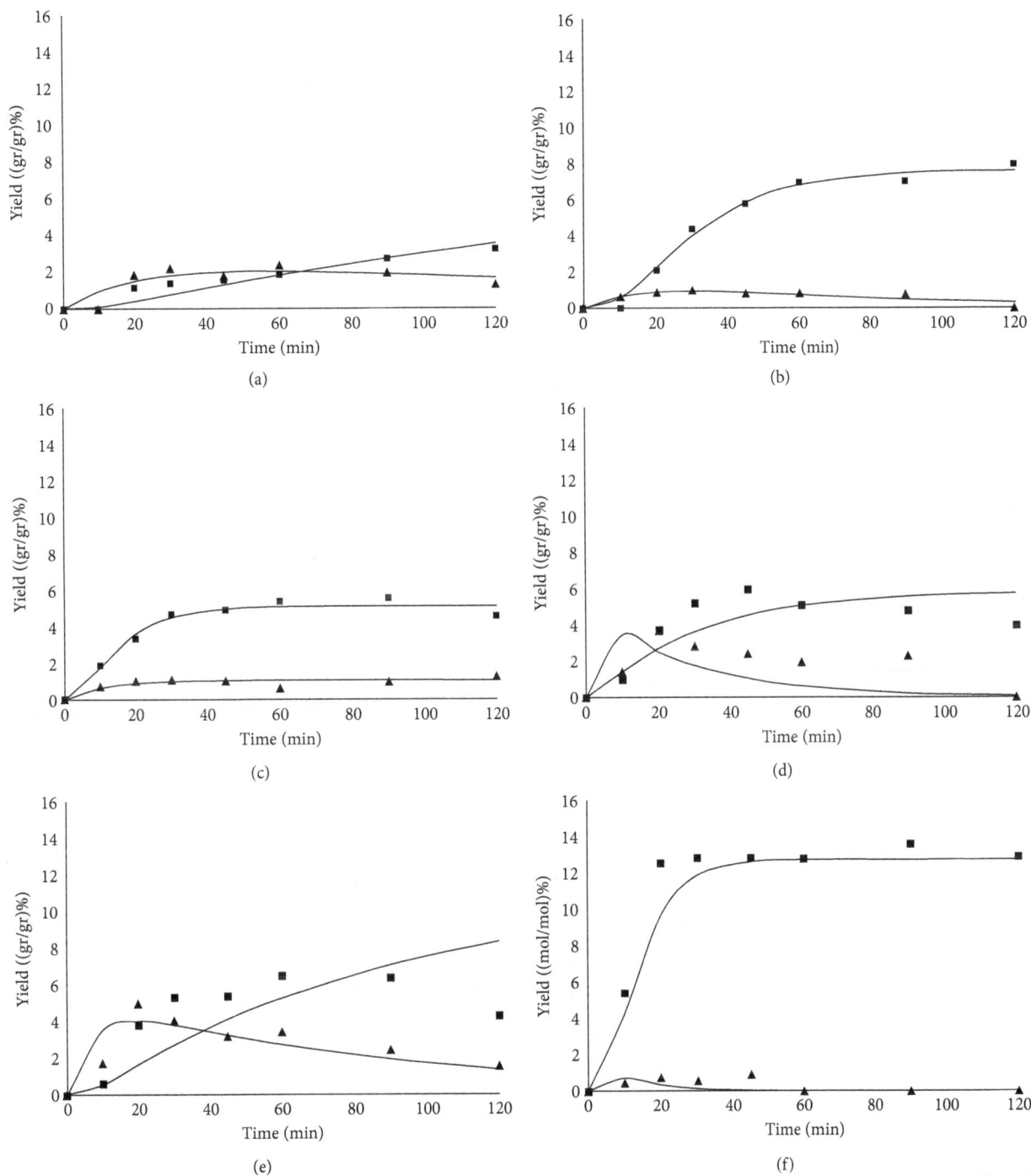

FIGURE 6: LA and furfural simultaneous optimization results between experimental data (■, LA; ▲, furfural) and simultaneous kinetic model (lines): (a) 150°C and 0.5 M; (b) 150°C and 1 M; (c) 160°C and 0.5 M; (d) 160°C and 1 M; (e) 170°C and 0.5 M; and (f) 170°C and 1 M.

comprised of amorphous fibres that are easily degraded in the biomass hydrolysis process.

The acid reaction order in the simultaneous kinetic model of LA and furfural production showed higher values than that in the separate kinetic model. This suggested that acid had an important role as a catalyst in the hydrolysis reaction of pretreated POEFBs and LA-furfural production.

According to Table 5, the activation energy of LA and furfural acid production from simultaneous kinetic model

calculation had the same trend as the separate kinetic model calculation. In separate and simultaneous kinetic models, the activation energy of LA production ($E_a k_{HMF}$) increased with higher acid catalyst concentration and the activation energy of furfural production ($E_a k_{XYN}$) decreased with higher acid concentration.

Simultaneous kinetic model is better than separate kinetic models for evaluating kinetic parameters in LA and furfural production. Glucan and xylan in separate kinetic

TABLE 5: Kinetic parameters for separate and simultaneous LA and furfural production.

| Reaction rate constants | Activation energy (E_a) (kJ/mol) | | | | Acid reaction order n (–) | |
| | Separate | | Simultaneous | | | |
	0.5 M	1 M	0.5 M	1 M	Separate	Simultaneous
k_{GLN}	132.37	108.48	25.95	61.18	0.78	4.75
k_{GLC1}	212.40	119.49	99.27	60.36	2.00	3.51
k_{GLC2}	188.31	62.12	18.38	4.25	1.66	1.37
k_{HMF}	42.55	56.08	33.27	83.97	1.15	2.18
k_{XYN}	76.76	59.22	171.74	88.81	1.02	2.37
k_{FUR}	9.89	77.08	98.21	175.14	1.64	1.35

models of LA and furfural were combined into one kinetic model (Figure 2) because glucan and xylan are fractions in POEFBs; this is the main reason simultaneous kinetic model was developed in this study. The sum of square error (SSE) of the optimization result between the experimental data and the models in the kinetic parameter estimation was lower in the simultaneous kinetic model (14.32) compared to those in the separate kinetic models (88.23 for LA and 44.04 for furfural). Therefore, the simultaneous kinetic model is better than separate kinetic models.

4. Conclusion

This study evaluated separate and simultaneous kinetic models of LA and furfural production from POEFBs that were pretreated with ammonia. A kinetic experiment was performed using a pressurized reactor at a temperature of 150–170°C, with a sulfuric acid catalyst at concentrations of 0.5 M and 1 M. In the kinetic experiments, the greatest LA yield was 52.1 mol%, which was obtained at a reaction temperature of 170°C, after a 90 min reaction, using an acid concentration of 1 M. The highest furfural yield was 27.94 mol%, which was obtained at a temperature of 170 C, after a 20 min reaction, using an acid concentration of 0.5 M. SAA pretreatment affected activation energy in glucose degradation reactions and favoured direct conversion of hemicellulose to furfural. Based on the evaluation of the kinetic parameters, the simultaneous kinetic model has been shown to have the same trends as the separate kinetic models in LA and furfural production: the activation energy of LA production ($E_a k_{HMF}$) increases with higher acid catalyst concentration, and the activation energy of furfural production ($E_a k_{XYN}$) decreases with higher acid concentration. Higher reaction temperature and acid concentration will increase LA production. However, higher acid concentration can reduce furfural production because of furfural decomposition. Based on the lower sum of square error (SSE) of the optimization result between the experimental data and the models in the kinetic parameter estimation, simultaneous kinetic model is better to calculate kinetic parameters of LA and furfural production than separate kinetic models.

Nomenclature

A: Preexponential factor (s^{-1})
C_5:   C5 sugar concentration (mol/L)
C_6:   C6 sugar concentration (mol/L)
$C_{furfural}$:   Furfural concentration (mol/L)
C_{glucan}:   Glucan concentration (mol/L)
$C_{glucose}$:   Glucose concentration (mol/L)
C_{HMF}:   5-Hydroxylmethylfurfural concentration (mol/L)
C_{HUM}:   Humin concentration (mol/L)
C_{LA}:   Levulinic acid concentration (mol/L)
C_{RES}:   Resinification product concentration (mol/L)
C_{xylan}:   Xylan concentration (mol/L)
E_a: Activation energy (kJ/mol)
FA: Formic acid
FAME: Fatty acid methyl esters
GVL: γ-Valerolactone
H_2SO_4:   Sulfuric acid
HCl: Hydrochloric acid
HMF: 5-Hydroxylmethylfurfural
HPLC: High-performance liquid chromatography
k: Reaction rate constant (s^{-1})
k_{FUR}:   Furfural reaction rate constant (s^{-1})
k_{GLC1}:   Glucose reaction rate constant 1 (s^{-1})
k_{GLC2}:   Glucose reaction rate constant 2 (s^{-1})
k_{GLN}:   Glucan reaction rate constant (s^{-1})
k_{HMF}:   5-Hydroxylmethylfurfural reaction rate constant (s^{-1})
k_{XLN}:   Xylan reaction rate constant (s^{-1})
LA: Levulinic acid
n: Acid reaction order
NREL: National Renewable Energy Laboratory
POEFBs: Palm oil empty fruit bunches
R: Gas constant (kJ/mol·°C)
SAA: Soaking in aqueous ammonia
SEM: Scanning electron microscopy
SSE: Sum of square error
T: Temperature (°C)
Y_F: Furfural yield (mol%)
Y_{LA}:   Levulinic acid yield (mol%)
$Y_{n,experiment}$:   Experimental yield for every sampling time (mol/mol)%
$Y_{n,predicted}$:   Model yield predicted for every sampling time (mol/mol)%.

Conflicts of Interest

The authors declare that they have no conflicts of interest.

Acknowledgments

The authors acknowledge financial support from Indonesia Estate Crop Fund for Palm Oil (BPDPKS; Research Grant no. Peng-01/DPKS.4/2015); from the Ministry of Research, Technology and Higher Education of the Republic of Indonesia through the World Class Professor (WCP) Program (no. 168. A10/D2/KP/2017); and from the USAID through the SHERA program—Centre for Development of Sustainable Region (CDSR).

References

[1] Y. H. Oh, I. Y. Eom, J. C. Joo et al., "Recent advances in development of biomass pretreatment technologies used in biorefinery for the production of bio-based fuels, chemicals and polymers," *Korean Journal of Chemical Engineering*, vol. 32, no. 10, pp. 1945–1959, 2015.

[2] S. H. Mood, A. H. Golfeshan, M. Tabatabaei et al., "Lignocellulosic biomass to bioethanol, a comprehensive review with a focus on pretreatment," *Renewable and Sustainable Energy Reviews*, vol. 27, pp. 77–93, 2013.

[3] X. Ye and Y. Chen, "Kinetic study of enzymatic hydrolysis of paulownia by diluted acid, alkali, and ultrasonic-assisted alkali pretreatments," *Biotechnology and Bioprocess Engineering*, vol. 20, no. 2, pp. 242–248, 2015.

[4] S. Saka, M. V. Munusamy, M. Shibata, Y. Tono, and H. Miyafuji, "Chemical constituents of the different anatomical parts of the oil palm (*Elaeis guineensis*) for their sustainable utilization," in *Proceedings of the JSPS-VCC Group Seminar 2008 on Natural Resources and Energy Environment*, Kyoto, Japan, November 2008.

[5] S. K. Maity, "Opportunities, recent trends and challenges of integrated biorefinery: part 1," *Renewable and Sustainable Energy Reviews*, vol. 43, pp. 1427–1445, 2015.

[6] H. Rabemanolontsoa and S. Saka, "Various pretreatments of lignocellulosics," *Bioresource Technology*, vol. 199, pp. 83–91, 2016.

[7] J. Domanski, S. Borowski, O. M. Mikolajczyk, and P. Kubacki, "Pretreatment of rye straw with aqueous ammonia for conversion to fermentable sugar as a potential substrates in biotechnological processes," *Biomass and Bioenergy*, vol. 91, pp. 91–97, 2016.

[8] B. S. Harish, M. J. Ramaiah, and K. B. Uppuluri, "Bioengineering strategies on catalysis for the effective production of renewable and sustainable energy," *Renewable and Sustainable Energy Reviews*, vol. 51, pp. 533–547, 2015.

[9] S. Choi, C. W. Song, J. H. Shin, and S. Y. Lee, "Biorefineries for the production of top building block chemicals and their derivatives," *Metabolic Engineering*, vol. 28, pp. 223–239, 2015.

[10] X. Lin, Q. Huang, G. Qi et al., "Adsorption behavior of levulinic acid onto microporous hyper-cross-linked polymers in aqueous solution: equilibrium, thermodynamic, kinetic simulation and fixed-bed column studies," *Chemosphere*, vol. 171, pp. 231–239, 2017.

[11] X. Zheng, Z. Zhi, X. Gu, X. Li, R. Zhang, and X. Lu, "Kinetic study of levulinic acid production from corn stalk at mild temperature using FeCl₃ as catalyst," *Fuel*, vol. 187, pp. 261–267, 2017.

[12] A. Morone, M. Apte, and R. A. Pandey, "Levulinic acid production from renewable waste resource: bottlenecks, potential remedies, advancements and applications," *Renewable and Sustainable Energy Reviews*, vol. 51, pp. 548–565, 2015.

[13] J. R. H. Panjaitan and M. Gozan, "Formic acid production from palm oil empty fruit bunches," *International Journal of Applied Engineering Research*, vol. 12, no. 14, pp. 4382–4390, 2017.

[14] K. Yan, C. Jarvis, J. Gu, and Y. Yan, "Production and catalytic transformation of levulinic acid: a platform for speciality chemicals and fuels," *Renewable and Sustainable Energy Reviews*, vol. 51, pp. 986–997, 2015.

[15] S. Kang and J. Yu, "An intensified reaction technology for high levulinic acid concentration from lignocellulosic biomass," *Biomass and Bioenergy*, vol. 95, pp. 214–220, 2016.

[16] K. Yan, G. Wu, T. Lafleur, and C. Jarvis, "Production, properties and catalytic hydrogenation of furfural to fuel additives and value-added chemicals," *Renewable and Sustainable Energy Reviews*, vol. 38, pp. 663–676, 2014.

[17] S. Z. Amraini, L. P. Ariyani, H. Hermansyah et al., "Production and characterization of cellulase from *E. coli* EgRK2 recombinant based on oil palm empty fruit bunches," *Biotechnology and Bioprocess Engineering*, vol. 22, no. 3, pp. 287–295, 2017.

[18] C. Chang, X. Ma, and P. Cen, "Kinetic studies on wheat straw hydrolysis to levulinic acid," *Chinese Journal of Chemical Engineering*, vol. 17, no. 5, pp. 835–839, 2009.

[19] C. Chang, X. Ma, and P. Cen, "Kinetics of levulinic acid formation from glucose decomposition at high temperature," *Chinese Journal of Chemical Engineering*, vol. 14, no. 5, pp. 708–712, 2006.

[20] S. X. Chin, C. H. Chia, Z. Fang, S. Zakaria, X. K. Li, and F. Zhang, "A kinetic study on acid hydrolysis of oil palm empty fruit bunch fibers using a microwave reactor system," *Energy and Fuels*, vol. 28, no. 4, pp. 2589–2597, 2014.

[21] K. Dussan, B. Girisuta, D. Haverty, J. J. Leachy, and M. H. B. Hayes, "Kinetics of levulinic acid and furfural production from Miscanthus x giganteus," *Bioresource Technology*, vol. 149, pp. 216–224, 2013.

[22] B. Girisuta, K. Dussan, H. Haverty, J. J. Leahy, and M. H. B. Hayes, "A kinetic study of acid catalysed hydrolysis of sugar cane bagasse to levulinic acid," *Chemical Engineering Journal*, vol. 217, pp. 61–70, 2013.

[23] B. Girisuta, B. Danon, R. Manurung, L. P. B. M. Janssen, and H. J. Heeres, "Experimental and kinetic modelling studies on the acid-catalysed hydrolysis of the water hyacinth plant to levulinic acid," *Bioresource Technology*, vol. 99, no. 17, pp. 8367–8375, 2008.

[24] B. Girisuta, P. B. M. Janssen, and H. J. Heeres, "Green chemicals: a kinetic study on the conversion of glucose to levulinic acid," *Chemical Engineering Research and Design*, vol. 84, no. 5, pp. 339–349, 2006.

[25] V. S. Lacerda, J. B. L Sotelo, A. C. Guimaraes et al., "A kinetic study on microwave-assisted conversion of cellulose and lignocellulosic waste into hydroxymethylfurfural/furfural," *Bioresource Technology*, vol. 180, pp. 88–96, 2015.

[26] Q. Wei, Z. S. Ping, X. Q. Li, R. Z. Wei, and Y. Y. Jie, "Degradation kinetics of xylose and glucose in hydrolysate containing dilute sulfuric acid," *Chinese Journal of Process Engineering*, vol. 8, p. 6, 2008.

[27] N. Zulkiple, M. Y. Maskat, and O. Hassan, "Pretreatment of oil palm empty fruit fiber (OPEFB) with aqueous ammonia for high production of sugar," *Procedia Chemistry*, vol. 18, pp. 155–161, 2016.

[28] A. Sluiter, R. Ruiz, C. Scarlata, J. Sluiter, and D. Templeton, "Determine of structure carbohydrates and lignin in biomass: laboratory analytical procedures (LAP)," Technical Report NREL/TP-510e42618, National Renewable Energy Laboratory, Golden, CO, USA, 2008.

[29] A. Sluiter, A. R. Ruiz, C. Scarlata, J. Sluiter, and D. Templeton, "Determine of ash in biomass: laboratory analytical procedures (LAP)," Technical Report NREL/TP-510e42622, National Renewable Energy Laboratory, Golden, CO, USA, 2005.

[30] T. H. Kim, F. Taylor, and K. B. Hicks, "Bioethanol production from barley hull using SAA (soaking in aqueous ammonia) pretreatment," *Bioresource Technology*, vol. 99, no. 13, pp. 5694–5702, 2008.

[31] S. W. Pryor, B. Karki, and N. Nahar, "Effect of hemicellulase addition during enzymatic hydrolysis of switchgrass pretreated by soaking in aqueous ammonia," *Bioresource Technology*, vol. 123, pp. 620–626, 2012.

[32] C. M. Xanthopoulou, E. Jurado, I. V. Skiadas, and H. N. Gavala, "Effect of aqueous ammonia soaking on the methane yield and composition of digested manure fibers applying different ammonia concentrations and treatment durations," *Energies*, vol. 7, no. 7, pp. 4157–4168, 2014.

[33] G. Antonopoulou, H. N. Gavala, I. V. Skladas, and G. Lyberatos, "The effect of aqueous ammonia soaking pretreatment on methane generation using different lignocellulosic biomasses," *Waste and Biomass Valorization*, vol. 6, no. 3, pp. 281–291, 2015.

[34] P. Daorattanachai, N. Viriya-empikul, N. Laosiripojana, and K. Faungnawakij, "Effects of kraft lignin on hyrolysis/dehydration of sugars, cellulosic and lignocellulosic biomass under hot compressed water," *Bioresource Technology*, vol. 144, pp. 504–512, 2013.

[35] A. K. Chandel, F. A. F. Antunes, M. B. Silva, and S. S. Silva, "Unraveling the structure of sugarcane bagasse after soaking in concentrated aqueous ammonia (SCAA) and ethanol production by Scheffersomyces (pichia) stipitis," *Biotechnology for Biofuels*, vol. 6, p. 102, 2013.

[36] B. Danon, L. V. D. Aa, and W. D. Jong, "Furfural degradation in a diluted acidic and saline solution in the presence of glucose," *Carbohydrate Research*, vol. 375, pp. 145–152, 2013.

[37] K. Lamminpaa, J. Aloha, and J. Tanskanen, "Acid-catalysed xylose dehydration into furfural in the presence of kraft lignin," *Bioresource Technology*, vol. 177, pp. 94–101, 2015.

11

Synthesis of a New Copper-based Supramolecular Catalyst and its Catalytic Performance for Biodiesel Production

Fei Chang ⓘ,[1] Chen Yan,[2] and Quan Zhou[3]

[1]Institute of Comprehensive Utilization of Plant Resources, Kaili University, Kaili 556011, China
[2]An Shun City People's Hospital, People's Hospital Republic of China, An Shun 561000, China
[3]Pharmaceutical and Bioengineering College, Hunan Chemical Vocational Technology College, Zhuzhou, Hunan 412000, China

Correspondence should be addressed to Fei Chang; feichang1980@126.com

Guest Editor: Shunmugavel Saravanamurugan

A new copper-based supramolecular (β-cyclodextrins, β-CD) catalyst was synthesized and used for transesterification of *Xanthium sibiricum* Patr oil to biodiesel. This catalyst exhibited high activity (88.63% FAME yield) in transesterification under the ratio of methanol-oil: 40 : 1; catalyst dosage: 8 wt.%; reaction temperature: 120°C; and reaction time: 9 h. The XRD, SEM, TEM, XPS, and BET characterization results showed that Cu-β-CD catalyst was amorphous and had clear mesoporous structure (17.2 nm) as compared with the native β-CD. This phenomenon is attributed to the coordination of Cu and β-CD.

1. Introduction

With the rapid socioeconomic development, the demand for petrochemical energy is on the increase. At the same time, the shortage of energy and environmental pollution have become the focus [1, 2]. Biodiesel is a good substitute for petrochemical diesel because of its sustainability, biodegradability, and cleanability [3]. Biodiesel, also known as fatty acid monoester, mainly including fatty acid methyl esters (FAME) and fatty acid ethyl esters (FAEE), is typically prepared via esterification or transesterification reactions of animal and vegetable oils with methanol or ethanol in the presence of an acidic and/or basic catalyst [4]. The reaction processes can be divided into homogeneous and heterogeneous ones depending on the type of catalysts, and researchers are more inclined to heterogeneous research for its advantages such as simple steps, easy postprocessing, and less pollution [5, 6]. Heterogeneous catalysts mainly include inorganic acid salts, solid heteropoly acids, metal oxides [7, 8], zeolites [9], and hydrotalcites [10]. Among them, the single and mixed metal oxides were studied by numerous studies due to their environment-friendly, cheap, and efficient catalytic characteristics, which were generally prepared by coprecipitation, sol-gel, impregnation, and hydrothermal methods [11]. In particular, the metal oxides composed of Ca, Mg, and Al were extensively illustrated to be active for biodiesel production [12–14]. However, Cu-based catalysts used for efficient biodiesel preparation have been rarely reported so far.

In this report, a new Cu-based supramolecular catalyst was prepared from $CuSO_4 \cdot 5H_2O$ and β-CD by simple organic synthesis and was applied to biodiesel synthesis. The results showed that the catalyst had obvious mesoporous structure and good catalytic activity. The results of this study fill the gaps of copper-based catalysts for biodiesel production.

2. Experiments

2.1. Materials. *Xanthium sibiricum* Patr oil was extracted with the reported method [15]. Pure fatty acid methyl esters were purchased from Sigma (USA). β-CD was purchased from Hongchang Pharmaceutical Reagent Co., Ltd., Xi'an. Anhydrous methanol, NaOH, and $CuSO_4 \cdot 5H_2O$ are analytically pure (AR) and purchased from Chemical Reagent Co., Ltd., Tianjin.

2.2. Catalyst Preparation.

2.2. Catalyst Preparation. According to previous reports [16, 17], 2.5 g β-CD and 0.8 g NaOH were dissolved into 50 mL distilled water and stirred to completely dissolve at room temperature, and then 50 mL aqueous solution of 0.5 g $CuSO_4 \cdot 5H_2O$ was gradually added at room temperature under magnetic stirring for 1.5 h and filtered. Upon completion, 500 mL ethanol was added to the filtrate, and a precipitate formed, which was filtered and washed with absolute ethanol to give a neutral precipitate. The attained solid was further dried at 80°C for 5 h.

2.3. Catalyst Characterization. TGA analysis was recorded by NETZSCH STA 429 instrument. XRD patterns were measured with the Bruker D8 advanced X-ray diffractometer (XRD) with Cu Kα radiation (λ = 0.154 nm) at 40 kV and 30 mA with a step size of 0.02. The surface morphologies of the catalysts were characterized via FEI inspect F50 type scanning electron microscope (SEM). The internal structure of catalysts was analyzed by the FEI Tecnai G2 F20 S-TWIN 200 kV transmission electron microscope (TEM). XPS analysis was conducted using the Thermo Scientific ESCALAB 250Xi spectrometer employing a monochromatic Al Kα X-ray source (hν = 1486.8 eV) and 500 μm test spot area, 15 kv test tube voltage, 10 mA tube current, and 2×10^{-9} mbar analysis room floor vacuum. The Brunauer–Emmett–Teller (BET) surface areas were measured by N_2 adsorption/desorption apparatus (Micromeritics ASAP 2020), and the pore size and pore volume distributions were calculated using the Barrett–Joyner–Halenda (BJH) model.

2.4. Product Analysis. The appropriate amount of *X. sibiricum* Patr oil, catalyst, and methanol were added into a 25 mL glass three-necked flask with a condensing means and placed in a an oil bath (120°C) with magnetic stirring for a certain time. After the reaction completion, the reaction mixture was cooled down and filtered, while the excess methanol was removed by rotary evaporation. Hereafter, the FAME contents of the samples were determined by the gas chromatography (GC, Agilent 6890 GC), and the FAME contents were calculated according to the methods reported in [18].

3. Results and Discussions

3.1. Catalyst Characterization

3.1.1. TGA Analysis. The TGA analysis results of the Cu-β-CD catalyst are shown in Figure 1. It can be seen that the weight loss of the Cu-β-CD catalyst mainly included three stages, namely, loss of water (50–150°C), catalyst decomposition (150–300°C), and complete decomposition of the catalyst (300–800°C). Evidently, this catalyst was stable until the temperature of around 150°C.

3.1.2. XRD. Usually, the catalytic activity is closely related to the morphology of the catalyst. The catalytic effect of the amorphous material was generally better than the crystal

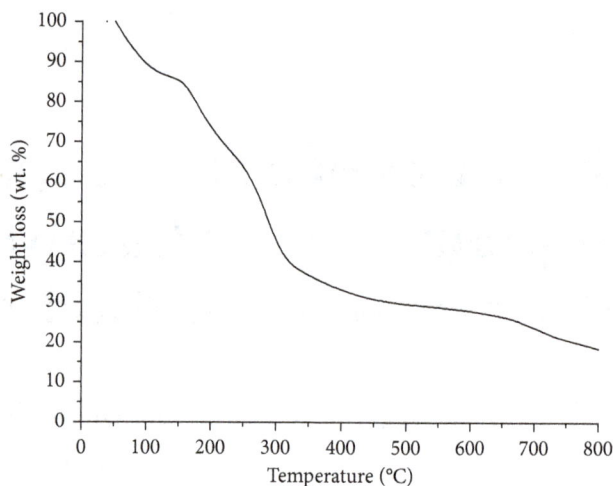

FIGURE 1: TGA curves of Cu-β-CD catalyst.

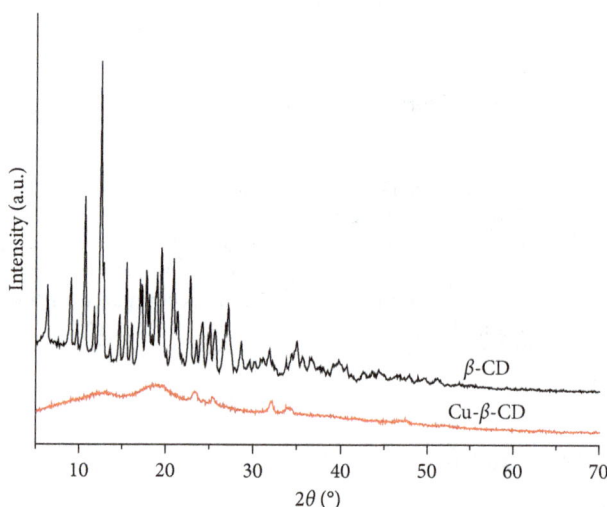

FIGURE 2: XRD patterns of pure β-CD and Cu-β-CD catalyst.

counterpart [19, 20]. XRD patterns of β-CD and Cu-β-CD are shown in Figure 2, it could be clearly seen that the single β-CD had distinct diffraction peaks, belonging to crystal state material. However, Cu-β-CD did not show significant diffraction peaks but appeared as wave packets. So, the structures were greatly changed when the copper ions were involved, which changed its morphology and increased its specific surface area (Figures 3–5), while improving its catalytic activity. This is consistent with the experimental results (Table 1).

3.1.3. XPS. The valence of copper ions and structure of the complex were determined by XPS spectra (Figure 6). As can be seen from Figures 6(a) and 6(b), Cu ions existed in the Cu-β-CD catalyst. C1s might be divided into three signals in Figure 6(a), namely, C-C (284.7 eV), C-O (286.4 eV), and C=O (287.9 eV), respectively. In addition, it can be seen from Figure 6(b) that the Cu's basic binding energy was 933.3 ev (Cu2p3/2) and 953.6 ev (Cu2p1/2). Therefore, Cu^{2+},

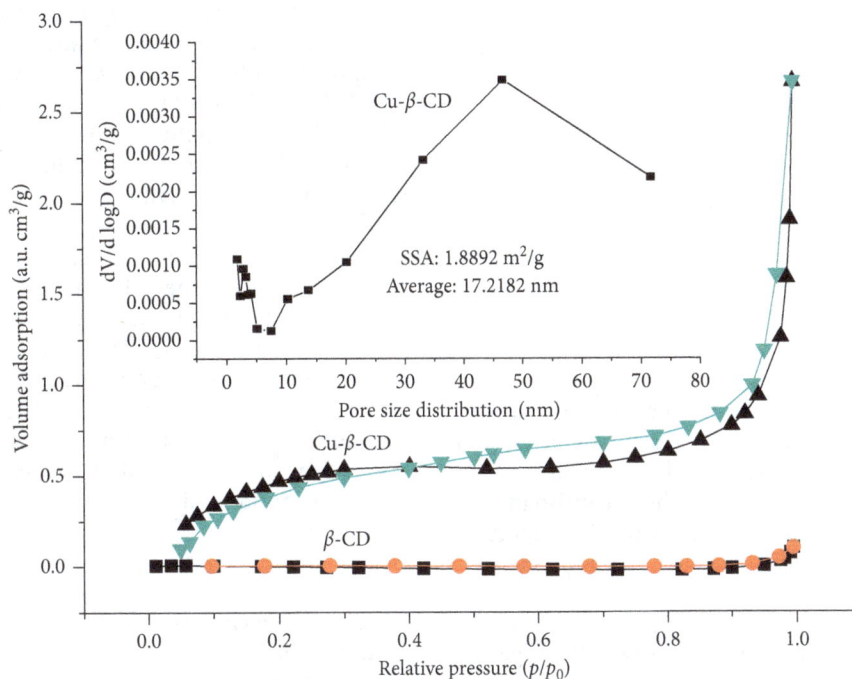

FIGURE 3: N$_2$ adsorption-desorption isotherms and pore size distribution of β-CD and Cu-β-CD.

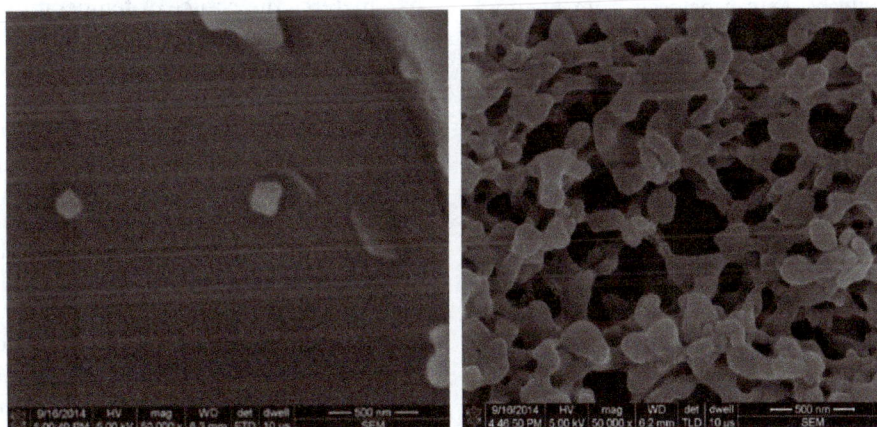

FIGURE 4: SEM images of pure (a) β-CD and (b) Cu-β-CD catalysts.

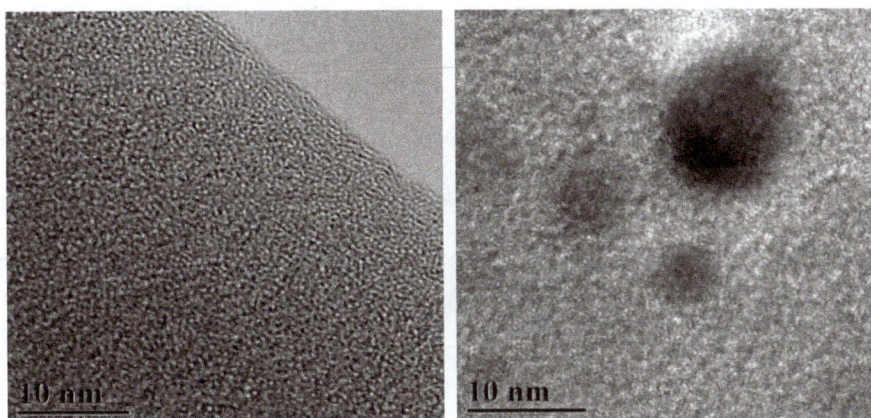

FIGURE 5: TEM images of pure (a) β-CD and (b) Cu-β-CD samples.

TABLE 1: The catalytic activity of the catalysts.

Entry	Catalyst	FAME%
1	β-CD (native)	nd
2	$Cu(SO_4)_2 \cdot 5H_2O$	nd
3	$Cu(SO_4)_2 \cdot 5H_2O + \beta$-CD	
4	(simple physical mixture)[a]	nd
5	CuO	nd
6	Cu-β-CD	88.6

Condition of reaction: the ratio of methanol-oil: 40 : 1; the amount of catalyst: 8 wt.%; reaction temperature: 120°C; reaction time: 9 h; [a]physical mixing; nd: not detected; FAME%: the data of biodiesel production percentages.

copper, is predominantly present in this complex, and coordination compounds were formed such as $CuCO_3$ and CuO. For this study, it can be deduced that a similar C-O-Cu bond existed in the Cu-β-CD catalyst. This is consistent with previous reports [21, 22] and FT-IR (Figure 1, supporting information (available here).

3.1.4. N_2 Adsorption-Desorption Isotherm.
The specific surface area (SSA) and pore size are also the main factors that affect the activity of the catalyst. So, SSA and pore size distribution of the β-CD and Cu-β-CD were studied via N_2 adsorption-desorption isotherm and calculated by BET and BJH methods, respectively. As can be seen from Figure 3, β-CD did not display apparent hysteresis loops, but hysteresis ring closure point of the Cu-β-CD appeared at $p/p_0 = 0.4$. In addition, the dramatic increase trend in the high-pressure section indicated that it belongs to the type IV isotherms and type H4 hysteresis ring [22, 23]. These results demonstrated that β-CD had no distribution of pores and the Cu-β-CD possessed slit hole formed by multilayer structure, and its average pore size is 17.2 nm. Those were consistent with SEM and TEM studies. Apart from this, the SSA of Cu-β-CD catalyst was 1.9 m^2/g, which is much larger than that of β-CD (0.1 m^2/g) [24].

3.1.5. SEM and TEM.
The morphology of the catalyst is typically correlated to its activity directly [25]. In order to understand the structure of Cu-β-CD, the catalyst was characterized by SEM and TEM, and the results are shown in Figures 4 and 5. The surface of native β-CD is smooth (Figure 4), and the obvious pore structure cannot be observed (Figure 5), but the Cu-β-CD showed multihole structure and heterogeneous mesoporous structure (Figure 4). Furthermore, a uniform worm-like duct structure of the Cu-β-CD was also observed (Figure 5). So, it can be concluded that the Cu-β-CD is a porous mesoporous material, and it can be inferred that Cu-β-CD catalyst has a larger specific surface area (SSA) than β-CD, which was confirmed by the BET test results. As we all know, the catalyst with porous mesoporous structures, small particles, and large SSA can improve the activity of the catalyst [26, 27], and the Cu-β-CD should have a high catalytic activity. Accordingly, the results of catalytic performance of the catalysts are shown in Table 1.

3.2. Catalytic Performance of the Catalysts.
Catalytic performance of the relevant catalysts is shown in Table 1 (supporting information). As can be seen from Table 1, β-CD (native), $Cu(SO_4)_2 \cdot 5H_2O$, $Cu(SO_4)_2 \cdot 5H_2O + \beta$-CD (simple physical mixture), and CuO did not show catalytic activity (Table 1, entries 1–5). In contrast, the Cu-β-CD showed a higher activity (FAME yield: 88.6%, Table 1, entry 6) under 40 : 1 methanol-oil ratio, 8 wt.% catalyst load, 120°C reaction temperature, and 9 h reaction time. In combination with the relevant card results that can be determined its catalytic activity, it can be deduced that the superior activity of Cu-β-CD is mainly due to the Cu^{2+} and β-CD which formed the Cu-OH bonds, and the Cu^{2+} may act as electrophilic species to activate ester. Furthermore, the Cu-OH bonds act as nucleophilic species to attack the carbon of the ester, and two synergies may weaken the ester bond and make $-OCH_3$ attack ester bonds easily [17].

3.3. Effect of Single Factor on the FAME Content.
In order to optimize the biodiesel catalytic process of the Cu-β-CD, reaction temperature, methanol/oil molar ratio, catalyst loading, and reaction time were studied, respectively. The results are shown in Figure 7, and in most chemical reactions, reaction temperature is one of the most important parameters. The choice of temperature has a direct effect on the reaction rate and product yield. As can be seen from Figure 7(a), the FAME content is only 20% at 65°C, but it increased with the increase of temperature. When the temperature reached 120°C, the maximum yield is obtained, while continuing to increase the temperature to 140°C leads to no change in the FAME. Figure 7(b) shows the effect of the molar ratio of methanol to oil in the reaction system. When the methanol-oil molar ratio is 10 : 1–50 : 1, it is proportional to the yield of FAME. As the methanol-oil molar ratio is 40 : 1 and 50 : 1, the yield of FAME was 88.39% and 89.11%, respectively. It can be considered that the increase of the ratio of methanol-oil yields of FAME can be neglected. Taking into account the catalyst concentration and cost, the methanol-oil molar ratio need not be further increased; therefore, the optimal molar ratio of methanol to oil is 40 : 1 in this reaction. Such a high molar ratio of methanol to oil is related to the characteristic of β-CD having alcoholicity [28]. The catalyst is the most critical factor in transesterification, and Figure 7(c) shows the yield of FAME under 2 wt.%–8 wt.% catalyst; the content of FAME is lowest with 2 wt.% catalyst amount, and with the increase of the amount of catalyst, the yield of FAME also increases. The yield of FAME reached its maximum when increasing to 8 wt.%. Therefore, the optimal catalyst loading should be chosen to be 8 wt.% for the cost problem. The reaction time is also a key factor affecting the reaction result. The impact of reaction time on the yield of FAME is shown in Figure 7(d). It can be seen from the Figure 7(d) that the conversion rate of FAME reached the maximum after 9 h. This shows that 88.63% FAME conversion was received under the optimized reaction conditions of 40 : 1 molar

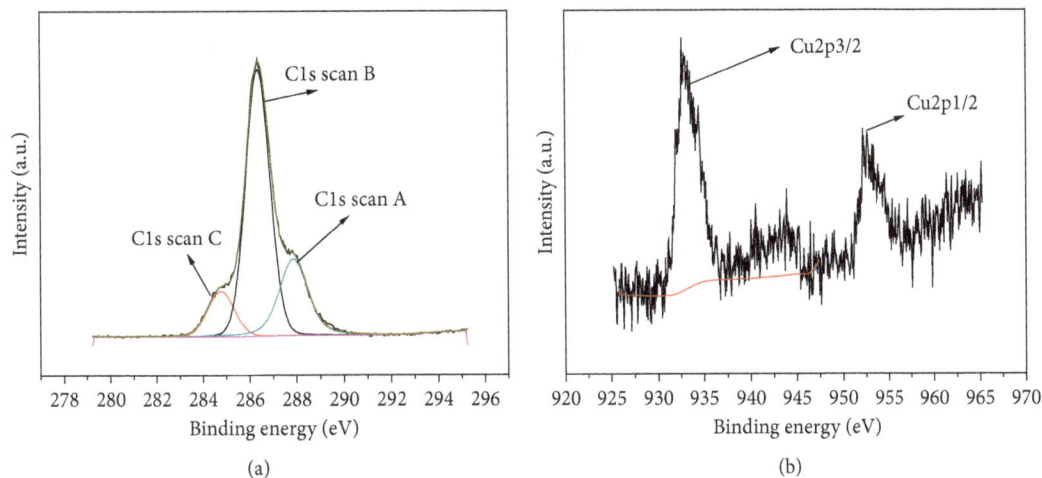

FIGURE 6: The XPS spectra of Cu-β-CD catalyst.

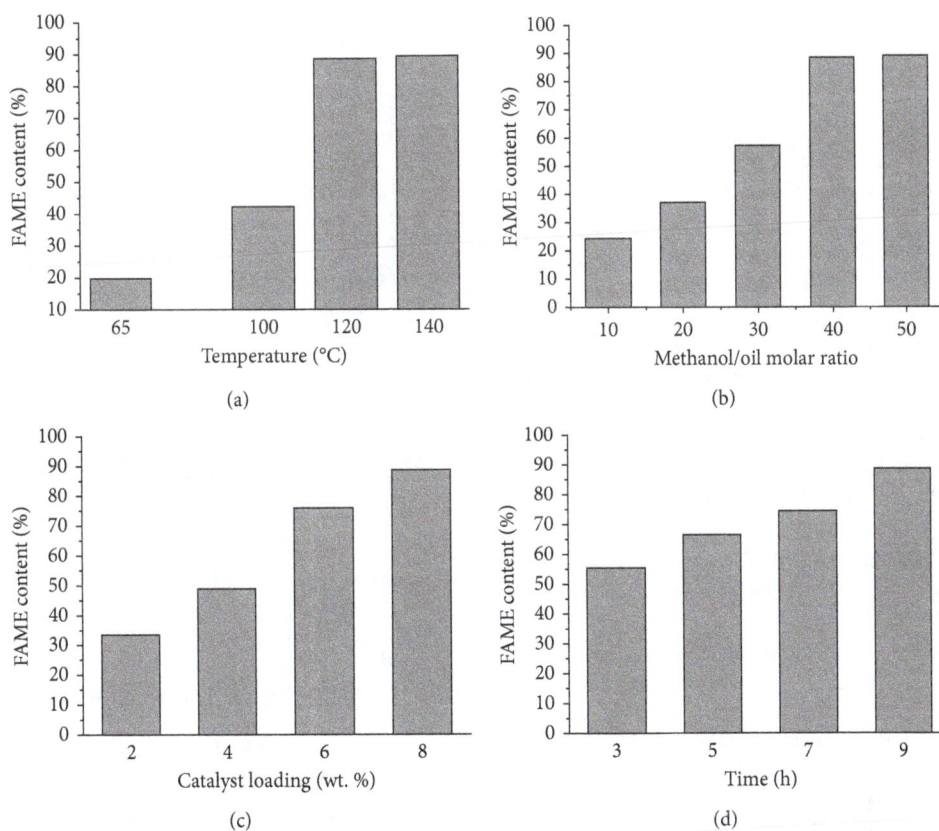

FIGURE 7: The effect of single factor on the FAME content. (a) Effect of temperature on FAME content (methanol/oil molar ratio = 40 : 1, CA = 8 wt.%, t = 9 h). (b) Effect of methanol/oil molar ratio on FAME content (CA = 8 wt.%, T = 120°C, t = 9 h). (c) Effect of catalyst loading on FAME content (methanol/oil molar ratio = 40 : 1, T = 120°C, t = 9 h). (d) Effect of time on FAME content (methanol/oil molar ratio = 40 : 1, T = 120°C, CA = 8 wt.%).

ratio of methanol/oil, 8 wt.% Cu-β-CD amount, 120°C, and 9 h.

Overall, Cu-β-CD was stable until around 150°C, which was a mesoporous material having a large SSA (1.8892 m^2/g) compared with β-CD (0.11 m^2/g), and its activity lies in the synergy of β-CD and copper.

4. Conclusions

The Cu-β-CD was prepared by a simple method, which was found to be a kind of uniform worm-like duct and porous mesoporous structured material. It was successfully applied to biodiesel production, giving 88.63% FAME conversion under optimal conditions. This study further demonstrated that Cu^{2+} and β-CD in the catalyst played a synergistic catalytic role, greatly improving the activity of Cu-based catalyst in transesterification.

Disclosure

Fei Chang and Chen Yan contributed equally to this work.

Conflicts of Interest

The authors declare that they have no conflicts of interest.

Acknowledgments

This work was financially supported by Guizhou Provincial State University S&T Technology Joint Fund Program (nos. LH[2015]7762 and LH [2015]7752).

References

[1] Y. He, S. Wang, and K. K. Lai, "Global economic activity and crude oil prices: a cointegration analysis," *Energy Economics*, vol. 32, no. 4, pp. 868–876, 2010.

[2] S. N. Dodić, S. D. Popov, J. M. Dodić, J. A. Ranković, and Z. Z. Zavargo, "Biomass energy in Vojvodina: market conditions, environment and food security," *Renewable and Sustainable Energy Reviews*, vol. 14, no. 2, pp. 862–867, 2010.

[3] M. Balat, "Production of bioethanol from lignocellulosic materials via the biochemical pathway: a review," *Energy Conversion and Management*, vol. 52, no. 2, pp. 858–875, 2011.

[4] A. D. Lele, K. Anand, and K. Narayanaswamy, *Surrogates for Biodiesel: Review and Challenges*, Springer, Singapore, 2017.

[5] M. R. Avhad and J. M. Marchetti, "A review on recent advancement in catalytic materials for biodiesel production," *Renewable and Sustainable Energy Reviews*, vol. 50, pp. 696–718, 2015.

[6] H. H. Mardhiah, H. C. Ong, H. H. Masjuki, S. Lim, and H. V. Lee, "A review on latest developments and future prospects of heterogeneous catalyst in biodiesel production from non-edible oils," *Renewable and Sustainable Energy Reviews*, vol. 67, pp. 1225–1236, 2017.

[7] F. Chang, Q. Zhou, H. Pan et al., "Solid mixed-metal-oxide catalysts for biodiesel production: a review," *Energy Technology*, vol. 2, no. 11, pp. 865–873, 2014.

[8] N. F. Balsamo, K. Sapag, M. I. Oliva, G. A. Pecchi, G. A. Eimer, and M. E. Crivello, "Mixed oxides tuned with alkaline metals to improve glycerolysis for sustainable biodiesel production," *Catalysis Today*, vol. 279, no. 2, pp. 209–216, 2017.

[9] S. Manadee, O. Sophiphun, N. Osakoo et al., "Identification of potassium phase in catalysts supported on zeolite NaX and performance in transesterification of *Jatropha* seed oil," *Fuel Processing Technology*, vol. 156, pp. 62–67, 2017.

[10] K. G. Georgogianni, A. K. Katsoulidis, P. J. Pomonis, G. Manos, and M. G. Kontominas, "Transesterification of

[11] H. V. Lee, J. C. Juan, T. Y. Yun Hin, and H. C. Ong, "Environment-friendly heterogeneous alkaline-based mixed metal oxide catalysts for biodiesel production," *Energies*, vol. 9, no. 8, p. 611, 2016.

[12] G. Tao, Z. Hua, Z. Gao, Y. Zhu, Y. Z. Chen, and J. Shi, "KF-loaded mesoporous Mg-Fe bi-metal oxides: high performance transesterification catalysts for biodiesel production," *Chemical Communications*, vol. 49, no. 73, pp. 8006–8008, 2013.

[13] J. R. Mercury, A. H. De Aza, and P. Pena, "Synthesis of $CaAl_2O_4$ from powders: particle size effect," *Journal of the European Ceramic Society*, vol. 25, no. 14, pp. 3269–3279, 2005.

[14] V. Mandić and S. Kurajica, "The influence of solvents on sol–gel derived calcium aluminate," *Materials Science in Semiconductor Processing*, vol. 38, pp. 306–313, 2015.

[15] F. Chang, M. A. Hanna, D. J. Zhang et al., "Production of biodiesel from non-edible herbaceous vegetable oil: *Xanthium sibiricum* Patr," *Bioresource Technology*, vol. 140, pp. 435–438, 2013.

[16] Y. Matsui, T. Kurita, M. Yagi, T. Okayama, K. Mochida, and Y. Date, "The formation and structure of Copper (II) complexes with cyclodextrins in an alkaline solution," *Bulletin of the Chemical Society of Japan*, vol. 48, no. 7, pp. 2187–2191, 1975.

[17] F. Chang, Q. Zhou, H. Pan, X. F. Liu, H. Zhang, and S. Yang, "Efficient production of biodiesel from Xanthium sibiricum Patr oil via supramolecular catalysis," *Renewable Energy*, vol. 111, pp. 556–560, 2017.

[18] H. Pan, X. F. Liu, H. Zhang, K. Yang, S. Huang, and S. Yang, "Multi-SO₃H functionalized mesoporous polymeric acid catalyst for biodiesel production and fructose-to-biodiesel additive conversion," *Renewable Energy*, vol. 107, pp. 245–252, 2017.

[19] L. M. Correia, R. M. A. Saboya, N. de Sousa Campelo et al., "Characterization of calcium oxide catalysts from natural sources and their application in the transesterification of sunflower oil," *Bioresource Technology*, vol. 151, pp. 207–213, 2014.

[20] W. Suryaputra, I. Winata, N. Indraswati, and S. Ismadji, "Waste capiz (Amusium cristatum) shell as a new heterogeneous catalyst for biodiesel production," *Renewable Energy*, vol. 50, pp. 795–799, 2013.

[21] E. Norkus, G. Grinciene, T. Vuorinen, E. Butkus, and R. Vaitkus, "Stability of a dinuclear Cu(II)–β-Cyclodextrin complex," *Supramolecular Chemistry*, vol. 15, no. 6, pp. 425–431, 2003.

[22] E. Norkus, G. Grinciene, T. Vuorinen, and R. Vaitkus, "Cu(II) ion complexation by excess of β-cyclodextrin in aqueous alkaline solutions," *Journal of Inclusion Phenomena*, vol. 48, no. 3-4, pp. 147–150, 2004.

[23] Y. Chen, Y. Cao, Y. Suo, G. P. Zheng, X. X. Guan, and X. C. Zheng, "Mesoporous solid acid catalysts of 12-tungstosilicic acid anchored to SBA-15: characterization and catalytic properties for esterification of oleic acid with methanol," *Journal of the Taiwan Institute of Chemical Engineers*, vol. 51, pp. 186–192, 2015.

[24] Y. Luo, Z. Mei, N. Liu, H. Wang, C. Han, and S. He, "Synthesis of mesoporous sulfated zirconia nanoparticles with high surface area and their applies for biodiesel production as effective catalysts," *Catalysis Today*, vol. 298, pp. 99–108, 2017.

[25] Y. H. Taufiq-Yap, H. V. Lee, M. Z. Hussein, and R. Yunus, "Calcium-based mixed oxide catalysts for methanolysis of *Jatropha* curcas oil to biodiesel," *Biomass and Bioenergy*, vol. 35, no. 2, pp. 827–834, 2011.

[26] Z. Helwani, M. R. Othman, N. Aziz, W. J. N. Fernando, and J. Kim, "Technologies for production of biodiesel focusing on green catalytic techniques: a review," *Fuel Processing Technology*, vol. 90, no. 12, pp. 1502–1514, 2009.

[27] D. Y. C. Leung, X. Wu, and M. K. H. Leung, "A review on biodiesel production using catalyzed transesterification," *Applied Energy*, vol. 87, no. 4, pp. 1083–1095, 2010.

[28] A. Buvari, J. Szejtli, and L. Barcza, "Complexes of short-chain alcohols with β-cyclodextrin," *Journal of Inclusion Phenomena*, vol. 1, no. 2, pp. 151–157, 1983.

12

Polyelectrolyte Complexation versus Ionotropic Gelation for Chitosan-based Hydrogels with Carboxymethylcellulose, Carboxymethyl Starch, and Alginic Acid

Elizabeth Henao,[1] Ezequiel Delgado,[2] Héctor Contreras,[2] and Germán Quintana (iD)[1]

[1]Grupo Pulpa y Papel, Facultad de Ingeniería Química, Universidad Pontificia Bolivariana, Sede Central Medellín, Circular 1 No. 70-01, Medellín, Colombia
[2]Departamento de Madera, Celulosa y Papel (DMCyP), Universidad de Guadalajara, Km. 15.5 Carretera Guadalajara-Nogales Las Agujas, 45020 Zapopan, JAL, Mexico

Correspondence should be addressed to Germán Quintana; german.quintana@upb.edu.co

Academic Editor: Donald L. Feke

The preparation of gels by charge interaction methods has been extensively studied, but it is not yet clear how these methods influence gel characteristics. The objective of this work was to study differences in morphology and surface charge of hydrogels prepared by ionotropic gelation, polyelectrolyte complexation, and a combination of both methods. Thus, the anionic charge was provided by carboxymethylcellulose (CMC), carboxymethylated starch (CMS), and alginic acid (AA); calcium chloride ($CaCl_2$) and chitosan (CS) were used for the ionotropic gelation and polyelectrolyte complexation, respectively. Those materials are commercially available, have low toxicity, and are widely used in the area. These compounds interact through physical crosslinks, which are affected by physical changes of the medium. Our results showed that these two methods produced changes in the morphology of the hydrogels. CMC gels exhibited larger pores in the presence of $CaCl_2$. In polyelectrolyte complexation, CMS produced an increased agglomeration of particles, while the addition of $CaCl_2$ to AA generated dispersed particles of size in the order of millimeters. Mixing both ionotropic gelation and polyelectrolyte complexation methods yielded gels of varied charge (568 mV for CMC, 502 mV for CMS, and 1713 mV for AA). FTIR spectra of the hydrogels showed interactions between the different polymeric compounds, being the greatest changes between 1250 and 1600 cm^{-1}, due possibly to the replacement of Na by Ca at crosslinking points. Therefore, the method of gel preparation employed had a major influence on the size and pore distribution, parameters which in turn influence encapsulation and drug delivery in these systems.

1. Introduction

Hydrogels are three-dimensional networks capable of absorbing large quantities of water or biological fluids, therefore, they attract interest in medicine [1]. Polysaccharides have been studied during the last years as raw materials in the encapsulation of drugs forming a mesh structure by charge interaction between anions and cations. Polysaccharides have peculiar properties like good solubility in aqueous environments, high stability, null toxicity, biocompatibility, biodegradability, and have the ability to encapsulate drugs [2–4]. In drug delivery applications, polysaccharides possessing hydroxyl, carboxyl, and amino groups are bioadhesives and could increase the residence time [3].

Several types of hydrogels have been prepared by noncovalent crosslinking, incorporating anionic polymers, such as carboxymethylcellulose, carboxymethylated starch or alginic acid, on cationic polymers like chitosan (polyelectrolyte complexation) [5–10]. Also, the same anionic polymers have been used with $CaCl_2$ (ionotropic gelation) [11, 12] or a mix of polymers with the salt [13, 14].

In ionotropic gelation, hydrogels are produced due to the ability of polyelectrolytes to crosslink with counterions, forming a meshwork structure of ionically crosslinked. The polyelectrolyte complexation technique forms hydrogels by the addition of one polyelectrolyte to another polyelectrolyte [15] having opposite charge.

There have been several proposals to generate hydrogels with a particular structure. Here, the method chosen plays an important role in the morphology. Some methods that show great promise as a tool for hydrogel manufacture are ionotropic gelation and polyelectrolyte complexation. Each method has some advantages and limitations.

This paper compares the ionotropic gelation, polyelectrolyte complexation, and the mixture of both procedures, on the morphology of gels prepared from carboxymethylcellulose, carboxymethyl starch, and alginic acid with chitosan.

2. Materials and Methods

2.1. Materials. Carboxymethylcellulose sodium salt (CMC) average Mw ~250,000 and degree of substitution 0.7 (CAS 9004-32-4), chitosan (CS) medium molecular weight (75–85% deacetylation) (CAS 9012-76-4), and alginic acid (AA) of medium viscosity (CAS 9005-38-3) are products of Sigma-Aldrich. Carboxylated starch (CMS) was purchased from DFE Pharma. Calcium chloride dihydrate (CAS 10035-04-8) was supplied by Carlo Erba.

2.2. Hydrogel Formation. Polymer solutions were prepared at such concentrations that were able to pass through the needle and formed a separate drop without problems of fluency [16].

2.2.1. Cationic Substance Concentrations. CS concentration was fixed in 0.01% w/v (dissolved in acetic acid 1% v/v) [17]. $CaCl_2$ was used at 1 and 4.9% w/v dissolved in water (in previous work, lower concentrations did not produce a gel).

2.2.2. Anionic Polymer Concentrations. CMC solutions were prepared at 2% w/v, CMS at 3.2% w/v, and AA at 3 and 0.5% w/v. For ionotropic gelation only, concentrations of 1.6% w/v for CMC and 5.6 for CMS were used. In order to solubilize the polymers, it was necessary to leave each solution under magnetic stirring for 24 hours. $CaCl_2$ was solubilized under magnetic agitation for 2 hours approximately.

2.2.3. Ionotropic Gelation. Each anionic polymer was dissolved in deionized water (to the desired concentration), and approximately 10 mL of this solution was dropped using a programmable syringe pump (New Era Pump System Inc.), passing through a needle (21 G) from a plastic syringe into a beaker containing 10 mL of the $CaCl_2$ solution; this system was kept under continuous stirring (600 rpm) for 10 minutes at room temperature. Then, the mix was stirred at the same conditions for 20 minutes more, and the hydrogel was kept 24 hours at room temperature to reach the equilibrium. Afterward, the sample was centrifuged at 3300 rpm for 30 minutes and the precipitate was removed from the supernatant. Finally, the gels were washed with distilled water and separated again by centrifugation.

2.2.4. Polyelectrolyte Complexation. The procedure followed a previously reported methodology [18, 19]. Briefly, a CS solution in acetic acid 1% w/v was used as a cationic agent,

instead of $CaCl_2$. The mix was stirred at the same conditions for 20 minutes, and later, the hydrogel was kept 24 hours at room temperature until the equilibrium was reached. The sample was then centrifuged at 3300 rpm for 30 minutes and the precipitate was removed from the supernatant. Finally, the gels were washed with distilled water and separated again by centrifugation.

A change to evaluate the effect of the aggregation order was made for the CMC polyelectrolyte complexation, dropping CS 0.016% w/v onto a CMC 0.8% w/v solution.

2.2.5. Preparation of Gels by Simultaneous Polyelectrolyte Complexation and Ionotropic Gelation. A mix of both procedures was carried on by solubilizing the $CaCl_2$ in the CS solution (2 hours of agitation), and then dropping the CMC on this mixture. The mix was stirred at the same conditions for 20 minutes more, and the hydrogel was kept 24 hours at room temperature to reach the equilibrium. Thus, the sample was centrifuged at 3300 rpm for 30 minutes, and the precipitate was removed from the supernatant, washed with distilled water and separated again by centrifugation.

Three gel samples were prepared for each system in order to compare results.

2.3. Morphological Characterization of the Gels. Images of the wet gels were obtained by an optical microscope (Leica DMREB, software LAS, version 7.1. Camera Leica DFC320). A drop of the sample was put on a slide with a drop of water to disperse the gel before the observation. Then, a coverslip was put upon the sample and was observed.

The surface morphology and internal structure of lyophilized gels were observed by a scanning electron microscope (SEM; HITACHI TM-1000). The samples were cooled with liquid nitrogen and freeze-dried for 24 hours, then stored and characterized.

2.4. Zeta Potential. Zeta potential of the preparations was measured after equilibrium (around 24 hours after the preparation) was established. The zeta potential was read with a Zetasizer Nano ZS90 with DS170 cuvettes, using acetic acid as a solvent for chitosan and hydrogels made by both polyelectrolyte complexation and the mix of ionotropic gelation and polyelectrolyte complexation. Water was used as a solvent for the polymers and hydrogels made by ionotropic gelation. Three readings of zeta potential values were made for each sample.

2.5. FTIR Characterization of Gels. FTIR spectra of freeze-dried samples were recorded with a PerkinElmer FTIR spectrometer, model spectrum GX with an attachment of attenuated total reflectance (ATR) and crystal diamond, taking spectra between 4000 and 600 cm^{-1} with a resolution of 4.00 cm^{-1} and 16 scannings.

3. Results

It has been suggested that polymers containing carboxyl groups are better than sulfated polymers for the encapsulation

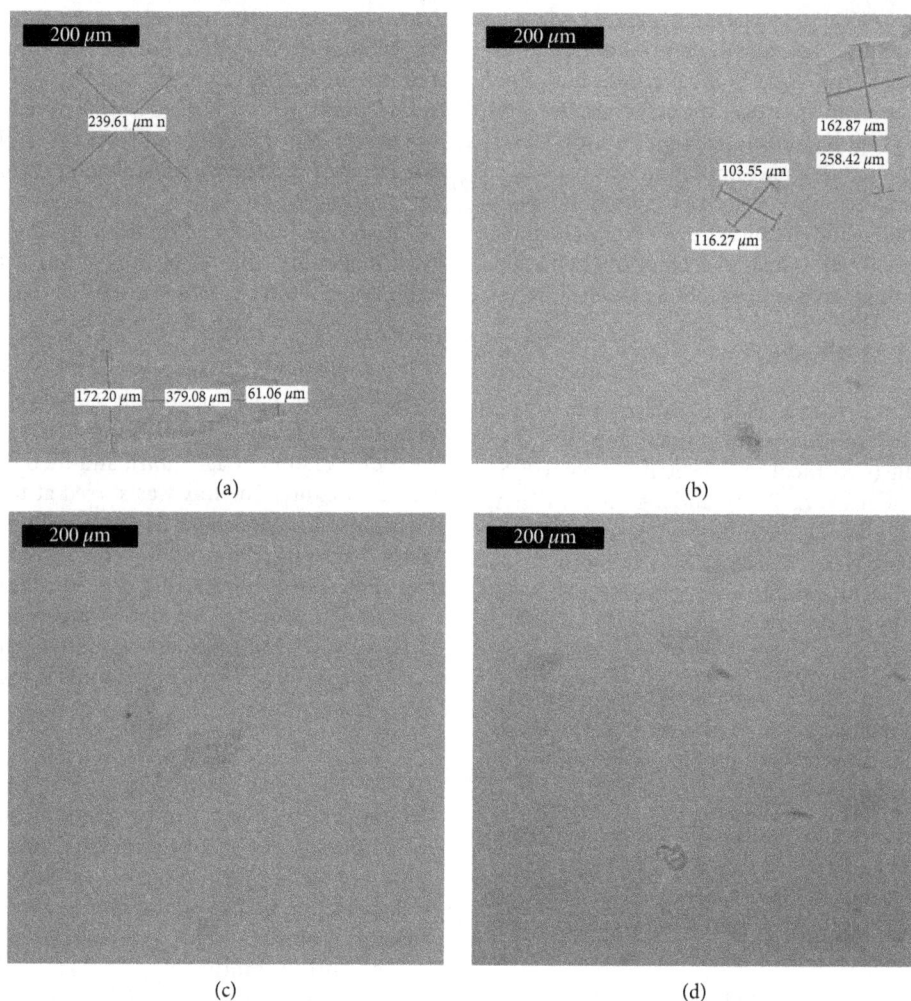

FIGURE 1: Optical micrographs of hydrogels made from CMC 2%: (a) ionotropic gelation with $CaCl_2$ 1% w/v; (b) ionotropic gelation with $CaCl_2$ 4.9% w/v; (c) polyelectrolyte complexation with CS 0.01% w/v; (d) combination of ionotropic gelation and polyelectrolyte complexation ($CaCl_2$ 1% w/w). Scale (a, b) 200x and (c, d) 100x.

and controlled release of drugs [20]. All polyanions used in the investigation have carboxyl groups [20, 21]. The mixture of the different polyanions with the CS produced a hydrogel at a pH lower than 4, and the morphology of hydrogels is very sensitive to the preparation procedures [16, 22, 23]. In this study, differences which are related to the method used in each polymer gel were observed from SEM images. Gulrez et al. [21] claim that, at acidic pHs, the mechanism promotes hydrogen bonding, which induces a decrease in the solubility in water and results in the formation of an elastic hydrogel.

3.1. Carboxymethylcellulose- (CMC-) Based Hydrogels. Optical microscopy images of CMC hydrogels showed that mixing the polymer with $CaCl_2$ yielded conical and spiral-like particles without any visible agglomeration. On the contrary, the polyelectrolyte complexation method generates membranes in the form of sponges of different sizes, which could be reduced by the simultaneous use of the two methods of hydrogel preparation.

Micrographs obtained with an optical microscope are reported in Figure 1 for a CMC concentration of 2%, to

compare hydrogels made of CMC by different methods. Ionotropic gelation generated a transparent hydrogel, which became cloudy when the $CaCl_2$ concentration was increased. Micrographs showed that CMC hydrogels appearance changed with the method used for their production. Conical and spiral shapes were observed by using ionotropic gelation, while polyelectrolyte complexation produced a sponge-like membrane. On the contrary, the mix of procedures yielded a cloudy and viscous suspension (as $CaCl_2$ concentration increased) with elongated particles.

Many studies about gelation with CMC and $CaCl_2$ are reported in the literature, but few microscope images are showed and analyzed. Dhanaraju et al. [22], studied the ionotropic gelation of Na-CMC and Na-alginate at 5% fixed concentration of $CaCl_2$, finding that the mean size of hydrogel beads increased at higher concentration of both polymers. Huei et al. [24] showed, with optical images, that polyelectrolyte complexation of CMC/gelatin ionically crosslinked using aqueous ferric.

Hosny et al. [23] found that the size of CMC gel beads was affected by the polymer concentration (i.e., 3% w/v); however, more homogeneous forms were obtained at

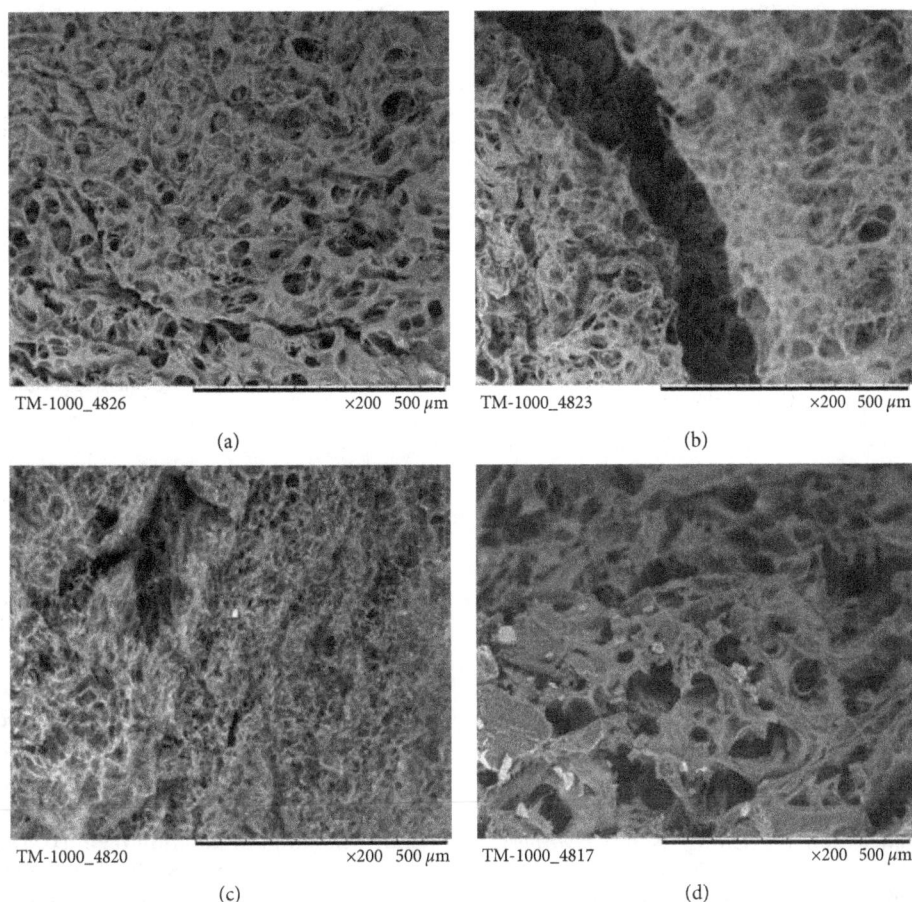

FIGURE 2: SEM images for hydrogels made from CMC 2%: (a) ionotropic gelation with $CaCl_2$ 1% w/w; (b) ionotropic gelation with $CaCl_2$ 4.9% w/v; (c) polyelectrolyte complexation with CS 0.01% w/v; (d) combination of ionotropic gelation and polyelectrolyte complexation ($CaCl_2$ 1% w/w). Scale 200x.

concentrations of 2%, attributing the increased size to the increment in viscosity of the polymer solutions.

Figure 2 shows SEM images for hydrogels made from CMC 2% by different methods. All methods produced sponge-like particles with an observable macroporous structure. In the case of ionotropic gelation, the thickness of the membrane seemed to increase at lower calcium chloride concentrations. The polyelectrolyte complexation gave a membrane with smaller pores while combining the ionotropic gelation and complexation methods produced a thicker membrane with larger pores. SEM images of both ionotropic gelation and polyelectrolyte complexation showed traces apparently of calcium salt not solubilized.

From SEM images, it is evident that the combination of methods rendered a network of larger pores and a denser network, while the ionotropic gelation generated smaller pores and a less dense network. Polyelectrolyte complexation produced even smaller pores and a spongy appearance of the membrane. Dhanaraju et al. [22] only reported SEM images where the observed porous membranes are similar to the sponges obtained in this study.

In order to examine the effect of varying $CaCl_2$ concentration on hydrogels made by ionotropic gelation, the CMC concentration was decreased to reach 1.6%, and

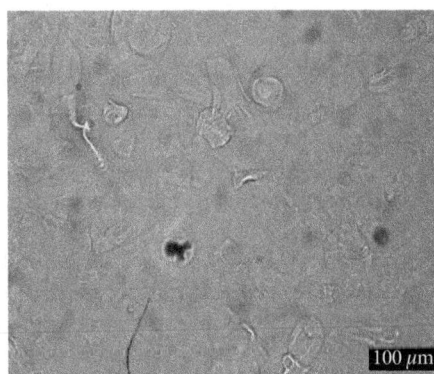

FIGURE 3: Optical micrographs of a hydrogel sample made by ionotropic gelation from CMC 1.6% and $CaCl_2$ 4.9% w/v. Scale 50x.

different concentrations of $CaCl_2$ were used (4.9, 7.35 y 14.7% w/v) to guarantee the formation of gel beads (Figure 3).

All $CaCl_2$ concentration levels resulted in particles with truncated-conical and spiral shape in the micrometer range, remaining as a sponge when the gel was lyophilized. The amount and size of these particles increased with increased quantities of the $CaCl_2$ present in the solution, suggesting that the more $CaCl_2$ concentration, the more gel was

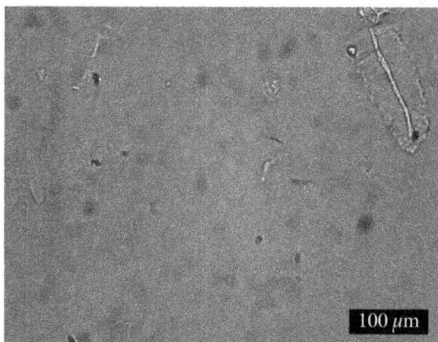

FIGURE 4: Optical micrographs of a hydrogel sample made by polyelectrolyte complexation of CS 0.016% w/v over CMC 0.8% w/v. Scale 50x.

produced. The shape noted in these hydrogels was not observed in the polyelectrolyte complexation of CMC and CS, as well as in the combination of polyelectrolyte complexation and ionotropic gelation of CMC, CS, and CaCl$_2$.

Similar particles were obtained in the case of polyelectrolyte complexation, only when CS 0.016% w/v was added drop by drop on CMC 0.8% w/v (changing the sequence of polymers addition). However, the recovery of the precipitate after centrifuging was difficult, due to its smaller size (Figure 4).

3.2. Carboxymethylated Starch- (CMS-) Based Hydrogels.

Optical micrographs are shown in Figure 5 at a CMS concentration of 3.2%. The resulting particles did not appear to change noticeably with the gel-forming method employed. The ionotropic gelation generated more individualized particles, while agglomeration was detected in polyelectrolyte complexation. The agglomeration process seemed to decrease at lower calcium chloride concentrations, as well as possibly due to interaction with CS chains in the combined procedure.

The increase in the concentration of both CMS and CaCl$_2$ in the ionotropic gelation did not produce major differences in the gel particle structures (Figure 6).

Figure 7 shows SEM images of hydrogel samples made of CMS 3.2% by different methods. All methods studied generated particles similar to raw starch, and only in the case of the polyelectrolyte complexation, it was observed in structures that could be caused by agglomeration of particles.

It is known that CMS gels in water generating particles [25]. Ionotropic gelation and the combination of gel-forming methods also yielded these types of particles. Moreover, polyelectrolyte complexation (without added salt) produced CMS hydrogels with an increased particle agglomeration, a fact confirmed by SEM images which showed amorphous macrostructures of apparently linked particles.

The difference observed between ionotropic gelation and coacervation is the initial agglomeration of the particles of CMS. According to the literature, when the particles agglomerate, they form networks between them, generating cavities attributed to the hydrophilic groups that tend to remain on the surface of the polymer. These cavities are suitable for drug loading and release [26].

Mihaela Friciu et al. [27] indicate that the largest granules observed in the CMS hydrogel are probably due to the association of numerous small particles forming larger granules. Figures 7(c) and 7(d) containing CS show particles with rough surfaces and larger than those obtained by ionotropic gelation, due to an alteration of the structure of the starch by the association of hydrogen between hydroxyl groups that promote repulsion and leads to reorganization of the network. The complex formation shows a greater amount of ionic interactions that lead to the formation of agglomerates of small particles that cover the larger granules as happens in the complexation of CMS with lecithin [27]. The formation of these large particles could help drug storage.

3.3. Alginic Acid- (AA-) Based Hydrogels.

Optical micrographs of hydrogels made from AA (0.5%) employing different techniques are reported in Figure 8. Polyelectrolyte complexation of AA and CS produced an important increment in the particle size, growing from micrometers to a millimeter order, compared to the particles generated by the ionotropic gelation method.

By increasing the AA concentration, particles with a larger size and more defined shape were obtained by means of the polyelectrolyte complexation and the combination of methods (Figure 9).

The most significant differences observed were that with alginic acid and salt addition the mixture generates spheres in the order of millimeters (the shape of the falling drop) for both ionotropic gelation and the combination of ionotropic gelation and polyelectrolyte complexation, suggesting that the cationic substances are able to envelop the AA polyanion. The blending of AA with Ca$^+$ ions has been widely studied for the formation of hydrogels by crosslinking [4], and most studies about polyelectrolyte complexation involve AA and CS [23].

Hosny et al. [23] reported that spheres of the millimetric order were formed by ionotropic gelation of alginic acid with aluminum ions. Dhanaraju et al. [22] also suggested that gel particles increased in size with a higher concentration of AA because the viscosity increased the size of the falling drop. For polyelectrolyte complexation, a few small points are observed. Hosny et al. [23] indicated as well that, in polyelectrolyte complexation, higher concentrations of AA (3%) yielded less homogeneous spheres with less stability while decreasing the concentration produced spheres of different sizes. Spheres of the millimetric order could be obtained by polyelectrolyte complexation by increasing the CS concentration.

In another study, Blemur et al. [16] found that the ionotropic gelation of AA with CaCl$_2$ formed a porous structure, while chitosan interacted with alginate to form microspheres which exhibited porosity like alginate.

Mi et al. [28] found that chitosan and alginate beads with higher crosslink density were spherical and had a relatively gross and an even cross-section and affirm that the particles consist of an inner core and an outer layer; those results are similar to the observed in this work. When the cationic charge was increased, particles with a size in the order of mm were obtained which had an even surface. Similarly, the

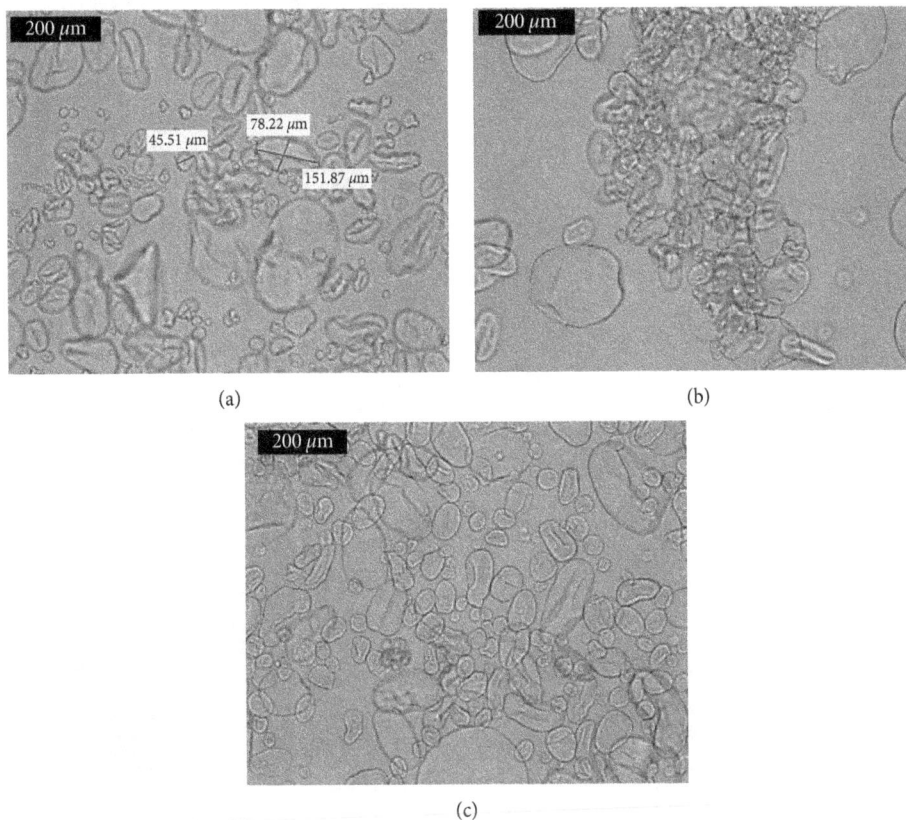

(a)

(b)

(c)

FIGURE 5: Comparison of hydrogel beads made from CMS 3.2% by (a) ionotropic gelation with $CaCl_2$ 4.9% w/v, (b) polyelectrolyte complexation with CS 0.01% w/v, and (c) combination of ionotropic gelation and polyelectrolyte complexation ($CaCl_2$ 1% w/w). Scale 100x.

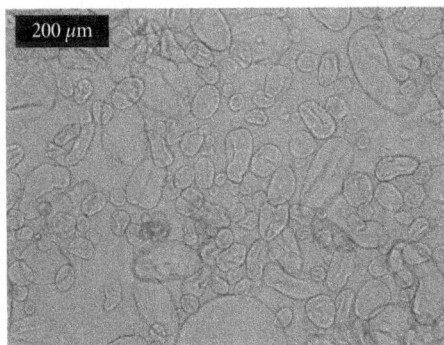

FIGURE 6: Optical micrograph of a hydrogel sample made by ionotropic gelation of CMS 5.6% and $CaCl_2$ 4.9%. Scale 50x.

complexation of alginate with chitosan also has porous walls (data not shown), and Xu et al. [29] showed that the mixture of AA with CS forms a random network that decreases pore formation with the increase in AA.

Finally, it was concluded that each polymer combination produces gels with different pore sizes that could modify the performance of the gels in drug delivery applications. When the superficial pores are in contact with water, they fill with water and facilitate the initial diffusion of the drug, controlled by the dissolution of the solute in the water of the pores and by its diffusion [30]. Besides, when the gel is hydrophilic, a progressive swelling occurs and produces changes in the shape and size of the pores with a drug

diffusion through them, then the release will be both through the pores filled with water and through the swollen polymer.

Release decreases with pore size [31], depending on the porosity of the hydrogel, the size of the drug, and the chemical properties of each, and the drug will diffuse slowly into the gel. Diffusion is regulated by movement through the polymer matrix or by mass erosion of the hydrogel to as it decomposes. Environmentally sensitive hydrogels effectively open their pores for drug diffusion [32]. Then, it can be proposed that, for CMC- and AA-based hydrogel, the release occurs in both ways and the liberation could be slower for hydrogel obtained by polyelectrolyte complexation, and in the case of CMS-based hydrogel, the release is by swelling without dependence of the mechanisms of the formation.

3.4. Zeta Potential of Hydrogels.
Table 1 shows the effect of the salt (calcium chloride) on the charge interaction of hydrogels. Calcium chloride has a smaller size than CS and can reach more sites for interaction with the anionic polymer, thus reducing the negative charge of the CMS or AA polymers.

In the study, it was also found that there was a greater equivalence of charge when the two methods of gel formation are combined. Sadeghi et al. [33] suggested that zeta potential is an indicator of charge, which is available on the gel surface, and the values reflect the charge that is not neutralized. They obtained higher negative zeta potential

FIGURE 7: SEM micrographs of hydrogel samples made of CMS 3.2% by (a) ionotropic gelation with CaCl$_2$ 1% w/w, (b) ionotropic gelation with CaCl$_2$ 4.9% w/v, (c) polyelectrolyte complexation with CS 0.01% w/v, and (d) ionotropic gelation and polyelectrolyte complexation (CaCl$_2$ 1% w/w). Scale (a, b) 400x and (c, d) 200x.

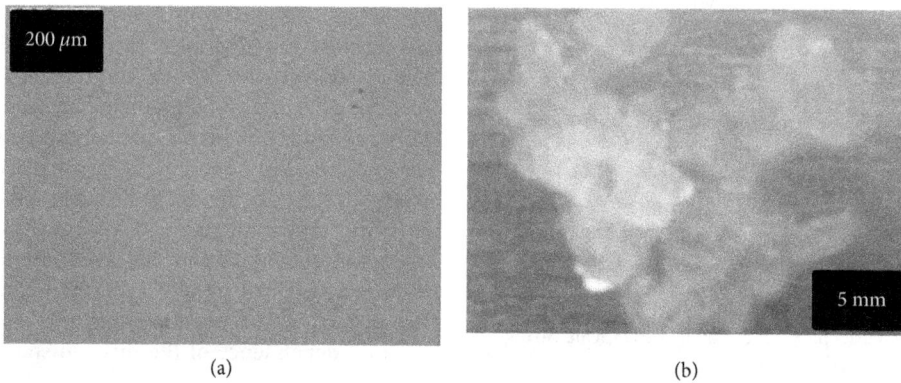

FIGURE 8: Micrographs of hydrogel samples made of AA 0.5% by (a) ionotropic gelation with CaCl$_2$ 1% w/v and (b) polyelectrolyte complexation with CS 0.01% w/v.

values for particles made by the complex polyelectrolyte method than with ionotropic gelation because the salt (cationic electrolyte present in this system) gives more positive charge, in order to neutralize more negative charges.

Then, according to Sadeghi et al. [33], polyelectrolyte complexation reveals that not all anionic polymer charges were balanced, probably due to the low concentration of chitosan. The addition of cationic electrolyte (such as in the combination of methods) decreases the magnitude of negative charges and suggests an increased molecular interaction for the resulting zeta potential values that are close to zero.

Particles with positive zeta potential values are more stable. According to Lamarra et al. [34], a higher negative charge gives larger and less stable particles, which was also observed in this work. On the contrary, negative zeta potential values seem to propitiate aggregation. It is known that positive zeta potential keeps a thicker double-layer that appears to prevent aggregation [35].

(a)

(b)

(c)

FIGURE 9: Optical microscope images of hydrogel samples made of AA 3% by (a) ionotropic gelation with $CaCl_2$ 1% w/v, (b) polyelectrolyte complexation with CS 0.01% w/v, and (c) combination of ionotropic gelation and polyelectrolyte complexation ($CaCl_2$ 1% w/w).

TABLE 1: Zeta potential for samples made by polyelectrolyte complexation and ionotropic gelation-polyelectrolyte complexation.

Anionic polymer/CS ratio	Zeta potential (mV) ionotropic gelation + polyelectrolyte complexation	Zeta potential (mV) polyelectrolyte complexation
CMC/CS, 200	−394	−962
CMS/CS, 320	−134	−636
AA/CS, 50	330	−786
AA/CS, 300	−37	−1750

3.5. *FTIR of Hydrogels.* FTIR spectra showed interactions between the cationic and anionic polymers in the polyelectrolyte complexation. The same was noted between the anionic polymers and calcium chloride for the ionotropic gelation and between the three components for the combination of ionotropic gelation and polyelectrolyte complexation methods.

Figures 10 and 11 show the IR spectra of hydrogel samples prepared from CMC and CMS by ionotropic gelation, polyelectrolyte complexation, and a combination of both. Both samples showed membrane-like structures in the micrometer order.

The characteristic absorption bands of chitosan -OH bond are at $3352 \, cm^{-1}$. At $2918 \, cm^{-1}$ there is the signal of $-CH_2$ groups. Also, the characteristics bands at 1640, 1570, and $1460 \, cm^{-1}$, belonging to vibration of carbonyl bonds (C=O) of the amide group and protonated amine group (NH_3^+) can be seen. Signals of C-H bonds are identified at 1420 and $1380 \, cm^{-1}$. The bands at 1300 and $1250 \, cm^{-1}$ correspond to C-N stretch, while the bands at 1150, 1040, and $1030 \, cm^{-1}$

belong to C-O group (COH, COC, and CH_2OH). The band at $1150 \, cm^{-1}$ is attributed to asymmetric vibrations of CO resulting from the deacetylation of chitosan. The band at $890 \, cm^{-1}$ is related to the glycosidic bonds [36].

Figure 10 shows bands for CMC at $1570 \, cm^{-1}$, $1412 \, cm^{-1}$, and $1315 \, cm^{-1}$, which are assigned to asymmetrical COO-stretching, symmetrical stretching, and C-H bending, respectively, indicating the presence of the carboxymethyl ether group. The band at $3315 \, cm^{-1}$ corresponds to -OH stretch. Also, the band at $2918 \, cm^{-1}$ can be assigned to CH- stretching of the $-CH_2$ groups. Bands at 1080, 1040, and $1030 \, cm^{-1}$ belong to C-O-C and C-O characteristic of polysaccharides. The band at $890 \, cm^{-1}$ is attributed to the glycosidic bond and the saccharide structure [37].

The CMC/CS complex has the characteristic bands of both chitosan and carboxymethylcellulose, namely, glycosidic bonds, OH, C=O, CH_2, and amide (Figure 10), which is evidence of the formation of the complex. The CMC/$CaCl_2$ gel has the characteristic bands of both carboxymethylcellulose and $CaCl_2$, and finally, the CMC/CS/$CaCl_2$ spectrum

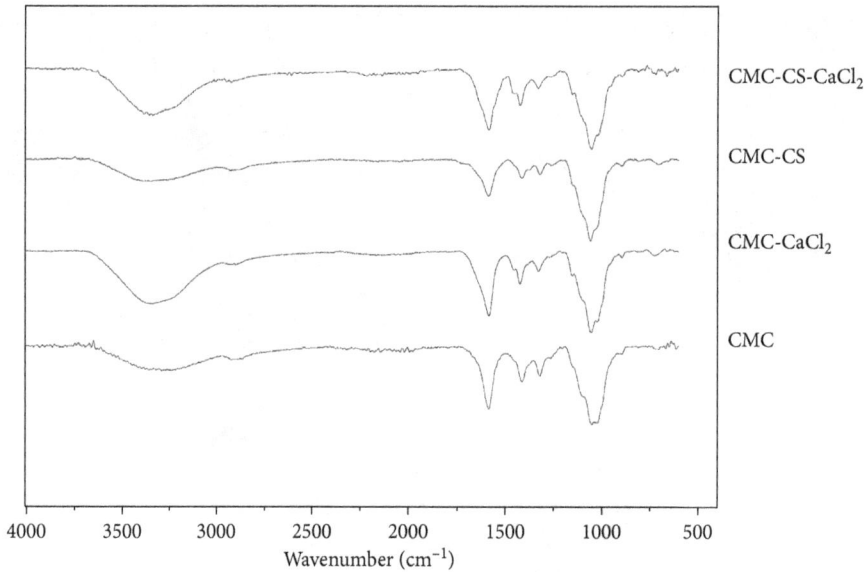

FIGURE 10: FTIR Spectra of hydrogels made from CMC (a) 2% by (b) ionotropic gelation with CaCl$_2$ 1% w/v, (c) polyelectrolyte complexation with CS 0.01% w/v, and (d) combination of ionotropic gelation and polyelectrolyte complexation mixing.

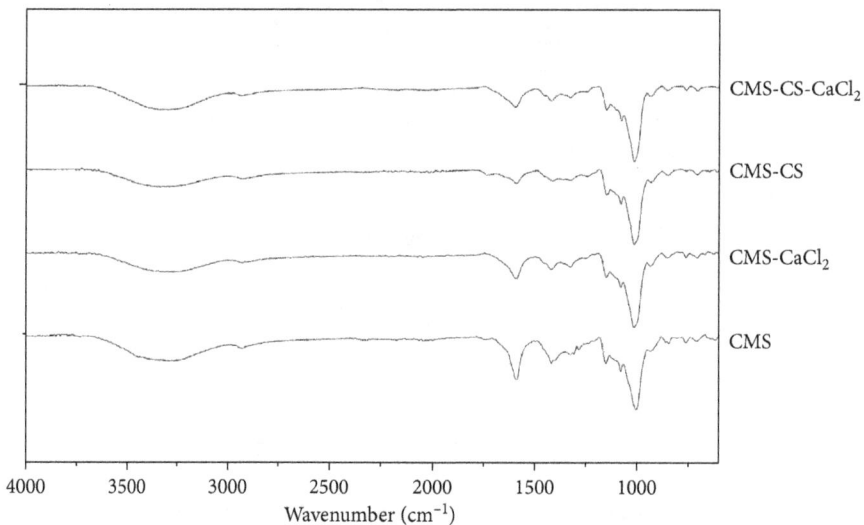

FIGURE 11: FTIR Spectra from hydrogel samples made of CMS (a) 3.2% by (b) ionotropic gelation with CaCl$_2$ 1% w/v, (c) polyelectrolyte complexation with CS 0.01% w/v, and (d) combination of ionotropic gelation and polyelectrolyte complexation.

shows characteristic of all compounds, indicating that all groups of interest are present.

The CMC/CS gel has all the characteristic bands of CMC, and also a new band in 1725 cm^{-1} due to C=O stretch; carbonyl groups are very sensitive to atoms and groups nearby, so the shift of this signal is due to the different chemical environment of the carbonyl group in the complex. The spectrum shows characteristic bands of chitosan 1640, 1380, and 1150 cm^{-1}. The signal at 1040 cm^{-1} is more attenuated than the band at 1030 cm^{-1} probably due to the same changes in the environment.

Interaction of CS polymer chains in complexation with CMS brings major differences in 3300 cm^{-1}, which suggests a rearrangement of hydrogen bonds when forming the complex, and also between 1700 and 1250 cm^{-1} bands which probably results from the new environment of the C=O. For ionotropic CMC gels, respective to the CMC alone spectra, the presence of calcium chloride increases the 3300 cm^{-1} band, makes a new band at 1490 cm^{-1} appear, decreases the 1300 cm^{-1} band, makes appear a peak in 1200 cm^{-1}, and reduces the 1100 cm^{-1} band (Figure 10). By comparison to the CMC spectrum, CS in the polyelectrolyte complexation reduces the 1620 cm^{-1} band, makes the peak at 1300 cm^{-1} more defined due to the increased concentration of C=O bands contributed by the carboxymethylated starch; also CS causes the band at 1150 cm^{-1} to disappear and also attenuates the band at 1050 cm^{-1}.

The FTIR spectrum of CMS shows peaks similar to those of CMC. Additionally, it has a weak absorption at 1730 cm^{-1}, probably due to the C=O stretch and asymmetrical COO-stretching of carboxylic acids groups present in carboxylated starch molecules [38]. Wang et al. [39] report that CMS has a strong peak around $1625\ cm^{-1}$ belonging to carboxylic groups and also another around $1421\ cm^{-1}$ of OH groups, as well as another at $1300\ cm^{-1}$ due to CH.

For CMS/CS in polyelectrolyte complexation gels, there is a noticeable band at $1700\ cm^{-1}$, which can be due to C=O bonds in a different chemical environment. The addition of calcium chloride and its interaction with CS gives a band at $1450\ cm^{-1}$; both hydrogels have bands at 1300 and $1320\ cm^{-1}$ characteristic of CMS (Figure 11).

The CMS/CS complex polyelectrolyte has more characteristic bands of carboxymethylated starch than those of chitosan (Figure 11). The spectrum shows a decrease in the signal at $3400\ cm^{-1}$ attributed to the -OH bonds; also the band at $1730\ cm^{-1}$ was shifted to $1700\ cm^{-1}$. A new band begins to be noticed at $1650\ cm^{-1}$ attributed to N-H bending, and the band at $1590\ cm^{-1}$ related to COO- groups of CMS is smaller. Bands at 1420 and $1315\ cm^{-1}$ from the CMS are also observed, which overlapped the amine bands of CS. Besides, there is an additional band at $1250\ cm^{-1}$, probably belonging to C-N stretching of chitosan. All these signals and their changes are evidence of the interaction between CMA and CS, strongly related to the new chemical environment of the polymeric chains and their functional groups in the formation of the complexes.

Regarding the interactions between the polymers and calcium chloride in the different gel preparations, the major differences were observed at $3300\ cm^{-1}$ and between 1700 and $1250\ cm^{-1}$. The OH, C=O, and C-N bonds are the functional groups mostly affected by the interactions of the polymer chains in the system. For CMC gels in relation to the CMC spectra, the formation of a salt increased the $3300\ cm^{-1}$ band, made the $1490\ cm^{-1}$ band appear, decreased the $1300\ cm^{-1}$ band, made a peak in $1200\ cm^{-1}$ appear, and reduced the signal at $1100\ cm^{-1}$ (Figure 11). In the case of CMS, differences were noticeable in the appearance of a peak at $1450\ cm^{-1}$ and the overlapping of bands between 1320 and $1300\ cm^{-1}$. Cho et al. [40] reported that stretching vibrations of O-H bonds from alginate, calcium chloride, resveratrol, and alginate microspheres that appeared in the range of $3207.59-3486\ cm^{-1}$ were changed after forming resveratrol-loaded cyclodextrin nanosponges complexes. The O-H stretching peak of alginate shifted to lower wavenumbers when resveratrol was encapsulated into alginate microspheres, suggesting that a molecular interaction between alginate and the drug was formed by hydrogen bonding. In the present work, changes between 1600 and $1250\ cm^{-1}$ bands in spectra obtained of gels made by ionotropic gelation and a combination of ionotropic gelation and polyelectrolyte complexation can be assigned to the replacement of Na+ by Ca++, which may indicate points of crosslinking due to the interaction of the components [40].

4. Conclusions

This contribution provides a more specific knowledge regarding the influence of three gel-forming methods, namely, ionotropic gelation, polyelectrolyte complexation, and a combination of both, on the morphology, size, and potential zeta of the hydrogels formed.

Charge interaction between one of the anionic polymer (CMC, AA, or CMS) and CS, $CaCl_2$, or a mixture of both, produce hydrogels with different morphological characteristics dependent on the method used. Regarding the size of hydrogel particles, in case of the ionotropic gelation, the addition of a salt, such as calcium chloride, yields definite individualized particles with AA generating particles in the range of millimeters being smaller when $CaCl_2$ is using without CS. The small size of calcium ions allows a more efficient diffusion to neutralize anionic charges along the anionic polymer chains. The polyelectrolyte complexation produces particles that tend to agglomerate. This is due probably to the difficulty of cationic and anionic polymeric chains to establish proper electrostatic interactions. The combination of ionotropic gelation and polyelectrolyte complexation techniques yields zeta potential values that approach to a neutralization of charges. The presence of calcium chloride guarantees a more balanced interaction between the negative and positive charge of polymers.

FTIR characteristic bands of the individual polymers employed in the different gel systems were reported. Shifting of these bands, mostly to higher wavenumbers, can be regarded as evidence of the molecular interactions taken place between their polymeric chains when forming the gel complexes.

Features like shape and variation of gel pores may influence negatively or positively the process of encapsulation and drug delivery, being crucial when choosing the type of hydrogel to encapsulate active substances.

Conflicts of Interest

The authors declare that they have no conflicts of interest.

Acknowledgments

The authors are grateful to the Alianza del Pacífico México for the financial support for a research stay at the Departamento de Madera Celulosa y Papel of the Universidad de Guadalajara. The authors also acknowledge the academic support provided by Colciencias (Colombia) for this research.

References

[1] J. Siepmann, R. A. Siegel, and M. J. Rathbone, *Fundamentals and Applications of Controlled Release Drug Delivery*, Springer, Berlin, Germany, 2012.

[2] A. S. Hoffman, "Hydrogels for biomedical applications," *Advanced Drug Delivery Reviews*, vol. 54, no. 1, pp. 3–12, 2002.

[3] Z. Liu, Y. Jiao, Y. Wang, C. Zhou, and Z. Zhang, "Polysaccharides-based nanoparticles as drug delivery systems," *Advanced Drug Delivery Reviews*, vol. 60, no. 15, pp. 1650–1662, 2008.

[4] C. Alvarez-Lorenzo, B. Blanco-Fernandez, A. M. Puga, and A. Concheiro, "Crosslinked ionic polysaccharides for stimuli-sensitive drug delivery," *Advanced Drug Delivery Reviews*, vol. 65, no. 9, pp. 1148–1171, 2008.

[5] M. R. Saboktakin, R. M. Tabatabaie, A. Maharramov, and M. A. Ramazanov, "Synthesis and in vitro evaluation of carboxymethyl starch–chitosan nanoparticles as drug delivery system to the colon," *International Journal of Biological Macromolecules*, vol. 48, no. 3, pp. 381–385, 2011.

[6] E. Assaad, Y. J. Wang, X. X. Zhu, and M. A. Mateescu, "Polyelectrolyte complex of carboxymethyl starch and chitosan as drug carrier for oral administration," *Carbohydrate Polymers*, vol. 84, no. 4, pp. 1399–1407, 2011.

[7] C. Calinescu and M. A. Mateescu, "Carboxymethyl high amylose starch: Chitosan self-stabilized matrix for probiotic colon delivery," *European Journal of Pharmaceutics and Biopharmaceutics*, vol. 70, no. 2, pp. 582–589, 2008.

[8] F. Bigucci, A. Abruzzo, B. Vitali et al., "Vaginal inserts based on chitosan and carboxymethylcellulose complexes for local delivery of chlorhexidine: preparation, characterization and antimicrobial activity," *International Journal of Pharmaceutics*, vol. 478, no. 2, pp. 456–463, 2015.

[9] H. Fukuda, "Polyelectrolyte complexes of Chitosan with sodium carboxymethylcellulose," *Bulletin of the Chemical Society of Japan*, vol. 53, no. 4, pp. 837–840, 1980.

[10] T. Gotoh, K. Matsushima, and K.-I. Kikuchi, "Preparation of alginate–chitosan hybrid gel beads and adsorption of divalent metal ions," *Chemosphere*, vol. 55, no. 1, pp. 135–140, 2004.

[11] R. Bilgainya, F. Khan, and S. Mann, "Spontaneous patterning and nanoparticle encapsulation in carboxymethylcellulose/alginate/dextran hydrogels and sponges," *Materials Science and Engineering: C*, vol. 30, no. 3, pp. 352–356, 2010.

[12] P. Poonam, D. Chavanke, and W. Milind, "A review on ionotropic gelation method: novel approach for controlled gastroretentive gelispheres," *International Journal of Pharmaceutics and Pharmaceutical Sciences*, vol. 4, pp. 27–32, 2012.

[13] L. Wang, E. Khor, and L.-Y. Lim, "Chitosan–alginate–CaCl$_2$ system for membrane coat application," *Journal of Pharmaceutical Sciences*, vol. 90, no. 8, pp. 1134–1142, 2001.

[14] M. Tavakol, E. Vasheghani-Farahani, and S. Hashemi-Najafabadi, "The effect of polymer and CaCl$_2$ concentrations on the sulfasalazine release from alginate-N,O-carboxymethyl chitosan beads," *Progress in Biomaterials*, vol. 2, no. 1, p. 10, 2013.

[15] J. S. Patil, M. V. Kamalapur, S. C. Marapur, and D. V. Kadam, "Ionotropic gelation and polyelectrolyte complexation: the novel techniques to design hydrogel particulate sustained, modulated drug delivery system: a review," *Digest Journal of Nanomaterials and Biostructures*, vol. 5, no. 1, pp. 241–248, 2010.

[16] L. Blemur, T. Canh Le, L. Marcocci, P. Pietrangeli, and M. A. Mateescu, "Carboxymethyl starch/alginate microspheres containing diamine oxidase for intestinal targeting," *Biotechnology and Applied Biochemistry*, vol. 63, no. 3, pp. 344–353, 2016.

[17] M. N. Ravi Kumar, "Nano and microparticles as controlled drug delivery devices," *Journal of Pharmacy and Pharmaceutical Sciences*, vol. 3, no. 2, pp. 234–258, 2000.

[18] P. Severino, C. F. da Silva, M. A. da Silva, M. H. A. Santana, and E. B. Souto, "Chitosan cross-linked pentasodium tripolyphosphate micro/nanoparticles produced by ionotropic gelation," *Sugar Tech*, vol. 18, no. 1, pp. 49–54, 2016.

[19] S. Al-Musa, D. Abu Fara, and A. A. Badwan, "Evaluation of parameters involved in preparation and release of drug loaded in crosslinked matrices of alginate," *Journal of Controlled Release*, vol. 57, no. 3, pp. 223–232, 1999.

[20] E. Russo, F. Selmin, S. Baldassari et al., "A focus on mucoadhesive polymers and their application in buccal dosage forms," *Journal of Drug Delivery Science and Technology*, vol. 32, pp. 113–125, 2016.

[21] S. K. H. Gulrez, S. Al-Assaf, and G. O. Phillips, "Hydrogels: methods of preparation, characterisation and applications," in *Progress in Molecular and Environmental Bioengineering-from Analysis and Modeling to Technology Applications*, A. Carpi, Ed., p. 5, InTech, Rijeka, Croatia, 2011.

[22] M. Dhanaraju, V. D. Sundar, S. NandhaKumar, and K. Bhaskar, "Development and evaluation of sustained delivery of diclofenac sodium from hydrophilic polymeric beads," *Journal of Young Pharmacists*, vol. 1, no. 4, pp. 301–304, 2009.

[23] E. A. Hosny, A. R. M. Al-Helw, and M. A. Al-Dardiri, "Comparative study of in-vitro release and bioavailability of sustained release diclofenac sodium from certain hydrophilic polymers and commercial tablets in beagle dogs," *Pharmaceutica Acta Helvetiae*, vol. 72, no. 3, pp. 159–164, 1997.

[24] G. O. S. Huei, S. Muniyandy, T. Sathasivam, A. Kumar, and P. Janarthanan, "Iron cross-linked carboxymethyl cellulose-gelatin complex coacervate beads for sustained drug delivery," *Chemical Papers*, vol. 70, no. 2, pp. 243–252, 2016.

[25] M. Lemieux, P. Gosselin, and M. A. Mateescu, "Carboxymethyl high amylose starch as excipient for controlled drug release: mechanistic study and the influence of degree of substitution," *International Journal of Pharmaceutics*, vol. 382, no. 1-2, pp. 172–182, 2009.

[26] G. Huang, J. Gao, Z. Hu, J. V. S. John, B. C. Ponder, and D. Moro, "Controlled drug release from hydrogel nanoparticle networks," *Journal of Controlled Release*, vol. 94, no. 2-3, pp. 303–311, 2004.

[27] M. Mihaela Friciu, T. Canh Le, P. Ispas-Szabo, and M. A. Mateescu, "Carboxymethyl starch and lecithin complex as matrix for targeted drug delivery: I. Monolithic mesalamine forms for colon delivery," *European Journal of Pharmaceutics and Biopharmaceutics*, vol. 85, no. 3, pp. 521–530, 2013.

[28] F.-L. Mi, H.-W. Sung, and S.-S. Shyu, "Drug release from chitosan–alginate complex beads reinforced by a naturally occurring cross-linking agent," *Carbohydrate Polymers*, vol. 48, no. 1, pp. 61–72, 2002.

[29] Y. Xu, C. Zhan, L. Fan, L. Wang, and H. Zheng, "Preparation of dual crosslinked alginate–chitosan blend gel beads and in vitro controlled release in oral site-specific drug delivery system," *International Journal of Pharmaceutics*, vol. 336, no. 2, pp. 329–337, 2007.

[30] R. W. Korsmeyer, R. Gurny, E. Doelker, P. Buri, and N. A. Peppas, "Mechanisms of solute release from porous hydrophilic polymers," *International Journal of Pharmaceutics*, vol. 15, no. 1, pp. 25–35, 1983.

[31] P. Horcajada, A. Rámila, J. Pérez-Pariente, and R. Vallet-Regí, "Influence of pore size of MCM-41 matrices on drug delivery rate," *Microporous and Mesoporous Materials*, vol. 68, no. 1-3, pp. 105–109, 2004.

[32] N. Bhattarai, J. Gunn, and M. Zhang, "Chitosan-based hydrogels for controlled, localized drug delivery," *Advanced Drug Delivery Reviews*, vol. 62, no. 1, pp. 83–99, 2010.

[33] A. M. M. Sadeghi, F. A. Dorkoosh, M. R. Avadi, P. Saadat, M. Rafiee-Tehrani, and H. E. Junginger, "Preparation, characterization and antibacterial activities of chitosan, N-trimethyl chitosan (TMC) and N-diethylmethyl chitosan (DEMC) nanoparticles loaded with insulin using both the ionotropic gelation and polyelectrolyte complexation

methods," *International Journal of Pharmaceutics*, vol. 355, no. 1-2, pp. 299–306, 2008.

[34] J. Lamarra, S. Rivero, and A. Pinotti, "Design of chitosan-based nanoparticles functionalized with gallic acid," *Materials Science and Engineering: C*, vol. 67, pp. 717–726, 2016.

[35] M. R. Avadi, A. M. M. Sadeghi, N. Mohammadpour et al., "Preparation and characterization of insulin nanoparticles using chitosan and Arabic gum with ionic gelation method," *Nanomedicine: Nanotechnology, Biology and Medicine*, vol. 6, no. 1, pp. 58–63, 2010.

[36] S. M. L. Silva, C. R. C. Braga, M. V. L. Fook, C. M. O. Raposo, L. H. Carvalho, and E. L. Canedo, "Application of infrared spectroscopy to analysis of chitosan/clay nanocomposites," in *Infrared Spectroscopy–Materials Science, Engineering and Technology*, T. M. Theophanideseditor, Ed., p. 524, InTech, Rio de Janeiro, RJ, Brazil, 2012.

[37] J. Wang and P. Somasundaran, "Adsorption and conformation of carboxymethyl cellulose at solid-liquid interfaces using spectroscopic, AFM and allied techniques," *Journal of Colloid and Interface Science*, vol. 291, no. 1, pp. 75–83, 2005.

[38] L. P. Massicotte, W. E. Baille, and M. Mateescu, "Carboxylated high amylose starch as pharmaceutical excipients: structural insights and formulation of pancreatic enzymes," *International Journal of Pharmaceutics*, vol. 356, no. 1-2, pp. 212–223, 2008.

[39] S. Wang, X. Sun, F. You, H. Dai, S. Mao, and J. Wang, "Application of cationic modified carboxymethyl starch as a retention and drainage aid in wet-end system," *Bioresources*, vol. 7, no. 3, pp. 3870–3882, 2012.

[40] A. R. Cho, Y. G. Chun, B. K. Kim, and D. J. Park, "Preparation of alginate–CaCl$_2$ microspheres as resveratrol carriers," *Journal of Materials Science*, vol. 49, no. 13, pp. 4612–4619, 2014.

13

Upgrading of Carbohydrates to the Biofuel Candidate 5-Ethoxymethylfurfural (EMF)

Xiaofang Liu ⓘ **and Rui Wang**

Guizhou Engineering Research Center for Fruit Processing, Food and Pharmaceutical Engineering Institute, Guiyang University, Guiyang 550005, China

Correspondence should be addressed to Xiaofang Liu; liuxfzap@163.com

Guest Editor: Masaru Watanabe

5-Ethoxymethylfurfural (EMF), one of the significant platform molecular derivatives, is regarded as a promising biofuel and additive for diesel, owing to its high energy density ($8.7\,kWh \cdot L^{-1}$). Several catalytic materials have been developed for the synthesis of EMF derived from different feedstocks under relatively mild reaction conditions. Although a great quantity of research has been conducted over the past decades, the unsatisfactory production selectivity mostly limited to the range 50%–70%, and the classic fructose used as the substrate restricted its application for fuel manufacture in large scale. To address these production improvements, this review pays attention to evaluate the activity of various catalysts (e.g., mineral salts, zeolites, heteropolyacid-based hybrids, sulfonic acid-functionalized materials, and ionic liquids), providing potential research directions for the design of novel catalysts for the achievement of further improved EMF yields.

1. Introduction

Diminishing fossil reserves and growing environmental problems have determined research for sustainable, green, and environmentally benign resources for liquid fuels and chemicals [1–3]. Biomass is widely available, inexpensive, and a CO_2-neutral source of carbon, the catalytic conversion of which to platform chemicals has potential to substitute the products from nonrenewable fossil sources [4–6]. A large number of strategies have been investigated for the conversion of carbohydrates in lignocellulosic biomass into chemicals and fuels [7–9]. The choice of an appropriate catalyst plays a significant role in observing high conversion and selectivity to the target chemicals in a sustainable, green, and economic process [10].

The main purpose of this review is to evaluate the activity of various catalysts (e.g., mineral salts, zeolites, heteropolyacid-based hybrids, sulfonic acid group-functionalized materials, and ionic liquids) with different catalytic effects and functional groups for the production of EMF from HMF,

fructose, glucose, and other carbohydrates under the applied reaction conditions, providing potential research directions for the design of novel catalysts for the achievement of further improved EMF yields. Fossil fuels and the derived chemicals have been produced from the limited natural sources. Increasing demand for limited fossil fuels and environmental degradation is gradually severe. Thus, replacing the fossil fuels with alternative and sustainable energy sources is imperative [11, 12]. Biomass is regarded as the only sustainable source of organic carbon compounds that have been suggested as the ideal equivalent to petroleum for the synthesis of fuels and chemicals. Biomass is widely existing and available as a proper feedstock, whose production estimates 1.0×10^{11} tons per year. Extensive research and studies have been conducted to produce biofuels and biodegradable products from biomass [13–15].

Among multiple furan derivatives derived from biomass, EMF is recognized as an excellent additive for regular diesel with promising properties as follows:

(i) With high boiling point (508 K) in comparison with diesel fuel.

(ii) High energy density (30.3 MJ·L^{-1}) is similar to regular gasoline (31.9 MJ·L^{-1}) and is compared to diesel fuel (33.6 MJ·L^{-1}), which is notably higher than ethanol (21 MJ·L^{-1}) [16].

(iii) Due to its low toxicity, EMF can be used as a flavor and aroma ingredient in the food beverage industries [17–21].

(iv) To blend EMF with diesel fuel in a diesel engine reduces the formation of particulate contamination, SO$_2$ emissions, and soot. Meanwhile, EMF-blended fuel can make the engine run smoothly for hours [11, 12, 22, 23].

(v) Ethers that included EMF have a high cetane number [16], which is a very important factor for combustion performance and emission.

(vi) No hydrogenation step is required in EMF production, which is an advantage over other fuel additives (e.g., DMF) obtained from HMF production.

Considering the fact that EMF plays a significant role as a fuel candidate, diesel fuel additive [24], FDCA, or cyclopentenones precursor [25, 26], it is vital to review their different production aspects with detailed attention. Thus, recent progress in EMF synthesis is summarized in this review.

2. Synthesis of EMF

In order to produce EMF, HMF should participate in the etherification reaction with ethanol in the presence of an acid catalyst. Fructose or glucose may also be used as an initiator feed in large scale because their price is considerably low compared to fructose and HMF. Production of EMF from disaccharides, polysaccharides, and biomass through a "one-pot" approach is economically desirable due to the fact that the costs of saccharification, sugar isolation, and purification can be eliminated. However, the yield of synthesized EMF decreases significantly when HMF or fructose has not been used as the feed, and the possible reaction pathways for the synthesis of EMF from carbohydrates are presented in Scheme 1. The proposed reaction mechanism of hexose conversion to EMF is shown in Scheme 2.

Another method to produce EMF from different feedstocks is to follow a multistep mechanism. In the first step of the mechanism, an intermediate is produced, and then the intermediate is converted to EMF at very high yield. A representative example of the multistep method is found in the preparation of halomethylfurfural, namely, 5-chloromethylfurfural (CMF), where the reaction of biomass, polysaccharides, or C6 sugars with HCl leads to the production of CMF, and then, EMF is produced by nucleophilic substitution of CMF with ethanol [27]. Herein, we reviewed and compared different acid catalysts and their efficacies on the production yields of EMF from various feedstocks. Optimal reaction conditions for the case were presented.

2.1. Homogeneous Mineral Salts. Soluble catalysts or the homogeneous catalysts possessed excellent catalytic performance and the fast reaction rate, the existent ions of which are highly dispersed through the medium to enhance the availability of active sites with reactants. Generally, mineral salts acted as Lewis acid in favor of converting the glucose-based substrates to synthesize EMF. However, the separation and recycle wasted energy are not catered to the green and sustainable development.

Additionally, the heterogeneous catalytic reaction caused lower corrosion for the environment and equipment. Therefore, much more researchers concentrated on solid acid catalysts and enhanced their performances and activities [28–30]. Reaction temperature plays a crucial role in carbohydrates transformation and EMF synthesis, and the temperature ranges from 70°C to 160°C. The major byproducts existing in the system were EL, 5-(ethoxymethyl) furfural diethylacetal (EMFDEA) [17], and 5,5′(oxybis(methylene))bis-2-furfural (OBMF) [31].

Metal chloride always is a commercially available Lewis acid with low toxicity and a high catalytic activity for the conversion of hexoses. Series of metal chlorides have been introduced for the production of EMF (Scheme 3). Initially, Liu et al. [32] examined various inorganic salts to promote fructose into EMF with NH$_4$Br, CuCl$_2$·4H$_2$O, and NiCl$_2$·6H$_2$O giving low yields of EMF. Among those mentioned above, NH$_4$Cl demonstrated superior catalytic activity with 42% total yield of HMF and EMF under optimum reaction condition. Simultaneously, Yang et al. [33] chose the AlCl$_3$·6H$_2$O as a catalyst for carbohydrates including glucose, sucrose, maltose, cellobiose, starch, and cellulose and converted to EMF in ethanol/water binary solvents. The higher total furans yield (included HMF and EMF) of 57% was obtained, and a moderate EMF yield of 40% was derived from sucrose. However, the conversion efficiencies of other carbohydrates to furan were low. The highest EMF yield through single step reaction from HMF was proved to be 92.9% catalyzed by AlCl$_3$ at 100°C for 5 h by Liu et al. [31]. With the same carbohydrates conversion system, Yu et al. [34] performed the use of large-scale common metal salts as catalysts for the preparation of EMF to understand the catalytic mechanism. According to the research, AlCl$_3$ gave good catalytic activity for producing EMF from fructose at 140°C, while CuSO$_4$ and Fe$_2$(SO$_4$)$_3$ provided comparable EMF yields at 110 to 120°C, and the latter favored the EL formation.

To enhance the catalytic performance, Jia et al. [35] explored combinations of AlCl$_3$ with different cocatalysts such as B(OH)$_3$ and BF$_3$·(Et)$_2$O or halide salts such as NaF, NaCl, and NaBr. The result of BF$_3$·(Et)$_2$O/AlCl$_3$·6H$_2$O proved to be significantly superior to other catalysts; 55.0%, 45.4%, and 23.9% high yields of EMF derived from fructose, inulin, and sucrose were achieved, respectively, under the optimized conditions. Later, Zhou et al. [36] studied the reactivity of FeCl$_3$, CrCl$_2$, GeCl$_4$, IrCl$_3$, and so on for the production of EMF. The presented toxicity, high price, and instability drawbacks of other metal chlorides affect the further exploration. In the presence of FeCl$_3$, a maximum EMF yield of 30.1% was obtained in the mixture solvent composed of [Bmim]Cl and ethanol.

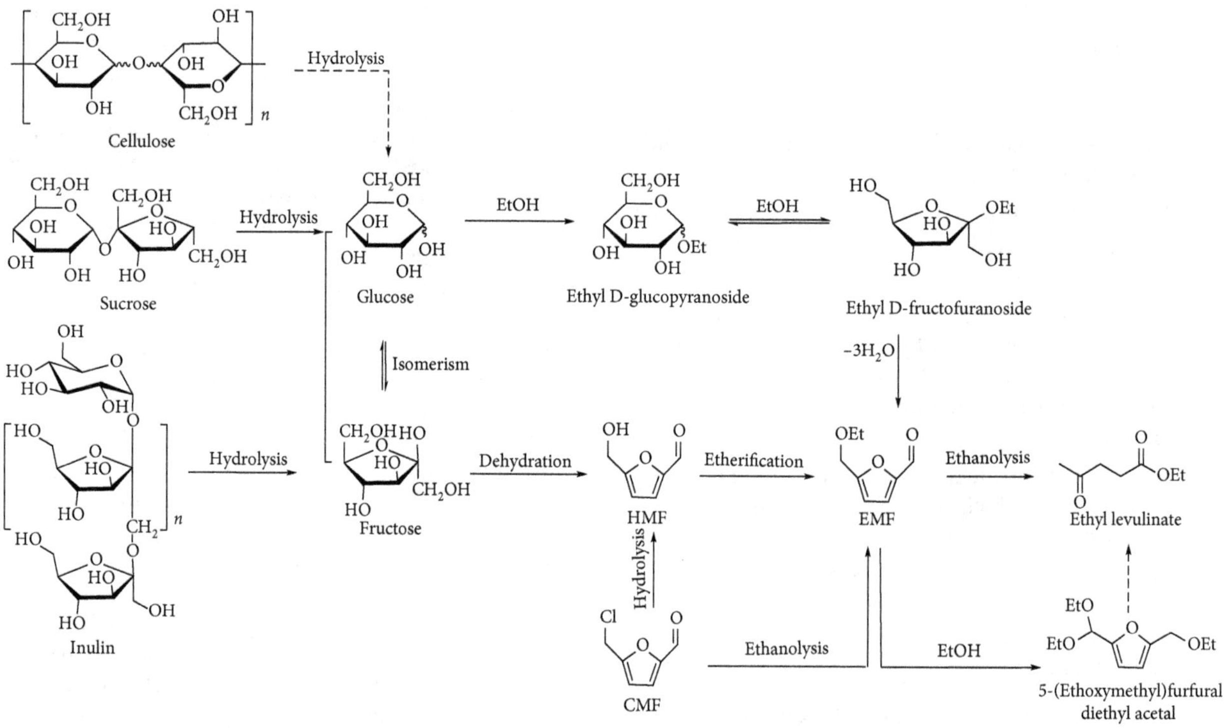

SCHEME 1: Reaction pathways for the synthesis of EMF from carbohydrates.

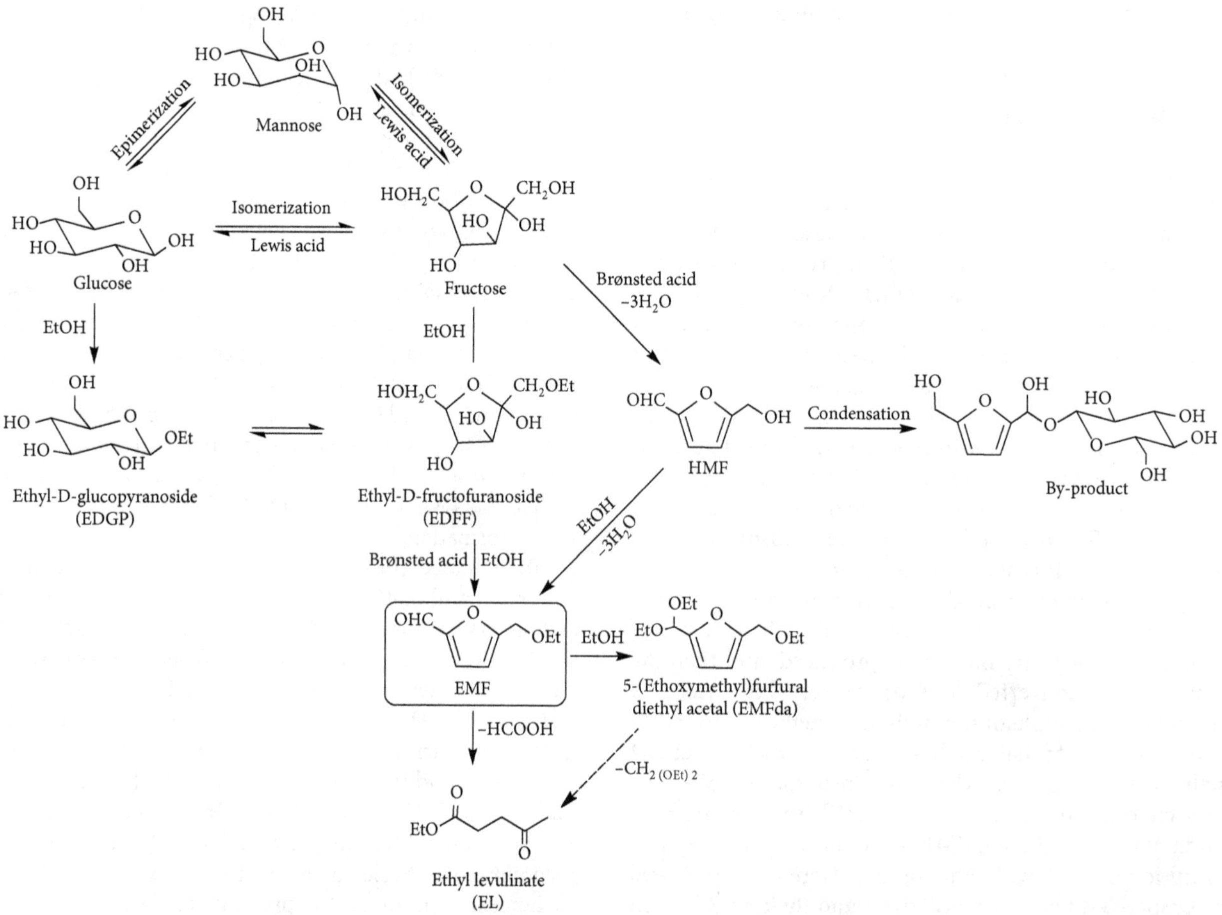

SCHEME 2: Proposed mechanism for one-pot conversion of hexose to EMF.

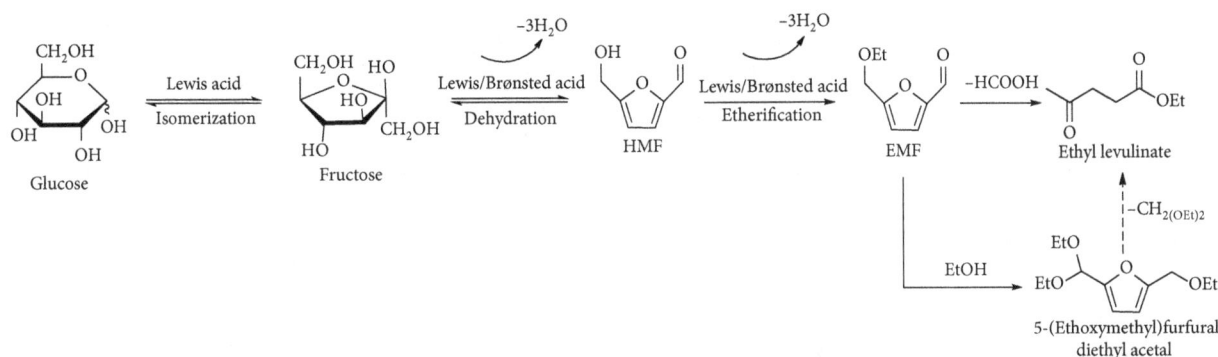

SCHEME 3: Reaction pathways for the conversion of hexose to EMF catalyzed by Lewis acid.

2.2. Heterogeneous Catalysts

2.2.1. Zeolitic Catalysts.
Zeolites with the highly dispersed and uniform channel microporous structure typically restrict the formation of large and unwanted by-product for the EMF preparation system. Meanwhile, the micropores also limit the effective diffusion of HMF and EMF due to the molecular sizes and small pore openings of zeolites [37, 38]. Expansion of the pore sizes needed to be further explored for the application in more areas.

The enlarged pore sizes play an important role for these materials. For instance, Lanzafame et al. [29] conducted the efficient etherification of HMF catalyzed by mesoporous Al-MCM-41 (with different Si/Al ratios), zirconia, or sulfated zirconia immobilized on SBA-15. Strong Lewis acid sites ZrO_2 and isolated Al^{3+} sites were explained to be beneficial for EMF production. In addition, strong Brønsted acid sites led to the formation of ethyl 4-oxopentanoate. Investigation on mesoporous aluminosilicates by further introducing aluminum into the framework Al-TUD-1 (Si/Al ratio: 21) was carried out by Neves et al. [39]. Based on the unique and regular channel of MCM-41, Che et al. [30] illustrated a nanosphere catalyst with highly dispersed $H_4SiW_{12}O_{40}$/ MCM-41 for the etherification of the hydroxyl group of HMF. Comparing the catalytic performance with p-TSA, H_2SO_4, Amberlyst-15, and H_3PO_4, $H_4SiW_{12}O_{40}$/MCM-41 showed 84.1% selectivity to EMF with 92.0% conversion of HMF under mild conditions. Meanwhile, the intermediate 5,5′(oxybis(methylene))bis-2-furfural could also be converted to target product EMF. They found that, with the strength of acids and heteropoly anion effects, the effective catalytic performance of $H_4SiW_{12}O_{40}$/MCM-41 was obtained. In this regards, Liu et al. [40] developed MCM-41-HPW with different dosages of phosphotungstic acid (HPW) supporting on MCM-41. The optimal 40 wt.% MCM-41-HPW achieved a high EMF yield of 83.4%, converted from HMF at 100°C for 12 h.

Inspired by the published results that tin-containing zeolite (Sn-BEA) was effective for the glucose-fructose-HMF conversion, Lew et al. [41] further employed Sn-BEA combined with Amberlyst-131 for one-pot glucose-to-fructose-to-HMF-to-EMF conversion, giving 31% yield of EMF. Li et al. [42] found DeAl-H-beta could provide a moderate EMF yield (43%) from sucrose, and the EMF yield

was improved up to 50% by a one-pot two-step method. However, poor catalytic performance was achieved using cellobiose as a substrate for EMF. Lewis et al. [43] explored a series of zeolites including Hf-, Zr-, Ti-, Ta-, Nb-, and Sn-Beta that could promote the etherification reaction of HMF. Under batch conditions, Ta-Beta and Al-Beta showed comparable EMF yields of 56% and 41%, respectively.

Product selectivity depends on the control of active sites strength. Barbera et al. [44] introduced NH_4-exchanged zeolites, and the NH_4 was used to block the strong sites, thus preventing the secondary reactions. The results proved that NH^{4+} effectively increased the product selectivity of NH_4-BEA catalyst for HMF etherification reaction. Recently, Bai et al. [45] exhibited the glucose conversion to EMF catalyzed by multifunctional MFI-Sn/Al zeolite with dual meso-/ microporosity and dual Lewis and Brønsted acidity. Combination of Lewis acidic Sn and Al sites and Brønsted acidic Al-O(H)-Si sites was efficient for the cascade isomerization-dehydration-etherification reaction, affording 44% EMF yield.

2.2.2. Heteropolyacid-Based Hybrid Catalysts.
Heteropolyacids have several advantages such as strong Brønsted acidity, tunable acid-base properties, and high proton mobility. But, they tend to dissolve in water and polar solvents and possess low surface area and low thermal stability, which limits their application in catalytic conversion. Supporting a solid support is a method that has been used by different groups.

Originally, Yang et al. [46] investigated the fructose-EMF transformation in a mixed ethanol/THF (tetrahydrofuran) medium catalyzed by $H_3PW_{12}O_{40}$ (HPW). The microwave offered EMF in yield of 76% under the optimum reaction condition, and the cosolvent THF significantly increased fructose conversion. Wang et al. [47] compared the activity of HPW, phosphomolybdic acid (HPM), $AlCl_3$, H_3PO_4, and Amberlyst-15 for production of EMF from fructose in an ethanol/DMSO mixture. Results indicated that HPW and HPM exhibited superior catalytic activities than $AlCl_3$, H_3PO_4, and Amberlyst-15, and the former formed 64% EMF within 130 min at 140°C. In this regard, Ren et al. [48] exchanged H^+ ion of HPW with Ag^+ and achieved Ag_1H_2PW, which presented a high activity and EMF yield of 88.7%. Further replacing another H^+ in the

Ag$_1$H$_2$PW, the acid strength of the catalyst decreased. Ren et al. used Ag$_1$H$_2$PW catalyst for fructose dehydration and HMF etherification, and a relatively high yield of EMF (69.5%) was obtained.

Stability and efficient separation are crucial criteria to be the catalyst support. Addition of heteropolyacid into Fe$_3$O$_4$@SiO$_2$, simultaneously with the introduction of the silica layer could modify the MNPs-formed inert surface. Fe$_3$O$_4$@SiO$_2$-HPW catalyzed the reaction of HMF with ethanol, and therefore, the activity increased dramatically and the production of EMF yield reached 83.6% [49]. Wang et al. [50] also applied Fe$_3$O$_4$@SiO$_2$-HPW for EMF production from HMF to obtain the same effect, which was in accordance with the results of Liu and Zhang [51]. Besides, Fe$_3$O$_4$@SiO$_2$-HPW achieved an EMF yield of 54.8% derived from fructose.

Organic-inorganic hybrid materials become the hotspot to be the catalyst, for instance, HPA-based [MIMBS]$_3$PW$_{12}$O$_{40}$ hybrid catalyst reported by Liu et al. [52] for the preparation of EMF. Combination of IL with H$_3$PW$_{12}$O$_{40}$ enhanced the EMF yield to 90.7% with 98.1% conversion of HMF under the optimum reaction condition at 70°C within 24 h. Wang et al. [53] studied the uniform nanospheric hybrids by controlling the molar ratio of phosphotungstic acid (HPW) and pyridine (PY) or trimethylamine (TEA) in different nanosizes. Particularly, PY-PW-1 showed an excellent catalytic activity with 90% of EMF, which attribute to the relatively strong acidity and regular pore size. With the prepared nanosphere PY-PW-1, fructose was also converted to EMF in one-pot reaction process and obtained a moderate yield of 55% with the enhanced temperature at 120°C.

The synthesized K-10 clay-HPW not only avoided the HPW being dissolved into polar organic solvents and expressed the high activity. Liu et al. [54] used 30 wt.% of K-10 clay-HPW for HMF-EMF transformation. Under the optimum reaction conditions (100°C, 10 h), catalytic activity was increased to 91.5% yield, whereas for one-pot fructose conversion to EMF, 61.5% yield of EMF was gained. The prepared catalyst with high stability could be reused several times without loss of catalytic performance. MCM-41 with uniform channel structure and pore size was suitable to support 12-tungstophosphoric acid that allowed the one-pot conversion of fructose directly into EMF, leading to 42.9% EMF by 40 wt.% MCM-41-HPW [40].

Apart from K-10 clay and MCM-41, MOF-based [Cu-BTC][HPM] (NENU-5) was also efficient for the etherification of HMF to EMF with up to 68.4% yield under normal pressure and optimized conditions (Figure 1). Benefiting from the unique structure, MOFs can provide adsorption sites for HPM to avoid HPM dissolving into solvents, making the catalysts be recyclable with 55% EMF yield [55].

The addition of cosolvent exhibits improvement in EMF yield as compared with the single solvent, and enhanced yield of EMF ranges from 5 to 20 percentages. Xu et al. [56] explored the ethanol/n-hexane mixture solvent indicating that the introduction of n-hexane could enhance the yield of EMF to 66.3% with optimum volume ratio (ethanol:n-hexane, 6:4), conducted at 120°C within 180 min. With the ethanol-DMSO mixture, Li et al. [57] also developed acid-base bifunctional hybrid nanosphere catalyst Lys/PW that possessed an optimal reaction activity of 76.6% EMF yield from fructose, implying that the base sites of the catalysts play an important role in increasing EMF stability.

Combination of Lewis acid AlCl$_3$·6H$_2$O and Brønsted solid acid PTSA-POM proved cooperative effect for glucose transformation into EMF and generated 30.6% EMF under the ethanol-water (9:1) system. [58].

The appropriate supports play a vital role in heteropolyacid as the active center producing EMF. Partial substitution of protons is another effective approach to change the property of heteropolyacid for better catalytic activity. Raveendra et al. [59] reported the cesium-exchanged silicotungstic acid to enhance the surface area and acidity and finally obtained 91% EMF yield with Cs$_2$STA at 120°C for 2.5 h.

2.2.3. Sulfonic Acid-Functionalized Catalysts.

The strong acidity of sulfonic acid groups can be immobilized onto the diverse supporter by sulfonation reactions and hence gain multiple sulfonic acid-functionalized catalysts. Simultaneously, the acidic density can be flexibly controlled by the dosage of sulfonic sources. Considering the green and sustainable development request, loss of element S was not catered for the present development.

Magnetic carriers to load sulfonic acid groups can be easily separated by a permanent magnet and recycled for times without loss of activity attracting much attention of scientists. Zhang et al. [60] chose Fe$_3$O$_4$@SiO$_2$ for preparation supporting a high yield (up to 89.3%) of EMF converted from HMF. Magnetic nanoparticles (MNPs) and amorphous carbon were selected to immobilize sulfonic acid groups (−SO$_3$H) which act as a solid acid catalyst for EMF production [61]. Successfully, Fe$_3$O$_4$@C-SO$_3$H shared active catalytic properties in HMF-EMF, fructose-EMF, sucrose-EMF, and inulin-EMF systems, and considerable yields of 88.4%, 67.8%, 33.2%, and 58.4% were obtained, respectively. Among those, the low yield of EMF was produced from sucrose owing to the composition of sucrose, which contains one fructose and glucose unit, and glucose cannot be converted to EMF directly. The more deep-seated research was carried out by Fe$_3$O$_4$@SiO$_2$-SH-Im-HSO$_4$ [62] and facilitated by it, indicating that EMF yields derived from inulin (56.1%) and fructose were consistent while glucose always promoted to ethyl glucoside. Soon afterward, Wang et al. [63] illustrated the smooth conversion of sucrose and inulin with superior catalytic performance offering 53.6% and 26.8% EMF yields, respectively, at 140°C for 24 h. Excellent catalytic performance of OMC-SO$_3$H indicated its promising application for biomass conversion into value-added chemicals and liquid fuels.

Considering the sustainable and green development trend, the same magnetic Fe$_3$O$_4$@C-SO$_3$H prepared from wheat straw biomass afforded comparable EMF yield of 64.2% in DMSO-ethanol binary solvent. Bearing multiple

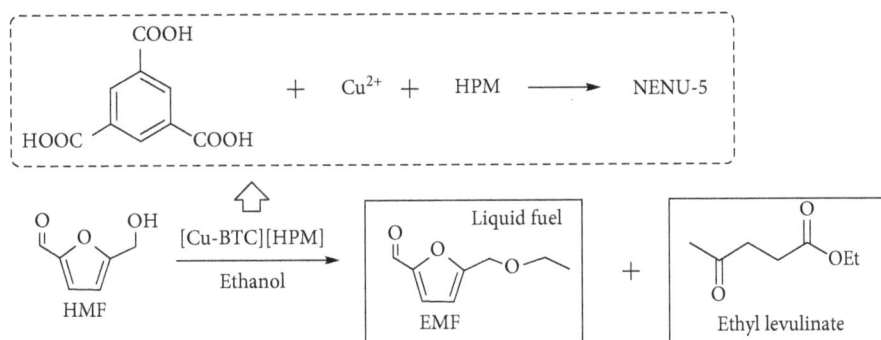

FIGURE 1: Schematic of the catalyst preparation and catalytic procedure.

-COOH, -SO$_3$H, and -OH groups, the authors described the progress of fructose or HMF to EMF in relatively high yields with 64.2% and 85.6% [64].

Grafting sulfonic acid onto the inorganic insoluble supports is a technique to prepare high-performance heterogeneous solid acidic catalysts. Liu and Zhang [51] studied the grafting of sulfonic acid onto the mesoporous silica and utilized for effective etherification of HMF for EMF with 83.8% under the optimal conditions (100°C, 10 h), while the high yields of 63.1% and 60.7% were synthesized by one-pot transformation of fructose and inulin. However, the aldose-based carbohydrates such as glucose were mainly turned to ethyl D-glucopyranoside with a high yield of 91.7%. The unique properties such as 2D structure, high stability, and high surface areas decided GO as a promising catalytic material that was oxidized by Hummers' method clarified the engineered catalytic performance for the conversion of HMF into EMF with high yield of EMF (92%), while EMF yield was decreased when fructose replaced HMF as a starting material [65] (Figure 2). Under mixture solvent ethanol-DMSO in 3 : 7, furan selectivity enhanced to 71% EMF yield and 34%, and 66% EMF yields were achieved when sucrose and inulin were used as the substrates. Partially reducing GO, amorphous carbon black (CB), and carbon nanotubes (CNTs) by sulfuric acid (S-RGO) were conducted for the synthesis of EMF converted from HMF. The catalytic activity of S-RGO proved to be superior to S-CB, S-CNTs, and even the classic Amberlyst-15 [66].

Besides, new solid acid catalysts obtained by sulfonating natural biopolymers met the green trend for biomass conversion to high value-added chemicals. Liu et al. [67] developed the cellulose sulfuric acid for the etherification of HMF for the synthesis of EMF and observed high yield (84.4%) under optimized condition. Similarly, application of glucose-derived magnetic solid acid, glu-Fe$_3$O$_4$-SO$_3$H, also acted as an etherification catalyst and generated 92% isolated yield EMF [68]. With the optimal 50 wt.% loading rate of glu-Fe$_3$O$_4$-SO$_3$H, EMF from fructose was effectively produced by 81% yield.

Amberlyst-15 as one of the standard solid acid catalysts was applied in various reaction systems and first explored by Zhu et al. [69] for the HMF-5-methoxymethylfurfural (MMF) conversion in low-boiling point solvent. The presence of cosolvent THF enhanced the contact of -SO$_3$H of Amberlyst-15 leading to promoting the catalytic progress

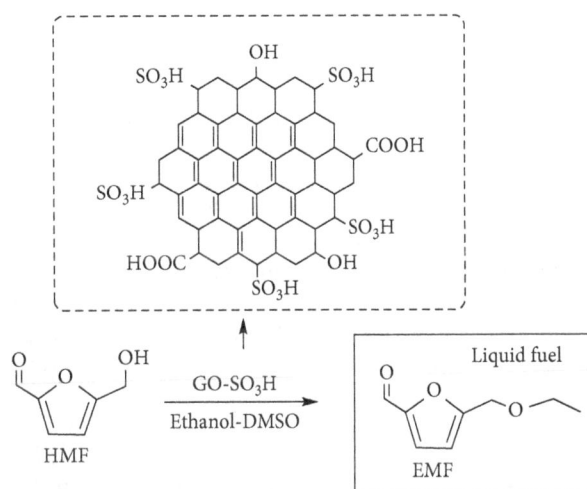

FIGURE 2: The conversion of HMF to EMF catalyzed by GO-SO$_3$H.

and accumulating the target molecules. Because of short of the high surface area and the limited contact of active sites and substrate, the catalytic performance of Amberlyst-15 was dissatisfactory. MOF provided large enough specific surface area and acidic density, and MIL-101-SO$_3$H(100) showed a better catalytic performance with 89.2% conversion of fructose and 67.7% EMF yield in the ethanol and THF mixture.

Well-defined structure and uniform distribution of active sites catalyst arenesulfonic acid-modified SBA-15 Ar-SO$_3$H-SBA-15 was studied for the conversion of fructose to EMF in binary ethanol-DMSO solvent in optimizing reaction conditions (e.g., temperature, catalyst loading, and DMSO concentration), demonstrating the maximum EMF yield of 63.4% [70]. Li et al. [71] generated a series of SO$_3$H-functionalized polymer solid acid catalysts for successive fructose dehydration and HMF etherification, and an EMF yield of 72.8% was achieved with the optimal catalyst poly(VMPS)-PW at 110°C within 10 h.

2.2.4. Ionic Liquid Catalysts. Due to the tailored design, good thermal and chemical stability, low melting point, and good solubility of acidic ionic liquids, the investigation of

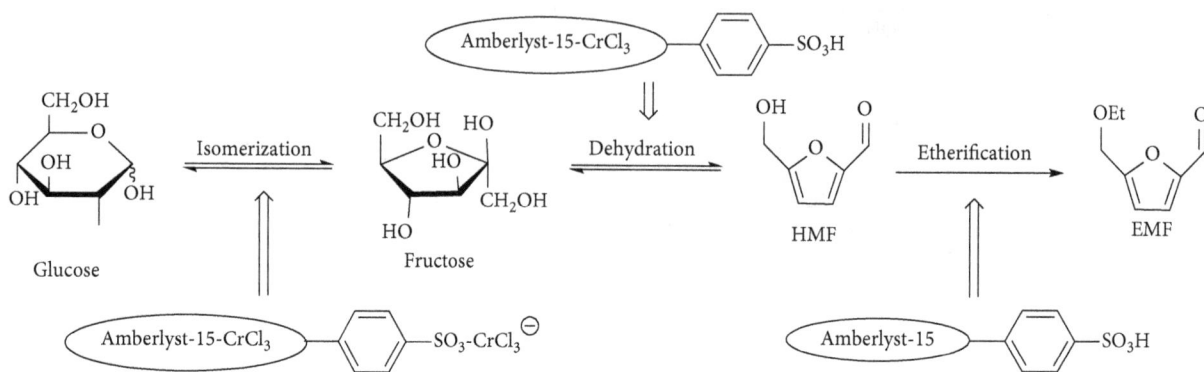

FIGURE 3: Schematic illustration of the catalysis progress.

them attracts much attention. The problem of high cost and recycle is urgently to be solved.

Kraus and Guney [72] appointed the sulfonic acid-functionalized IL to catalyze the conversion of fructose for the EMF fabrication in a one-pot procedure without the addition of solvent or acid catalysts. The application of 1-butyl-3-(3-sulfopropyl)-imidazolium chloride (4) and 1-methyl-3-(3-sulfopropyl)-imidazolium chloride (5) for EMF provided comparable 55% and 54% yields under the mild condition (100°C, 80 min), and the novel biphasic medium assembly of hexanes and IL enhanced the EMF yields to a certain extent.

The Brønsted acidic IL $[DMA]^+[CH_3SO_3]^-$ was effective for fructose-EMF conversion with a relatively high yield of 64%, while one-pot cellulose transformation gave 22% yield of EMF [73]. Alam et al. [37] further showed EMF production directly from Foxtail and Red nut sedge weeds in presence of $[DMA]^+[CH_3SO_3]^-$, which was better than $[NMP]^+[CH_3SO_3]^-$ when converting foxtail to EMF.

Except the ionic liquid catalysts, functionalized ionic liquids can also act as the reaction medium to afford efficient EMF yield. Guo et al. [74] chose hydrogen sulfate ionic liquid as a homogeneous catalyst and mixed with ethanol as a reaction solvent for efficient preparation of EMF. Catalyzed by the $[C_4mim][HSO_4]$-ethanol system, high yield of 83% EMF was observed from fructose at 130°C within 20 min, which could be attributed to the acidity of $[C_4mim]$ $[HSO_4]$, the viscosity of the mixture system, and the formed hydrogen bonds between $[C_4mim][HSO_4]$, ethanol solvent, and fructose.

Due to the superior physical and chemical properties and benefits of acidic ILs, an effective progress for EMF derived from carbohydrates by deep eutectic solvent (DES) mixture as a solvent was proposed [75]. 77.3% EMF yield was gained with fructose as substrate catalyzed by Amberlyst-15, and more excellent catalytic performance afforded a high EMF yield of 46.7% obtained from glucose with $CrCl_3$-modified Amberlyst-15 (Figure 3).

2.2.5. *Others.* Except for the mentioned catalysts in the previous section, there exist many other effective materials such as Zr-Mont, Co(x)Pc, $MoO_2Cl_2(H_2O)_2$, and a

combination of metal chloride with resins which could be also used for carbohydrates transformation to EMF.

Acidic Zr-montmorillonite (Zr-Mont) catalyst was utilized for synthesis mixture of EMF and 2-(diethoxymethyl)-5-(ethoxymethyl)furan by Shined and Rode [76], and the latter can turn into EMF by conducting in water with the same Zr-Mont. Under the optimized reaction conditions, EMF was observed in the highest yields of 91% by etherification HMF with ethanol. Yadav et al. [77] investigated the use of cobalt (I, II, III) phthalocyanine (Co(x)Pc) for the isomerization of glucose to fructose, for fructose dehydration, and in the subsequent etherification. It was noted that Co(x)Pc was effective for the direct conversion of carbohydrates to EMF in [EMIM]Cl ionic liquid, and Co(III) Pc exhibited much higher catalytic activity. The levulinate could be reduced by the waste and basic additive oil shale ash, which is formed by burning oil shale in power plants, while it was efficient for the preparation of EMF with high yield and purity [78]. The final product EMF syntheses starting from 5-bromomethylfurfural (BMF) afforded 88% yield without further purification which reacted under room temperature for 17 h in 96% ethanol.

Considering the excellent catalytic property for the reduction reaction, high valent oxomolybdenum complexes were investigated for the synthesis of EMF by one-pot fructose conversion in ethanol/THF (5:2) mixture and optimal yield of 53% was achieved within 17 h at 120°C. With the efficient catalyst, 40% and 23% yield of EMF were produced from inulin and sucrose, respectively [79].

Marine carbohydrate agar derived from red algae was reported for production of EMF in the presence of [EMIM] Cl, $CrCl_2$, and Dowex resin mixture, afford isolate 3.9 g of EMF and EL (EMF to EL mole ratio: 5:2) from 10 g of agar [80]. Similarly, biomass wheat straw was directly converted by alcoholysis reaction in 94% (w/w) ethanol with 30 mM H_2SO_4 at 200°C, and yields for EMF and by-product EL were 20% and 25%, respectively [81].

3. Conclusions

As mentioned above, various Lewis and Brønsted acid catalysts have been conducted to convert HMF, fructose, and inulin in high yields without extra costs for isolation and purification. It is worth noting that HMF is not affordable as a feed for EMF production regardless of the high yields of 70%–95% reported for HMF etherification to EMF by different groups. The technoeconomic analysis carried out by Torres et al. [82] estimated the HMF price of 2.16 \$/kg based on the process proposed by Roman-Leshkov et al. [83]. EMF synthesis from fructose at a desirable yield of 50%–70%, but high reaction temperature (90–130°C) and retention time (6–24 h), is needed compared to the HMF cases. In general, EMF production derived from sucrose did not gain high yields, because of the fact that glucose obtained from sucrose conversion was not converted to EMF in good yields. Nevertheless, the fructose polymer inulin exhibited better performance and produced EMF at high yields. One-pot transformation of cellulose or biomass to EMF led to low yields, due to their complex structures with the low glucose conversion to EMF. Ethanol/THF or ethanol/DMSO mixture is applied as a reaction medium for EMF synthesis from fructose. In a series of reactions, EMF may also participate in the rehydration reaction to produce EL.

Conflicts of Interest

The authors declare that there are no conflicts of interest regarding the publication of this paper.

Acknowledgments

This work was financially supported by the Joint Science and Technology Funds of the Youth Growth S&T Personnel Foundation of Guizhou Education Department (No. KY [2018]292), the Special Funding of Guiyang Science and Technology Bureau and Guiyang University (GYU-KYZ [2018]01-12), and the Technical Talent Support Program of Guizhou Education Department (No. KY[2018]069).

References

[1] E. L. Kunkes, D. A. Simonetti, R. M. West, J. C. Serrano-Ruiz, C. A. Gartner, and J. A. Dumesic, "Catalytic conversion of biomass to monofunctional hydrocarbons and targeted liquid-fuel classes," *Science*, vol. 322, no. 5900, pp. 417–421, 2008.

[2] D. R. Dodds and R. A. Gross, "Chemicals from biomass," *Science*, vol. 318, no. 5854, pp. 1250-1251, 2007.

[3] P. Gallezot, "Conversion of biomass to selected chemical products," *Chemical Society Reviews*, vol. 41, no. 4, pp. 1538–1558, 2012.

[4] R. J. Putten, J. C. Waal, E. Jong, C. B. Rasrendra, H. J. Heeres, and J. G. Vries, "Hydroxymethylfurfural, a versatile platform chemical made from renewable resources," *Chemical Reviews*, vol. 113, no. 3, pp. 1499–1597, 2013.

[5] J. C. Serrano-Ruiz, R. Luque, and A. Sepúlveda-Escribano, "Transformations of biomass-derived platform molecules: from high added-value chemicals to fuelsvia aqueous-phase processing," *Chemical Society Reviews*, vol. 40, no. 11, pp. 5266–5281, 2011.

[6] K. Yan, G. Wu, T. Lafleur, and C. Jarvis, "Production, properties and catalytic hydrogenation of furfural to fuel additives and value-added chemicals Renewable Sustainable," *Renewable and Sustainable Energy Reviews*, vol. 38, pp. 663–676, 2014.

[7] J. B. Binder and R. T. Raines, "Simple chemical transformation of lignocellulosic biomass into furans for fuels and chemicals," *Journal of the American Chemical Society*, vol. 131, no. 5, pp. 1979–1985, 2009.

[8] J. N. Chheda, G. W. Huber, and J. A. Dumesic, "Liquid-phase catalytic processing of biomass-derived oxygenated hydrocarbons to fuels and chemicals Angew," *Angewandte Chemie International Edition*, vol. 46, no. 38, pp. 7164–7183, 2007.

[9] Y. Zhang, Z. Xue, J. Wang et al., "Controlled deposition of Pt nanoparticles on Fe_3O_4@carbon microspheres for efficient oxidation of 5-hydroxymethylfurfural," *RSC Advances*, vol. 6, no. 56, pp. 51229–51237, 2016.

[10] E. Taarning, I. S. Nielsen, K. Egeblad, R. Madsen, and Christensen, "Chemicals from renewables: aerobic oxidation of furfural and hydroxymethylfurfural over gold catalysts," *ChemSusChem*, vol. 1, no. 1-2, pp. 75–78, 2008.

[11] H. Omidvarborna, A. Kumar, and D. S. Kim, "Recent studies on soot modeling for diesel combustion," *Renewable and Sustainable Energy Reviews*, vol. 48, pp. 635–647, 2015.

[12] H. Omidvarborna, A. Kumar, and D. S. Kim, "Characterization of particulate matter emitted from transit fueled with B20 in idle modes," *Journal of Environmental Chemical Engineering*, vol. 2, no. 4, pp. 2335–2342, 2014.

[13] G. W. Huber, S. Iborra, and A. Corma, "Synthesis of transportation fuels from biomass: chemistry, catalysts, and engineering," *Chemical Reviews*, vol. 106, no. 9, pp. 4044–4098, 2006.

[14] T. Werpy, G. Petersen, A. Aden, J. Bozell, J. Holladay, and J. White, *Top Value Added Chemicals from Biomass. Volume 1-Results of Screening for Potential Candidates from Sugars and Synthesis Gas*, Department of Energy Washington DC, Washington, DC, USA, 2004.

[15] B. Kamm, P. R. Gruber, and M. Kamm, *Biorefineries-Industrial Processes and Products*, John Wiley & Sons, Hoboken, NY, USA, 2007.

[16] M. J. Murphy, J. D. Taylor, and R. L. McCormick, "Compendium of experimental cetane number data," NREL/SR-540-36805, National Renewable Energy Laboratory, Golden, Colorado, 2004.

[17] J. S. Câmara, M. A. Alves, and J. C. Marques, "Changes in volatile composition of Madeira wines during their oxidative ageing," *Analytica Chimica Acta*, vol. 563, no. 1-2, pp. 188–197, 2006.

[18] B. Vanderhaegen, H. Neven, S. Coghe, K. J. Verstrepen, H. Verachtert, and G. Derdelinckx, "Evolution of chemical and sensory properties during aging of top-fermented beer," *Journal of Agricultural and Food Chemistry*, vol. 51, no. 23, pp. 6782–6790, 2003.

[19] M. S. Pérez-Coello, M. A. González-Viñas, E. García-Romero, M. C. Díaz-Maroto, and M. D. Cabezudo, "Influence of storage temperature on the volatile compounds of young white wines," *Food Control*, vol. 14, no. 5, pp. 301–306, 2003.

[20] S. H. Oliveira, P. P. Guedes, B. P. Machado, T. Hogg, J. C. Marques, and J. S. Câmara, "Impact of forced-aging process on Madeira wine flavor," *Journal of Agricultural and Food Chemistry*, vol. 56, no. 24, pp. 11989–11996, 2008.

[21] I. Cutzach, P. Chatonnet, and D. Dubourdieu, "Study of the formation mechanisms of some volatile compounds during the aging of sweet fortified wines," *Journal of Agricultural and Food Chemistry*, vol. 47, no. 7, pp. 2837–2846, 1999.

[22] G. J. M. Gruter and F. Dautzenberg, "Method for the synthesis of 5-hydroxy-methylfurfural ethers and their use," U.S. Patent/0082304 A1, 2011.

[23] H. Omidvarborna, A. Kumar, and D. S. Kim, "Variation of diesel soot characteristics by different types and blends of biodiesel in a laboratory combustion chamber," *Science of The Total Environment*, vol. 544, pp. 450–459, 2016.

[24] B. Liu and Z. H. Zhang, "One-pot conversion of carbohydrates into furan derivatives via furfural and 5-hydroxylmethylfurfural as intermediates," *ChemSusChem*, vol. 9, no. 16, pp. 1–23, 2016.

[25] M. E. Janka, D. M. Lange, M. C. Morrow et al., "Oxidation process to produce a crude and/or purified carboxylic acid product," US patent 2012/0302770A1, 2012.

[26] A. Bredihhin, S. Luiga, and L. Vares, "Application of 5-ethoxymethylfurfural (EMF) for the production of cyclopentenones," *Synthesis*, vol. 48, no. 23, pp. 4181–4188, 2016.

[27] S. Alipour, H. Omidvarborna, and D. S. Kim, "A review on synthesis of alkoxymethyl furfural, a biofuel candidate," *Renewable and Sustainable Energy Reviews*, vol. 71, pp. 908–926, 2017.

[28] M. Balakrishnan, E. R. Sacia, and A. T. Bell, "Etherification and reductive etherification of 5-(hydroxymethyl) furfural: 5-(alkoxymethyl) furfurals and 2,5-bis (alkoxymethyl) furans as potential bio-diesel candidates," *Green Chemistry*, vol. 14, no. 6, pp. 1626–1634, 2012.

[29] P. Lanzafame, D. M. Temi, S. Perathoner et al., "Etherification of 5-hydroxymethyl-2-furfural (HMF) with ethanol to biodiesel components using mesoporous solid acidic catalysts," *Catalysis Today*, vol. 175, no. 1, pp. 435–441, 2011.

[30] P. H. Che, F. Lu, J. J. Zhang et al., "Catalytic selective etherification of hydroxyl groups in 5-hydroxymethylfurfural over $H_4SiW_{12}O_{40}$/MCM-41 nanospheres for liquid fuel production," *Bioresource Technology*, vol. 119, pp. 433–436, 2012.

[31] B. Liu, Z. H. Zhang, K. , C. Huang, and Z. F. Fang, "Efficient conversion of carbohydrates into 5-ethoxymethylfurfural in ethanol catalyzed by $AlCl_3$," *Fuel*, vol. 113, pp. 625–631, 2013.

[32] J. T. Liu, Y. Tang, K. G. Wu, C. F. Bi, and Q. Cui, "Conversion of fructose into 5-hydroxymethylfurfural(HMF) and its derivatives promoted by inorganic salt in alcohol," *Carbohydrate Research*, vol. 350, pp. 20–24, 2012.

[33] Y. Yang, C. W. Hu, and M. M. Abu-Omar, "Conversion of glucose into furans in the presence of $AlCl_3$ in an ethanol-water solvent system," *Bioresource Technology*, vol. 116, pp. 190–194, 2012.

[34] X. Yu, X. Y. Gao, R. L. Tao, and L. C. Peng, "Insights into the metal salt Catalyzed 5-ethoxymethylfurfural synthesis from carbohydrates," *Catalysts*, vol. 7, no. 6, pp. 182–192, 2017.

[35] X. Q. Jia, J. P. Ma, P. H. Che et al., "Direct conversion of fructose-based carbohydrates to 5-ethoxymethylfurfural catalyzed by $AlCl_3 \cdot 6H_2O/BF_3(Et)_2O$ in ethanol," *Journal of Energy Chemistry*, vol. 22, no. 1, pp. 93–97, 2013.

[36] X. M. Zhou, Z. H. Zhang, B. Liu, Q. Zhou, S. G. Wang, and K. J. Deng, "Catalytic conversion of fructose into furans using $FeCl_3$ as catalyst," *Journal of Industrial and Engineering Chemistry*, vol. 20, no. 2, pp. 644–649, 2014.

[37] M. I. Alam, S. De, S. Dutta, and B. Saha, "Solid-acid and ionic-liquid catalyzed one-pot transformation of biorenewable substrates into a platform chemical and a promising biofuel," *RSC Advances*, vol. 2, no. 17, pp. 6890–6896, 2012.

[38] S. Saravanamurugan and A. Riisager, "Solid acid catalysed formation of ethyl levulinate and ethyl glucopyranoside from mono-and disaccharides," *Catalysis Communications*, vol. 17, pp. 71–75, 2012.

[39] P. Neves, M. M. Antunes, P. A. Russo et al., "Production of biomass-derived furanic ethers and levulinate esters using heterogeneous acid catalysts," *Green Chemistry*, vol. 15, no. 12, pp. 3367–3376, 2013.

[40] A. Q. Liu, Z. H. Zhang, Z. F. Fang, B. Liu, and K. C. Huang, "Synthesis of 5-ethoxymethylfurfural from 5-hydroxymethylfurfural and fructose in ethanol catalyzed by MCM-41 supported phosphotungstic acid," *Journal of Industrial and Engineering Chemistry*, vol. 20, no. 4, pp. 1977–1984, 2014.

[41] C. M. Lew, N. Rajabbeigi, and M. Tsapatsis, "One-pot synthesis of 5-(ethoxymethyl)furfural from glucose using Sn-BEA and Amberlyst catalysts," *Industrial & Engineering Chemistry Research*, vol. 51, no. 14, pp. 5364–5366, 2012.

[42] H. Li, S. Saravanamurugan, S. Yang, and A. Riisager, "Direct transformation of carbohydrates to the biofuel 5-ethoxymethylfurfural by solid acid catalysts," *Green Chem*, vol. 18, no. 3, pp. 726–734, 2016.

[43] J. D. Lewis, S. V. Vyver, A. J. Crisci et al., "A continuous flow strategy for the coupled transfer hydrogenation and etherification of 5-(hydroxymethyl)furfural using lewis acid zeolites," *ChemSusChem*, vol. 7, no. 8, pp. 2255–2265, 2014.

[44] K. Barbera, P. Lanzafame, S. Perathoner et al., "HMF etherification using NH_4-exchanged zeolites," *New Journal of Chemistry*, vol. 40, no. 5, pp. 4300–4306, 2016.

[45] Y. Y. Bai, L. Wei, M. F. Yang et al., "Three-step cascade over a single catalyst: synthesis of 5-(ethoxymethyl)furfural from glucose over a hierarchical lamellar multi-functional zeolite catalyst," *Journal of Materials Chemistry A*, vol. 6, no. 17, pp. 7693–7705, 2018.

[46] Y. Yang, M. M. Abu-Omar, and C. W. Hu, "Heteropolyacid catalyzed conversion of fructose, sucrose, and inulin to 5-ethoxymethylfurfural, a liquid biofuel candidate," *Appl. Energy*, vol. 99, pp. 80–84, 2012.

[47] H. L. Wang, T. S. Deng, Y. X. Wang, Y. Q. Qi, X. L. Hou, and Y. L. Zhu, "Efficient catalytic system for the conversion of fructose into 5-ethoxymethylfurfural," *Bioresource Technology*, vol. 136, pp. 394–400, 2013.

[48] Y. Ren, B. Liu, Z. Zhang, and J. Lin, "Silver-exchanged heteropolyacid catalyst (Ag_1H_2PW): an efficient heterogeneous catalyst for the synthesis of 5-ethoxymethylfurfural from 5-hydroxymethylfurfural and fructose," *Journal of Industrial and Engineering Chemistry*, vol. 21, pp. 1127–1131, 2015.

[49] L. H. Reddy, J. L. Arias, J. Nicolas, and P. Couvreur, "Magnetic nanoparticles: design and characterization, toxicity and biocompatibility, pharmaceutical and biomedical applications," *Chemical Reviews*, vol. 112, no. 11, pp. 5818–5178, 2012.

[50] S. G. Wang, Z. H. Zhang, B. Liu, and J. L. Li, "Silica coated magnetic Fe_3O_4 nanoparticles supported phosphotungstic acid: a novel environmentally friendly catalyst for the synthesis of 5-ethoxymethylfurfural from 5-hydroxymethylfurfural and fructose," *Catalysis Science & Technology*, vol. 3, no. 8, pp. 2104–2112, 2013.

[51] B. Liu and Z. H. Zhang, "One-pot conversion of carbohydrates into 5-ethoxymethylfurfural and ethyl D-glucopyranoside in ethanol catalyzed by a silica supported sulfonic acid catalyst," *RSC Advances*, vol. 3, no. 30, pp. 12313–12319, 2013.

[52] B. Liu, Z. H. Zhang, and K. J. Deng, "Efficient one-pot synthesis of 5-(ethoxymethyl) furfural from fructose catalyzed by a novel solid catalyst," *Industrial & Engineering Chemistry Research*, vol. 51, no. 47, pp. 15331–15336, 2012.

[53] Z. W. Wang, H. Li, C. J. Fang, W. F. Zhao, T. T. Yang, and S. Yang, "Simply assembled acidic nanospheres for efficient production of 5-ethoxymethylfurfural from 5-hydroxymethylfurfural and fructose," *Energy Technology*, vol. 5, no. 11, pp. 2046–2054, 2017.

[54] A. Q. Liu, B. Liu, Y. M. Wang, R. S. Ren, and Z. H. Zhang, "Efficient one-pot synthesis of 5-ethoxymethylfurfural from fructose catalyzed by heteropolyacid supported on K-10 clay," *Fuel*, vol. 117, pp. 68–73, 2014.

[55] Z. H. Wang and Q. W. Chen, "Conversion of 5-hydroxymethylfurfural into 5-ethoxymethylfurfural and ethyl levulinate catalyzed by MOF-based heteropolyacid materials," *Green Chemistry*, vol. 18, no. 21, pp. 5884–5889, 2016.

[56] G. Z. Xu, B. L. Chen, Z. B. Zheng, K. Li, and H. G. Tao, "One-pot ethanolysis of carbohydrates to promising biofuels: 5-ethoxymethylfurfural and ethyl levulinate," *Asia-Pacific Journal of Chemical Engineering*, vol. 12, no. 4, pp. 527–535, 2017.

[57] H. Li, K. S. Govind, R. Kotni, S. Shunmugavel, A. Riisager, and S. Yang, "Direct catalytic transformation of carbohydrates into 5-ethoxymethylfurfural with acid-base bifunctional hybrid nanospheres," *Energy Conversion and Management*, vol. 88, pp. 1245–1251, 2014.

[58] H. S. Xin, T. W. Zhang, W. Z. Li et al., "Dehydration of glucose to 5-hydroxymethylfurfural and 5-ethoxymethylfurfural by combining Lewis and Brønsted acid," *RSC Advances*, vol. 7, no. 66, pp. 41546–41551, 2017.

[59] G. Raveendra, A. Rajasekhar, M. Srinivas, P. S. S. Prasad, and N. Lingaiah, "Selective etherification of hydroxymethylfurfural to biofuel additivesover Cs containing silicotungstic acid catalysts G," *Applied Catalysis A: General*, vol. 520, pp. 105–113, 2016.

[60] Z. H. Zhang, Y. M. Wang, Z. F. Fang, and B. Liu, "Synthesis of 5-ethoxymethylfurfural from fructose and inulin catalyzed by a magnetically recoverable acid catalyst," *ChemPlusChem*, vol. 79, no. 2, pp. 233–240, 2014.

[61] Z. L. Yuan, Z. H. Zhang, J. D. Zheng, and J. T. Lin, "Efficient synthesis of promising liquid fuels 5-ethoxymethylfurfural from carbohydrates," *Fuel*, vol. 150, pp. 236–242, 2015.

[62] S. S. Yin, J. Sun, B. Liu, and Z. H. Zhang, "Magnetic material grafted cross-linked imidazolium based polyionic liquids: an efficient acid catalyst for the synthesis of promising liquid fuel 5-ethoxymethylfurfural from carbohydrates," *Journal of Materials Chemistry A*, vol. 3, no. 9, pp. 4992–4999, 2015.

[63] J. M. Wang, Z. H. Zhang, S. W. Jin, and X. Z. Shen, "Efficient conversion of carbohydrates into 5-hydroxylmethylfurfan and 5-ethoxymethylfurfural over sufonic acid-functionalized mesoporous carbon catalyst," *Fuel*, vol. 192, pp. 102–107, 2017.

[64] Y. Yao, A. Gu, Y. Wang, H. J. Wang, and W. Li, "Magnetically recoverable carbonaceous material: an efficient catalyst for the synthesis of 5-hydroxymethylfurfural and 5-ethoxymethylfurfural from carbohydrates," *Russian Journal of General Chemistry*, vol. 86, no. 7, pp. 1698–1704, 2016.

[65] H. L. Wang, T. S. Deng, Y. X. Wang et al., "Graphene oxide as a facile acid catalyst for the one-pot conversion of carbohydrates into 5-ethoxymethylfurfural," *Green Chemistry*, vol. 15, no. 9, pp. 2379–2383, 2013.

[66] M. M. Antunes, P. A. Russo, P. V. Wiper et al., "Sulfonated graphene oxide as effective catalyst for conversion of 5-(hydroxymethyl)-2-furfural into biofuels," *ChemSusChem*, vol. 7, no. 3, pp. 804–812, 2014.

[67] B. Liu, Z. H. Zhang, and K. C. Huang, "Cellulose sulfuric acid as a bio-supported and recyclable solid acid catalyst for the synthesis of 5-hydroxymethylfurfural and 5-ethoxymethylfurfural from fructose," *Cellulose*, vol. 20, no. 4, pp. 2081–2089, 2013.

[68] R. S. Thombal and V. H. Jadhav, "Application of glucose derived magnetic solid acid for etherification of 5-HMF to 5-EMF, dehydration of sorbitol to isosorbide, and esterification of fatty acids," *Tetrahedron Letters*, vol. 57, no. 39, pp. 4398–4400, 2016.

[69] H. Zhu, Q. Cao, C. H. Li, and X. Mu, "Acidic resin-catalysed conversion of fructose into furan derivatives in low boiling point solvents," *Carbohydrate Research*, vol. 346, no. 13, pp. 2016–2018, 2011.

[70] G. Morales, M. Paniagua, J. A. Melero, and J. Iglesias, "Efficient production of 5-ethoxymethylfurfural from fructose by sulfonic mesostructured silica using DMSO as co-solvent," *Catalysis Today*, vol. 279, pp. 305–316, 2017.

[71] H. Li, Q. Y. Zhang, and S. Yang, "Catalytic cascade dehydration-etherification of fructose into 5-ethoxymethylfurfural with SO3H-functionalized polymers," *International Journal of Chemical Engineering*, vol. 2014, Article ID 481627, 7 pages, 2014.

[72] G. A. Kraus and T. Guney, "A direct synthesis of 5-alkoxymethylfurfural ethers from fructose via sulfonic acid-functionalized ionic liquids," *Green Chemistry*, vol. 14, no. 6, pp. 1593–1596, 2012.

[73] S. De, S. Dutta, and B. Saha, "One-pot conversions of lignocellulosic and algal biomass into liquid fuels," *ChemSusChem*, vol. 5, no. 9, pp. 1826–1833, 2012.

[74] H. X. Guo, X. H. Qi, Y. Y. Hiraga, T. M. Aida, and R. L. Smith, "Efficient conversion of fructose into 5-ethoxymethylfurfural with hydrogen sulfate ionic liquids as co-solvent and catalyst," *Chemical Engineering Journal*, vol. 314, pp. 508–514.

[75] M. Zuo, K. Le, Y. C. Feng et al., "An effective pathway for converting carbohydrates to biofuel 5-ethoxymethylfurfural via 5-hydroxymethylfurfural with deep eutectic solvents (DESs)," *Industrial Crops and Products*, vol. 112, pp. 18–23, 2018.

[76] S. Shinde and C. Rode, "Cascade reductive etherification of bioderived aldehydes over Zr-based catalysts," *ChemSusChem*, vol. 10, no. 20, pp. 4090–4101, 2017.

[77] K. K. Yadav, S. Ahmad, and S. M. S. Chauhan, "Elucidating the role of cobalt phthalocyanine in the dehydration of carbohydrates in ionic liquids," *Journal of Molecular Catalysis A: Chemical*, vol. 394, pp. 170–176, 2014.

[78] I. Viil, A. Bredihhin, U. Mäeorgb, and L. Vares, "Preparation of potential biofuel 5-ethoxymethylfurfural and other 5-alkoxymethylfurfurals in the presence of oil shale ash," *RSC Advances*, vol. 4, no. 11, pp. 5689–5693, 2014.

[79] J. G. Pereira, S. C. A. Sousa, and C. A. Fernandes, "Direct conversion of carbohydrates into 5-ethoxymethylfurfural (EMF) and 5-hydroxymethylfurfural (HMF) catalyzed by oxomolybdenum complexes," *ChemistrySelect*, vol. 2, no. 16, pp. 4516–4521, 2017.

[80] B. Kim, J. Jeong, S. Shin et al., "Facile single-step conversion of macroalgal polymeric carbohydrates into biofuels," *ChemSusChem*, vol. 3, no. 11, pp. 1273–1275, 2010.

[81] R. J. Grisel, J. C. van der Waal, E. de Jong, and W. J. Huijgen, "Acid catalysed alcoholysis of wheat straw: towards second generation furan-derivatives," *Catalysis Today*, vol. 223, pp. 3–10, 2014.

[82] A. I. Torres, P. Daoutidis, and M. Tsapatsis, "Continuous production of 5-hydroxymethylfurfural from fructose: a design case study," *Energy & Environmental Science*, vol. 3, no. 10, pp. 1560–1572, 2010.

[83] Y. Román-Leshkov, J. N. Chheda, and J. A. Dumesic, "Phase modifiers promote efficient production of hydroxymethylfurfural from fructose," *Science*, vol. 312, no. 5782, pp. 1933–1937, 2006.

14

Decomposition of the Methylene Blue Dye using Layered Manganese Oxide Materials Synthesized by Solid State Reactions

M. E. Becerra,[1,2,3,4,5] **A. M. Suarez** ⓘ**,**[3,5] **N. P. Arias** ⓘ**,**[4,6] **and O. Giraldo** ⓘ[1,3,4]

[1]*Departamento de Física y Química, Facultad de Ciencias Exactas y Naturales, Universidad Nacional de Colombia, Sede Manizales 170003, Colombia*
[2]*Departamento de Ingeniería Química, Facultad de Ingeniería y Arquitectura, Universidad Nacional de Colombia, Sede Manizales 170003, Colombia*
[3]*Laboratorio de Materiales Nanoestructurados y Funcionales, Facultad de Ciencias Exactas y Naturales, Universidad Nacional de Colombia, Sede Manizales 170003, Colombia*
[4]*Grupo de Investigación en Procesos Químicos, Catalíticos y Biotecnológicos, Universidad Nacional de Colombia, Sede Manizales 170003, Colombia*
[5]*Departamento Química, Facultad de Ciencias Exactas y Naturales, Universidad de Caldas, Manizales 17003, Colombia*
[6]*Facultad de Ciencias e Ingeniería, Universidad de Boyacá, Sogamoso, Colombia*

Correspondence should be addressed to O. Giraldo; ohgiraldoo@unal.edu.co

Academic Editor: Donald L. Feke

The modulation in the synthesis parameters of layered manganese oxides allowed us to produce materials with different AC conductivities. These conductivities were correlated with the catalytic performance of the materials in the decomposition of methylene blue, as a model of electron transfer reactions. The manganese oxides were prepared by thermal reduction of $KMnO_4$ at 400°C and 800°C where one sample was heated at 1°C/min and the other was heated at 10°C/min. The materials were characterized by atomic absorption, average oxidation states of manganese, X-ray diffraction, thermogravimetric analysis, and scanning electron microscopy. The results indicate that, by increasing the synthesis temperature, both the lamellar arrangement and the crystal size increased, while the Mn^{4+} amount in the material decreased. Furthermore, it was observed that as the conductivity increases for the materials, the catalytic performance also increases. Therefore, a direct correlation between the conductivity and catalytic performance can be established. For example, the layered manganese oxides material synthesized at 400°C, using a heating rate of 10°C/min, showed the highest AC conductivity and had the best performance in the degradation of methylene blue. Finally, we propose a general mechanism for understanding how manganese oxides behave as catalysts that produce oxidizing species from H_2O_2 which degrades methylene blue. Our proposed mechanism takes into consideration the state of aggregation of the catalyst, the availability of Mn^{4+}, and the electrical conductivity.

1. Introduction

Today's world requires efficient, clean, and environmentally friendly processes. In this context, manganese oxide base materials are an alternative, since by changing the synthesis parameters it is possible to modulate several characteristics: structure, composition, morphology, and electrical conduction [1–3], among others. These features allow designing materials with applications in the primary and secondary batteries [4], supercapacitors [5], catalytic processes [5–7],

and the degradation of dyes [8], among others. A recent critical review [9] accounts for the environmental catalytic applications of the Mn-based oxides and recognizes them as one of the most promising catalysts.

A simple route for obtaining birnessite-type layered manganese oxide is the thermal reduction of $KMnO_4$ [6]. Birnessite consists of MnO_6 octahedrons where the Mn atoms are present as Mn^{4+} and Mn^{3+} ions. The presence of Mn^{3+} ions generates an excess negative charge on the layers, which is offset by cations, commonly Na^+ or K^+, and a monolayer of water molecules in the interlayer region [10, 11]. These materials show ionic conduction, due to the movement of the interlaminar cation, and the electronic conduction of (hopping) electrons in the layers [1–3]. The characterization of manganese oxides by AC conductivity is a nondestructive experimental technique, which can provide relevant information on the potential performance of these materials in redox reactions, which can be a useful tool and easy to apply for the design of heterogeneous catalysts for applications that are likely to spread to several redox reactions.

To correlate the conductivity with the catalytic potential of birnessite in electron transfer reactions, its performance in the degradation of methylene blue (chloride 3,7-bis (dimethylamino)phenothiazine-5-inio) was evaluated as a model reaction. The methylene blue (MB) degradation has been studied widely by using Mn-doped g-C_3N_4 [12], ferrimagnetic materials [13], nickel oxide nanoparticles [14], Au/ZnO [15], and titanium oxides [16, 17], among others. In contrast, few studies report the use of manganese oxides (MnOx) for the MB degradation. Some of those report the use of hydrogen peroxide as an oxidizing agent to generate the reactive oxygen species (ROS), responsible for degrading MB up to the mineralization to carbon dioxide, sulfate, and nitrate [18, 19].

The reaction mechanisms proposed for the MB degradation using MnOx and H_2O_2 [19, 20] suggest a reaction between hydrogen peroxide and the surface of MnOx giving the formation of ROS mainly singlet oxygen ($^1O_2$) and free radicals such as $OH^{\bullet}$ and $O_2^{\bullet-}$. However, we did not find a general mechanism for electron transfer reactions that consider the catalyst aggregation state, the Mn^{4+} availability, the reactive oxygen species formation, and electrical conductivity, as relevant features in the catalytic activity of the birnessite materials. In the present study, we propose a general mechanism that considers all the experimental parameters including the conductivity.

2. Materials and Methods

2.1. Preparation of the Samples.
The samples for this study were synthesized by thermal reduction of $KMnO_4$ (Merck, 99%) [6] at a heating rate of 1°C and 10°C/min. The final temperatures were 400°C and 800°C. Once calcinated, the materials were washed with distilled and deionized water (DDW) until the pH of 9.50 was obtained, and finally, the materials were dried at 60°C for 48 hours. The materials obtained are named according to the final synthesis temperature and heating rate as follows: 4R1 and 8R1 for the first set and 4R10 and 8R10 for the second set.

2.2. Characterization

2.2.1. Atomic Absorption Spectroscopy (AA).
The elemental analysis was performed on a Thermo Series S4 atomic absorption spectrophotometer. About 100 mg of the sample powder was taken for analysis and dissolved in 10.0 mL solution of 37% HCl : DDW in a 2 : 1 ratio, and it was then heated to obtain 50% of the initial volume. 1.0 mL of lanthanum chloride at 1.0% was added, and the final volume

was adjusted with DDW at 100.0 mL for Mn and K content determination.

2.2.2. Average State of Manganese Oxidation (AOS).
The analysis of each sample was performed in duplicate, following the method reported by Glover et al. [21]. Briefly, to determine the total content of Mn, approximately 40 mg of material was dissolved in 10.0 mL of 37% HCl and 10.0 mL of DDW; then, they were heated until the solution became transparent, and the volume was adjusted to 100.0 mL. 100.0 mL of a saturated solution of $Na_2P_2O_7 \cdot 10H_2O$ was added to 10.0 mL of the prepared sample, the pH was adjusted to about 7.00 with 37% HCl, and it was titrated by potentiometry with $KMnO_4$ 0.101 M up to potential jump higher than 100 mV. To determine the available oxygen, approximately 40 mg of the sample was dissolved in 15.0 mL of a 0.10 M $(NH_4)Fe(SO_4)_2 \cdot 6H_2O$ (FAS) (Merck 98%) solution acidified with H_2SO_4 at 98% and titrated with 0.101 M $KMnO_4$ until it turned pink. 1.0 mL of FAS was used as a blank under the same conditions described above. The average oxidation state of manganese was calculated like this AOS = (total moles of O/total moles of Mn) × 2.

2.2.3. X-Ray Diffraction (XRD).
The XRD patterns from the powdered samples were obtained at room temperature in a Rigaku MiniFlex II diffractometer equipped with a radiation source of Cu $K\alpha$ ($\lambda = 1.5406$ Å) at 30 kV and 15 mA in the continuous mode. Data were taken from 3 to 70° in 2$\grave{e}$ with an accuracy of 0.01° in 2$\grave{e}$ at a scan rate of 0.2°/min and a step size of 0.02° in 2θ. The estimated crystal size was performed using the Debye–Scherrer equation [22]:

$$T = \frac{0.9\lambda}{\sqrt{(FWHM)_M^2 - (FWHM)_s^2} * \cos\theta}, \qquad (1)$$

where λ is the Cu Ka radiation and $(FWHM)_M$ and $(FWHM)_s$ are the full-width at half maximum of the more intense diffraction peak in the sample and silicon standard, respectively.

2.2.4. Thermogravimetric Analysis (TGA).
Thermogravimetric analysis was performed on a TA Instruments, TGA Q500 model, with a sensitivity of 0.1 μg, a resolution of ± 0.1°C, and an accuracy of 0.01%. The measurement was doing over approximately 10 mg of the sample and analyzed under N_2 with a flow of 100.0 mL·min^{-1}, in a temperature range of 25°C to 800°C, at a heating rate of 10°C min^{-1}.

2.2.5. Scanning Electron Microscopy (SEM).
The micrographs of the materials were taken on a QUANTA 250 FEI microscope with a tungsten electron source, with a resolution of 3.0 nm at 30 kV. The samples were deposited on a carbon tape and analyzed in the high vacuum mode with a power voltage range between 10.00 kV and 15.00 kV and magnifications of 10000x.

2.2.6. AC Conductivity at Room Temperature.

The AC conductivity of the powder materials was measured at 21°C and 50% relative humidity in accordance with the methodology proposed by Arias et al. [1] in a SOLARTRON 1260 equipment with a SOLARTRON 1296 dielectric interface and a 1296-4A test cell equipped with two bronze electrodes in a parallel arrangement in two-point configuration. The distance between the working and the reference electrodes was 1.57 mm. To prevent border effects and eddy currents, the working electrode, with an effective diameter of 20 mm, has a guard ring. The acquisition of impedance data was performed using the Z-plot software version 3.3, in a frequency range of 10 MHz to 0.1 Hz with a voltage amplitude of 100 mV rms. The analyses were performed in duplicate. For treating the data, a Z-view software version 3.3 (Scribner Association) was used.

The real-AC conductivity was found from the impedance data using

$$\sigma' = \left(\frac{d}{A}\right)\frac{Z'}{Z'^2 + Z''^2} = \left(\frac{d}{A}\right) * Y', \qquad (2)$$

where d is the thickness of the sample (cm), A is the effective electrode area (cm^2), Z' is the real component of the impedance, Z'' is the imaginary part of the complex impedance, d/A is the geometric factor of the sample, and Y' is the conductance.

2.3. Catalytic Test: Degradation of Methylene Blue.

The degradation of methylene blue at room temperature was performed by the method reported by Zhang et al. [19], which is described briefly: 150.0 mL of methylene blue at 30 ppm and 30.0 mg of the MnOx material were stirred for 20 minutes, and then 6.8 mL of H$_2$O$_2$ at 30% were added. The concentration of methylene blue was monitored at baseline and in the following reaction times: 15, 30, 60, and 120 minutes. For the tests, two aliquots of 1.0 mL were taken and centrifuged for 4 minutes at 5000 rpm. The concentration of methylene blue was measured by UV-Vis spectroscopy at $\lambda = 665$ nm.

To determine the effect of the superoxide anion in the degradation of the dye [19], the same reaction was carried out in the presence of 100 mM of gallic acid.

3. Results and Discussion

3.1. X-Ray Diffraction (XRD).

The diffraction patterns from the set of synthesized materials heated at 1°C/min heating rate (Figure 1(a)) show the characteristic diffraction peak of birnessite-type layered manganese oxide [6, 23] around 12.30° (7.19 Å). There were no crystallographic phase changes of the layer structure as it was determined from XRD patterns. However, in the range of 30° to 70° (Figure 1(b)), the characteristic peaks of certain lamellar stacking faults appear in the samples that have been heated to 400°C. Thus, the presence of diffraction peaks at 36.50°, 37.26°, 41.98°, 65.56°, and 66.90° in the 4R1 material has been reported for a turbostratic stacking fault in manganese oxides [24]. In 8R1, the peaks located at 36.02° (2.49 Å), 38.24° (2.35 Å), 40.88° (2.20 Å), 44.28° (2.04 Å), 53.08° (1.72 Å), and 65.30° (1.42 Å) were characteristic of hexagonal-type birnessite [24, 25], and weak peaks at 35.18° (2.55 Å), 37.42° (2.40 Å), 40.06° (2.25 Å), 43.70°

(2.07 Å), 51.04° (1.79 Å), 52.44° (1.74 Å), and 64.06° (1.45 Å) were typical of orthogonal-type birnessite [25]. These changes indicate that the structural order increases as the temperature increases. The XRD patterns for the synthesized materials at 10°C/min heating rate are shown in Figure 1(c). There was no evidence of significant changes in the lamellar arrangement (Figure 1(d)), compared with the materials synthesized at 1°C/min (Figure 1(b)).

These figures show that materials that were heated at the same temperature have similar structure independent of the rate at which they heated. In contrast, the estimated crystal size was affected by both the synthesis temperature (Table 1) and the heating rate. At higher temperatures and lower heating rate, larger crystal size was obtained.

3.2. Chemical Composition, Average Oxidation State, Thermal Stability, and Structural Formulas Determination.

As it can be seen in Table 1, the content of K$^+$ does not vary significantly in the analyzed samples, whereas the Mn^{4+} content varies with both the temperature and the heating rate.

In the materials calcined at 800°C, a lower content of Mn^{4+} is observed in comparison with those that were calcined at 400°C. The material 4R10 showed the highest content of Mn^{4+}. The AOS for Mn is greater for 4R1 and 4R10 than for 8R1 and 8R10. These values were also higher than the average oxidation state values reported for birnessite-type materials obtained by low-temperature oxidation-reduction methods [1, 26]. The thermal reduction of KMnO$_4$ generated materials with higher content of Mn^{4+} at lower temperatures [2, 27]. The content of Mn^{4+} has been reported as a critical surface species in similar materials [18, 19, 28–30] due to the fact that it initiates the decomposition reactions of NOx, SOx [30], and dyes on the surface. Thermal stability studies through TGA showed that the mass losses in the two sets of synthesized materials (Figures 2(a) and 2(b)) up to 150°C were associated with physisorbed water, between 150°C and 250°C with structural water. Finally, for temperatures above 250°C, the mass losses were associated with a structural change involving the release of oxygen, as it has been reported in previous studies [6]. In general, the thermal stability of the material increases with the synthesis temperature, and no significant differences were observed concerning the heating rate.

With the information obtained by the atomic absorption analysis, the average oxidation, and thermal gravimetric analysis, the different structural formulas were calculated (Table 1) using a modification of the equation reported by Gaillot et al. [23, 31] in accordance with the following equation:

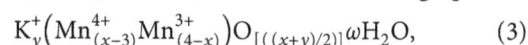

$$K_y^+\left(Mn_{(x-3)}^{4+}Mn_{(4-x)}^{3+}\right)O_{[((x+y)/2)]}\omega H_2O, \qquad (3)$$

where x is the average oxidation state of Mn, y is the K/Mn ratio, and w is the water content calculated from the mass loss of up to 250°C under a nitrogen atmosphere.

3.3. Morphological Analysis by Scanning Electron Microscopy (SEM).

The morphology of the materials (Figure 3) shows particle aggregates of different sizes, with a sponge morphology.

(a)

♣ Turbostratic
♦ Hexagonal
● Orthogonal

(b)

(c)

♣ Turbostratic
♦ Hexagonal
● Orthogonal

(d)

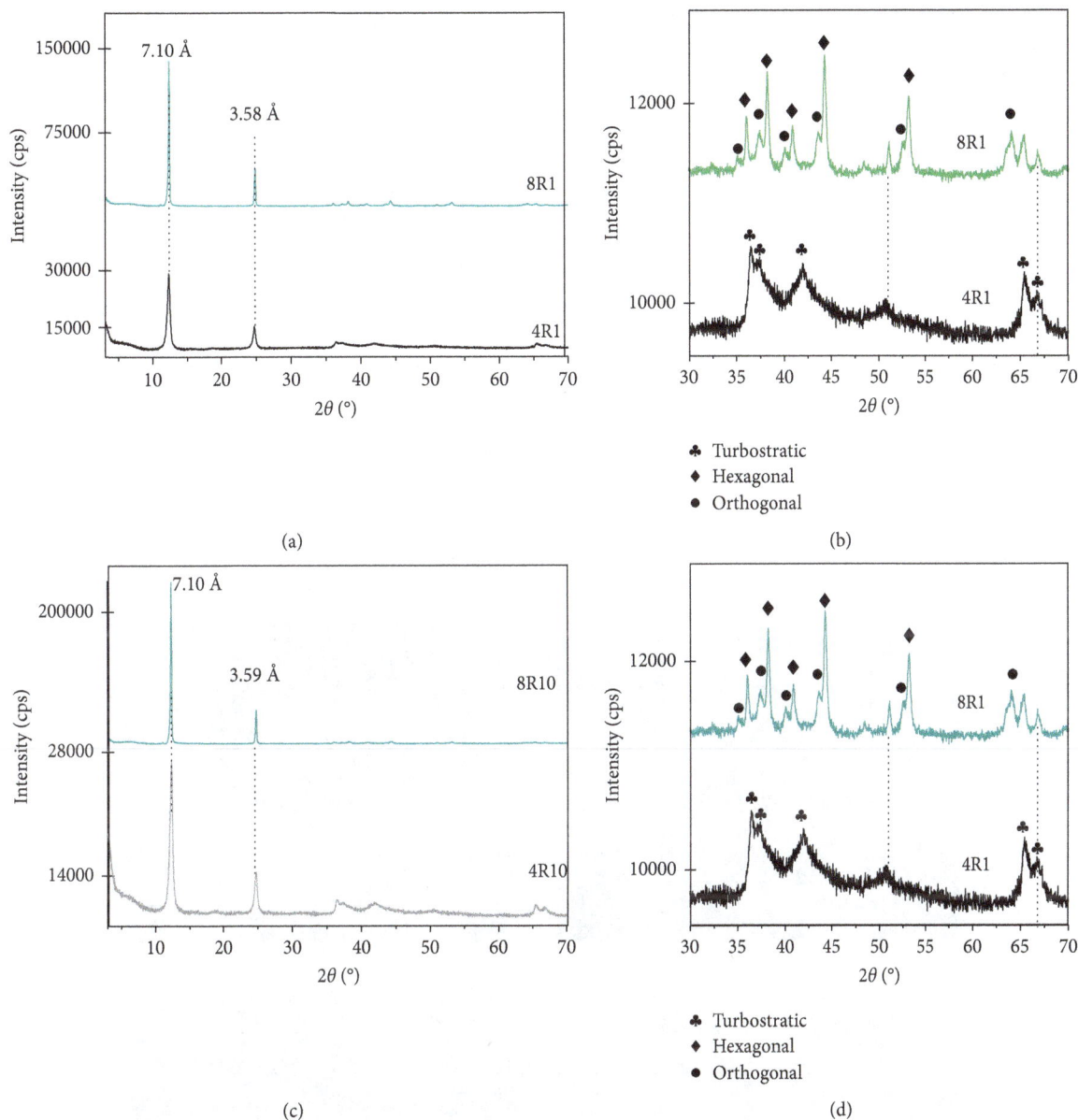

FIGURE 1: X-ray patterns for birnessites synthesized at 400°C and 800°C: (a) materials at 1°C/min heating rate, (b) zoom between 30° and 70° (2θ) for the materials at 1°C/min heating rate, (c) materials at 10°C/min heating rate, and (d) zoom between 30° and 70° (2θ) for the materials at 10°C/min heating rate.

TABLE 1: Physicochemical parameters and structural formulas of the synthesized materials.

Material	K/Mn	AOS[a]	Mn^{4+}	Mn^{3+}	Structural formula	Crystal size (nm)	$\sigma'(\omega)$ at 0.1 Hz (μS)	$\sigma'(\omega)$ at 7.94 MHz (μS)
4R1	0.20	3.85	0.85	0.15	$K^+_{0.20}(Mn^{4+}_{0.85}Mn^{3+}_{0.15})O_{2.02}0.660H_2O$	14	5.64	21.60
8R1	0.23	3.82	0.82	0.18	$K^+_{0.23}(Mn^{4+}_{0.82}Mn^{3+}_{0.18})O_{2.03}0.512H_2O$	117	2.41	16.40
4R10	0.21	3.89	0.89	0.11	$K^+_{0.21}(Mn^{4+}_{0.89}Mn^{3+}_{0.11})O_{2.05}0.660H_2O$	9	13.60	41.20
8R10	0.20	3.82	0.82	0.18	$K^+_{0.20}(Mn^{4+}_{0.82}Mn^{3+}_{0.18})O_{2.01}0.592H_2O$	67	1.10	12.90

[a]Average oxidation state of Mn.

The larger particles in 8R1 and 8R10 materials take the form of a mesh with different spacings between crystals.

These results showed a relationship between aggregate size and both temperature synthesis and heating rate, as it has been reported for ceramic materials like perovskites [32]. The morphological changes were consistent with the increase in the crystal size as it was estimated by the XRD (Table 1).

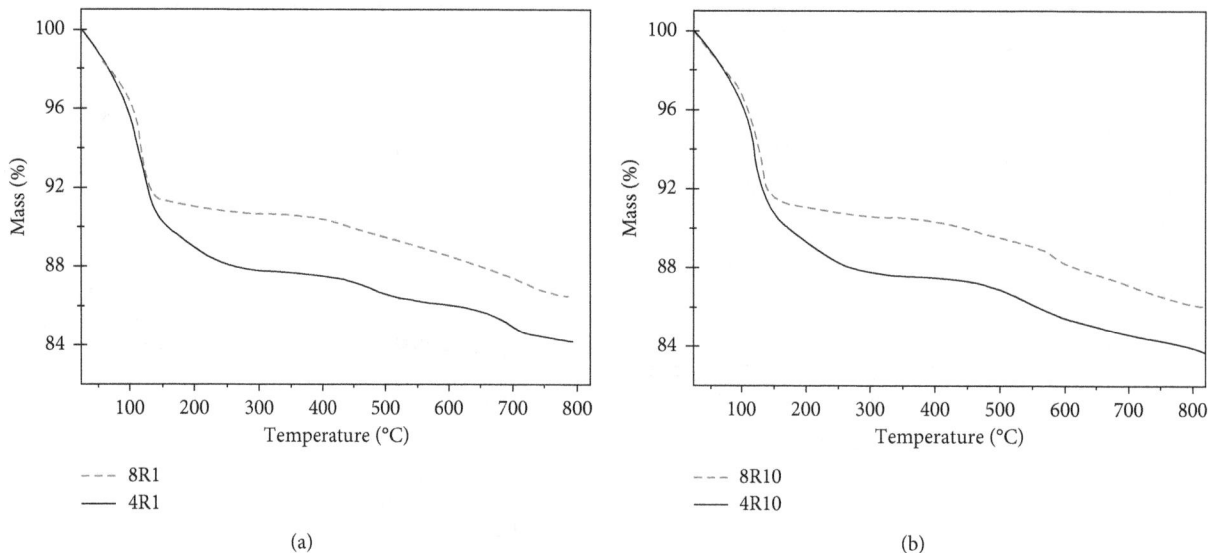

FIGURE 2: Thermograms for the materials synthesized at a heating rate of (a) 1°C/min and (b) 10°C/min.

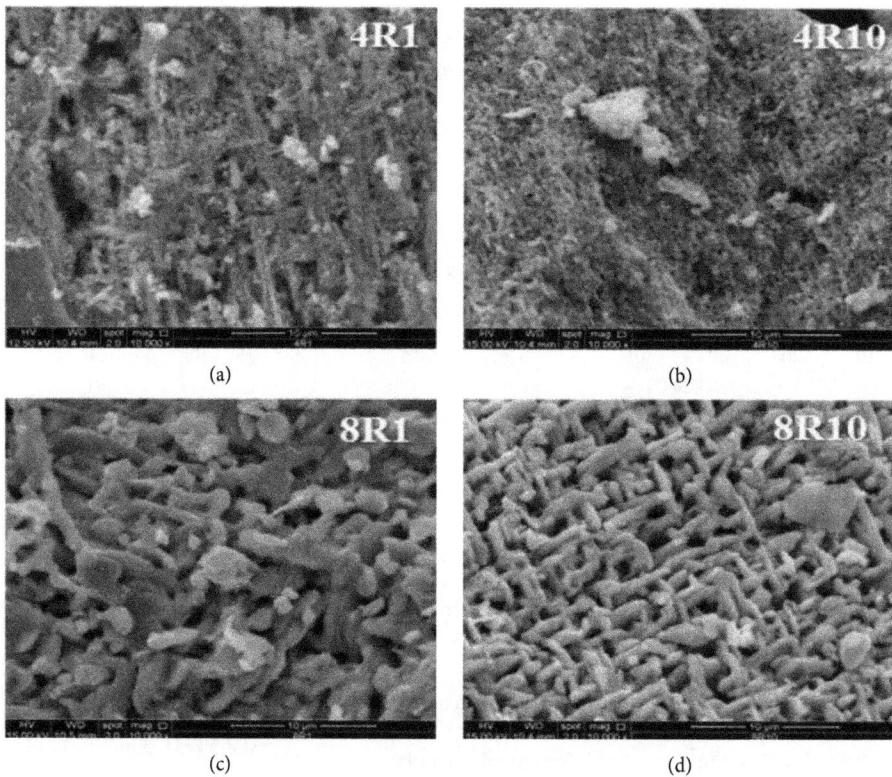

FIGURE 3: SEM micrographs of the materials synthesized at 400°C and 800°C and heating rates of 1°C/min and 10°C/min in both cases.

3.4. AC Conductivity at Room Temperature.

The comparative study of the conductivity of synthesized materials (Figure 4) showed that those obtained at 10°C/min had conductivities up to 1 order of magnitude higher than materials obtained at 1°C/min. In materials synthesized at 1°C and 10°C/min, the real component of the complex conductivity in the low-frequency region (0.1 Hz to 10^3 Hz) decreases as the synthesis temperature increases. Above 1000 Hz, the conductivity depends more on the frequency, according to the "Universal Johnscher's Law" [33], while at low frequencies, it is less dependent, which suggests DC conductivity. This phenomenon is similar to the one reported for birnessite synthesized by thermal reduction routes [2] and soft chemistry routes [1, 3]. For frequencies up to 1×10^5 Hz, in both sets of materials, σ' increases at lower temperature synthesis, and at higher frequencies, the conductivity become similar, having values in the range of semiconductors [34–36]. The increase in conductivity with the frequency

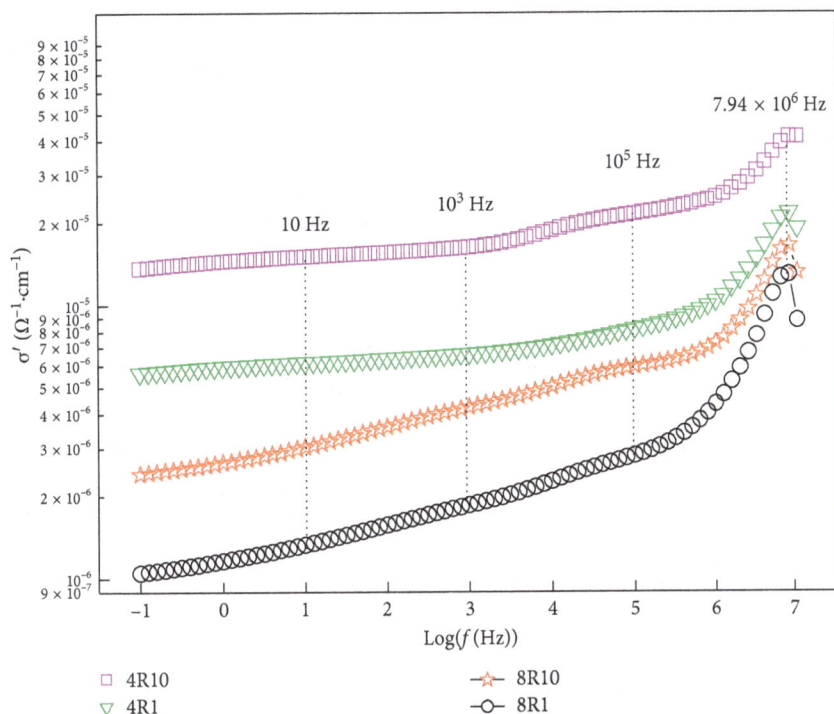

FIGURE 4: Effect of the synthesis temperature and the heating rate on the conductivity of the materials.

indicates a joint movement of charge carriers and changes in the conduction mechanisms [1, 2], mainly dominated by short-range conduction.

The conductivity as a function of the preparation temperature (Figure 4) showed that, at 400°C, the materials were more conductive, thereby suggesting that the lamellar arrangement and the average oxidation state of manganese affect the conduction mechanisms as it was reported by Arias et al. [1, 2]. On the contrary, the presence of Mn^{4+}, understood as the amount of surface Mn^{4+} per nm of the crystal size (Table 1) as well as the stacking faults found by XRD (Figure 1), favors the electron *"hopping"* mechanism [1, 37] and therefore the electrical conduction. Moreover, the high heating rate used in the synthesis of these materials favors the generation of structural defects and microcracks [38] that modified the electrical conduction pathways. Also, both the variation in the lamellar ordering [24] and the birnessite crystal size alter this conduction routes. Consistent with the above, it was found that the conduction process was more favorable in 4R10. This material had the largest lamellar disorder, Mn^{4+} content, and the smallest crystal size.

3.5. Catalytic Degradation of Methylene Blue. With the aim to correlate the conductivity results with the catalytic performance, the catalytic degradation of methylene blue (MB) was used as the reaction model. Therefore, the synthesized materials were tested in the MB degradation reaction using hydrogen peroxide as an oxidizing agent. The 4R10 material exhibits the best performance in this reaction as it can be seen in Figure 5(a). This result was consistent with its highest conductivity (Figure 4), the content of Mn^{4+}, and the smallest crystal size (Table 1) among all materials studied as

it was discussed above. The literature reports that the materials with smaller crystal size exhibit better catalytic performance [38–41], which can be correlated with the increased surface area, necessary for reactions that occur mainly on the surface of the material [28, 39] like in those of MB degradation.

To study the maximal degradation of MB achieved by the 4R10 material, the reaction was monitored up to 1000 min. Between 120 and 300 min, the degradation percentage of MB was 10% higher than that obtained at 60 min (Figure 5(b)). At 600 min, a 63.0% of degradation was reached. After 600 minutes, no significant increase was observed in the disappearance of MB. When only H_2O_2 was used in the MB degradation, the percentage of the demise of MB was 9.3% at 120 min of reaction and 6.4% using only 4R10 (inset in Figure 5(b)), while the MB degradation with 4R10 in the presence of H_2O_2 was 50%. These observations suggest that MB degradation occurs through the formation of reactive oxygen species (ROS) from H_2O_2 over MnOx such as $O_2^{\bullet-}$ and $OH^{\bullet}$ [19, 42, 43]. The formation of the $OH^{\bullet}$ radical finally promotes the degradation of the organic dye until its complete mineralization and protonation of the peroxide HO_2^{-} anion which regenerates the H_2O_2, and it is shown as

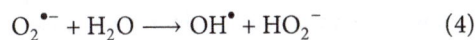

$$O_2^{\bullet-} + H_2O \longrightarrow OH^{\bullet} + HO_2^{-} \qquad (4)$$

$$HO_2^{-} + H^{+} \longrightarrow H_2O_2 \qquad (5)$$

To elucidate if $O_2^{\bullet-}$ radical participate in this reaction, an additional experiment was done. It consisted of the addition of gallic acid, a scavenger specific for the radical $O_2^{\bullet-}$ [44]. Figure 5(c) shows a reduction to 16.1% in the percentage of degradation of MB at 120 minutes of reaction. These results

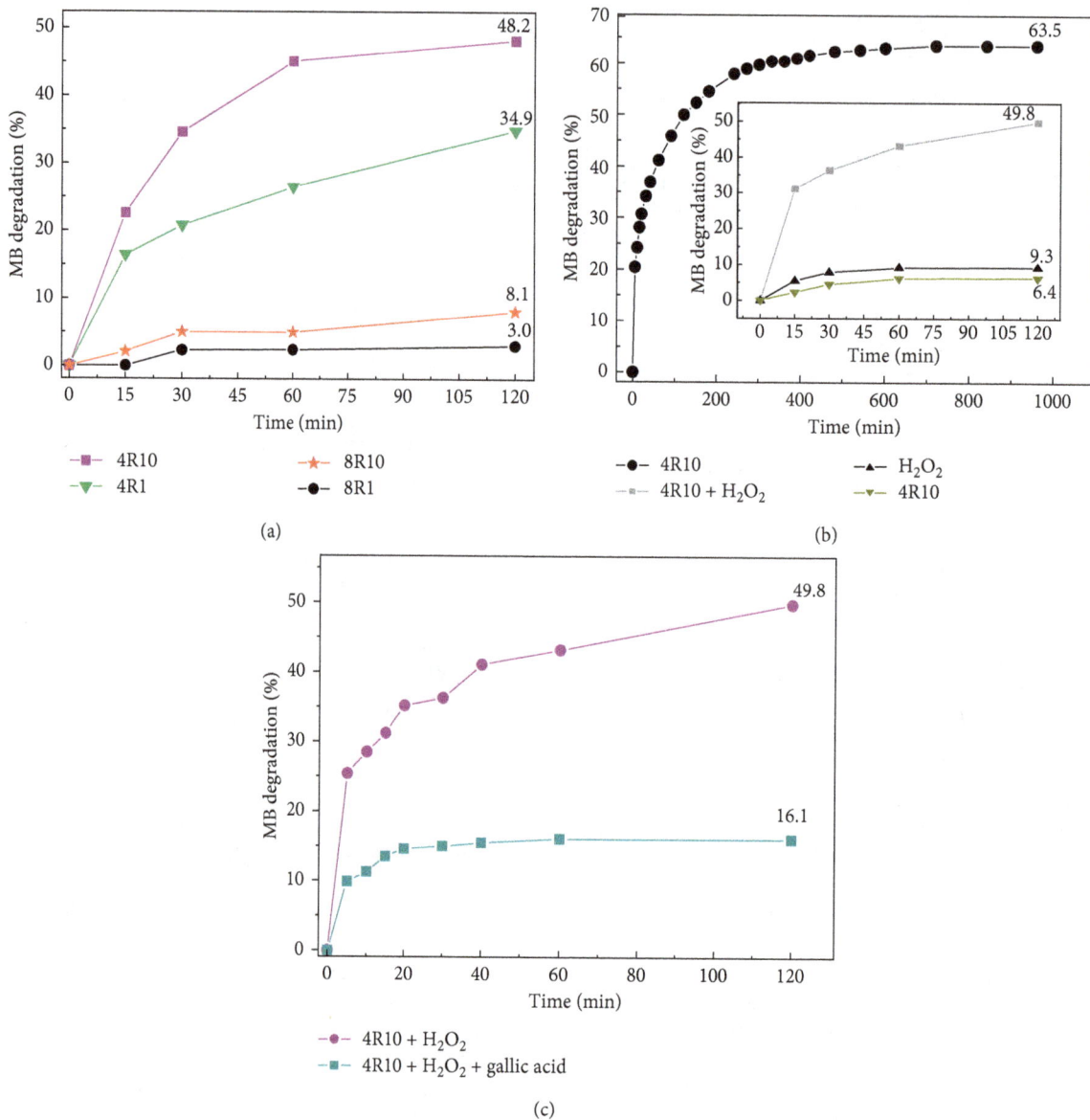

FIGURE 5: MB degradation. (a) Reactions using H_2O_2 as an oxidizing agent. (b) 4R10 material using H_2O_2 as an oxidizing agent. Insert: comparison with the reaction blanks up to 120 min. (c) Effect of the gallic acid addition on the MB degradation for the 4R10 material.

confirm the participation of these radicals in the mechanism of MB degradation.

3.6. Proposed Mechanism Involving Electron Transfer.
Oxidative degradation of organic compounds, especially methylene blue, has been reported as a surface reaction that generates reactive oxygen species [45, 46]. The oxidation of methylene blue through Fenton-type reactions involves the generation of free radicals from an oxidizing substance and from a material that can potentially provide electrons [45, 47]. In this case, hydrogen peroxide is decomposed by the manganese oxide to generate highly reactive oxygen species [18, 19, 45, 48].

The reaction between the birnessite and the hydrogen peroxide involves the adsorption of hydrogen peroxide over the material surface. The hydrogen peroxide should be adsorbed by an acid-Lewis site, with the Mn^{4+} being fundamental versus the

Mn^{3+} because of its most acidic characteristic. Finally, an electronic transfer from the hydrogen peroxide to the manganese (4+) led to the oxidation of the hydrogen peroxide to form a reactive oxygen species [19].

According to the results obtained in this study and considering the availability of Mn^{4+} in the surface of the material, Figure 6(a) shows the reaction mechanism for the formation of ROS from H_2O_2 and the layered material, mainly considering the presence of the Mn^{4+}/Mn^{3+} [19] system. The described mechanism suggests acid-base reactions on the surface of the layered material, which has active sites of Mn^{4+}. Initially, an acid-base reaction occurs between one of the oxygen molecules of H_2O_2 and Mn^{4+}, followed by an electrostatic attraction between the hydrogen molecules of H_2O_2 and the layered material, which is well known to have a negative charge [1, 10]. Once this

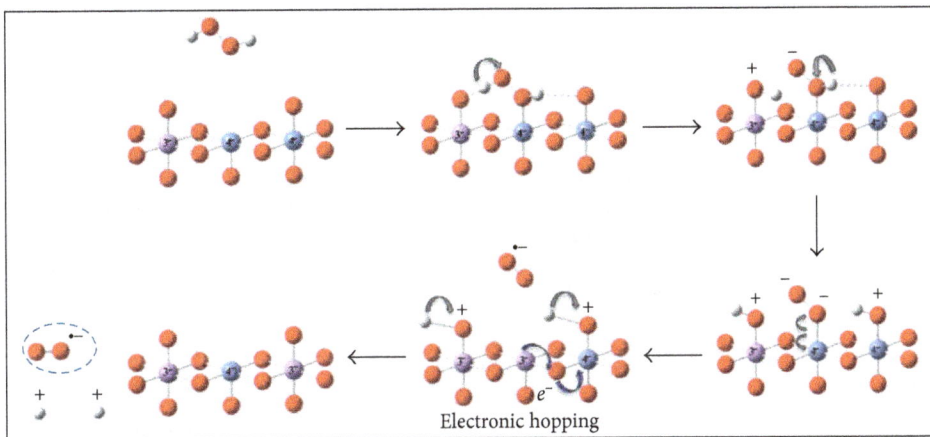

(a)

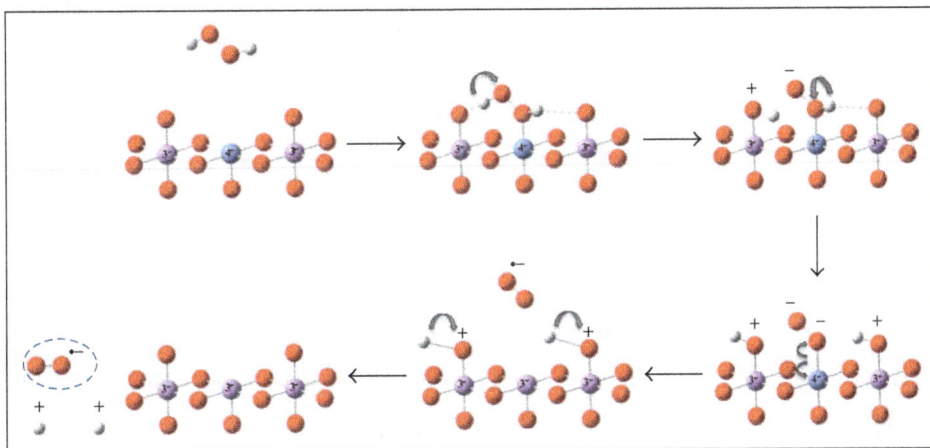

(b)

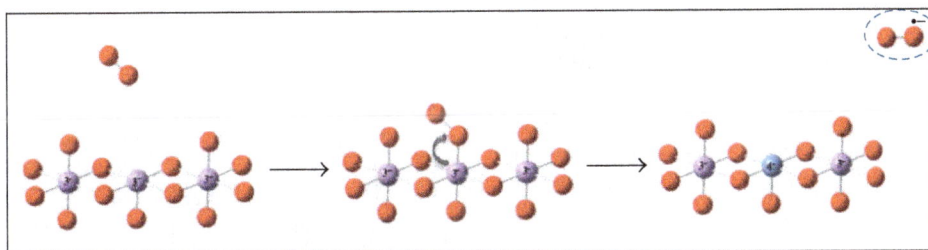

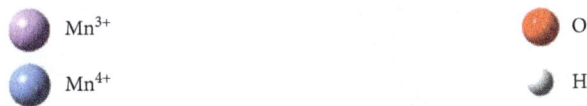

(c)

FIGURE 6: Continued.

$$O_2^{\bullet-} + H_2O \longrightarrow HO_2^- + OH^{\bullet}$$

$$HO_2^- + H^+ \longrightarrow H_2O_2$$

(d)

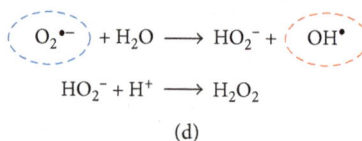

FIGURE 6: Proposed mechanism for the formation of the superoxide and hydroxyl radicals in MB degradation. (a) Hydrogen peroxide adsorption at an Mn^{4+} active site (oxygen vacant) with a posterior electron transfer to the Mn^{4+} active site and formation of a superoxide radical and an Mn^{3+} inactive site. There is an electron transfer process from an Mn^{3+} ion to an adjacent Mn^{4+} ion via an oxobridge (*electronic hopping*) to finally regenerate the Mn^{4+} active site. (b) A similar step as in (a) but in this case without regeneration of the active side. (c) Oxidation of Mn^{3+} with O_2 at the oxygen vacant site to restore the Mn^{4+} active site and superoxide radical. (d) Formation of hydroxyl radical ($OH^{\bullet}$) and regeneration of hydrogen peroxide.

interaction is established, a homolytic rupture is suggested in the bond formed by Mn^{4+} and the oxygen of the peroxide, with the consequent reduction from Mn^{4+} to Mn^{3+} with the formation of the superoxide radical $O_2^{\bullet-}$. The original active site would be regenerated by the transfer of an electron from Mn^{3+}, which has remained after the reaction, to an adjacent Mn^{4+} ion through the oxobridge, as illustrated in Figure 6(a). The electron transfer process from Mn^{3+} to Mn^{4+} via an oxobridge (electronic hopping) is well known in perovskite-type manganese oxides [49]. The described process should be facilitated in those materials, which have a higher electronic conductivity, thus helping to regenerate the active sites efficiently and thus increasing the catalytic activity of the material. It is suggested that this mechanism is most likely in materials with high oxidation state, that is, the highest content of Mn^{4+}.

Another possible scenario would be given when the active sites of Mn^{4+} are adjacent only to Mn^{3+} ions. As shown in Figure 6(b), the reaction mechanism would be like that described above, except that the active sites that have been reduced to Mn^{3+} would no longer be so easily regenerated. For this purpose, it is proposed that an oxygen molecule can oxidize Mn^{3+} to Mn^{4+} and generate a superoxide radical, as it is illustrated in Figure 6(c). Finally, Figure 6(d) shows the reactions for the formation of hydroxyl radical from superoxide radical and the regeneration of hydrogen peroxide.

The proposed mechanism is consistent with experimental data and explains the trends found between conductivity, the average oxidation state of Mn, and catalytic activity of the studied materials. Unlike mechanisms reported in the literature, this mechanism not only collects experimental observations but also considers explicitly the surface sites on which the reaction may be occurring, indicating the state of aggregation of the catalyst and the heterogeneous nature of the process.

4. Conclusions

We obtained birnessites whose conductivity differed significantly as a result of variations in the synthesis temperature and heating rate. A direct relationship between catalyst conductivity and MB degradation performance was observed. The 4R10 material which had a smaller crystal size, higher content of Mn^{4+}, and higher conductivity showed the largest percentage of MB degradation. A general mechanism was proposed for understanding how manganese oxides behave as catalysts that produce oxidizing species from H_2O_2 that degrade methylene blue. Our proposed mechanism takes into consideration the following experimental observations: the availability of Mn^{4+}, the electrical conductivity, and the heterogeneous nature of the process. This relationship can be useful and easy to apply for the design of heterogeneous catalysts for applications that are likely to spread beyond the studied reaction.

Conflicts of Interest

The authors declare that there are no conflicts of interest regarding the publication of this paper.

Acknowledgments

The work at the Universidad Nacional de Colombia, Manizales Campus, was supported by Facultad de Ciencias Exactas y Naturales. The authors also acknowledge Laboratorio de Química (Atomic Absorption Analysis), Laboratorio de Magnetismo y Materiales Avanzados (Thermal and Calorimetric Analysis), and Laboratorio de Microscopia Electronica de Barrido (SEM) at the Universidad de Caldas. A. M. Suarez also acknowledges Facultad de Ciencias Exactas y Naturales, Universidad Nacional de Colombia-Manizales, for a research internship.

References

[1] N. P. Arias, M. T. Dávila, and O. Giraldo, "Electrical behavior of an octahedral layered OL-1-type manganese oxide material," *Ionics*, vol. 19, no. 2, pp. 201–214, 2013.

[2] N. P. Arias, M. E. Becerra, and O. Giraldo, "Caracterización eléctrica de un óxido de manganeso laminar tipo birnesita," *Revista Mexicana de Física*, vol. 61, pp. 380–387, 2015.

[3] O. Giraldo, N. P. Arias, and M. E. Becerra, "Electrical properties of TiO_2-pillared bidimensional manganese oxides," *Applied Clay Science*, vol. 141, pp. 157–170, 2017.

[4] Y. Tang, S. Zheng, Y. Xu, X. Xiao, H. Xue, and H. Pang, "Advanced batteries based on manganese dioxide and its composites," *Energy Storage Materials*, vol. 12, pp. 284–309, 2018.

[5] H. R. Barai, A. N. Banerjee, and S. W. Joo, "Improved electrochemical properties of highly porous amorphous

manganese oxide nanoparticles with crystalline edges for superior supercapacitors," *Journal of Industrial and Engineering Chemistry*, vol. 56, pp. 212–224, 2017.

[6] M. E. Becerra, N. P. Arias, O. H. Giraldo, F. E. López Suárez, M. J. Illán Gómez, and A. Bueno López, "Soot combustion manganese catalysts prepared by thermal decomposition of $KMnO_4$," *Applied Catalysis B: Environmental*, vol. 102, no. 1-2, pp. 260–266, 2011.

[7] S. Dey, G. C. Dhal, D. Mohan, and R. Prasad, "Low-temperature complete oxidation of CO over various manganese oxide catalysts," *Atmospheric Pollution Research*, vol. 9, no. 4, pp. 755–763, 2018.

[8] S. Das, A. Samanta, and S. Jana, "Light-assisted synthesis of hierarchical flower-like MnO_2 nanocomposites with solar light induced enhanced photocatalytic activity," *ACS Sustainable Chemistry and Engineering*, vol. 5, no. 10, pp. 9086–9094, 2017.

[9] H. Xu, N. Yan, Z. Qu et al., "Gaseous heterogeneous catalytic reactions over Mn-based oxides for environmental applications: a critical review," *Environmental Science and Technology*, vol. 51, no. 16, pp. 8879–8892, 2017.

[10] S. L. Brock, N. Duan, Z. R. Tian, O. Giraldo, H. Zhou, and S. L. Suib, "A review of porous manganese oxide materials," *Chemistry of Materials*, vol. 10, no. 10, pp. 2619–2628, 1998.

[11] A. C. Thenuwara, S. L. Shumlas, N. H. Attanayake et al., "Intercalation of cobalt into the interlayer of birnessite improves oxygen evolution catalysis," *ACS Catalysis*, vol. 6, no. 11, pp. 7739–7743, 2016.

[12] J.-C. Wang, C.-X. Cui, Q.-Q. Kong et al., "Mn-doped g-C_3N_4 nanoribbon for efficient visible-light photocatalytic water splitting coupling with methylene blue degradation," *ACS Sustainable Chemistry and Engineering*, vol. 6, no. 7, pp. 8754–8761, 2018.

[13] Y.-Y. Yang, M.-Q. He, M.-X. Li, Y.-Q. Huang, T. Chi, and Z.-X. Wang, "Ferrimagnetic copper-carboxyphosphinate compounds for catalytic degradation of methylene blue," *Inorganic Chemistry Communications*, vol. 94, pp. 5–9, 2018.

[14] G. Jayakumar, A. Albert Irudayaraj, and A. Dhayal Raj, "Photocatalytic degradation of methylene blue by nickel oxide nanoparticles," *Materials Today: Proceedings*, vol. 4, no. 11, pp. 11690–11695, 2017.

[15] L. Wolski, A. Walkowiak, and M. Ziolek, "Formation of reactive oxygen species upon interaction of Au/ZnO with H_2O_2 and their activity in methylene blue degradation," *Catalysis Today*, 2018, In press.

[16] Z. M. Abou-Gamra and M. A. Ahmed, "Synthesis of mesoporous TiO_2–curcumin nanoparticles for photocatalytic degradation of methylene blue dye," *Journal of Photochemistry and Photobiology B: Biology*, vol. 160, pp. 134–141, 2016.

[17] F. Azeez, E. Al-Hetlani, M. Arafa et al., "The effect of surface charge on photocatalytic degradation of methylene blue dye using chargeable titania nanoparticles," *Scientific Reports*, vol. 8, no. 1, p. 7104, 2018.

[18] T. Sriskandakumar, N. Opembe, C.-H. Chen, A. Morey, C. King'ondu, and S. L. Suib, "Green decomposition of organic dyes using octahedral molecular sieve manganese oxide catalysts," *Journal of Physical Chemistry A*, vol. 113, no. 8, pp. 1523–1530, 2009.

[19] L. Zhang, Y. Nie, C. Hu, and X. Hu, "Decolorization of methylene blue in layered manganese oxide suspension with H_2O_2," *Journal of Hazardous Materials*, vol. 190, no. 1–3, pp. 780–785, 2011.

[20] I. A. Salem and M. S. El-Maazawi, "Kinetics and mechanism of color removal of methylene blue with hydrogen peroxide catalyzed by some supported alumina surfaces," *Chemosphere*, vol. 41, no. 8, pp. 1173–1180, 2000.

[21] D. Glover, B. Schumm, and A. Kozowa, *Handbook of Manganese Dioxides Battery Grade*, International Battery Materials Association, Cleveland, OH, USA, 1989.

[22] L. Alexander and H. P. Klug, "Determination of crystallite size with the X-ray spectrometer," *Journal of Applied Physics*, vol. 21, no. 2, pp. 137–142, 1950.

[23] A.-C. Gaillot, D. Flot, V. A. Drits, A. Manceau, M. Burghammer, and B. Lanson, "Structure of synthetic K-rich birnessite obtained by high-temperature decomposition of $KMnO_4$. I. Two-layer polytype from 800°C experiment," *Chemistry of Materials*, vol. 15, no. 24, pp. 4666–4678, 2003.

[24] V. A. Drits, B. Lanson, and A.-C. Gaillot, "Birnessite polytype systematics and identification by powder X-ray diffraction," *American Mineralogist*, vol. 92, no. 5-6, pp. 771–788, 2007.

[25] V. A Drits, E. Silvester, A. Gorshkov, and A. Manceau, "Structure of synthetic monoclinic Na-rich birnessite and hexagonal birnessite: I. Results from X-ray diffraction and selected-area electron diffraction," *American Mineralogist*, vol. 82, no. 9-10, pp. 946–961, 1997.

[26] C.-H. Chen and S. L. Suib, "Control of catalytic activity via porosity, chemical composition, and morphology of nanostructured porous manganese oxide materials," *Journal of the Chinese Chemical Society*, vol. 59, no. 4, pp. 465–472, 2012.

[27] H.-J. Cui, J.-W. Shi, and M.-L. Fu, "Synthesis, and catalytic activity of magnetic cryptomelane-type manganese oxide nanotubes," *Journal of Cluster Science*, vol. 23, no. 3, pp. 607–614, 2012.

[28] Y.-C. Son, V. D. Makwana, A. R. Howell, and S. L. Suib, "Efficient, catalytic, aerobic oxidation of alcohols with octahedral molecular sieves," *Angewandte Chemie International Edition*, vol. 40, no. 22, pp. 4280–4283, 2001.

[29] M. Ilyas, M. Saeed, M. Sadiq, and M. Siddique, "Mixed-valence manganese oxide catalysed oxidation of benzyl alcohol and cyclohexanol in the liquid phase," *Progress in Reaction Kinetics and Mechanism*, vol. 39, no. 4, pp. 375–390, 2014.

[30] W. Yang, J. Zhang, Q. Ma, Y. Zhao, Y. Liu, and H. He, "Heterogeneous reaction of SO_2 on manganese oxides: the effect of crystal structure and relative humidity," *Scientific Reports*, vol. 7, no. 1, p. 4550, 2017.

[31] A.-C. Gaillot, V. A. Drits, A. Manceau, and B. Lanson, "Structure of the synthetic K-rich phyllomanganate birnessite obtained by high-temperature decomposition of $KMnO_4$: substructures of K-rich birnessite from 1000°C experiment," *Microporous and Mesoporous Materials*, vol. 98, no. 1–3, pp. 267–282, 2007.

[32] A. Ecija, K. Vidal, A. Larrañaga, L. Ortega-San-Martín, and M. I. Arriortua, "Synthetic methods for perovskite materials; structure and morphology," in *Advances in Crystallization Processes*, Y. Mastai, Ed., pp. 485–506, InTech, Den Haag, Netherlands, 2012, https://www.intechopen.com/books/advances-in-crystallization-processes/synthetic-methods-for-perovskite-materials-structure-and-morphology.

[33] A. K. Jonscher, "The 'universal' dielectric response," *Nature*, vol. 267, no. 5613, pp. 673–679, 1977.

[34] D. M. Sherman, "Electronic structures of iron(III) and manganese(IV) (hydr)oxide minerals: thermodynamics of photochemical reductive dissolution in aquatic environments," *Geochimica et Cosmochimica Acta*, vol. 69, no. 13, pp. 3249–3255, 2005.

[35] H. Sato, J.-I. Yamaura, T. Enoki, and N. Yamamoto, "Magnetism and electron transport phenomena of manganese

oxide ion exchanger with tunnel structure," *Journal of Alloys and Compounds*, vol. 262-263, pp. 443–449, 1997.

[36] R. N. De Guzman, A. Awaluddin, Y.-F. Shen et al., "Electrical resistivity measurements on manganese oxides with layer and tunnel structures: birnessites, todorokites, and cryptomelanes," *Chemistry of Materials*, vol. 7, no. 7, pp. 1286–1292, 1995.

[37] C. Stampfl and C. G. Van de Walle, "Energetics and electronic structure of stacking faults in AlN, GaN, and InN," *Physical Review B*, vol. 57, no. 24, pp. R15052–R15055, 1998.

[38] F. Schüth, "General principles for synthesis and modification of porous materials," in *Handbook of Porous Solids*, Wiley, Berlin, Germany, 2002.

[39] S. L. Suib, "Structure, porosity, and redox in porous manganese oxide octahedral layer and molecular sieve materials," *Journal of Materials Chemistry*, vol. 18, no. 14, pp. 1623–1631, 2008.

[40] J. Mondal, Q. T. Trinh, A. Jana et al., "Size-dependent catalytic activity of palladium nanoparticles fabricated in porous organic polymers for alkene hydrogenation at room temperature," *ACS Applied Materials and Interfaces*, vol. 8, no. 24, pp. 15307–15319, 2016.

[41] A. Singh, R. K. Hocking, S. L. Y. Chang et al., "Water oxidation catalysis by nanoparticulate manganese oxide thin films: probing the effect of the manganese precursors," *Chemistry of Materials*, vol. 25, no. 7, pp. 1098–1108, 2013.

[42] H. Zhou, Y. F. Shen, J. Y. Wang, X. Chen, C.-L. O'Young, and S. L. Suib, "Studies of decomposition of H_2O_2 over manganese oxide octahedral molecular sieve materials," *Journal of Catalysis*, vol. 176, no. 2, pp. 321–328, 1998.

[43] W. Zhang, H. Wang, Z. Yang, and F. Wang, "Promotion of H_2O_2 decomposition activity over β-MnO_2 nanorod catalysts," *Colloids and Surfaces A: Physicochemical and Engineering Aspects*, vol. 304, no. 1–3, pp. 60–66, 2007.

[44] A. G. Couto, C. A. L. Kassuya, J. B. Calixto, and P. R. Petrovick, "Anti-inflammatory, antiallodynic effects and quantitative analysis of gallic acid in spray dried powders from *Phyllanthus niruri* leaves, stems, roots and whole plant," *Revista Brasileira de Farmacognosia*, vol. 23, no. 1, pp. 124–131, 2013.

[45] K. Dutta, S. Mukhopadhyay, S. Bhattacharjee, and B. Chaudhuri, "Chemical oxidation of methylene blue using a Fenton-like reaction," *Journal of Hazardous Materials*, vol. 84, no. 1, pp. 57–71, 2001.

[46] A. K. M. Atique Ullah, A. K. M. Fazle Kibria, M. Akter, M. N. I. Khan, A. R. M. Tareq, and S. H. Firoz, "Oxidative degradation of methylene blue using Mn_3O_4 nanoparticles," *Water Conservation Science and Engineering*, vol. 1, no. 4, pp. 249–256, 2017.

[47] Y. Zhang, J. Shang, Y. Song et al., "Selective Fenton-like oxidation of methylene blue on modified Fe-zeolites prepared via molecular imprinting technique," *Water Science and Technology*, vol. 75, no. 3, pp. 659–669, 2017.

[48] K. M. Kwon, I. G. Kim, Y.-S. Nam et al., "Catalytic decomposition of hydrogen peroxide aerosols using granular activated carbon coated with manganese oxides," *Journal of Industrial and Engineering Chemistry*, vol. 62, pp. 225–230, 2018.

[49] A. S. Bhalla, R. Guo, and R. Roy, "The perovskite structure–a review of its role in ceramic science and technology," *Materials Research Innovations*, vol. 4, no. 1, pp. 3–26, 2000.

15

Synthesis and Characterization of Nanofiber of Oxidized Cellulose from Nata De Coco

Ditpon Kotatha⊕ and **Supitcha Rungrodnimitchai**⊕

Department of Chemical Engineering, Faculty of Engineering, Thammasat University, Khlong Luang, Pathum Thani 12120, Thailand

Correspondence should be addressed to Supitcha Rungrodnimitchai; supitcha@engr.tu.ac.th

Academic Editor: Michael Harris

Oxidized cellulose (OC) nanofiber was successfully prepared from the dry sheet of Nata De Coco (DNDC) using the mixture system of HNO_3/H_3PO_4–$NaNO_2$ for the first time. The carboxyl content of the OC was investigated at different conditions (HNO_3/H_3PO_4 ratios, reaction times, and reaction temperatures). The results revealed that the carboxyl content of the OC increased along with the reaction time, which yielded 0.6, 14.8, 17.5, 20.9, 21.0, and 21.0% after 0, 6, 12, 36, and 48 hours, respectively. The reaction yields of the OC ranged between 79% and 85% when using HNO_3/H_3PO_4 ratio of 1 : 3, 1.4% wt of $NaNO_2$ at 30°C at different reaction times. From the structural analysis, the OC products showed a nanofibrous structure with a diameter of about 58.3–65.4 nm. The Fourier transform infrared spectra suggested the formation of carboxyl groups in the OC after oxidation reaction. The crystallinity and crystalline index decreased with an increase of reaction time. The decrease of crystallinity from oxidation process agreed with the decrease of degree of polymerization from the hydrolysis of β-1,4-glycosidic linkages in the cellulose structure. The thermal gravimetric analysis results revealed that the OC products were less thermally stable than the raw material of DNDC. In addition, the OC products showed blood agglutinating property by dropping blood on the sample along with excellent antibacterial activity.

1. Introduction

Oxidized cellulose (OC), which could be obtained by partial oxidation of the hydroxymethyl groups on the hydroglucose rings to produce cellulose containing carboxyl group (Figure 1), is an important biocompatible polymer. It has hemostatic property and has been widely used in medical researches and surgery; for example, it was used as hemostatic agents in cardiac surgery [1], thyroid surgery [2], laparoscopic surgery for small uterine perforation [3], and it was used for tubal hemorrhage hemostasis during laparoscopic sterilization [4]. Moreover, it can also be used as drug delivery materials [5] and heavy metal adsorbents [6, 7].

Due to special applications of the OC, many researchers have been trying to synthesize OC, for example, by reactions with oxidizing reagents such as nitrogen dioxide [8, 9], the mixture of nitric acid, phosphoric acid, and sodium nitrite

[10, 11], or reaction with relatively less unstable nitroxyl radical reagents such as 2,2,6,6-tetramethylpiperidine *N*-oxyl (TEMPO) [12–14].

However, almost all of reports use cellulose from the plant or viscose fiber with a diameter about 10–20 μm [15] as raw material. It was reported that Nata De Coco is a nanofiber of bacterial cellulose with about 30–70 nm in diameter [16–18] and able to absorb 200 times of water of its original weight [19]. Furthermore, the nanofiber was reported for its super functionalities for various applications such as optically and transparently functional materials, tissue engineering scaffolds, food package, catalysts, textiles, or surface coatings [20–24]. Therefore, if Nata De Coco is used as a raw material for OC, it possibly can produce higher oxidation efficiency or different properties than previous reports.

Nowadays, Nata De Coco is a well-known food product, which is served as a dessert by cutting in the form of cubes

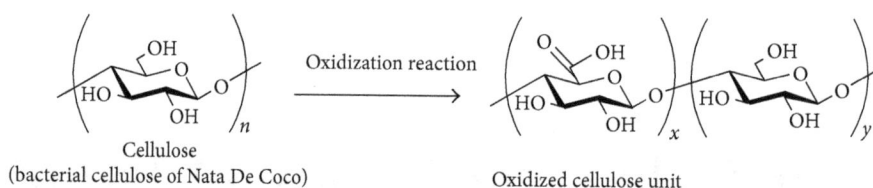

Cellulose
(bacterial cellulose of Nata De Coco)

Oxidized cellulose unit

FIGURE 1: Oxidation reaction of cellulose.

and boiling in syrup. The Nata De Coco is produced by the fermentation of coconut juice, which gels through the production of microbial cellulose that can produce acetic acid such as *Pseudomonas* sp., *Alcaligenes* sp., *Agrobacterium* sp., *Rhizobium* sp., and the most common one is *Acetobacter xylimum* [25]. Moreover, it can be produced from the other sources of sugar such as fruit juices (orange, pineapple, apple, Japanese pear, and grape juices) [26], coconut milk, milk serum, and molasses which is a by-product of sugarcane refining. It is a natural nanofiber which was produced at low costs. If Nata De Coco can be used as raw materials to synthesize OC nanofiber, then nanofibers of other cellulose derivatives would be able to be prepared from Nata De Coco as well.

In this work, the OC nanofiber was synthesized from the dry sheet of Nata De Coco (DNDC) for the first time by a mixture of the HNO_3/H_3PO_4–$NaNO_2$ system. The properties and characterization of the novel OC and its blood agglutination and antibacterial activity properties were discussed.

2. Materials and Methods

2.1. Materials. Nata De Coco of food grade was purchased from a local market in Pathumthani province, Thailand. The nature cotton linter was purchased from a cotton mill in Sakonnakhon province, Thailand. Nitric acid (65.0%), orthophosphoric (85.0%), sodium nitrite (97.0%), calcium acetate (99.0%), sodium hydroxide (98.0%), and ethanol (99.9%) were purchased from Carlo Erba. Copper(II)-ethylenediamine complex (1 M solution in water) was purchased from Acros. All chemicals were of reagent grade or analytical grade and used as received.

2.2. Preparation of Raw Material. Nata De Coco (1,600 g) in the form of 1 cm cubes was washed in distilled water. Subsequently, it was cut and ground in a cooking mixture for 3 minutes and was filtered for removing water. Then, the solid residue was poured into a tray (34 × 25 cm) and was dried at 60°C for 48 hours. After that, the DNDC was cut to 1 × 1 cm size and then used as a raw material.

2.3. Synthesis of OC. Approximately 2.0 g of dry Nata De Coco was added to 50 mL of HNO_3/H_3PO_4 mixture in a glass bottle with the ratios of 1:1, 2:1, 1:2, 1:3, and 1:4 (v/v). Subsequently, $NaNO_2$ was added with amounts of 1.0, 1.4, 1.8, and 2.2% (w/v). The reddish fumes instantaneously occurred. The glass bottle was capped and let to react at 10, 20, 30, and 40°C for 6, 12, 24, 36, and 48 hours, respectively,

with a shaking speed of 100 rpm in a water bath shaker. The obtained samples were washed with distilled water until the filtrate reached a pH of about 4. It was then washed with 50% v/v ethanol and finally washed with distilled water again. The OC was air dried at room temperature before characterization and the % yield was determined using the following equation:

$$\% \text{ yield} = \frac{\text{dry weight of the OC (g)}}{\text{dry weight of Nata De Coco (g)}} \times 100. \quad (1)$$

2.4. Determination of Carboxyl Content. The carboxyl content was measured in accordance with the United States Pharmacopeia (USP23-NF18) [27]. The sample (0.5 g) with a known amount of the moisture content was added in 50 mL of 2% calcium acetate solution and shaken in the water bath shaker with a shaking speed of 100 rpm for 12 hours. After that, the mixture was titrated with 0.1 N·NaOH standard solutions by using phenolphthalein as an indicator. The volume of the consumed NaOH was corrected by the blank. The carboxyl content was determined using the following equation:

$$\% \text{ carboxyl content} = \frac{N \times V \times MW_{-COOH}}{m} \times 100, \quad (2)$$

where N is the normality of NaOH, V is the volume of NaOH used in titration, MW_{-COOH} is the molecular weight of carboxyl group, and m is the weight of the dry testing sample (mg).

2.5. Fourier Transform Infrared Spectrometry (FT-IR). Fourier transform infrared imaging microscopy (Perkin Elmer; model: Spectrum Spotlight 300) was used for identification of functional groups of the obtained samples by micro-attenuated total reflectance technique. All FT-IR spectra were scanned with a resolution of 1 cm^{-1} in the range of 4000–400 cm^{-1}.

2.6. Morphology Analysis. The scanning electron microscopy (SEM, Hitashi model S-3400N) was used for analysis of the morphology of the fiber at 10 kV with 10 k magnification, and the diameter and size distribution of the fiber were analyzed by using image visualization software (Image-Pro Plus). The samples were coated with gold vapor before observation.

2.7. X-Ray Diffraction (XRD) Analysis. XRD patterns of the obtained samples were recorded using an X-ray

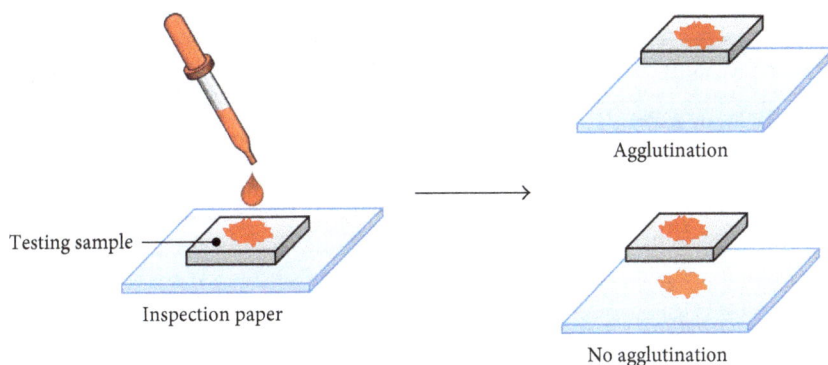

FIGURE 2: A schematic diagram of the blood agglutination test.

diffractometer (PANalytical, model: X'Pert Pro). The samples were scanned from 3° to 40° of 2θ in 0.02° step per 0.5 seconds, and the operating voltage and current was 40 kV and 30 mA, respectively. The radiation was Ni-filtered Cu-Kα radiation of wavelength 1.54 Å. The crystallinity degree (X_c) and crystalline index (CrI) of the obtained samples were estimated by following the literatures [11, 28] as per the following equation:

$$X_c = \frac{S_c}{S_c + S_a} \times 100,$$

$$\text{CrI} = \frac{I_{(0\,0\,2)} - I_{AM}}{I_{(0\,0\,2)}} \times 100,$$

(3)

where S_c is the sum of all crystal areas and S_a is the amorphous area. $I_{(0\,0\,2)}$ is the maximum intensity of the (0 0 2) lattice diffraction which is the peak around 22° of 2θ and I_{AM} is the diffraction intensity of the amorphous peak around 18° of 2θ.

2.8. Degree of Polymerization (DP).

The degrees of polymerization (DP) of the obtained samples were estimated by the intrinsic viscosity method using a Cannon–Fanske calibrated viscometer (size 75) according to the standard test method for intrinsic viscosity of cellulose (ASTM D1795-96) [29]. The intrinsic viscosity measurement was performed as follows. The sample with a known amount of moisture content varying between 0.2 and 1.0 g was suspended in 25 mL of distilled water in a 125 mL Erlenmeyer flask. The suspension was flushed with nitrogen to remove entrapped air from the sample. After that, 25 mL of 1.0 M copper(II)-ethylenediamine complex (Cuen solution) was added into the suspension and continued flushing with nitrogen about 2 min. The Erlenmeyer flask was then closed with a stopper and shaken in water bath shaker for 24 hours with a shaking speed of 100 rpm. After the suspension completely dissolved in the solution, 7 mL of solution was added to the viscometer in a water bath at $25 \pm 0.5°C$. When the solution reached temperature equilibrium with the bath (in about 5 minutes), the efflux time for the solution between the two marks was measured. The same procedure was employed to measure the efflux time for the 0.5 M Cuen solution.

The intrinsic viscosity [η] was calculated using the Martin equation:

$$\log\left(\frac{\eta_{rel} - 1}{c}\right) = \log[\eta] + k[\eta]c,$$

(4)

where $\log[(\eta_{rel} - 1)/c]$ is plotted against c and the straight line is extrapolated through the point to c = 0. The intercept gives log [η], where c is the concentration of the sample in g/100 mL and η_{rel} is the relative viscosity calculated by dividing the efflux time of solution by the efflux time of 0.5 M Cuen solution as given in the following relationship:

$$\eta_{rel} = \frac{\eta_{solution}}{\eta_{solvent}} = \frac{\text{efflux time}_{solution}}{\text{efflux time}_{solvent}}$$

(5)

The DP was determined from the following equation:

$$\text{DP} = [\eta]190.$$

(6)

2.9. Thermal Gravimetric Analysis (TGA).

The thermal gravimetric analyzer (Mettler Toledo; model: TGA/SDTA 851^e) was used for analysis of the thermal property of the obtained samples. The samples were run at a heating rate of 20°C/min in the range 25–1000°C under the nitrogen atmosphere.

2.10. Agglutination of Blood on the OC.

The blood agglutination test was adapted from the procedure of Khatei et al. [30], Jarujamrus et al. [31], and Li et al. [32]. The testing samples of DNDC and the obtained OC were cut to 0.5×0.5 cm sizes and were placed on an inspection paper. Then, a drop of blood sample was placed on the testing samples and was let to stand for 10 minutes. If the blood stain was not found on the inspection paper after removing the testing sample, it can be concluded that the testing sample have the agglutinated blood property. A schematic diagram of the blood agglutination test is shown in Figure 2.

2.11. Antibacterial Activity.

The antibacterial activity of DNDC and the obtained OC was evaluated according to the standard test method for determining the antimicrobial activity of immobilized antimicrobial agents under dynamic control conditions (ASTM E2149) [33]. Gram-negative

TABLE 1: The conditions for synthesis of OC.

Sample name	Material	Ratios of HNO₃/H₃PO₄ (v/v)	Amount of NaNO₂ (%w/v)	Temperature (°C)	Reaction time (hours)
A0 (DNDC)	DNDC (raw material)	—	—	—	—
A1		1:1			
A2		2:1			
A3	DNDC	1:2	1.4	30	24
A4 (OC-24 hrs)		1:3			
A5		1:4			
B1			1.0		24
B2	DNDC	1:3	1.8	30	
B3			2.2		
C1				10	
C2				20	
C3	DNDC	1:3	1.4	40	24
C4				50	
D1 (OC-6 hrs)					6
D2 (OC-12 hrs)					12
D3 (OC-36 hrs)	DNDC	1:3	1.4	30	36
D4 (OC-48 hrs)					48
F0	Natural cotton linter (raw material)	—	—	—	—
F1					6
F2					12
F3	Natural cotton linter	1:3	1.4	30	24
F4					36
F5					48

bacterium, *Escherichia coli* ATCC 25922, was used as a test organism. Briefly, the testing samples and a control were placed in sterile glass flasks. *Escherichia coli* (ATCC 25922) was grown overnight and was diluted to 1.59×10^5 CFU/mL, and then it was added to the flask. Bacteria recovery at the start of the test ($t = 0$) was determined. The flasks were then shaken for 1 hour after which a small volume of the suspension is reanalyzed for the activity of the leached antimicrobial treatment. Plates were incubated at 37°C for 24 hours, and any resulting colonies were counted. The antimicrobial activity was determined in accordance with the standard.

3. Results and Discussion

3.1. Reaction Conditions. In this study, OC was synthesized from DNDC via a mixture of the HNO_3/H_3PO_4–$NaNO_2$ system with different conditions. In addition, the natural cotton linter was also used for synthesis of OC in order to investigate the effect of fiber size of oxidation efficiency. The varied conditions in this experiment are shown in Table 1.

3.2. Carboxyl Content. First of all, the effect of ratios of HNO_3/H_3PO_4 on the carboxyl content was investigated. The carboxyl content of the OC products by using different ratios of HNO_3/H_3PO_4, which are the reactions performed at 30°C and 1.4% (w/v) of $NaNO_2$ for 24 hours, are shown in Figure 3. All of reactions yielded OC products after

the oxidation reaction with high % carboxyl content (14.10–20.93%). The reaction which was obtained by $HNO_3/H_3PO_4 = 1:3$ (A4) gave the maximum of carboxyl content as 20.93%, and then the reaction with HNO_3/H_3PO_4 ratio of $1:3$ was chosen for the condition in the further steps.

The effect of amount of $NaNO_2$ on the carboxyl content of the OC is shown in Figure 4. The reactions were performed at 30°C for 24 hours by using $HNO_3/H_3PO_4 = 1:3$ of acid mixture with different amounts of $NaNO_2$. The result showed that the different amounts of $NaNO_2$ gave no significant difference. Moreover, the more vigorous reactions were observed when the amount of $NaNO_2$ was larger than 1.4% w/v (more of reddish fume and bubble were generated), and the bubble overflowed the glass bottle when more than 2.2% w/v of $NaNO_2$ was added. The amount of $NaNO_2$ did not have large effect on the generation of nitrogen oxide because it worked as an initiator. Therefore, condition with 1.4% $NaNO_2$ was chosen for the condition in the further reaction.

The effect of temperature on the carboxyl content of the OC is shown in Figure 5. Each reaction was performed at different temperatures by using $HNO_3/H_3PO_4 = 1:3$ of acid mixture and 1.4% (w/v) of $NaNO_2$ for 24 hours. The results showed that the carboxyl content increased when the reaction temperature increased from 10 to 30°C, suggesting that an increase of oxidation rate resulted from an increase of temperature. However, the carboxyl content decreased when the temperature rose up from 30 to 50°C, which was possibly caused by hydrolysis of

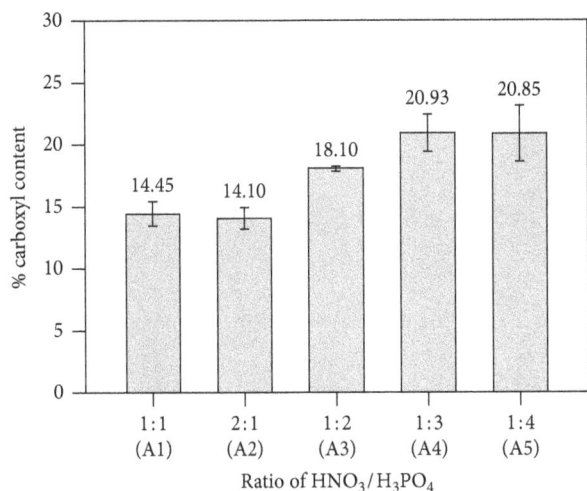

FIGURE 3: Effect of different ratios of HNO_3/H_3PO_4 on the carboxyl content of the OC. The reactions were performed at 30°C and 1.4% (w/v) of $NaNO_2$ for 24 hours.

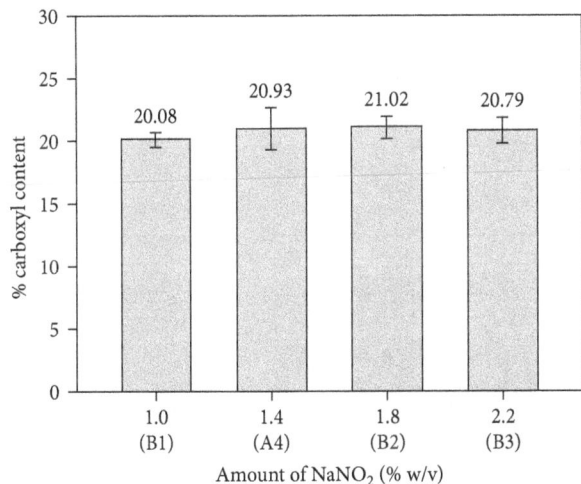

FIGURE 5: Effect of temperature on the carboxyl content of the OC. The reaction using $HNO_3/H_3PO_4 = 1:3$ of acid mixture and 1.4% (w/v) of $NaNO_2$ for 24 hours.

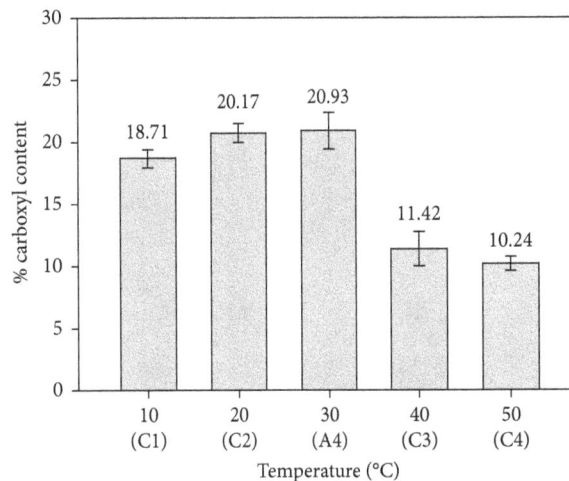

FIGURE 4: Effect of amount of $NaNO_2$ on the carboxyl content of the OC. The reactions were performed at 30°C for 24 hours by using $HNO_3/H_3PO_4 = 1:3$ of acid mixture.

cellulose from strong acid. From the results, it could be concluded that 30°C was the optimum temperature for the reaction, which yielded the maximum of carboxyl content.

The effect of reaction times on the carboxyl content and % yield of the OC is shown in Figure 6(a). The OC products were prepared at 30°C by using $HNO_3/H_3PO_4 = 1:3$ of acid mixture and 1.4% (w/v) of $NaNO_2$ for 6, 12, 24, 36, and 48 hours (the corresponding products were labelled as OC-6 hrs (D1), OC-12 hrs (D2), OC-24 hrs (A4), OC-36 hrs (D3), and OC-48 hrs (D4), resp.). The carboxyl content of the raw material of Nata De Coco was 0.57%. On the contrary, the carboxyl content of the OC products as OC-6 hrs (D1), OC-12 hrs (D2), OC-24 hrs (A4), OC-36 hrs (D3), and OC-48 hrs (D4) was 14.79, 17.49, 20.93, 20.96, and 20.98%, and the % yield was 85.34, 84.70, 83.19, 80.36, and 78.62%, respectively. The carboxyl content increased with an

increase of reaction time, while the % yield decreased with an increase of reaction time. However, after 24 hours, the results showed that the carboxyl content slightly increased (20.93 to 20.98% from 24 to 48 hours), whereas the % yield still linearly decreased. It suggested that the reaction at 30°C by using 1:3 HNO_3/H_3PO_4 of acid mixture and 1.4% (w/v) of $NaNO_2$ for 24 hours was the optimum condition for synthesis of the OC with high carboxyl content.

Moreover, the results from preparation of the OC from the natural cotton linter using the same condition as DNDC are shown in Figure 6(b). The tendency of the change of the carboxyl contents and % yields were similar to those of the OC products from DNDC, but the values were much less than those of the OC products from DNDC. From 6 to 48 hours of reaction times, the OC products from DNDC contained 14.79–20.98% of the carboxyl content and 85.3–78.6% of the % yield. On the contrary, the OC products from the natural cotton linter have the carboxyl content ranging from 8.60 to 10.77% and % yield ranging from 78.2 to 45.0%. The result implied that the diameter of the fiber has a great effect on carboxyl contents and % yields of the OC product. It was presumed that the mixture of acid (HNO_3/H_3PO_4 -$NaNO_2$) in this system is responsible for the oxidation reaction of cellulose, but excess amount of the acid mixture led to the hydrolysis reaction of cellulose. In DNDC case, the DNDC showed a nanofibrous structure with high surface for oxidation reaction so that almost all of acid mixture was involved in the oxidation reaction, while the fiber of natural cotton showed diameter larger than DNDC (see in morphology analysis) which is a small surface area for oxidation reaction so the amount of free mixture acid is very much remaining. It led to the hydrolysis reaction of the fiber, in which the % yield was consistent with the carboxyl content.

3.3. Fourier Transform Infrared Spectrometry (FT-IR) Characterization. The FT-IR spectra of microcrystalline

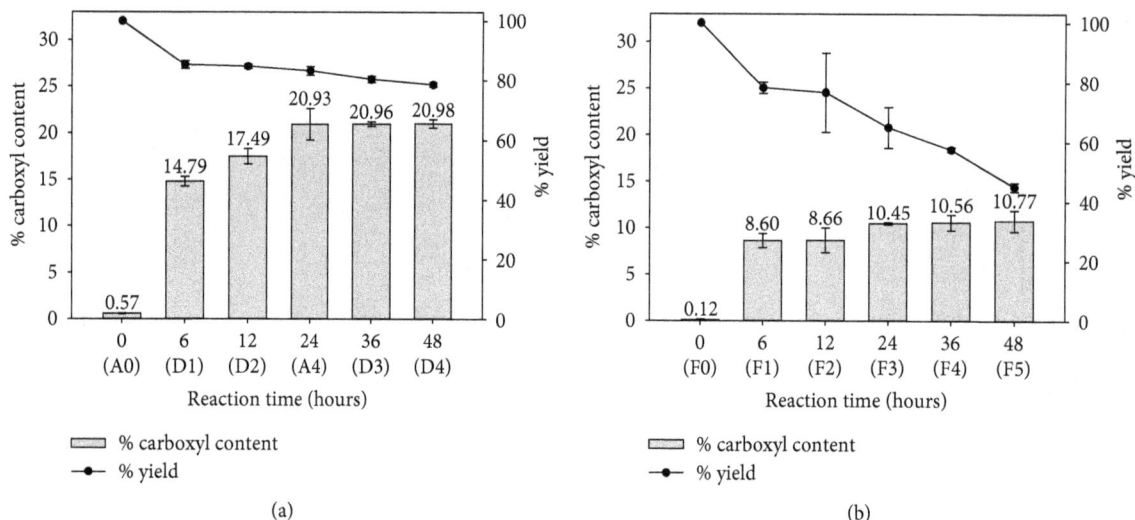

(a)

(b)

FIGURE 6: Effect of reaction time on the carboxyl content and % yield of the OC products from (a) DNDC and (b) natural cotton linter. The reaction was performed at 30°C by using 1 : 3 HNO_3/H_3PO_4 mixture and 1.4% (w/v) of $NaNO_2$.

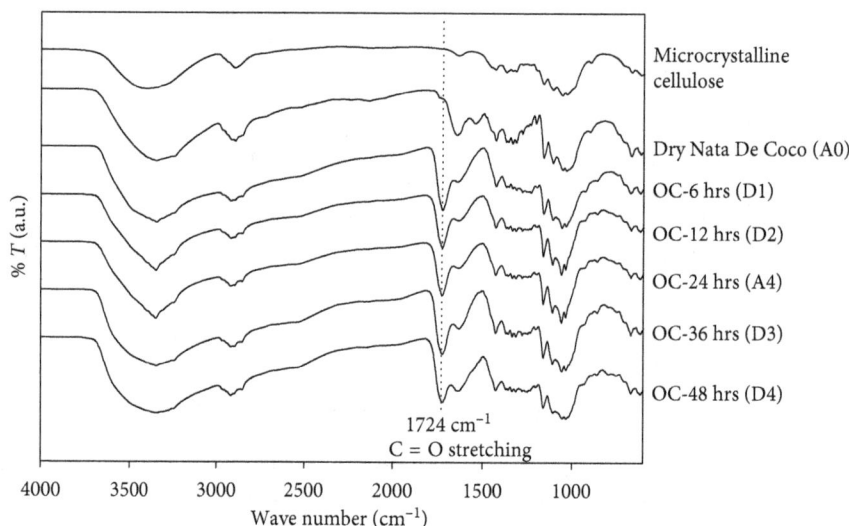

FIGURE 7: FT-IR spectra of microcrystalline cellulose, DNDC (A0), and the OC products.

cellulose (commercial), DNDC (A0), and the OC products are shown in Figure 7. The absorption peak at $3350 \, cm^{-1}$ corresponded to the stretching vibration of O–H, and the weak peaks at 2920 and $2850 \, cm^{-1}$ were induced by stretching vibration of–CH_2. The several peaks at 1430 to $1200 \, cm^{-1}$ were due to the stretching of C–H, and peak about 1060 and $1035 \, cm^{-1}$ were due to the stretching of C–O–C. These absorption peaks are characteristic of cellulose, while the absorption peak at $1640 \, cm^{-1}$ was related to the H_2O bending vibration of the adsorbed water of the material. The FT-IR spectra of the OC products were similar to those of microcrystalline cellulose and DNDC (A0), except that all of the OC products showed a new absorption peak at $1724 \, cm^{-1}$, which corresponded to the vibration of carbonyl group (–C=O), suggesting the formation of carboxyl groups in the products.

3.4. Morphology Analysis. In this research, the morphology of the fiber of the DNDC (A0) and the OC products were analyzed by scanning electron microscopy (SEM) at 10 kV with 10 and 35 k magnification and then, the diameter and size distribution of the fiber were analyzed by using image visualization software (Image-Pro Plus) from the SEM images.

The 10 k magnification of SEM images and diameter size distribution from SEM images of DNDC (A0) and the OC products are shown in Figure 8, and higher magnification of SEM images (35 k) of DNDC (A0) and OC-24 hrs (A4) are shown in Figure 9. SEM images of DNDC (A0) (Figure 8(a)) revealed a nanofibrous structure, and average diameter size was about 69.8 nm which is similar with the previous reports [16, 17]. Moreover, the diameter of the OC products as OC-6 hrs (D1), OC-12 hrs (D2), OC-24 hrs (A4), OC-36 hrs (D3),

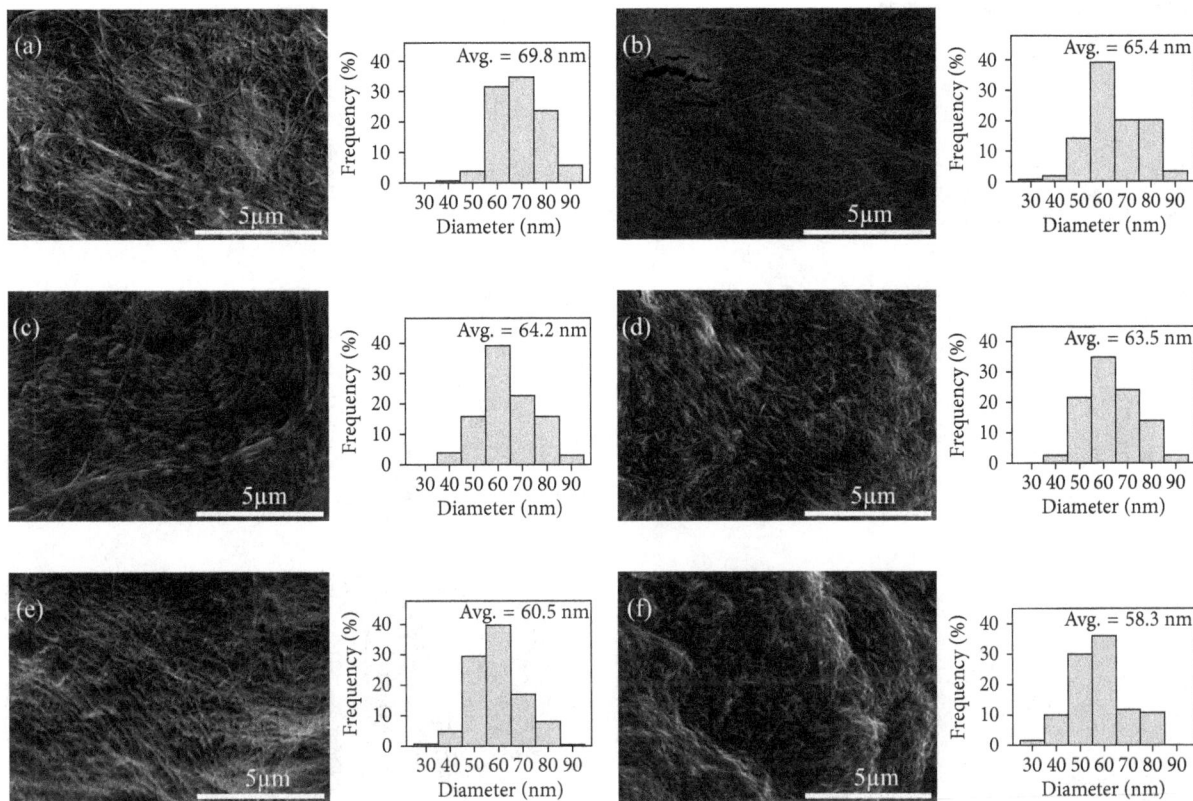

FIGURE 8: 10 k magnification of SEM images of (a) DNDC (A0), (b) OC-6 hrs (D1), (c) OC-12 hrs (D2), (d) OC-24 hrs (A4), (e) OC-36 hrs (D3), and (f) OC-48 hrs (D4) and their diameter size distribution.

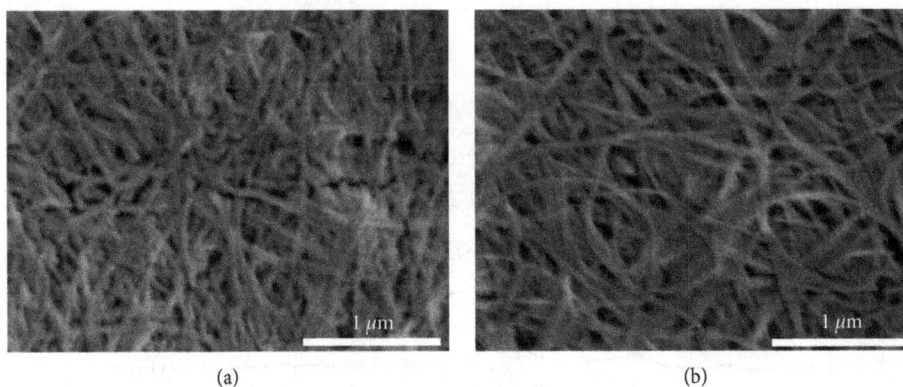

(a) (b)

FIGURE 9: 35 k magnification of SEM images of (a) DNDC (A0) and (b) OC-24 hrs (A4).

and OC-48 hrs (D4) was 65.4, 64.2, 63.5, 60.5, and 58.3 nm, respectively. The diameter size decreased with an increase of the reaction time because of the hydrolysis of the raw material by strong acid. This is a reason to support that the % yield of the OC decreased with an increase of reaction time.

On the contrary, the morphology of the fiber of the natural cotton linter and OC from cotton were also analyzed by SEM at 10 kV but using 100x magnifications. The SEM images and diameter size distribution from SEM images of the natural cotton linter and OC from natural cotton linter

are shown in Figure 10. The diameter of the natural cotton linter also decreased after oxidation reaction. The average diameter of the natural cotton linter was 17.3 μm, and their natural cotton linter oxidized for 6, 12, 24, 36, and 48 hours was 16.4, 16.4, 15.0, 14.3, and 12.7 μm, respectively. It should be noted that the diameter of the cellulose fiber of Nata De Coco was significantly smaller than that of the natural cotton linter.

3.5. X-Ray Diffraction Analysis. X-ray diffraction patterns of DNDC (A0) and the OC products are shown in Figure 11.

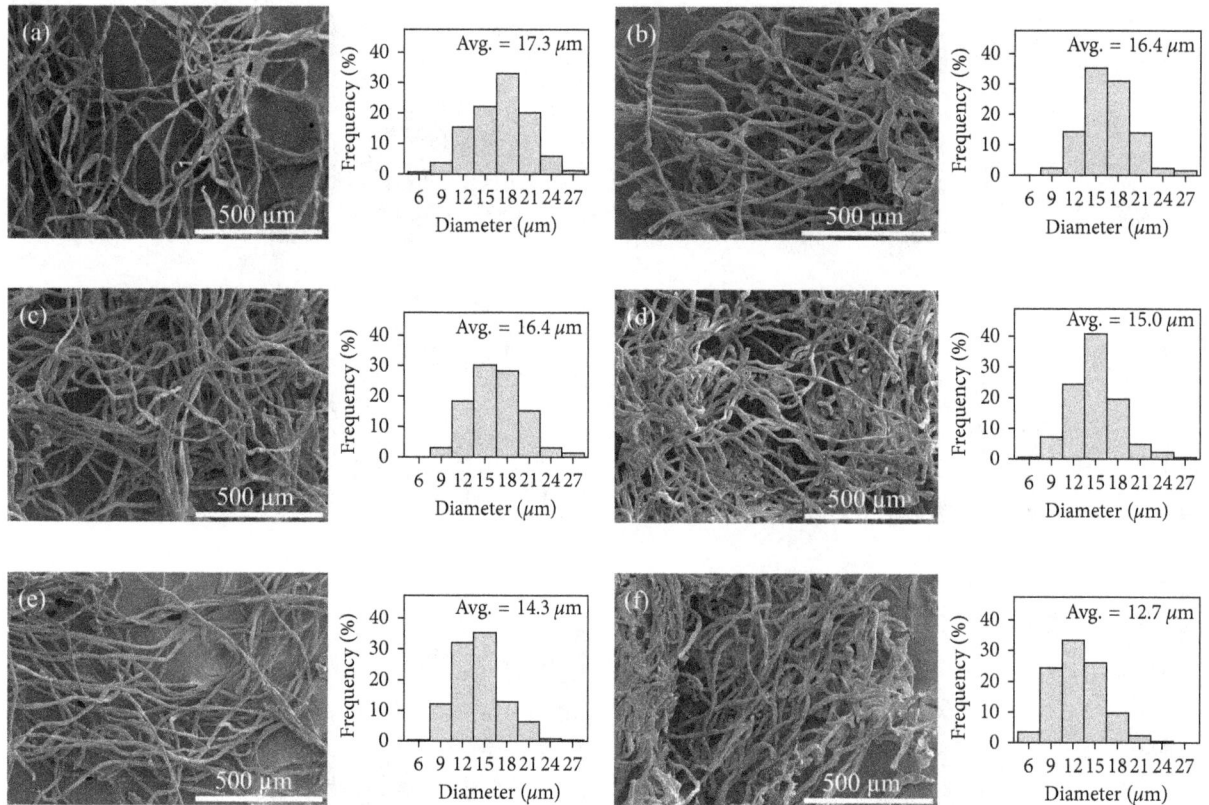

FIGURE 10: SEM images of (a) natural cotton linter and oxidized natural cotton linter for (b) F1, (c) F2, (d) F3, (e) F4, and (f) F5 and their diameter size distribution.

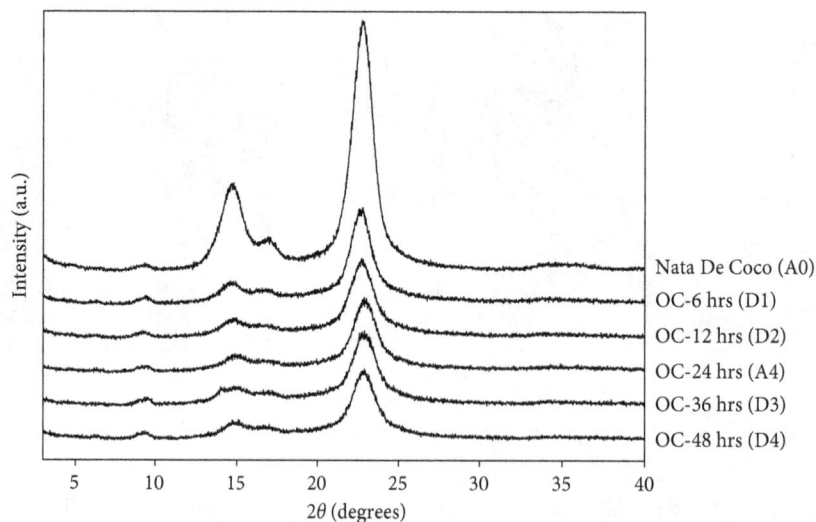

FIGURE 11: X-ray diffraction patterns of DNDC (A0) and the OC products.

The diffraction patterns of the OC product were similar with the raw material of DNDC (A0). All of the patterns showing three distinct peaks at $2\theta = 14.6°$, $16.8°$, and $22.9°$ were, respectively, attributed to the (1 0 1), (1 0 $\bar{1}$), and (0 0 2) reflection planes of the typical cellulose I structure [34, 35].

In addition, characteristic peaks, crystallinity (X_c), and crystalline index (CrI) are shown in Table 2. The location of

the characteristic peak of the OC products did not particularly change from the raw material of DNDC (A0). It could be concluded that the crystalline form did not change from HNO_3/H_3PO_4–$NaNO_2$ oxidation process. However, the crystallinity and crystalline index of the OC products showed remarkable decrease from the raw material of DNDC (A0). With an increase of reaction time, the decrease

TABLE 2: Characteristic peaks and crystalline parameters of DNDC (A0) and the OC products.

Sample	Location of characteristic peak, 2θ (degrees)			Crystallinity (%)	Crystalline index (%)
DNDC (A0)	22.84	16.76	14.63	89.29	92.65
OC-6 hrs (D1)	22.90	16.84	14.60	78.23	83.31
OC-12 hrs (D2)	22.75	16.90	14.54	77.20	80.32
OC-24 hrs (A4)	22.95	16.73	14.51	76.01	79.64
OC-36 hrs (D3)	22.92	16.79	14.54	74.94	79.56
OC-48 hrs (D4)	22.86	16.73	14.80	71.34	79.40

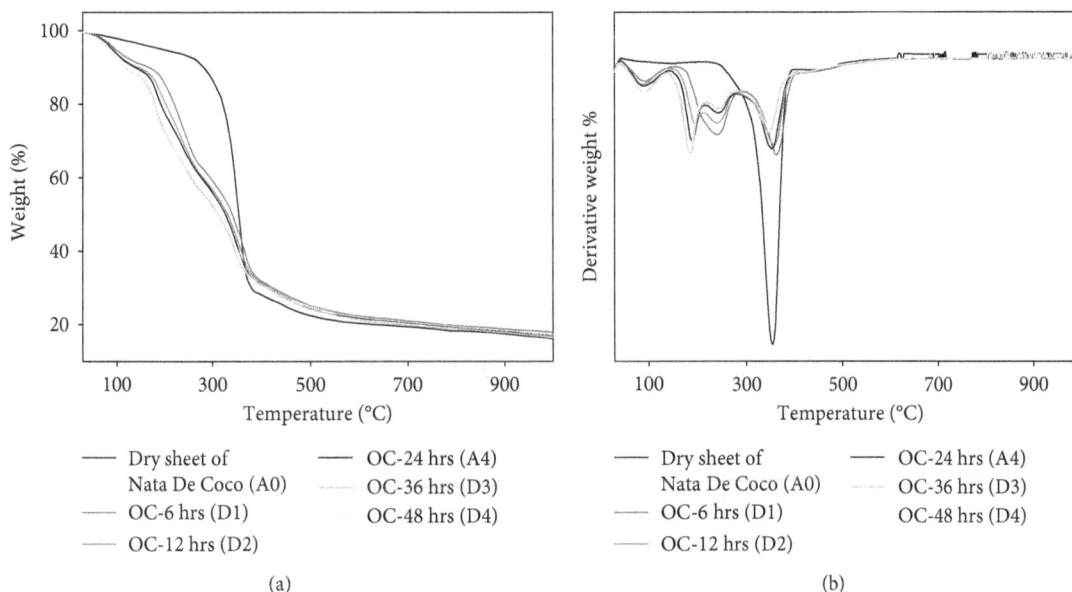

FIGURE 12: (a) TGA and (b) DGTA curves of DNDC (A0) and the OC products.

TABLE 3: Thermal gravimetric data of DNDC (A0) and the OC products.

Sample	Volatile content (%)	Degradation temperature (°C)		Weight loss at final stage (%)
		Initial stage	Final stage	
DNDC (A0)	5.74	348	—	77.99
OC-6 hrs (D1)	9.14	218	351	72.85
OC-12 hrs (D2)	9.68	186, 223	345	72.80
OC-24 hrs (A4)	9.87	182, 220	341	72.73
OC-36 hrs (D3)	11.73	178, 223	334	71.83
OC-48 hrs (D4)	11.79	176, 223	339	70.71

of crystallinity and crystalline index become slower because the first stage of oxidation reaction occurs on the amorphous region and crystalline region surfaces of cellulose. Then, the oxidation reaction sluggishly penetrates into the crystalline region and destroyed the crystalline structure of cellulose gradually [8].

3.6. Degree of Polymerization (DP). The DP of DNDC (A0) was about 1,126.6, while the DP of the OC products as OC-6 hrs (D1), OC-12 hrs (D2), OC-24 hrs (A4), OC-36 hrs

(D3), and OC-48 hrs (D4) was 94.6, 76.3, 74.1, 73.7, and 67.7, respectively. The DP of the OC products showed a significant decrease from the raw material of DNDC (A0) because of depolymerization of the cellulosic material by strong acid. Moreover, DP of the OC product decreased with increase of reaction time suggesting that the hydrolysis of the β-1,4-glycosidic linkages in the cellulose structure increased with an increase of reaction time.

3.7. Thermal Analysis. TGA and DGTA curves of DNDC (A0) and the OC products are shown in Figure 12, and thermal gravimetric data are also shown in Table 3. All samples showed a weight loss due to the volatile content on the surface evaporation between 40 and 125°C, and the volatile content increased with an increase of reaction time because the sample with higher carboxyl content had the greater affinity for molecules of water [10]. The initial degradation temperature of the raw material of DNDC (A0) was about 348°C, which is in the same range with degradation temperature of bacterial cellulose in the previous report [34–36], due to dehydration and decomposition of the cellulose molecules. The initial degradation temperature of the OC products was lower than that of DNDC (A0). Moreover, the decomposition patterns became more complex for OC products. The first step of degradation

(a)

(b)

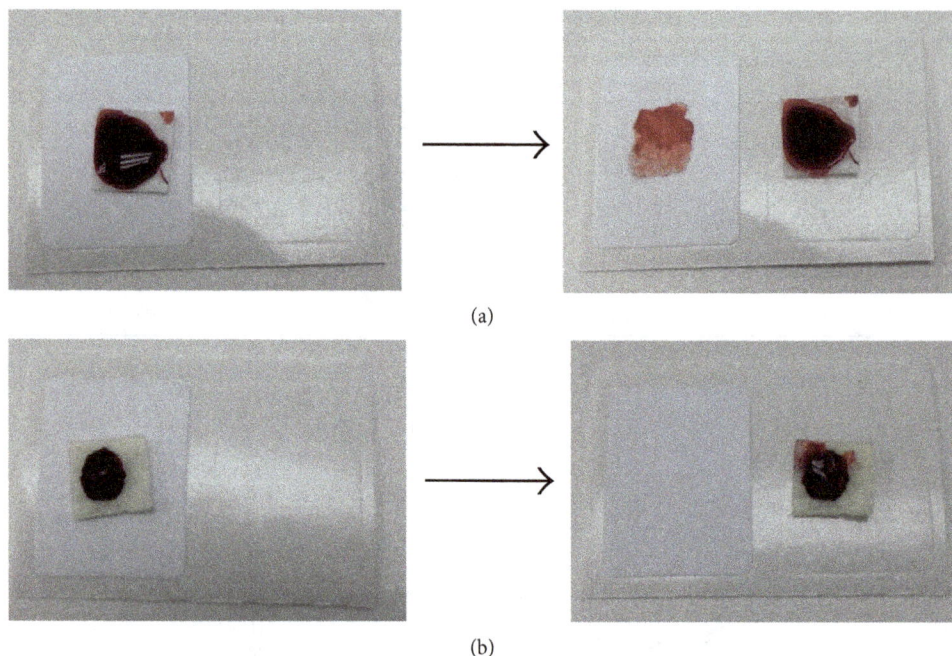

FIGURE 13: Agglutination of the blood test of (a) DNDC (A0) and (b) OC-24 hrs (A4).

temperature was approximately 176–223°C, and the final state was approximately 223–334°C due to the degradation of the remaining residue. The thermal stability and weight loss of the OC products are likely to decrease with an increase reaction time.

3.8. Agglutination of Blood Test. The results of agglutination of the blood test on DNDC (A0) and OC-24 hrs (A4) are shown in Figure 13. The thickness of DNDC (A0) and OC-24 hrs (A4) were 0.15 and 0.19 mm, respectively. The DNDC (A0) and OC-24 hrs (A4) were placed on an inspection paper. Then, a drop of blood sample was placed on the testing sample, and let to stand for 10 minutes. The blood stain was found on the inspection paper after removing the testing sample for DNDC (A0), while the blood stain was not found in OC-24 hrs (A4). Moreover, the blood sample on OC-24 hrs (A4) turned into dark brown or black gel. From the results, it could be concluded that the OC have the blood agglutinating property due to the activation of platelets and the artificial clot formation by combining between the Fe^{3+} ions in the blood and carboxyl groups of OC [9].

3.9. Antibacterial Activity. The antibacterial activity tests of DNDC (A0) and OC-24 hrs (A4) are shown in Table 4. The antibacterial performance of OC-24 hrs (A4), which was evaluated by means of Gram-negative bacterium *Escherichia coli* ATCC 25922 as representative, indicated that the reduction of CFU exceeds 99.99%, while the antibacterial activity property was not observed in DNDC (A0). It should be concluded that the OC possesses bactericidal activity [37].

TABLE 4: Antibacterial activity test of DNDC (A0) and the OC-24 hrs (A4).

Test organisms	Inoculum size (CFU/mL)	Sample	Number (CFU/mL)	% reduction
Escherichia coli ATCC 25922	1.59×10^5	DNDC (A0)	1.58×10^5	—
		OC-24 hrs (A4)	≤10	99.99

4. Conclusions

The results showed that the OC nanofiber was prepared from DNDC with high % yield by a reaction using a mixture of the HNO_3/H_3PO_4–$NaNO_2$ system. It was found that the OC from the reaction at 30°C by using $HNO_3/H_3PO_4 = 1 : 3$ of acid mixture and 1.4% (w/v) of $NaNO_2$ for 24 hours was the optimum condition for synthesis of the OC. The reaction yielded 83.19% of OC with 20.93% of carboxyl content. Moreover, the diameter of the fiber showed a great effect on the carboxyl content and % yield. It was found that the carboxyl content and % yield of the OC from DNDC (A0) was higher than those of the OC from the cotton linter at the same oxidizing condition.

Based on a structural analysis, the OC products showed a nanofibrous structure with a diameter about 58.3–65.4 nm, and the diameter of the fiber decreased gradually with increasing oxidation extent due to the hydrolysis of fiber by strong acid. The FT-IR spectra of the OC products clearly showed a new absorption peak at 1724 cm^{-1} from the vibration of –C=O, which suggested the formation of carboxyl groups after the oxidation reaction. The decrease of crystallinity from oxidation process agreed with the decrease of degree of polymerization from the hydrolysis of β-1,4-glycosidic linkages in the cellulose structure. The TG and DTG results of OC products showed a weight loss between 40 and 125°C due to the volatile content on the surface evaporation. The result revealed the OC products to be less thermally stable than the DNDC (A0).

In addition, the obtained products showed the blood agglutinating property by dropping blood on the sample test and also showed excellent antibacterial activity, thus the reduction of CFU exceeds 99.99% of *Escherichia coli* ATCC 25922. It shows that the novel OC nanofiber from Nata De Coco has a good potential for further applications such as the medical material or antibacterial material [38–40].

Conflicts of Interest

The authors declare that there are no conflicts of interest regarding the publication of this paper.

Acknowledgments

The authors gratefully acknowledge the financial support provided by Thammasat University Research Fund under the TU Research Scholar, Contract no. 57/2557.

References

[1] J. Barnard and R. Millner, "A review of topical hemostatic agents for use in cardiac surgery," *Annals of Thoracic Surgery*, vol. 88, no. 4, pp. 1377–1383, 2009.

[2] M. Amit, Y. Binenbaum, J. T. Cohen, and Z. Gil, "Effectiveness of an oxidized cellulose patch hemostatic agent in thyroid surgery: a prospective, randomized, controlled study," *Journal of the American College of Surgeons*, vol. 217, no. 2, pp. 221–225, 2013.

[3] J. B. Sharma, M. Malhotra, and P. Pundir, "Laparoscopic oxidized cellulose (surgical) application for small uterine perforations," *International Journal of Gynecology and Obstetrics*, vol. 83, no. 3, pp. 217–275, 2003.

[4] J. B. Sharma and M. Malhotra, "Topical oxidized cellulose for tubal hemorrhage hemostasis during laparoscopic sterilization," *International Journal of Gynecology and Obstetrics*, vol. 82, no. 3, pp. 221-222, 2003.

[5] L. Zhu, V. Kumar, and G. S. Banker, "Examination of oxidized cellulose as a macromolecular prodrug carrier: preparation and characterization of an oxidized cellulose-phenylpropanolamine conjugate," *International Journal of Pharmaceutics*, vol. 223, no. 1-2, pp. 35–47, 2001.

[6] N. Isobe, X. Chen, U. J. Kim et al., "TEMPO-oxidized cellulose hydrogel as a high-capacity and reusable heavy metal ion adsorbent," *Journal of Hazardous Materials*, vol. 260, pp. 195–201, 2013.

[7] T. Saito and A. Isogai, "Ion-exchange behavior of carboxylate groups in fibrous cellulose oxidized by the TEMPO-mediated system," *Carbohydrate Polymers*, vol. 61, no. 4, pp. 183–190, 2005.

[8] Y. D. Wu, J. M. He, Y. D. Huang, F. W. Wang, and F. Tang, "Oxidation of regenerated cellulose with nitrogen dioxide/carbon tetrachloride," *Fiber and Polymers*, vol. 13, no. 5, pp. 576–581, 2012.

[9] Y. Wu, J. He, W. Cheng et al., "Oxidized regenerated cellulose-based hemostat with microscopically gradient structure," *Carbohydrate Polymers*, vol. 88, no. 3, pp. 1023–1032, 2012.

[10] V. Kumar and T. Yang, "HNO₃/H₃PO₄–NANO₂ mediated oxidation of cellulose–preparation and characterization of bioabsorbable oxidized cellulose in high yields and with different levels of oxidation," *Carbohydrate Polymers*, vol. 48, no. 4, pp. 403–412, 2002.

[11] Y. Xu, X. Liu, X. Liu, J. Tan, and H. Zhu, "Influence of HNO₃/H₃PO₄–NANO₂ mediated oxidation on the structure and properties of cellulose fibers," *Carbohydrate Polymers*, vol. 111, pp. 955–963, 2014.

[12] P. Ma, S. Fu, H. Zhai, and C. Daneault, "Influence of TEMPO-mediated oxidation on the lignin of thermomechanical pulp," *Bioresource Technology*, vol. 118, pp. 607–610, 2012.

[13] T. Saito, Y. Okita, T. T. Nge, and J. Sugiyama, "TEMPO-mediated oxidation of native cellulose: microscopic analysis of fibrous fractions in the oxidized product," *Carbohydrate Polymers*, vol. 65, no. 4, pp. 435–440, 2006.

[14] G. Biliuta, L. Fras, M. Drobota et al., "Comparison study of TEMPO and phthalimide-N-oxyl (PINO) radicals on oxidation efficiency toward cellulose," *Carbohydrate Polymers*, vol. 91, no. 2, pp. 502–507, 2013.

[15] C. Ververis, K. Georghiou, N. Christodoulakis, P. Santas, and R. Santas, "Fiber dimensions, lignin and cellulose content of various plant materials and their suitability for paper product," *Industrial Crops and Products*, vol. 19, no. 3, pp. 245–254, 2004.

[16] N. Halib, M. C. I. Amin, and I. Ahmad, "Physicochemical properties and characterization of Nata de Coco from local food industries as a source of cellulose," *Sains Malaysiana*, vol. 41, no. 2, pp. 205–211, 2012.

[17] Y. Kaburagi, M. Ohoyama, Y. Yamaguchi et al., "Graphitization behavior of carbon nanofibers derived from bacteria cellulose," *Carbon*, vol. 55, pp. 371–374, 2013.

[18] E. L. Hult, S. Yamanaka, M. Ishihara, and J. Sugiyama, "Aggregation of ribbons in bacterial cellulose induced by high pressure incubation," *Carbohydrate Polymers*, vol. 53, no. 1, pp. 9–14, 2003.

[19] D. Klemm, D. Schumann, U. Udhardt, and S. Marsch, "Bacterial synthesized cellulose–artificial blood vessels for microsurgery," *Progress in Polymer Science*, vol. 26, no. 9, pp. 1561–1603, 2001.

[20] X. D. Cao, H. Dong, and C. M. Li, "New nanocomposite materials reinforced with flax cellulose nanocrystals in waterborne polyurethane," *Biomacromolecules*, vol. 8, no. 3, pp. 899–904, 2007.

[21] K. Das, D. Ray, C. Banerjee et al., "Physicomechanical and thermal properties of jute nanofiber reinforced biocopolyester composites," *Industrial and Engineering Chemistry Research*, vol. 49, no. 6, pp. 2775–2782, 2010.

[22] H. Dong, K. E. Strawheckera, J. F. Snydera et al., "Cellulose nanocrystals as a reinforcing material for electrospun poly (methyl methacrylate) fibers: formation, properties and nanomechanical characterization," *Carbohydrate Polymers*, vol. 87, no. 4, pp. 2488–2495, 2012.

[23] P. Lu and Y. L. Hsieh, "Cellulose nanocrystal-filled poly (acrylic acid) nanocomposite fibrous membranes," *Nanotechnology*, vol. 20, no. 41, p. 415604, 2009.

[24] H. Deng, X. Zhou, X. Wang et al., "Layer-by-layer structured polysaccharides film-coated cellulose nanofibrous mats for cell culture," *Carbohydrate Polymers*, vol. 80, no. 2, pp. 475–480, 2010.

[25] P. Ross, M. Raphae, and P. Moshe, "Cellulose biosynthesis and function in bacteria," *Microbiology Review*, vol. 55, pp. 35–38, 1991.

[26] A. Kurosumi, C. Sasaki, Y. Yamashita, and Y. Nakamura, "Utilization of various fruit juices as carbon source for production of bacterial cellulose by *Acetobacter xylinum* NBRC 13693," *Carbohydrate Polymers*, vol. 76, no. 2, pp. 333–335, 2009.

[27] USP, *United States Pharmacopeia 23/National Formulary 18: Oxidized Cellulose*, USP, Rockville, MD, USA, 1995.

[28] S. Park, J. O. Baker, M. E. Himmel, P. A. Parilla, and D. K. Johnson, "Cellulose crystallinity index: measurement techniques and their impact on interpreting cellulose performance," *Biotechnology and Biofuels*, vol. 3, no. 1, pp. 1–10, 2010.

[29] ASTM D 1795–96, *Standard Test Method for Intrinsic Viscosity of Cellulose*, ASTM, West Conshohocken, PA, USA, 2001.

[30] V. Khatri, K. Halász, L. V. Trandafilović et al., "ZnO-modified cellulose fiber sheets for antibody immobilization," *Carbohydrate Polymers*, vol. 109, pp. 139–147, 2014.

[31] P. Jarujamrus, J. Tian, X. Li et al., "Mechanisms of red blood cells agglutination in antibody-treated paper," *Analyst*, vol. 137, no. 9, pp. 2205–2210, 2012.

[32] M. Li, J. Tian, M. Al-Tamimi, and W. Shen, "Paper-based blood typing device that report patient's blood type "in writing"," *Angewandte Chemie International Edition*, vol. 51, no. 22, pp. 5497–5501, 2012.

[33] ASTM E2149, *Standard Test Method for Determining the Antimicrobial Activity of Antimicrobial Agent under Dynamic Control Conditions*, ASTM, West Conshohocken, PA, USA, 2013.

[34] Q. Gao, X. Shen, and X. Lu, "Regenerated bacterial cellulose fiber prepared by the NMMO·H$_2$O process," *Carbohydrate Polymers*, vol. 83, no. 3, pp. 1253–1256, 2011.

[35] X. Lu and X. Shen, "Solubility of bacteria cellulose in zinc chloride aqueous solutions," *Carbohydrate Polymers*, vol. 86, no. 1, pp. 239–244, 2011.

[36] J. A. Marins, B. G. Soares, H. S. Barud, and S. J. L. Ribeiro, "Flexible magnetic membranes based on bacterial cellulose and its evaluation as electromagnetic interference shielding material," *Materials Science and Engineering C*, vol. 33, no. 7, pp. 3994–4001, 2013.

[37] P. R. Murray, K. S. Rosenthal, G. S. Kobayashi, and M. A. Pfaller, *Bacteriology, Medical Biomicrobiology*, A Harcourt Health Sciences Company, London, UK, 4th edition, 2002.

[38] W. Huang, X. Li, Y. Xue et al., "Antibacterial multilayer films fabricated by LBL immobilizing lysozyme and HTCC on nanofibrous mats," *International Journal of Biological Macromolecules*, vol. 53, pp. 26–31, 2013.

[39] J. M. He, Y. D. Wu, F. W. Wang et al., "Hemostatic, antibacterial and degradable performance of the water-soluble chitosan-coated oxidized regenerated cellulose gauze," *Fiber and Polymers*, vol. 15, no. 3, pp. 504–509, 2014.

[40] A. Maleki, H. Movahed, and R. Paydar, "Design and development of a novel cellulose/gamma-Fe2O3/Ag nanocomposite: a potential green catalyst and antibacterial agent," *RSC Advances*, vol. 6, no. 17, pp. 13657–13665, 2016.

16

Densities, Apparent Molar Volume, Expansivities, Hepler's Constant, and Isobaric Thermal Expansion Coefficients of the Binary Mixtures of Piperazine with Water, Methanol, and Acetone at $T = 293.15$ to 328.15 K

Qazi Mohammed Omar (ID),[1] **Jean-Noël Jaubert** (ID),[2] **and Javeed A. Awan** (ID)[1]

[1]*Institute of Chemical Engineering and Technology, Faculty of Engineering and Technology, University of the Punjab, Lahore, Pakistan*
[2]*Université de Lorraine, Ecole Nationale Supérieure des Industries Chimiques, Laboratoire Réactions et Génie des Procédés (UMR CNRS 7274), 1 rue Grandville, 54000 Nancy, France*

Correspondence should be addressed to Jean-Noël Jaubert; jean-noel.jaubert@univ-lorraine.fr and Javeed A. Awan; javeedawan@ yahoo.com

Academic Editor: Gianluca Di Profio

The properties of 3 binary mixtures containing piperazine were investigated in this work. In a first step, the densities for the two binary mixtures (piperazine + methanol) and (piperazine + acetone) were measured in the temperature range of 293.15 to 328.15 K and 293.15 to 323.15 K, respectively, at atmospheric pressure by using a Rudolph research analytical density meter (DDM 2911). The concentration of piperazine in the (piperazine + methanol) mixture was varied from 0.6978 to 14.007 mol/kg, and the concentration of piperazine in the (piperazine + acetone) mixture was varied from 0.3478 to 1.8834 mol/kg. On the other hand, the density data for the (piperazine + water) mixture were taken from the literature in the temperature range of 298.15 to 328.15 K. In a second step, for the 3 investigated systems, the apparent molar volume (V_ϕ) and the limiting apparent molar volume (V_ϕ^0) at infinite dilution were calculated using the Redlich–Mayer equation. The limiting apparent molar volumes (V_ϕ^0) were used to study the influence of the solute-solvent and solute-solute interactions. The temperature dependency of the apparent molar volumes was used to estimate the apparent molar expansibility, Hepler's constant $(\partial^2 V_\phi^0 / \partial T^2)_\mathrm{p}$, and isobaric thermal expansion coefficients α_p.

1. Introduction

Information about the physical properties of solutions in the vast range of solute concentrations at different temperatures is greatly important for physicochemical processes (separation process, crystallization, vaporization, desalination, waste aqua treatment, environment protection, oil retrieval, etc.) and the natural environment [1, 2].

The apparent molar volumes are particularly relevant to determine the molecular interactions (solute to solute, solute to solvent, and solvent to solvent) happening in solutions [3]. Also, the apparent molar volumes of solutions at infinite dilution are useful to obtain information regarding solute to solvent and solvent to solvent interactions. However, the apparent molal volumes depend on strength of solution that

can be used for the determination of solute to solute interactions [4–6].

The thermophysical properties of piperazine + water is important for the design of gas processing technology [7] like in the treatment of natural gas having significant amount of H_2S and in processing of refinery waste gases as well as synthesis gas for manufacturing of NH_3, where solution of piperazine + water is used as a solvent for the removal of acidic gases (carbon dioxide and hydrogen sulfide). The highly effective removal of CO_2 from industrial gases can also be performed by mixing piperazine with an alcohol such as the 2-amino-2-methyl-1-propanol [7, 8], which suggests that the alcoholic solutions of piperazine are also important in many separation processes. As another example, the separation of o- and p-chlorobenzoic acids from their eutectic blend usually uses a mixture of (piperazine + methanol) [9] as the solvent. In chemical processes, piperazine is present with crude products and has to be separated. In such a case, acetone is classically used due to its stronger molecular interaction with piperazine as compared to the higher molecular weight ketone [10].

In conclusion, the thermophysical properties of the 3 binary systems (piperazine + water), (piperazine + acetone), and (piperazine + methanol) are involved in many separation processes and thus need to be known. Consequently, it was decided to measure the densities of the (piperazine + acetone) and (piperazine + methanol) systems in the temperature range of 293.15 to 328.1 K and 293.15 to 323.15 K, respectively, since they are not available in the open literature. The concentration of piperazine was varied from 0.6978 to 14.007 mol/kg and from 0.3478 to 1.8834 mol/kg for methanol and acetone, respectively. The density data for (piperazine + water) were taken from literature in the temperature range of 298.15 to 328.15 K. [11]. For the 3 investigated binary systems, the density data were used for the calculation of the apparent molar volume, limiting apparent molar volume, apparent molar expansivities, Hepler's constant, and isobaric thermal expansion coefficient.

2. Experimental Work

2.1. Materials. The chemicals used in this work are piperazine (purity ≥ 99%), methanol (≥99.4%), and acetone (≥99.8%). They were provided by Sigma-Aldrich (Germany) and were used without any further purification or treatment. The purity of these chemicals used along with their source and CAS number are tabulated in Table 1. The deionized water has been prepared in lab through alfa-pore machine (WAP-4).

2.2. Measurement of the Density. An analytical digital vibrating glass U-tube densitometer (DDM-2911, Rudolph) with an accuracy of 0.05 kg/m^3 was used to measure the density of the 2 mixtures: piperazine + methanol and piperazine + acetone. A schematic diagram of the used densitometer is illustrated in Figure 1. The binary mixtures were prepared by weight using a Sartorius analytical weight balance with an uncertainty of ±0.00029 g (the corresponding uncertainty in molality was ±0.0004 mol/kg). The densities of the pure solvents and their blends with piperazine were measured in a temperature range varying from 298.15 to 333.15 K. The calibration of the apparatus was conducted by comparing the density of air and water at 293.15 K and the barometric pressure. Air was provided through a suction tube filled with silica balls to ensure a provision of dry air, and double-distilled water was injected through a syringe into the density meter. Silica balls were regularly heated to remove the moisture content absorbed from the atmospheric air. Once calibrated, the U-tube densitometer was washed with distilled water and dried with acetone and air. The density data reported in this study are an average of at least three runs. To remove the air bubbles from the samples, all the solutions were sonicated by using a universal ultrasonic cleaner for 30 min. Later, the samples were stored in vials and placed into a desiccator for 10 minutes for proper mixing and settling. For each

TABLE 1: List of chemicals used in this work.

Chemicals	Purity	Source	CAS number
Piperazine	≥99%	Merck	110-85-0
Methanol	≥99.4%	Merck	67-56-1
Acetone	≥99.8%	Merck	67-64-1

measurement, the tube was washed with water and dried with acetone. During the measurements, the air pump was always turned off to avoid irregularities due to vibrations.

Table 2 shows the densities of the pure solvents (methanol and acetone) measured in this study in the temperature range of 293.1–328.15 K along with values reported in the literature. An average deviation of approximately 0.03% is observed between both sets of data, which suggests that our data are consistent with previously measured densities and that our equipment is reliable.

3. Results and Discussion

The densities of all three binary mixtures (piperazine + water), (piperazine + methanol), and (piperazine + acetone) as a function of the molality of piperazine and temperature are presented in Table 3 and plotted in Figures 2–4. The experiments cover the commercially significant concentration range of piperazine with water, methanol, and acetone, that is, concentrations that are important for industrial applications like the design of gas processing technology, liquid-liquid extraction, and leaching. More specifically, the mixtures of piperazine + methanol were prepared in a concentration range of 2.187 wt.% to 30.978 wt.% (0.6978 mol/kg to 14.007 mol/kg). Similarly, mixtures of piperazine + acetone were prepared in concentration range from 1.98 wt.% to 9.86 wt.% (0.3478 mol/kg to 1.8834 mol/kg). It is observed that the density of the mixture increases with an increase in the concentration of piperazine. However, the density decreases with an increase in the temperature.

The apparent molar volumes [26] (V_ϕ) (in m^3/mol) of piperazine were calculated from the densities of the solutions by using the equation given below:

$$V_\phi = \frac{M}{\rho} + \frac{\rho_0 - \rho}{m \cdot \rho \cdot \rho_0}, \quad (1)$$

where m is the molality of piperazine (mol/kg), ρ and ρ_0 are densities (in kg/m^3) of the solution and pure solvent, respectively, and M is the molar mass of piperazine (in kg/mol). The apparent molar volume (V_ϕ) of all three binary mixtures (piperazine + water), (piperazine + methanol), and (piperazine + acetone) calculated from Equation (1) as a function of molality of piperazine and temperature is tabulated in Table 4 and plotted in Figures 5–7 (denoted by markers). Figures 5–7 show that V_ϕ values rise with rise in temperature for each binary mixture, highlighting that the overall order of the structure is improved or increased in solution with rising temperature [27]. The influence of the molality depends on the studied system, that is, the apparent molar volumes may rise, decrease, or progress through a maximum. Our data were correlated with the Redlich–Mayer equation [28]:

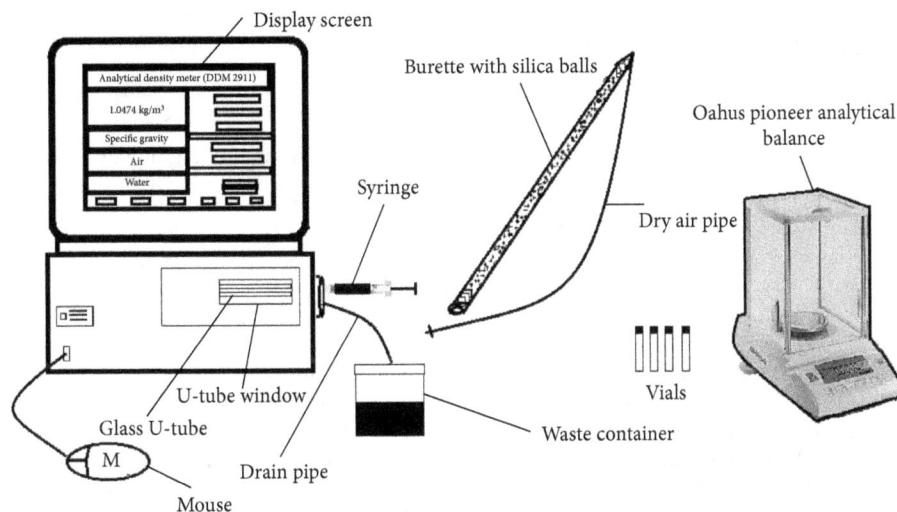

FIGURE 1: Schematic diagram of the experimental setup (analytical density meter (DDM 2911)).

TABLE 2: Comparison of the densities of the pure solvents measured in this study with those reported previously at various temperatures and at atmospheric pressure with standard uncertainties: u $(T) = \pm 0.01$ K u $(\rho) = \pm 0.1$ kg/m^3, u $(m) = \pm 0.0004$ mol/kg, and u $(P) = \pm 0.002$ atm.

				Density (ρ_0) kg/m^3			
		Methanol				Acetone	
T/K	This work	Lit. value	(Reference)	T/K	This work	Lit. value	(Reference)
293.15	791.6	791.9	(Papanastasiou and Ziogas [12])	293.15	789.9	790.02	(Kinart et al. [13])
		791.65	(Gonfa et al. [14])			790.355	(Janz and Tomkins [15])
298.15	786.9	786.884	(Anwar and Yasmeen [16])	298.15	784.2	784.45	(Kinart et al. [13])
		786.68	(Tu et al. [17])			784.638	(Janz and Tomkins [15])
303.15	782.2	782.158	(Anwar and Yasmeen [16])	303.15	778.5	778.7	(Fan et al. [18])
		781.9	(Tu et al. [19])			778.57	(Enders et al. [20])
308.15	777.2	777.2	(Tu et al. [17])	308.15	773.0	773.0	(Hafez and Hartland [21])
		777.414	(Anwar and Yasmeen [16])			773.065	(Janz and Tomkins [15])
313.15	772.4	772.3	(Tu et al. [19])	313.15	767.3	767.03	(Fan et al. [18])
		772.64	(Anwar and Yasmeen [16])			767.21	(Estrada-Baltazar et al [22])
318.15	767.5	767.6	(Tu et al. [17])	318.15	761.6	761.288	(Janz and Tomkins [15])
		767.844	(Anwar and Yasmeen [16])			761.3	(Hafez and Hartland [21])
323.15	762.7	762.7	(Bhuiyan and Uddin [23])	323.15	755.7	755.14	(Fan et al. [18])
		763.028	(Anwar and Yasmeen [16])			755.31	(Estrada-Baltazar et al [22])
328.15	757.7	759.2	(Cai et al. [24])			755.54	(Wu et al. [25])

$$V_\phi = V_\phi^0 + S_v \sqrt{m} + B_v \cdot m, \qquad (2)$$

where V_ϕ^0 is the limiting apparent molar volume of the piperazine mixtures and S_v and B_v are two regression parameters. Figures 6 and 7 highlight that our data are accurately correlated with such a simple model. The corresponding values of V_ϕ^0, S_v, and B_v are tabulated in Table 5. From this table, notably, the V_ϕ^0 values rise with rise in temperature for each binary mixture. As highlighted by [29], this behavior characterizes the presence of strong solute to solvent interactions that are strengthened with the rise in temperature. It is worth noting that this behavior was also observed for many systems. We can cite the (methanol + methyl acetate) system reviewed by [30], the (methanol + ethyl acetate), the (ethanol + methyl acetate) and (ethanol + ethyl acetate) systems studied by [29], the (methanol + isopropyl alcohol), the (methyl salicylate + DMSO) and (hydroxamic acid + DMSO) examined by [31]. The V_ϕ^0 values of the mixtures rise in the

following order: (piperazine + methanol) < (piperazine + water) < (piperazine + acetone), which could be due—as explained by [29]—to an enhancement in the strengths of the solute to solvent interactions. This enhancement results in an increase of contraction in the volume.

The values of S_v and B_v are also tabulated in Table 5. The S_v values are negative for (piperazine + acetone) and positive for the (piperazine + water) and (piperazine + methanol) systems at each temperature. The S_v value decreases with rise in temperature for each binary system. The strength of the solute to solute interactions of each system increases in the following order: (piperazine + methanol) > (piperazine + water) > (piperazine + acetone). The solute to solute interaction decreases with an increase in temperature for each binary system. The B_v values are negative for the (piperazine + methanol) and (piperazine + water) mixtures and positive for the (piperazine + acetone) mixture at each temperature. The B_v values rise with rise in temperature for each binary system. The negative B_v values show rise in solute to solute interactions for

TABLE 3: Densities (kg/m³) of the binary mixtures (piperazine + water), (piperazine + methanol), and (piperazine + acetone) as a function of the molality and temperature at atmospheric pressure. A global uncertainty calculation was performed on each point and the corresponding values are given in parentheses.

Temperature (K) Molality (mol/kg) of piperazine	293.15	298.15	303.15	308.15	313.15	318.15	323.15	328.15
	Binary system densities (kg/m³) with global uncertainty (kg/m³) in parentheses							
System (piperazine + water) (Muhammad et al. [11])								
0.9838	—	998.3	996.9	995.2	993.4	991.4	989.2	986.8
3.0226	—	999.4	997.9	996.3	994.4	992.4	990.1	987.8
6.4138	—	1001.3	999.8	998.0	996.1	994.0	991.7	989.3
System (piperazine + methanol) (this work)								
0.6978	805.8 (0.02)	801.2 (0.01)	796.4 (0.02)	791.3 (0.007)	786.5 (0.02)	781.7 (0.01)	776.6 (0.01)	771.3 (0.01)
1.4000	817.4 (0.04)	812.9 (0.02)	808.2 (0.03)	803.5 (0.01)	798.6 (0.02)	793.6 (0.01)	788.4 (0.02)	783.2 (0.01)
2.2148	829.6 (0.03)	825.2 (0.02)	820.5 (0.03)	815.7 (0.02)	810.9 (0.02)	805.9 (0.02)	800.8 (0.02)	795.6 (0.02)
3.0544	840.7 (0.04)	836.3 (0.02)	831.7 (0.04)	827.0 (0.04)	822.1 (0.03)	817.2 (0.03)	812.1 (0.03)	806.9 (0.02)
3.7151	849.0 (0.04)	844.6 (0.02)	840.0 (0.04)	835.3 (0.04)	830.5 (0.04)	825.6 (0.03)	820.5 (0.04)	815.4 (0.02)
5.1007	862.6 (0.05)	858.2 (0.03)	853.7 (0.04)	849.0 (0.05)	844.5 (0.04)	839.6 (0.04)	834.6 (0.04)	829.5 (0.03)
5.9003	870.3 (0.05)	866.0 (0.03)	861.5 (0.05)	856.9 (0.05)	852.1 (0.05)	847.2 (0.04)	842.2 (0.05)	837.1 (0.04)
7.4725	881.4 (0.06)	877.1 (0.03)	872.6 (0.05)	868.0 (0.06)	863.2 (0.06)	858.4 (0.05)	853.4 (0.06)	848.3 (0.04)
9.2198	893.9 (0.06)	889.6 (0.04)	885.2 (0.05)	880.6 (0.06)	875.8 (0.06)	871.0 (0.05)	866.1 (0.07)	861.1 (0.05)
14.007	912.6 (0.06)	908.4 (0.05)	903.9 (0.06)	899.3 (0.06)	894.6 (0.07)	889.6 (0.05)	884.2 (0.07)	879.3 (0.06)
System (piperazine + acetone) (this work)								
0.3478	792.4 (0.01)	786.5 (0.005)	780.8 (0.01)	774.9 (0.006)	768.8 (0.002)	762.7 (0.002)	756.3 (0.002)	—
0.8416	796.7 (0.02)	791.0 (0.02)	785.2 (0.02)	779.3 (0.02)	773.3 (0.008)	767.2 (0.01)	760.9 (0.01)	—
1.0951	799.3 (0.03)	793.5 (0.02)	787.7 (0.03)	781.6 (0.02)	775.5 (0.01)	769.1 (0.02)	762.7 (0.02)	—
1.3359	802.3 (0.03)	797.0 (0.03)	790.7 (0.03)	784.6 (0.03)	778.4 (0.02)	772.1 (0.02)	765.6 (0.02)	—
1.8834	807.5 (0.03)	802.0 (0.04)	795.9 (0.03)	789.9 (0.03)	784.1 (0.02)	778.2 (0.03)	772.1 (0.03)	—

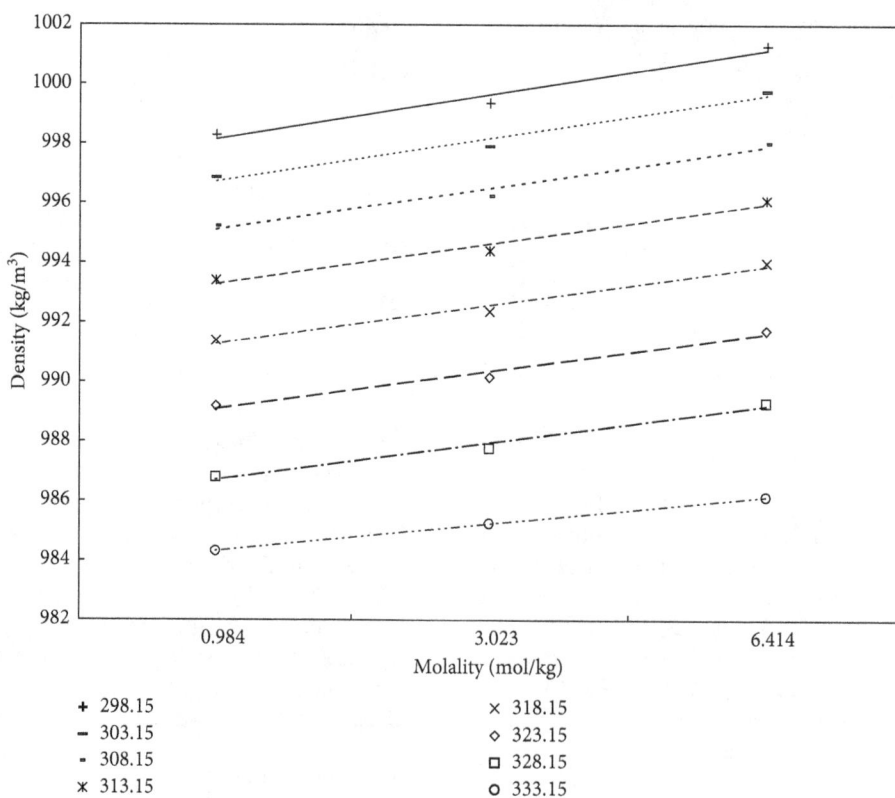

FIGURE 2: Density of the (piperazine + water) system as a function of piperazine molality at various temperatures.

mixtures that is, (piperazine + water) and (piperazine + methanol). The positive B_V values indicate strong solute to solute interactions for the (piperazine + acetone) system.

The temperature dependence of the limiting apparent molar volume (V_ϕ^0) can be expressed in terms of the following equation [26]:

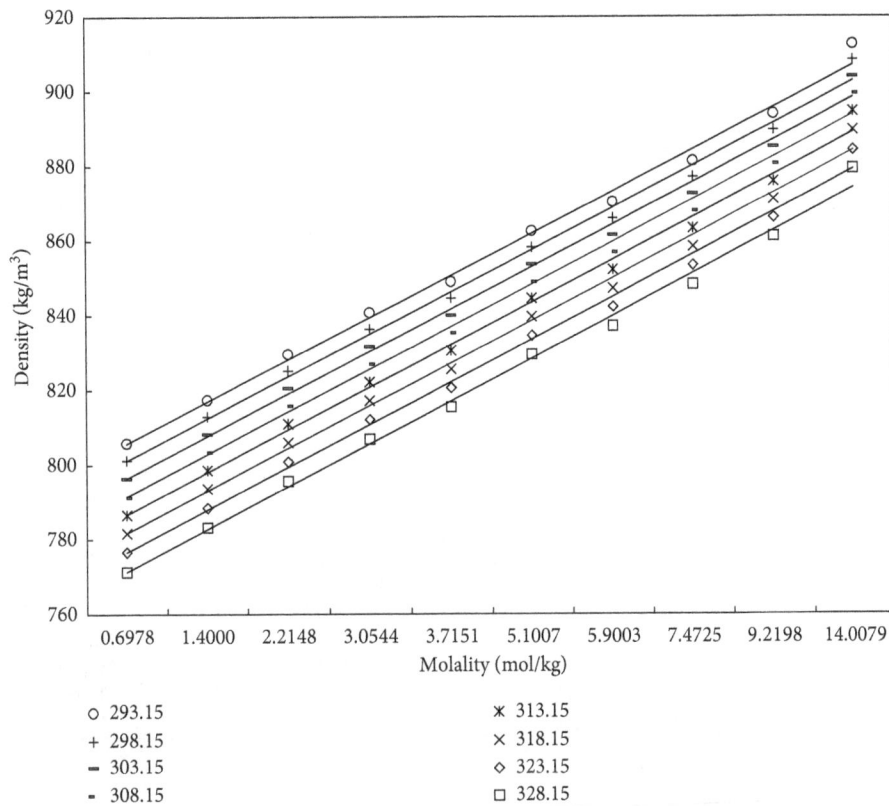

FIGURE 3: Density of the (piperazine + methanol) system as a function of piperazine molality at various temperatures.

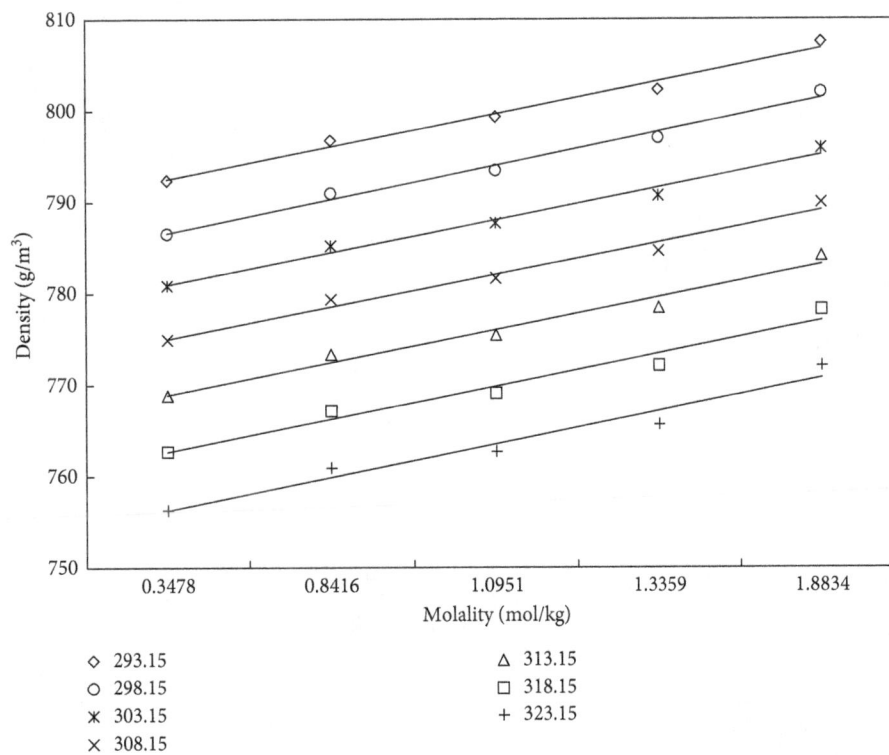

FIGURE 4: Density of the (piperazine + acetone) system as a function of piperazine molality at various temperatures.

TABLE 4: Apparent molar volume (m^3/mol) of the binary mixtures (piperazine + water), (piperazine + methanol), and (piperazine + acetone) as a function of the molality and temperature at atmospheric pressure with the standard uncertainty: u $(V_\phi) = \pm 0.35 \times 10^{-6}$ m^3/mol.

Temperature (K)	293.15	298.15	303.15	308.15	313.15	318.15	323.15	328.15
Molality (mol/kg) of piperazine	Binary system apparent molar volumes $10^6 \times V_\phi$ (m^3/mol)							
System (piperazine + water)								
0.9838	—	85.02	85.16	85.32	85.49	85.67	85.88	86.13
3.0226	—	85.43	85.57	85.72	85.89	86.08	86.28	86.50
6.4138	—	85.37	85.52	85.68	85.87	86.06	86.27	86.50
System (piperazine + methanol)								
0.6978	75.02	75.24	75.50	75.84	76.07	76.38	77.32	78.39
1.4000	76.96	77.05	77.18	77.15	77.48	77.97	78.69	79.32
2.2148	77.69	77.84	78.02	78.13	78.44	78.83	79.39	79.88
3.0544	78.29	78.46	78.65	78.78	79.11	79.48	79.98	80.41
3.7151	78.48	78.66	78.86	79.03	79.31	79.67	80.11	80.49
5.1007	79.49	79.70	79.93	80.13	80.30	80.66	81.07	81.45
5.9003	79.61	79.82	80.04	80.24	80.54	80.90	81.29	81.68
7.4725	80.52	80.75	81.00	81.23	81.55	81.89	82.29	82.68
9.2198	80.69	80.92	81.18	81.43	81.77	82.11	82.47	82.84
14.007	82.43	82.71	83.01	83.31	83.66	84.07	84.55	84.93
System (piperazine + acetone)								
0.3478	97.37	98.38	99.39	102.11	104.76	107.33	110.96	—
0.8416	95.20	95.88	96.72	98.09	99.43	100.86	102.50	—
1.0951	94.14	94.91	95.75	97.22	98.65	100.24	101.94	—
1.3359	92.75	92.69	94.15	95.49	96.78	98.16	99.71	—
1.8834	92.02	92.32	93.37	94.36	95.06	95.77	96.69	—

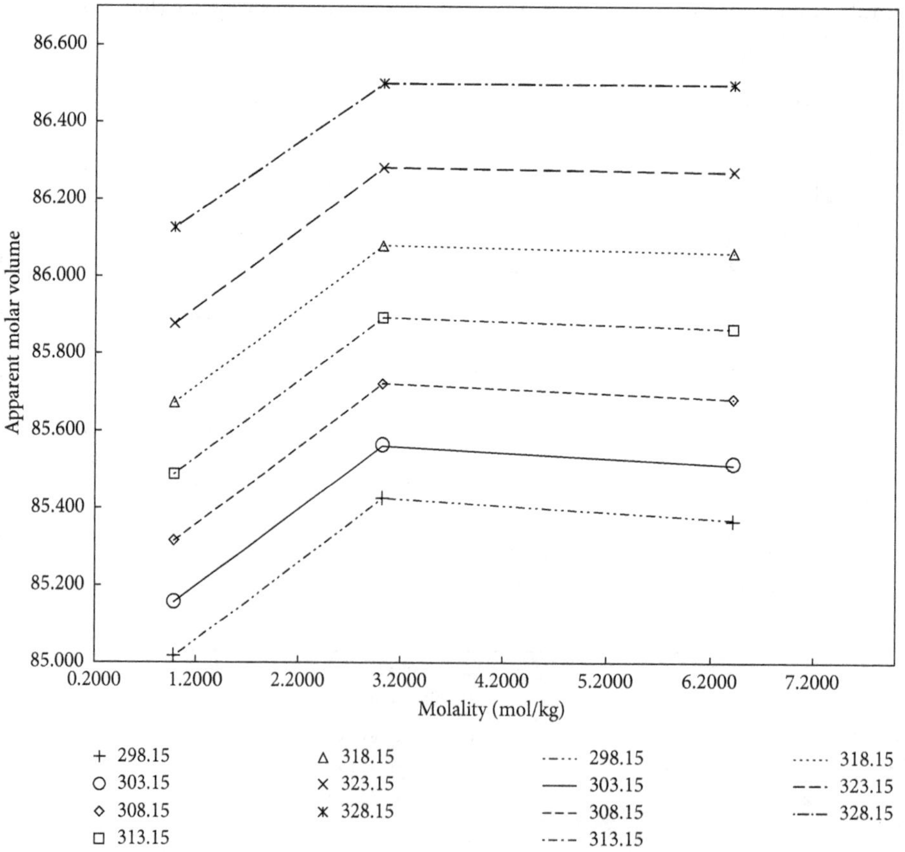

FIGURE 5: Apparent molar volume (V_ϕ) of the binary mixture (piperazine + water) as a function of piperazine molality at different temperatures.

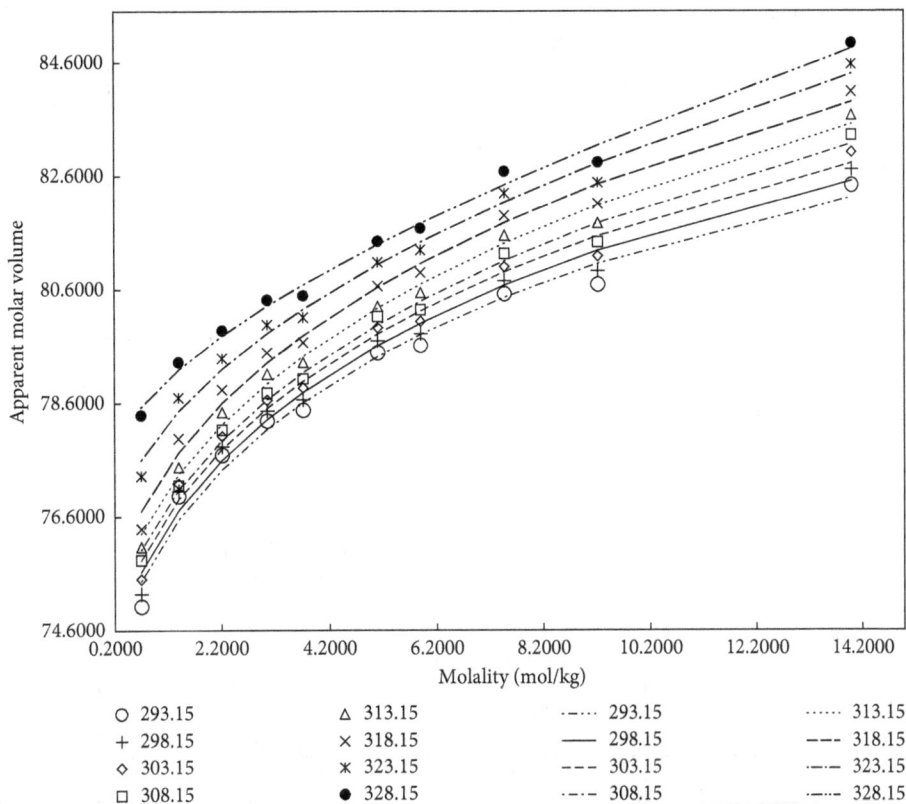

FIGURE 6: Apparent molar volume (V_ϕ) of the binary mixture (piperazine + methanol) as a function of piperazine molality at different temperatures. The correlations performed with the Redlich–Mayer equation are plotted as solid/dashed lines.

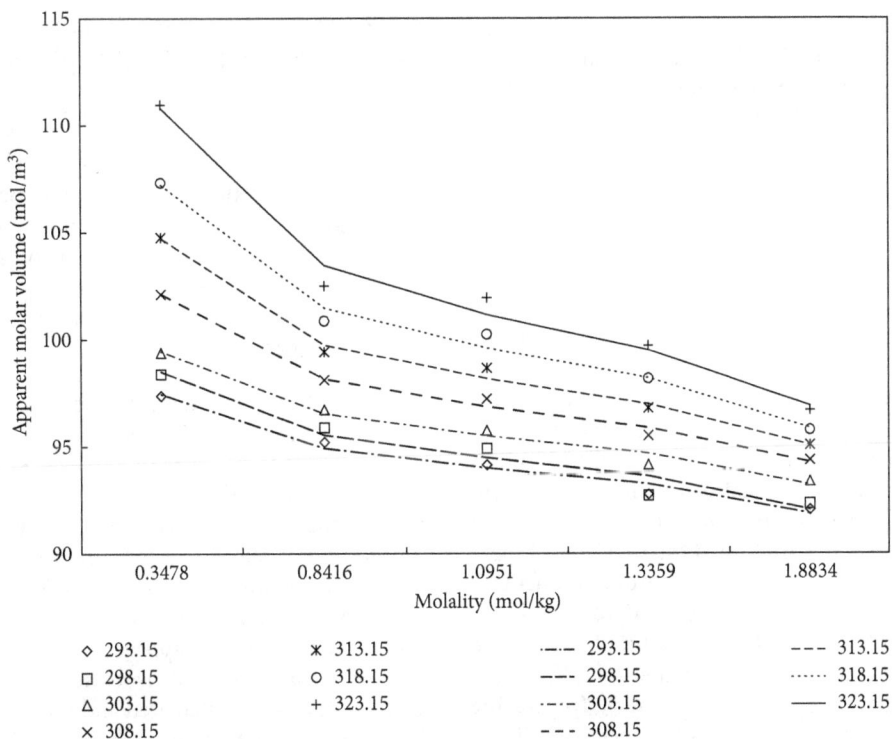

FIGURE 7: Apparent molar volume (V_ϕ) of the binary mixture (piperazine + acetone) as a function of piperazine molality at different temperatures. The correlations performed with the Redlich–Mayer equation are plotted as solid/dashed lines.

TABLE 5: Values of the limiting apparent molar volume (V_ϕ^0) along with the S_v and B_v parameters to be used in the Redlich–Mayer equation for each of the 3 binary systems as a function of temperature.

Temperature (K)	293.15	298.15	303.15	308.15	313.15	318.15	323.15	328.15
			System (piperazine + water)					
$10^6 \times V_\phi^0$ (m^3·mol^{-1})	—	83.78	83.94	84.11	84.30	84.49	84.71	85.07
$10^6 \times S_v$ (m^3·kg$^{1/2}$·mol$^{-3/2}$)	—	1.65	1.61	1.60	1.57	1.56	1.53	1.39
$10^6 \times B_v$ (m^3·kg·mol^{-2})	—	−0.40	−0.39	−0.39	−0.38	−0.37	−0.36	−0.33
			System (piperazine + methanol)					
$10^6 \times V_\phi^0$ (m^3·mol^{-1})	72.57	72.75	72.99	73.27	73.57	74.01	75.48	77.05
$10^6 \times S_v$ (m^3·kg$^{1/2}$·mol$^{-3/2}$)	3.72	3.69	3.63	3.47	3.45	3.38	2.58	1.67
$10^6 \times B_v$ (m^3·kg·mol^{-2})	−0.31	−0.29	−0.27	−0.22	−0.21	−0.20	−0.05	0.11
			System (piperazine + acetone)					
$10^6 \times V_\phi^0$ (m^3·mol^{-1})	102.68	104.72	105.89	111.82	117.01	121.20	129.50	—
$10^6 \times S_v$ (m^3·kg$^{1/2}$·mol$^{-3/2}$)	−9.60	−11.54	−12.17	−19.19	−24.51	−27.62	−37.74	—
$10^6 \times B_v$ (m^3·kg·mol^{-2})	1.26	1.68	2.14	4.66	6.21	6.67	10.19	—

TABLE 6: The limiting apparent molar expansibility (E_ϕ^0) and isobaric thermal expansion coefficient α_p.

Temperature (K) Binary mixtures	293.15	298.15	303.15	308.15	313.15	318.15	323.15	328.15
			$10^6 \times E_\phi^0$ (m^3·mol^{-1}·K^{-1})					
Piperazine + water	—	2.64	3.24	3.84	4.44	5.04	5.64	6.24
Piperazine + methanol	−0.04	0.03	0.10	0.16	0.23	0.29	0.35	0.41
Piperazine + acetone	25.38	46.78	68.18	89.58	110.98	132.38	153.78	—
			$10^3 \times \alpha_p$ (K^{-1})					
Piperazine + water	—	0.31	0.39	0.46	0.53	0.60	0.67	0.73
Piperazine + methanol	−0.39	0.28	0.95	1.62	2.28	2.93	3.52	4.08
Piperazine + acetone	2.47	4.47	6.44	8.01	9.48	10.92	11.88	—

$$V_\phi^0 = A + BT + CT^2, \qquad (3)$$

where A, B, and C are empirical parameters and T is the temperature. The limiting apparent molar expansibility (E_ϕ^0) can be obtained by differentiating Eq. (3) with respect to the temperature:

$$E_\phi^0 = \left(\frac{\partial V_\phi^0}{\partial T}\right) = B + 2CT, \qquad (4)$$

The (E_ϕ^0) values for each binary system are tabulated in Table 6 which gives important information related to the solute to solvent interactions [32]. Table 6 depicts that, at each temperature, the (E_ϕ^0) values are positive for all three binary systems and decrease with rise in temperature.

According to Hepler's theory [33] the so-called Hepler's constant, ($\partial^2 V_\phi^0/\partial T^2$), can be used to classify a solute into two categories, whether solute can act as a builder of structure or as a breaker of structure. If the ($\partial^2 V_\phi^0/\partial T^2$) value is positive, then the solute is favorable in development or making of structure. Conversely, if the ($\partial^2 V_\phi^0/\partial T^2$) value is negative, then the solute will act as a structure breaker. The values of ($\partial^2 V_\phi^0/\partial T^2$) are −0.33, −2.89, and −12.25 for the binary mixtures, that is, (piperazine + water), (piperazine + methanol), and (piperazine + acetone), respectively. Thus, piperazine acts as a structure breaker in solution. The proof for the effect of Hepler's constant on microscopic structure is discussed in the literature [34].

The isobaric thermal expansion coefficient, α_p, of the solute was calculated using the apparent molar volume and apparent molar expansibility at infinite dilution data:

$$\alpha_p = \frac{1}{V_\phi^0}\left(\frac{\partial V_\phi^0}{\partial T}\right) = \frac{E_\phi^0}{V_\phi^0}. \qquad (5)$$

The isobaric thermal expansion coefficient, α_p, is also tabulated in Table 6. A higher value of α_p was obtained for acetone, and lower value of α_p was obtained for water.

4. Conclusion

In this work, new density data for 2 binary mixtures (piperazine + methanol) and (piperazine + acetone) were measured from 293.15 K to 328.15 K. It was found that the density of both the mixtures increases with increase in temperature but decreases with increase in concentration. Similarly, V_ϕ values rise with increase in concentration of piperazine in methanol but decreases in case of acetone. Also, the apparent molar volume (V_ϕ), limiting apparent molar volume (V_ϕ^0), apparent molar expansibility (E_ϕ^0), Hepler's constant, and isobaric thermal expansion coefficient (α_p) were calculated and reported in this work. The limiting apparent molar volume V_ϕ^0 increases with temperature, which highlights the strong interactions of the solute with the solvent. The positive apparent molar expansibility (E_ϕ^0) decreases with temperature, which indicates that the interactions increase with a rise in the temperature of the solution. The negative values of the Helper's constant suggest that piperazine acts as a structure breaker in the solvent. The Redlich–Mayer equation was used to correlate the apparent molar volume with the standard deviation, $u(V_\phi) = \pm 0.35 \times 10^{-6}\,\text{mol/m}^3$.

Nomenclature

V_ϕ: Apparent molar volume
V_ϕ^0: Apparent molar volume at infinite dilution
ρ: Density of the solution
ρ_0: Density of the pure solvent
S_V: Empirical parameter of the apparent molar volume
B_V: Empirical parameter of the apparent molar volume
α_p: Isobaric thermal expansion coefficient
E_ϕ^0: Limiting apparent molar volume expansibility
M: Molar mass of the solute
m: Molality of the solute.

Conflicts of Interest

The authors declare that there are no conflicts of interest regarding the publication of this paper.

Acknowledgments

This research was performed as part of the employment of the authors. Consequently, University of the Punjab and University of Lorraine are warmly thanked. The work described in this paper has been presented at the Chisa 2018 International Conference that took place in Prague (Czech Republic) in August 2018.

References

[1] S. Gupta and J. D. Olson, "Industrial needs in physical properties," *Industrial & Engineering Chemistry Research*, vol. 42, no. 25, pp. 6359–6374, 2003.

[2] K. S. Pitzer, "Thermodynamics of natural and industrial waters," *Journal of Chemical Thermodynamics*, vol. 25, no. 1, pp. 7–26, 1993.

[3] F. J. Millero, "Molal volumes of electrolytes," *Chemical Reviews*, vol. 71, no. 2, pp. 147–176, 1971.

[4] Y. Marcus, *Ion Solvation*, Wiley, New York, NY, USA, 1985.

[5] F. J. Millero, *Water and Aqueous Solutions, Structure, Thermodynamics, and Transport Properties*, Chapter 13, R. A. Horne, Ed., Wiley Interscience, New York, NY, USA, 1972..

[6] R. Perkins, T. Andersen, J. O. M. Bockris, and B. E. Conway, *Modern Aspects of Electrochemistry*, Kluwer Academic Publishers, Norwell, MA, USA, 1969.

[7] A. Samanta and S. S. Bandyopadhyay, "Density and viscosity of aqueous solutions of piperazine and (2-amino-2-methyl-1-propanol + piperazine) from 298 to 333 K," *Journal of Chemical & Engineering Data*, vol. 51, no. 2, pp. 467–470, 2006.

[8] A. L. Kohl and R. Nielsen, *Gas Purif*, Gulf Professional Publishing, Houston, TX, USA, 1997.

[9] A. Lashanizadegan, N. S. Tavare, M. Manteghian, and D. M. Newsham, "Ternary phase equilibrium diagrams for o-and p-chlorobenzoic acids and their complex with piper-

[10] G. R. Bond, "Determination of piperazine as piperazine diacetate," *Analytical Chemistry*, vol. 32, no. 10, pp. 1332–1334, 1960.

[11] A. Muhammad, M. I. A. Mutalib, T. Murugesan, and A. Shafeeq, "Thermophysical properties of aqueous piperazine and aqueous (N-methyldiethanolamine + piperazine) solutions at temperatures (298.15 to 338.15) K," *Journal of Chemical & Engineering Data*, vol. 54, no. 8, pp. 2317–2321, 2009.

[12] G. E. Papanastasiou and I. I. Ziogas, "Physical behavior of some reaction media. 3. Density, viscosity, dielectric constant, and refractive index changes of methanol + dioxane mixtures at several temperatures," *Journal of Chemical & Engineering Data*, vol. 37, no. 2, pp. 167–172, 1992.

[13] C. M. Kinart, W. J. Kinart, and A. Ć. wiklińska, "Density and viscosity at various temperatures for 2-methoxyethanol+ acetone mixtures," *Journal of Chemical & Engineering Data*, vol. 47, no. 1, pp. 76–78, 2002.

[14] G. Gonfa, M. A. Bustam, N. Muhammad, and S. Ullah, "Density and excess molar volume of binary mixture of thiocyanate-based ionic liquids and methanol at temperatures 293.15–323.15 K," *Journal of Molecular Liquids*, vol. 211, pp. 734–741, 2015.

[15] G. J. Janz and R. P. T. Tomkins, *Nonaqueous Electrolytes Handbook*, Vol. 1, Academic Press, New York, NY, USA, 1972.

[16] N. Anwar and S. Yasmeen, "Volumetric, compressibility and viscosity studies of binary mixtures of [EMIM][NTf2] with ethylacetate/methanol at (298.15–323.15) K," *Journal of Molecular Liquids*, vol. 224, pp. 189–200, 2016.

[17] C. H. Tu, H. C. Ku, W. F. Wang, and Y. T. Chou, "Volumetric and viscometric properties of methanol, ethanol, propan-2-ol, and 2-methylpropan-2-ol with a synthetic C_{6+} mixture from 298.15 K to 318.15 K," *Journal of Chemical & Engineering Data*, vol. 46, no. 2, pp. 317–321, 2001.

[18] X. H. Fan, Y. P. Chen, and C. S. Su, "Densities and viscosities of binary liquid mixtures of 1-ethyl-3-methylimidazolium tetrafluoroborate with acetone, methyl ethyl ketone, and N-methyl-2-pyrrolidone," *Journal of the Taiwan Institute of Chemical Engineers*, vol. 61, pp. 117–123, 2016.

[19] C. H. Tu, S. L. Lee, and I. H. Peng, "Excess volumes and viscosities of binary mixtures of aliphatic alcohols (C1–C4) with nitromethane," *Journal of Chemical & Engineering Data*, vol. 46, no. 1, pp. 151–155, 2001.

[20] S. Enders, H. Kahl, and J. Winkelmann, "Surface tension of the ternary system water + acetone + toluene," *Journal of Chemical & Engineering Data*, vol. 52, no. 3, pp. 1072–1079, 2007.

[21] M. Hafez and S. Hartland, "Densities and viscosities of binary systems toluene-acetone and 4-methyl-2-pentanone-acetic acid at 20, 25, 35, and 45. degree. C," *Journal of Chemical & Engineering Data*, vol. 21, no. 2, pp. 179–182, 1976.

[22] A. Estrada-Baltazar, A. De León-Rodríguez, K. R. Hall, M. Ramos-Estrada, and G. A. Iglesias-Silva, "Experimental densities and excess volumes for binary mixtures containing propionic acid, acetone, and water from 283.15 K to 323.15 K at atmospheric pressure," *Journal of Chemical & Engineering Data*, vol. 48, no. 6, pp. 1425–1431, 2003.

[23] M. M. H. Bhuiyan and M. H. Uddin, "Excess molar volumes and excess viscosities for mixtures of N, N-dimethylformamide with methanol, ethanol and 2-propanol at different temperatures," *Journal of Molecular Liquids*, vol. 138, no. 1–3, pp. 139–146, 2008.

[24] D. Cai, J. Yang, H. Da, L. Li, H. Wang, and T. Qiu, "Densities

azine in methanol," *Journal of Chemical & Engineering Data*, vol. 45, no. 6, pp. 1189–1194, 2000.

and viscosities of binary mixtures N, N-dimethyl-N-(3-sulfopropyl) cyclohexylammonium tosylate with water and methanol at T = (303.15 to 328.15) K," *Journal of Molecular Liquids*, vol. 229, pp. 389–395, 2017.

[25] J. Y. Wu, Y. P. Chen, and C. S. Su, "The densities and viscosities of a binary liquid mixture of 1-n-butyl-3-methylimidazolium tetrafluoroborate, ([Bmim][BF4]) with acetone, methyl ethyl ketone and N, N-dimethylformamide, at 303.15 to 333.15 K," *Journal of the Taiwan Institute of Chemical Engineers*, vol. 45, no. 5, pp. 2205–2211, 2014.

[26] M. T. Zafarani-Moattar and H. Shekaari, "Apparent molar volume and isentropic compressibility of ionic liquid 1-butyl-3-methylimidazolium bromide in water, methanol, and ethanol at T = (298.15 to 318.15) K," *Journal of Chemical Thermodynamics*, vol. 37, no. 10, pp. 1029–1035, 2005.

[27] P. Khanuja, V. R. Chourey, and A. A. Ansari, "Apparent molar volume and viscometric study of glucose in aqueous solution," *Journal of Chemical and Pharmaceutical Research*, vol. 4, pp. 3047–3050, 2012.

[28] O. Redlich and D. M. Meyer, "The molal volumes of electrolytes," *Chemical Reviews*, vol. 64, no. 3, pp. 221–227, 1964.

[29] I. Bahadur, N. Deenadayalu, and D. Ramjugernath, "Effects of temperature and concentration on interactions in methanol+ ethyl acetate and ethanol+ methyl acetate or ethyl acetate systems: Insights from apparent molar volume and apparent molar isentropic compressibility study," *Thermochimica Acta*, vol. 577, pp. 87–94, 2014.

[30] I. Bahadur and N. Deenadayalu, "Apparent molar volume and isentropic compressibility for the binary systems {methyl-trioctylammoniumbis(trifluoromethylsulfonyl)imide + methyl acetate or methanol} and (methanol + methyl acetate) at T = 298.15, 303.15, 308.15 and 313.15 K and atmospheric pressure," *Journal of Solution Chemistry*, vol. 40, pp. 1528–1543, 2011.

[31] R. S. Sah, P. Pradhan, and M. N. Roy, "Solute–solvent and solvent–solvent interactions of menthol in isopropyl alcohol and its binary mixtures with methyl salicylate by volumetric, viscometric, interferometric and refractive index techniques," *Thermochimica Acta*, vol. 499, pp. 149–154, 2010.

[32] H. Shekaari, S. S. Mousavi, and Y. Mansoori, "Thermophysical properties of ionic liquid, 1-pentyl-3-methylimidazolium chloride in water at different temperatures," *International Journal of Thermophysics*, vol. 30, no. 2, pp. 499–514, 2009.

[33] L. G. Hepler, "Thermal expansion and structure in water and aqueous solutions," *Canadian Journal of Chemistry*, vol. 47, no. 24, pp. 4613–4617, 1969.

[34] P. K. Thakur, S. Patre, and R. Pande, "Thermophysical and excess properties of hydroxamic acids in DMSO," *Journal of Chemical Thermodynamics*, vol. 58, pp. 226–236, 2013.

17

Development of Flow-Through Polymeric Membrane Reactor for Liquid Phase Reactions: Experimental Investigation and Mathematical Modeling

Endalkachew Chanie Mengistie[1] and Jean-François Lahitte[2]

[1]*Bahir Dar Institute of Technology (BiT), Faculty of Chemical and Food Engineering, Bahir Dar University, 1920 Bahir Dar, Ethiopia*
[2]*Laboratoire de Genie Chimique, Universite Paul Sabatier, 118 route de Narbonne, Toulouse, France*

Correspondence should be addressed to Endalkachew Chanie Mengistie; endalk10@gmail.com

Academic Editor: Jose C. Merchuk

Incorporating metal nanoparticles into polymer membranes can endow the membranes with additional functions. This work explores the development of catalytic polymer membrane through synthesis of palladium nanoparticles based on the approaches of intermatrix synthesis (IMS) inside surface functionalized polyethersulfone (PES) membrane and its application to liquid phase reactions. Flat sheet PES membranes have been successfully modified via UV-induced graft polymerization of acrylic acid monomer. Palladium nanoparticles have been synthesized by chemical reduction of palladium precursor loaded on surface modified membranes, an approach to the design of membranes modified with nanomaterials. The catalytic performances of the nanoparticle incorporated membranes have been evaluated by the liquid phase reduction of p-nitrophenol using $NaBH_4$ as a reductant in flow-through membrane reactor configuration. The nanocomposite membranes containing palladium nanoparticles were catalytically efficient in achieving a nearly 100% conversion and the conversion was found to be dependent on the flux, amount of catalyst, and initial concentration of nitrophenol. The proposed mathematical model equation represents satisfactorily the reaction and transport phenomena in flow-through catalytic membrane reactor.

1. Introduction

The concept of membrane reactors (MRs), combining a membrane-based separation with a catalytic chemical reaction in one unit, dates back to the 1960s [1]. Since then, MRs played an important role in improving selectivity and yielding and enhancing conversion for thermodynamically limited chemical reactions in many chemical processes of industrial importance. In recent years, many approaches have been proposed to combine membrane properties with chemical reaction in order to intensify a process. These include extractor type, distributor, and contactor type. Depending on the type of membrane reactors, the membrane performs different functions [2, 3].

(1) Extractor Membrane Reactors. One of the products is continuously and selectively removed from the reaction mixture by the membrane. If the reaction is limited by the thermodynamic equilibrium, the conversion can be increased by removing one of the product components, so that the equilibrium shifts towards the desired product side [4, 5]. If the reaction rate of the undesired secondary reaction is higher than that of the primary reaction, the reaction selectivity can be significantly enhanced by removing the desired intermediate species. Particularly, the advantage of selectively removing the valuable product lies in avoiding further separation steps or reducing the separation units by increased product concentrations. Furthermore, if one of the products has inhibition effect, as in case of some fermentation, removing this product strongly improves the reactor productivity [6].

(2) Distributor Membrane Reactors. In this type of MR, one of the reactants is specifically added to the reaction mixture

across a membrane. The membrane can act as even distributor of the limiting reactant along the reactor to prevent side reactions and as upstream separation unit to selectively dose one component from a mixture. Controlled addition of oxygen in gas-phase partial oxidation of hydrocarbons in which the intermediate product reacts more intensively with oxygen than the reactants, in order to prevent total oxidation [7, 8], is the main application of this category of membrane reactor.

(3) Contactor Membrane Reactors. In this configuration, the reacting species are fed at different sides of the membrane and must diffuse through the catalytic layer to react. Therefore, the role of the membrane is to provide an interfacial contact area for the reacting streams but does not perform any selective separation. The two sides of membranes are used to bring reactants into contact and if the reaction rate is fast compared to the diffusion rates of the reactants, the reaction occurs in the catalytic layer in a way that prevents mixing of reactants. Due to higher surface area of membranes, a contactor mode can provide higher contact area between two different phases. Particularly, if one phase has lower solubility in the other phase, higher surface area contact between these phases can decrease the need of higher pressure that could have been applied to lower soluble component. Gas/liquid contactors and flow-through membrane reactors are important class of membrane contactors.

In flow-through catalytic membrane reactor (FTCMR) configuration, unselective porous catalytic membrane, either inherently catalytic or being made catalytic by impregnation of nanocatalysts, is applied in dead-end mode operation. The premixed reactants are forced to pass through the catalytic membrane. The function of the membrane is to create a reaction environment with intensive contact between the reactants and catalyst with short and controlled residence times and high catalytic activity. The main drawback in classical fixed-bed reactors is that the desired conversion is mainly limited by the pore diffusion. However, if reactants can flow convectively through the catalyst sites, the resulting intensive contact between reactants and catalyst can result in a high catalytic activity [9, 10]. Besides, this can avoid the problems derived from internal or external mass transfer resistance that may appear in a conventional fixed-bed reactor. In FTCMR, the reactants flow convectively through the membrane to catalyst sites, which in turn results in an intensive contact between the reactants and the catalyst, thereby leading to higher catalytic activity with negligible mass transport resistance. Furthermore, the introduction of convective flow can avoid undesired side reactions [1, 10].

Catalytic membranes (CMs), either inherently catalytic or being made catalytic by impregnation of nanocatalysts, are known for more than a decade. In spite of this fact, the development of CMs is still a major challenge. Majority of the catalytic membranes used in industries are inorganic (either ceramic or metal); for that reason, they can withstand harsh reaction conditions (high temperature, pressure, concentration, and corrosive chemicals). The main drawbacks of such CM materials are high cost and frangibility [5]. Because polymers are less expensive and more flexible than ceramics and metals, it is possible to use them in CM development instead of high cost metals and ceramics. However, majority of the polymers are only suitable for mild operation conditions. For this compensation, high reactive catalysts should be impregnated inside the polymer membrane matrix. In such cases, active catalysts can compromise the demand of higher temperature. Therefore, stabilization of active catalysts by encapsulating inside polymer membranes can help to boost and enhance the drawbacks of polymer membranes.

Metal nanoparticles (MNPs) have shown a great potential in different catalytic processes and are well known for their higher catalytic performances. Particularly, MNPs of transition metals are found to be efficient and selective catalysts for several types of catalytic reactions. This is due to higher percentage of surface atoms and associated quantum effects. However, metal nanoparticles lack chemical stability and mechanical strength. They exhibit extremely high pressure drop or head loss in fixed-bed column operation and are not found suitable for such systems [11]. Also, MNPs tend to aggregate; this phenomenon reduces their high surface area to volume ratio and subsequently reduces effectiveness. By appropriately dispersing metal nanoparticles into surface functionalized polymer membranes, many of these shortcomings can be overcome without compromising the properties of nanoparticles. Immobilizing MNPs on polymer membrane support, besides providing a mechanical strength, it offers an option to maintain their catalytic activities by preventing unnecessary growth and aggregation. Moreover, catalytic application of MNPs is the best alternative to efficiently utilize most expensive metals. Immobilization of MNPs on polymer membrane support is therefore best strategy to overcome the drawbacks of both polymers and MNPs [12]. The use of functionalized polymer membrane as a support and stabilizing media enables synthesizing nanocatalysts at the desired "point use" and will result in formation of catalytically active polymer membrane. Encapsulation of MNPs in polymers membranes offers also unique possibilities for enhancing accessibility of catalytic sites to reactants [13, 14].

In this research, we developed catalytic polymer membrane for liquid phase reactions by surface modification of flat sheet PES microfiltration membrane using UV-induced graft polymerization and synthesizing of stable palladium nanoparticle via intermatrix synthesis method inside surface functionalized PES membranes. The catalytic performance of the membrane was evaluated using reduction of *p*-nitrophenol (NP) as a model for liquid phase reaction.

2. Materials and Methods

2.1. Materials. The following chemicals and materials were used during the experiment: acrylic acid (AA), N-methyl-2-pyrrolidone (NMP), polyvinylpyrrolidone powder (PVP, M_w = 29,000), casting knife, acrylic acid (AA), N,N$'$-methylenebisacrylamide, 4-hydroxybenzophenone, and tetra-ammine palladium(II)chloride monohydrate. All compounds have been used without any purification and solutions were prepared with deionized water.

2.2. Membrane Preparation. Flat sheet PES microfiltration membranes were prepared via nonsolvent induced phase separation (NIPS) using a solution containing polyether-sulfone (18% wt) as polymer and N-methyl-2-pyrrolidone (62% wt) as solvent and polyvinylpyrrolidone (PVP 20% wt) as a pore former. The solution was casted using a casting knife with 350 μm thickness and is precipitated in a coagulation water bath at 18–20°C.

2.3. Membrane Functionalization.

2.3. Membrane Functionalization. Flat sheet PES microfiltration membranes were immersed for 3 minutes in 30 ml aqueous solution of AA monomer (25 wt%). After the immersion, samples were grafted using a simple photografting setup, containing quartz UV lamp. Exposure time was (5–20) minutes. Distance from the light source to the sample was adjusted to a minimum 6 cm, in order to avoid a possible heat up of the membrane. After grafting, samples were washed with deionized water in order to remove unreacted monomer. Dried samples were taken for surface analysis surface using attenuated total reflection Fourier transform infrared spectroscopy (ATR-FTIR, Thermo-Nicolet Nexus)

2.4. Precursor Loading and Intermatrix Synthesis of Pd-Nanoparticles. The synthesis of Pd-NPs inside the functionalized flat sheet PES polymer membrane matrix was carried out via intermatrix synthesis (IMS) method with procedures consisting of the following: (1) palladium salt [Pd(NH$_3$)$_4$Cl$_2$] which was loaded to the functionalized membrane which enables cation exchange between carboxylic groups of functionalized PES with [Pd(NH$_3$)$_4$]$^{2+}$ ions to took place and (2) subsequent chemical reduction by 0.1 M NaBH$_4$ solution: the cation exchange was performed by submerging grafted membrane in to palladium precursor solution (0.01 M Pd(NH$_3$)$_4$Cl$_2$) over night at room temperature. The synthesis can be summarized by the following sequential equations of ion exchange ((1) and (2)) and chemical reduction ((3) and (4)) [13]:

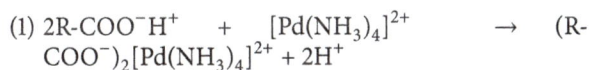

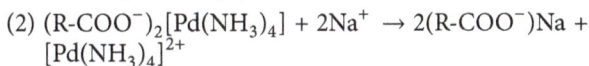

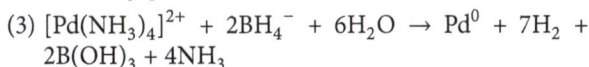

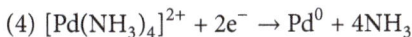

(1) $2R\text{-}COO^-H^+ \quad + \quad [Pd(NH_3)_4]^{2+} \quad \rightarrow \quad (R\text{-}COO^-)_2[Pd(NH_3)_4]^{2+} + 2H^+$

(2) $(R\text{-}COO^-)_2[Pd(NH_3)_4] + 2Na^+ \rightarrow 2(R\text{-}COO^-)Na + [Pd(NH_3)_4]^{2+}$

(3) $[Pd(NH_3)_4]^{2+} + 2BH_4^- + 6H_2O \rightarrow Pd^0 + 7H_2 + 2B(OH)_3 + 4NH_3$

(4) $[Pd(NH_3)_4]^{2+} + 2e^- \rightarrow Pd^0 + 4NH_3$

2.5. Palladium Content Measurement. The amount of palladium nanoparticle loaded to the functionalized membrane was determined by using inductively coupled plasma optical emission spectrometry (ICP-OES, Ultima 2, Horoba Jobin Yvon). 1 cm^2 sample of Pd loaded membrane was dissolved in aqua regia, which is a highly corrosive mixture of acids, for two days. The acid mixture was prepared by freshly mixing concentrated nitric acid (65%) and hydrochloric acid (35%) in a volume ratio of 1:3. It was then diluted in ultra-pure water so as to analyze in ICP.

2.6. Catalytic Performance Evaluation. The catalytic performance of Pd loaded flat sheet PES membrane was evaluated by the reduction of p-nitrophenol to p-aminophenol with

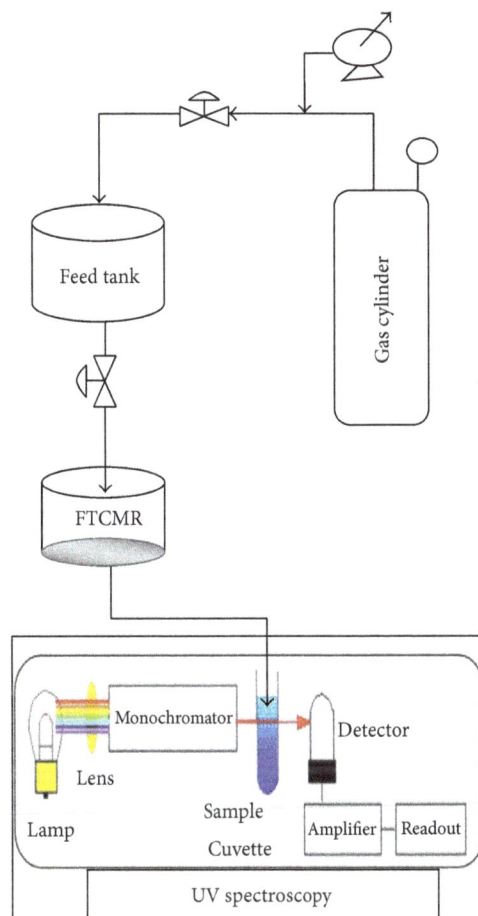

FIGURE 1: FTCMR experimental setup.

NaBH$_4$ in FTCMR setup shown in Figure 1. The reduction of p-nitrophenol in the presence of nanoparticles has been frequently used previously in order to evaluate the catalytic activity of different metal nanoparticles immobilized on membranes [15–17]. In addition, this reaction has been used to test the catalytic activity of metal nanoparticles immobilized in other carrier systems like core-shell. Previous work demonstrated that p-nitrophenol is reduced to aminophenol only in the presence of the catalyst; no reaction takes place in the absence of the nanoparticles [15].

In FTCMR configuration shown in Figure 2, a solution containing different concentrations of a mixture of p-nitrophenol and NaBH$_4$ was forced to pass through the membrane. An excess of NaBH$_4$ was used, so that the kinetics can be assimilated to a pseudo first order in terms of concentration of p-nitrophenol. The progress of the reaction was monitored by measuring the concentration of p-nitrophenol using UV-visible spectroscopy at ($\lambda = 400$ nm).

2.7. Mathematical Modeling. Convective mass transport is taking place in FTCMR, because there exists transmembrane pressure difference between the two sides of membrane, so as to enforce the reactants to pass through. In some membranes, such as nanofiltration membrane, the presence of convective flow by pressure difference enhances the diffusive driving

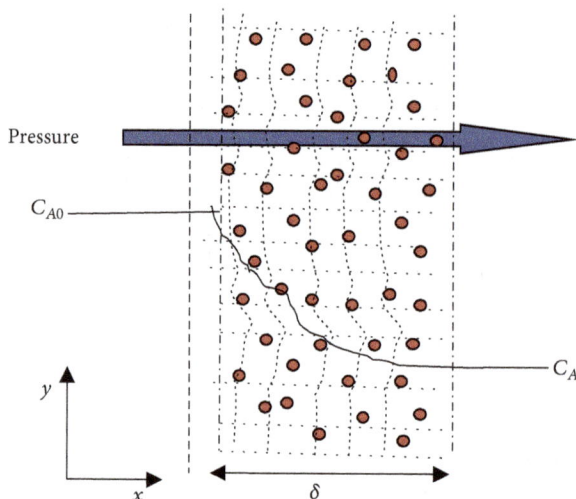

FIGURE 2: Scheme of one-dimensional mass transport across catalytic layer in flow-through membrane reactor (red particles are the catalysts, and C_{A0} and C_A are initial and final concentration of reactant A, resp.).

force. Hence investigating combined effect of the diffusive and convective flows is important in surface functionalized polymer membranes. The model takes into account simultaneous transport by convective and diffusive mass flow with chemical reaction. Emin et al. [17] have clearly shown that palladium nanoparticles have been only immobilized on the grafted poly (acrylic acid) layer. Analyzing the sample using energy-dispersive X-ray spectroscopy (EDX) showed that NPs were not found deep in the support membrane. This fact is considered here in modeling FTCMR.

The following assumptions are considered [18].

(i) Reaction occurs uniformly at every position within the catalyst layer.

(ii) Transport of reactants through the catalyst layer occurs both by diffusion and by convection. However the diffusion term does not contribute much to permeate flow rate.

(iii) Chemical reaction occurs on the interface of the catalytic particles so that the diffusion inside the dens nanoparticle is negligible.

(iv) The concentration changes only across the catalytic membrane layer (0.198 μm, from SEM measurement) and nanoparticles are immobilized and stabilized on the grafted layer.

The differential mass transport across a reactive membrane layer with constant transport parameters through combined mass transport with chemical reaction, which is represented in Figure 2, can be described by a continuity equation for Cartesian coordinate system. For reactant species A [18],

$$C\left(\frac{\partial X_A}{\partial t} + V_x\frac{\partial X_A}{\partial x} + V_y\frac{\partial X_A}{\partial y} + V_z\frac{\partial X_A}{\partial z}\right)$$
$$= C * D\left(\frac{\partial^2 X_A}{\partial x^2} + \frac{\partial^2 X_A}{\partial y^2} + \frac{\partial^2 X_A}{\partial z^2}\right) + r_A, \tag{1}$$

where C is the concentration, X_A is molar fraction, D is the diffusion coefficient in the membrane (m^2/hr), r_A is rate of reaction, and V is the convective velocity (m/hr) through the membrane. For steady state system with a convective flow in x direction, the above equation becomes

$$V_x\frac{dC_A}{dx} = D\frac{d^2C_A}{dx^2} + r_A. \tag{2}$$

For pseudo first-order reaction, $r_A = -K_{\text{app}} * C_A$, where K_{app} is the apparent kinetic constant and C_A is the concentration of reactant A. Substitution of rate of reaction to (2) gives

$$D\frac{d^2C_A}{dx^2} - V_x\frac{dC_A}{dx} - K_{\text{app}} * C_A = 0. \tag{3}$$

Defining and introducing the dimensionless length epsilon (ε) = x/δ, whose value varies from 0 to 1, where δ is catalytic membrane layer (0.198 μm) and x is the catalytic layer at any distance, substituting to (3) gives

$$\frac{d^2C_A}{d\varepsilon^2} - \frac{V * \delta}{D} * \frac{dC_A}{d\varepsilon} - \frac{K_{\text{app}}\delta^2}{D}C_A = 0. \tag{4}$$

Two dimensionless groups' Peclet number (P_e) and reaction modulus ($\emptyset^2$) are defined and inserted to (4), where

$$P_e = \frac{V * \delta}{D}$$
$$\emptyset^2 = \frac{K_{\text{app}}\delta^2}{D}. \tag{5}$$

The diffusion coefficient was predicted based on the variation of the actual measurement taken from the filtration unit and a model prediction that ignores the diffusion term (as if the entire process was controlled by the convective flow). The solution of (4) is

$$\frac{C_A}{C_{A0}} = \exp\left[\frac{P_e\varepsilon}{2}\right]$$
$$\cdot \left[\frac{\sinh(\varphi(1-\varepsilon)) * P_e/2 + \varphi\cosh[\varphi(1-\varepsilon)]}{(P_e/2)\sinh\varphi + \varphi\cosh\varphi}\right], \tag{6}$$

where $\varphi = \sqrt{(P_e^2/4 + \emptyset^2)}$.

We can predict the concentration distribution at the end of the membrane ($\varepsilon = 1$), as function of initial concentration (C_{A0}), Peclet number, and reaction modules with the following equation [18]:

$$\frac{C_A}{C_{A0}} = \frac{\varphi * \exp(P_e/2)}{(P_e/2)\sinh\varphi + \varphi\cosh\varphi}. \tag{7}$$

3. Results and Discussion

3.1. *Membrane Surface Functionalization.* Both unmodified PES and the UV-induced modified flat sheet membranes were characterized by ATR-FTIR. Figure 3 shows the spectra of the unmodified and modified membranes with AA monomer. As

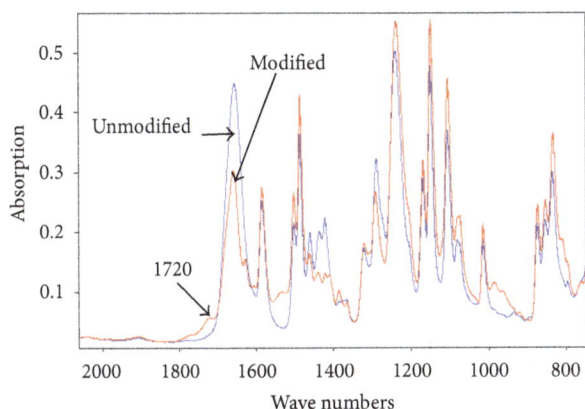

FIGURE 3: The ATR-FTIR spectra of unmodified and modified PES flat sheet membranes using UV grafting of AA (25 wt%) at 20 min UV batch irradiation time.

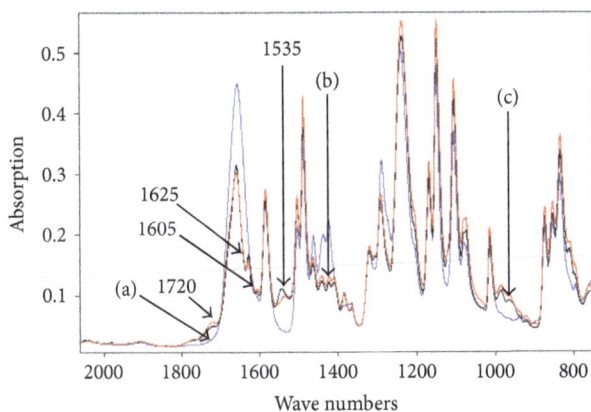

FIGURE 4: The ATR-FTIR spectra of the (a) unmodified flat sheet PES membrane and modified membranes using UV-induced grafting with (25 wt% AA); (b) 20; and (c) 15 minute UV irradiation time, respectively.

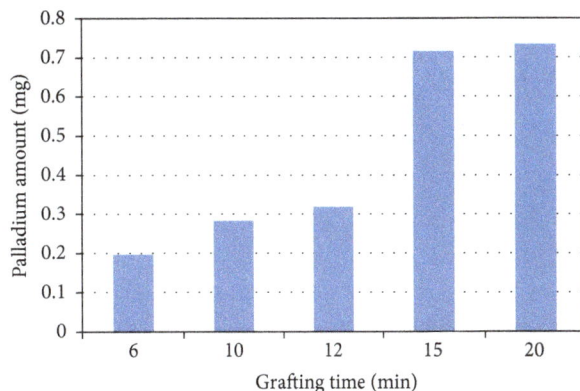

FIGURE 5: Palladium amount per membrane ($15.2 \, cm^2$) versus grafting time.

can be seen, UV-induced grafted membranes exhibit different ATR-FTIR spectra than the unmodified one. In addition to the typical PES bands of the unmodified membrane, the IR spectra of modified membrane show additional peak at $1720 \, cm^{-1}$, which corresponds to the carbonyl (C=O) group bands of COOH, which indicates the existence of poly(acrylic acid) chains [19] and assured that the monomer is successfully polymerized on the substrate PES. Also, with UV modification, some original absorbance peak intensity is decreased that can be contributed to increase coverage of the PES surface by poly(acrylic acid).

Apart from this, as can be seen from FTIR spectra of modified membrane in Figure 4, the intensity of the new peak at $1720 \, cm^{-1}$ increases with photografting reaction time which corresponds to the energy received during graft polymerization. The more the energy received by the monomer, the higher the degree of modification. In addition to the new absorption peak corresponding to the carbonyl group (C=O) of COOH, there are also other new small absorbance peaks appearing at 1625, 1605, and $1535 \, cm^{-1}$ for the modified

membranes. The small absorbance peak for the AA-modified membranes at these regions attributed to the presence of additives such as PVP in the original casting solution and cross linker in the monomer solution, as reported by Rahimpour [20] and Bernstein et al. [21]. Also, the disappearance of the original peaks at 1640–$1680 \, cm^{-1}$ for the modified membranes indicates that the modification was successful.

Membranes grafted at different times were used to synthesize Pd nanoparticle. ICP analysis in Figure 5 showed that the weight of Pd loading is increasing with grafting time (energy received). These data are in agreement with expected and FTIR results. At higher energy, the intensity of the modified functional group was stronger which can lead to better cation exchange with Pd precursor and thus higher Pd amount from the reduction.

3.2. Catalytic Performance Evaluation. The effect of the nitrophenol concentration in the feed on the conversion was investigated using different feed concentrations and the same amount of Pd at a room temperature. A closer look at Figure 6 clearly shows the effect of initial p-NP concentration on conversion. The conversion is higher for the lower p-NP initial concentration. In all of initial p-NP concentration ranges (i.e., 0.033–0.128 mM), the same trend was obtained, which strongly indicates that reaction on Pd nanoparticle is highly active dependant surface. For the same catalyst loading and permeate flux (i.e., approximately the same residence time and applied pressure within a membrane), the catalytic activity of membranes with lower initial concentration was higher than that of higher initial p-NP concentration.

Furthermore, the inverse proportionality of p-NP conversion with permeate flux confirmed that our Pd loaded membranes were not mass transfer limited; rather it was reaction limited. If membranes were mass transfer limited, we would not observe a decrease of conversion with increased fluxes. As the applied pressure in flow-through mode enforces reactants to have an intensive contact with catalyst, the reaction at the surface of the catalyst is the main limitation to such system. As a result of increasing pressure (flux), reactants permeate through the membrane without reacting;

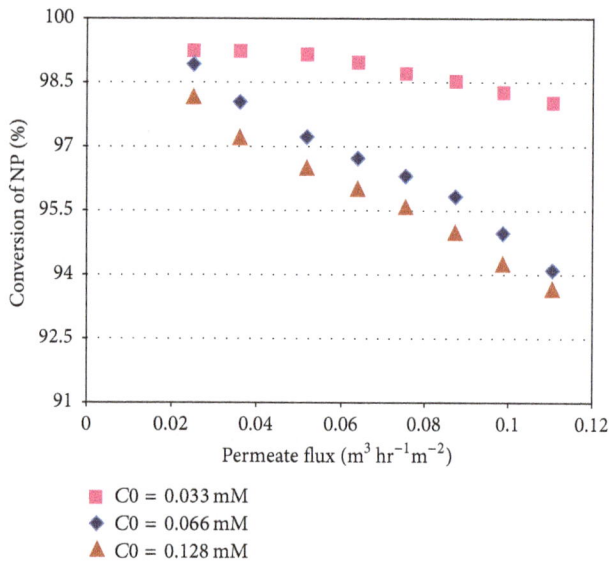

FIGURE 6: Conversion of nitrophenol (NP) in Pd loaded PES membrane versus flux for single pass in dead-end mode of filtration (0.283 mg of palladium).

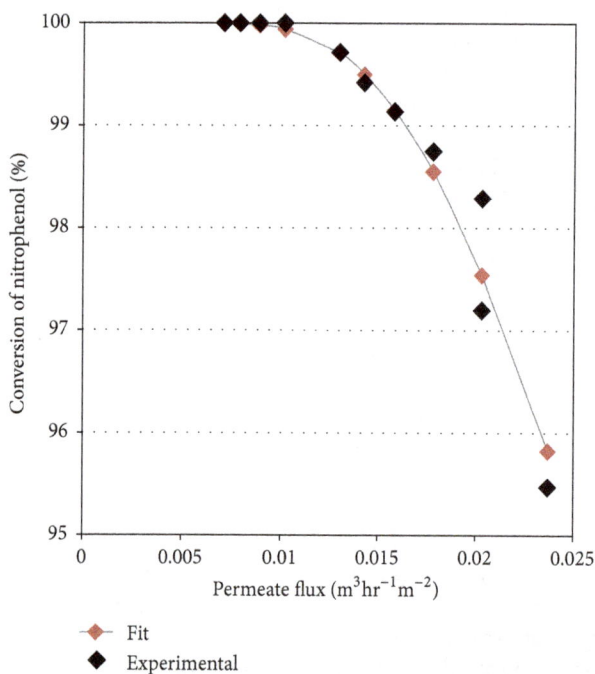

FIGURE 7: Plot of p-NP conversion versus flux. The curve (with red diamond) represents a first-order reaction model with a rate constant (K_{app}) of $300\,hr^{-1}$. Feed conditions: [p-nitrophenol] = $0.514\,mM$, [$NaBH_4$] = $14.38\,mM$, Pd = $0.733\,mg$, and catalytic layer = $0.198\,\mu m$.

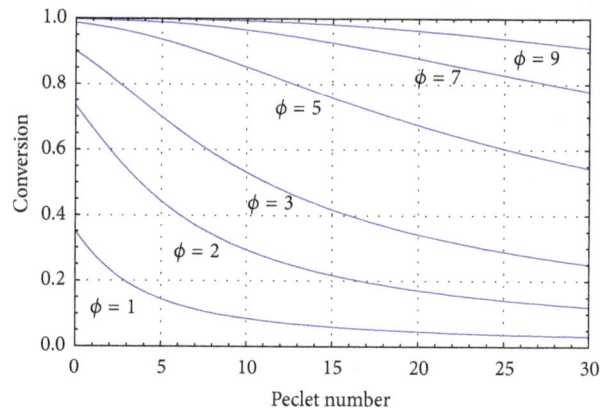

FIGURE 8: Simulation of conversion versus convective flow (Peclet number) at different values of reaction modulus.

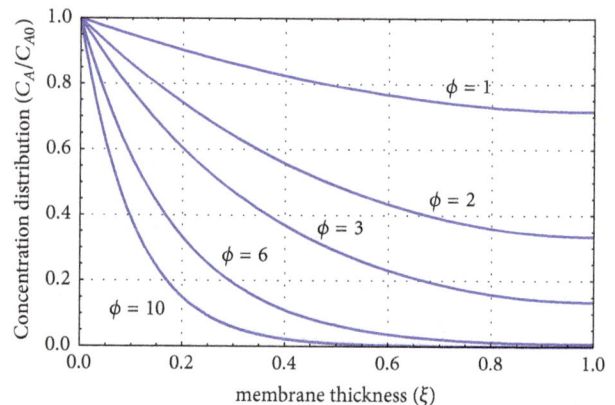

FIGURE 9: Simulation of concentration distribution across the catalytic layer at different values of reaction modulus as function of membrane thickness.

represents satisfactorily the reaction and transport phenomena in flow-through catalytic membrane reactor. From the same figure, it can be concluded that both convective and diffusive transport through the catalytic membrane layer contributed to the catalytic activity of the nanoparticle.

In order to understand better and check the effect of different parameters, simulations of the FTCMR were performed using Wolfram Mathematica 7, at different process parameters such as Peclet number (Pe), reaction modulus ($\emptyset^2$), catalytic membrane layer thickness (δ), and concentrations.

Simulation results of Figures 8 and 9 showed the effect of different process parameters such as Peclet number (Pe), reaction modulus ($\emptyset^2$), and catalytic membrane layer thickness (δ) on the outlet concentrations of the reactant.

4. Conclusion

In this work, surface modification of polyethersulfone membrane by UV-assisted grafting polymerization of acrylic

hence the conversion decreases. Similar results were reported by Dotzauer et al. [22] and Crock et al. [23].

Figure 7 shows the experimental plot points and model predictions according to the model equation (7) developed and it confirms well to a simple first-order kinetic model of the reaction. The proposed mathematical model equation

acid has been successfully done. Catalytically active and efficient Pd nanoparticle has been synthesized via intermatrix synthesis. As a model for liquid phase reaction, its catalytic performance was investigated by the reduction of aqueous p-nitrophenol to p-aminophenol with sodium borohydride as reductant. The catalytic activity of Pd embedded membrane was shown to be directly proportional to the palladium content in the nanocomposite. The catalytic activity of flow-through reactor for reduction of nitrophenol outperformed the batch mode of operation, as it was demonstrated by conversion comparison at the same initial nitrophenol concentration and weight of catalyst. This was attributed to the convective flow of reactants directly to the catalyst sites which can provide an intensive contact. An important asset in application of catalyst embedded membrane is the possibility of varying the flux to control the conversion. At lower and moderate range of fluxes, a complete conversion was achieved in FTCMR, but an increase in the flux decreased the conversion, because of insufficient contact time. The effect of initial nitrophenol concentration at the same flux and catalyst has been investigated. In all concentration ranges taken, a lower initial concentration had higher conversion, as higher concentration poses surface coverage of the catalyst. Further investigation of p-nitrophenol reduction reactions related to porosity of the grafted layer has to be done. The proposed mathematical model equation represents satisfactorily the reaction and transport phenomena in flow-through catalytic membrane reactor.

Conflicts of Interest

The authors would like to declare that there are no conflicts of interest regarding the publication of this paper.

References

[1] A. Julbe, D. Farrusseng, and C. Guizard, "Porous ceramic membranes for catalytic reactors - Overview and new ideas," *Journal of Membrane Science*, vol. 181, no. 1, pp. 3–20, 2001.

[2] T. Westermann, *Flow-Through Membrane Microreactor for Intensified Heterogeneous Catalysis*, 2009.

[3] T. Westermann and T. Melin, "Flow-through catalytic membrane reactors—principles and applications," *Chemical Engineering and Processing: Process Intensification*, vol. 48, no. 1, pp. 17–28, 2009.

[4] R. Dittmeyer, V. Höllein, and K. Daub, "Membrane reactors for hydrogenation and dehydrogenation processes based on supported palladium," *Journal of Molecular Catalysis A: Chemical*, vol. 173, no. 1-2, pp. 135–184, 2001.

[5] S. S. Ozdemir, M. G. Buonomenna, and E. Drioli, "Catalytic polymeric membranes: Preparation and application," *Applied Catalysis A: General*, vol. 307, no. 2, pp. 167–183, 2006.

[6] K. K. Sirkar, P. V. Shanbhag, and A. S. Kovvali, "Membrane in a reactor: A functional perspective," *Industrial & Engineering Chemistry Research*, vol. 38, no. 10, pp. 3715–3737, 1999.

[7] Y. Zhang, K. Su, F. Zeng, W. Ding, and X. Lu, "A novel tubular oxygen-permeable membrane reactor for partial oxidation of CH_4 in coke oven gas to syngas," *International Journal of Hydrogen Energy*, vol. 38, no. 21, pp. 8783–8789, 2013.

[8] M. A. Al-Juaied, D. Lafarga, and A. Varma, "Ethylene epoxidation in a catalytic packed-bed membrane reactor: Experiments and model," *Chemical Engineering Science*, vol. 56, no. 2, pp. 395–402, 2001.

[9] M. Pera-Titus, M. Fridmann, N. Guilhaume, and K. Fiaty, "Modelling nitrate reduction in a flow-through catalytic membrane contactor: Role of pore confining effects on water viscosity," *Journal of Membrane Science*, vol. 401-402, pp. 204–216, 2012.

[10] E. Nagy, "Mass transfer through a convection flow catalytic membrane layer with dispersed nanometer-sized catalyst," *Industrial & Engineering Chemistry Research*, vol. 49, no. 3, pp. 1057–1062, 2010.

[11] V. V. Pushkarev, Z. Zhu, K. An, A. Hervier, and G. A. Somorjai, "Monodisperse metal nanoparticle catalysts: Synthesis, characterizations, and molecular studies under reaction conditions," *Topics in Catalysis*, vol. 55, no. 19-20, pp. 1257–1275, 2012.

[12] K. Na, Q. Zhang, and G. A. Somorjai, "Colloidal Metal Nanocatalysts: Synthesis, Characterization, and Catalytic Applications," *Journal of Cluster Science*, vol. 25, no. 1, pp. 83–114, 2014.

[13] P. Ruiz, M. Muñoz, J. MacAnás, C. Turta, D. Prodius, and D. N. Muraviev, "Intermatrix synthesis of polymer stabilized inorganic nanocatalyst with maximum accessibility for reactants," *Dalton Transactions*, vol. 39, no. 7, pp. 1751–1757, 2010.

[14] J. Bastos-Arrieta, A. Shafir, A. Alonso, M. Muñoz, J. MacAnás, and D. N. Muraviev, "Donnan exclusion driven intermatrix synthesis of reusable polymer stabilized palladium nanocatalysts," *Catalysis Today*, vol. 193, no. 1, pp. 207–212, 2012.

[15] J. Macanás, L. Ouyang, M. Bruening, M. Muñoz, J. Remigy, and J. Lahitte, "Development of polymeric hollow fiber membranes containing catalytic metal nanoparticles," *Catalysis Today*, vol. 156, no. 3-4, pp. 181–186, 2010.

[16] L. Ouyang, D. M. Dotzauer, S. R. Hogg, J. MacAnás, J.-F. Lahitte, and M. L. Bruening, "Catalytic hollow fiber membranes prepared using layer-by-layer adsorption of polyelectrolytes and metal nanoparticles," *Catalysis Today*, vol. 156, no. 3-4, pp. 100–106, 2010.

[17] C. Emin, J.-C. Remigy, and J.-F. Lahitte, "Influence of UV grafting conditions and gel formation on the loading and stabilization of palladium nanoparticles in photografted polyethersulfone membrane for catalytic reactions," *Journal of Membrane Science*, vol. 455, pp. 55–63, 2014.

[18] C. Endalkachew and J.-F. Lahitte, *Development of Gas/Liquid Catalytic Membrane Reactor*, University of Zaragoza Library Online Thesis Catalog, 2014.

[19] B. Deng, J. Li, Z. Hou et al., "Microfiltration membranes prepared from polyethersulfone powder grafted with acrylic acid by simultaneous irradiation and their pH dependence," *Radiation Physics and Chemistry*, vol. 77, no. 7, pp. 898–906, 2008.

[20] A. Rahimpour, "UV photo-grafting of hydrophilic monomers onto the surface of nano-porous PES membranes for improving surface properties," *Desalination*, vol. 265, no. 1-3, pp. 93–101, 2011.

[21] R. Bernstein, E. Antón, and M. Ulbricht, "UV-photo graft functionalization of polyethersulfone membrane with strong polyelectrolyte hydrogel and its application for nanofiltration," *ACS Applied Materials & Interfaces*, vol. 4, no. 7, pp. 3438–3446, 2012.

[22] D. M. Dotzauer, J. Dai, L. Sun, and M. L. Bruening, "Catalytic membranes prepared using layer-by-layer adsorption of poly-electrolyte/metal nanoparticle films in porous supports," *Nano Letters*, vol. 6, no. 10, pp. 2268–2272, 2006.

[23] C. A. Crock, A. R. Rogensues, W. Shan, and V. V. Tarabara, "Polymer nanocomposites with graphene-based hierarchical fillers as materials for multifunctional water treatment membranes," *Water Research*, vol. 47, no. 12, pp. 3984–3996, 2013.

18

Mass and Heat Transport Models for Analysis of the Drying Process in Porous Media: A Review and Numerical Implementation

Hong Thai Vu [1] and **Evangelos Tsotsas** [2]

[1]*Department of Chemical Process Equipment, School of Chemical Technology, Hanoi University of Science and Technology, Hanoi, Vietnam*
[2]*Chair of Thermal Process Engineering, Faculty of Process and Systems Engineering, Otto-von-Guericke-University Magdeburg, Magdeburg, Germany*

Correspondence should be addressed to Hong Thai Vu; thai.vuhong@hust.edu.vn

Academic Editor: Bhaskar Kulkarni

The modeling and numerical simulation of drying in porous media is discussed in this work by revisiting the different models of moisture migration during the drying process of porous media as well as their restrictions and applications. Among the models and theories, we consider those are ranging from simple ones like the diffusion theory to more complex ones like the receding front theory, the model of Philip and de Vries, Luikov's theory, Krischer's theory, and finally Whitaker's model, in which all mass, heat transport, and phase change (evaporation) are taken into account. The review of drying models as such serves as the basis for the development of a framework for numerical simulation. In order to demonstrate this, the system of equations governing the drying process in porous media resulting from Whitaker's model is presented and used in our numerical implementation. A numerical simulation of drying is presented and discussed to show the capability of the implementation.

1. Introduction

The modeling and numerical simulation of drying in porous media is a topic of great interest in process engineering and has attracted the attention of research institutions for decades. The analysis of absorption or desorption of liquid in general, and of water in particular, finds its application not only in chemical engineering, food processing, and pharmaceuticals but also in electronic packaging, which is becoming more and more important in the area of the Internet of Things [1–8]. Over the years, many works were undertaken to understand the mass and heat transport phenomena during drying. In this work, we will revisit the different models of moisture migration during the drying of porous media. The restrictions and applications of these models will be reviewed and discussed.

Drying models were developed since the beginning of the twentieth century. Different methods and numerical solutions were presented and applied successfully for different porous media. A review of these methods until the 1980s was given by Fortes and Okos [9] and Bories [10]. Reviews of empirical, analytical, and numerical methods of drying until the 1990s can be found in the work of Tsotsas [11]. In what follows, we will examine the most relevant theories and models for drying of porous media and discuss the most suitable model that can today be used for numerical simulation. Amongst others, we will first consider the so-called diffusion theory, the receding front theory, the model of Philip and de Vries, Luikov's theory, and Krischer's theory. In particular, we will discuss Whitaker's model for drying and discuss the framework based on this model for numerical simulation of the drying process. Finally, we will present a numerical simulation using our implementation of Whitaker's model and discuss the sample drying characteristics in our numerical study. For details of the implementation of the model, refer [12].

2. The Different Theories of Drying in Porous Media

2.1. Diffusion Theory. Lewis and Sherwood are known as pioneers in developing mathematical drying models by applying the Fourier equation of heat conduction to the drying of solids. In this equation, temperature and thermal diffusivity were replaced by moisture and moisture diffusivity, respectively. Starting from the idea of Lewis, Sherwood [13] provided solutions of the diffusion equation. Sherwood showed that the moisture transport involves two independent processes: the evaporation of moisture at the solid surface and the internal diffusion of liquid to the surface. The following simple diffusion model, in which the diffusivity of liquid is constant, was used to calculate the moisture distribution in a solid during drying and compared with experimental data of some materials (e.g., slabs of wood, clay, and soap):

$$\frac{\partial X}{\partial t} = \delta_{\text{eff}} \nabla^2 X, \tag{1}$$

where X is vaguely defined as the moisture content, t represents time, and δ_{eff} can be considered as an effective diffusion coefficient and is determined experimentally.

At the end of the 1930s, Ceaglske and Hougen [14] and Hougen et al. [15] pointed out that the moisture distribution cannot be calculated correctly only from (1). It was noted that the moisture movement in a solid during drying is not only due to diffusion but also due to other mechanisms such as gravity, external pressure, capillarity, convection, and vaporization-condensation when a temperature gradient is applied. Experimental data were collected for different kinds of porous media (clay, paper pulp, sand, lead shot, porous brick, and wood) and compared with the numerical results obtained by Sherwood's model to show the limitation of his model and prove their criticism.

By using the capillary theory to describe the drying of granular materials (such as coarse, medium, and fine sand) and based on the collected experimental data, Ceaglske and Hougen [14] suggested that the effective diffusion coefficient δ_{eff} should be considered as varying during drying and proposed the following diffusion model:

$$\frac{\partial X}{\partial t} = \nabla \cdot \left(\delta_{\text{eff}} \nabla X \right). \tag{2}$$

The effective diffusion coefficient is now considered as a function of moisture content, temperature, material type, and drying history. In solving the diffusion equation, this parameter is usually taken as constant or in the form of linear, exponential, or polynomial functions of moisture content. One example of these functions was given by Suzuki and Maeda [16]:

$$\delta_{\text{eff}}(X) = (aX + b)^{n_e}, \tag{3}$$

where a, b, and n_e are the constant factors. Suzuki and Maeda also presented an approximation method to describe the moisture distribution within drying porous materials in which the effective diffusion coefficient is expressed as an exponential function of moisture content:

$$\delta_{\text{eff}}(X) = a \cdot e^{b \cdot X}, \tag{4}$$

where a and b are the constant factors as in (3). This is a common function used to describe the effective diffusion coefficient. Suzuki and Maeda then solved the nonlinear diffusion problem by using dimensionless variables for the diffusion coefficient, moisture content, time, and space. It was shown that the steady-flux model (pseudosteady state) is fairly accurate when it is employed in low moisture content drying. However, it was also found that the diffusion model leads to a noticeable error when it is applied to high moisture content drying. The reason is that, in this case, capillary pumping has strong effect, and this mechanism must be taken into account.

In order to determine the effective diffusion coefficient, in the past decades, a large number of works were carried out. For example, the works of Saravacos and Raouzeos [17], Jaros et al. [18], Mourad et al. [19], Ribeiro et al. [20], Li and Kobayashi [21], and Srikiatden and Roberts [22] focused on materials in the food industry and agriculture. Koponen [23] dealt with wood, Ketelaars et al. [24] studied clay, and Pel et al. [25] considered fired-clay brick, sand-lime brick, and gypsum. In developing the diffusion model, heat transfer was included in examining the problem, for example, in the work of Thijssen and Coumans [26]. A shortcut method was proposed to calculate drying rates in nonisothermal drying of particles and hollow spheres. In their work, heat transfer was taken into account in the calculation of the evaporation rate and the sample temperature. Shrinking and non-shrinking materials were considered. The method proposed by Thijssen and Coumans is based on the numerical solution of the diffusion equation with variable diffusion coefficient (as a function of moisture content) and based on the result (isothermal drying rate versus average moisture content) of drying experiments with a slab at different temperatures. By using this method, the information obtained from the isothermal drying experiments was applied to other geometries and nonisothermal conditions. Dimensionless variables (moisture content, time, space coordinate, and diffusion coefficient) were used in the study. Instead of solving numerically the nonlinear diffusion equation, a calculation procedure was applied in a step-by-step manner where the conditions for each new step were obtained from the experimental drying curves.

In order to measure moisture profiles during drying, some advanced methods such as scanning neutron radiography or nuclear magnetic resonance imaging (MRI) were used in a number of studies. Among others are the works of Blackband and Mansfield [27] on solid blocks of nylon, Schrader and Litchfield [28] on food gel, Pel et al. [29] on brick and kaolin clay, McDonald et al. [30] on sandstone and rock, and Koptyug et al. [31] on alumina pellets. By means of these advanced methods, the measured moisture profiles can be used directly to determine the effective diffusion coefficient. In the work of Schrader and Litchfield [28], MRI was used to measure moisture profiles in a cylinder of food gel during drying at room temperature. The measured profiles were then compared with the numerical results

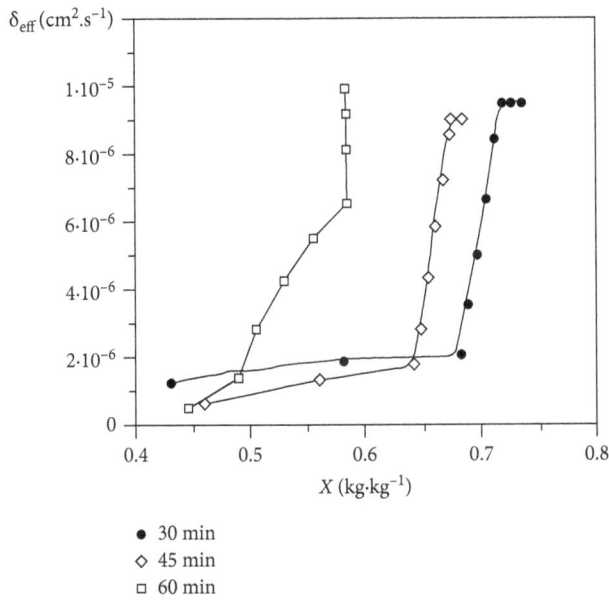

FIGURE 1: Determination of the effective diffusion coefficient for food gel based on the comparison of the MRI data with the diffusion model: effective diffusion coefficient as function of drying time and moisture content (Schrader and Litchfield [28]).

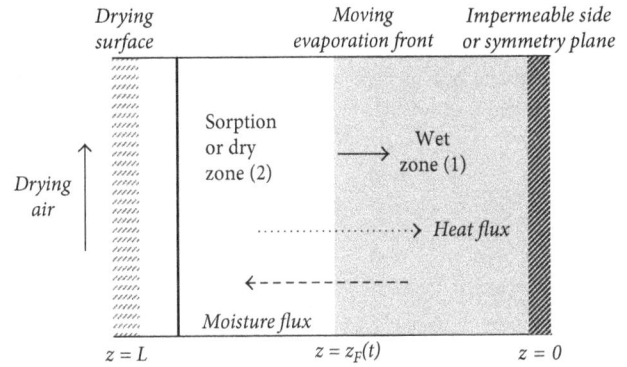

FIGURE 2: Receding front model.

calculated by the diffusion model, and in this way, the effective diffusion coefficient was computed. As an example, Figure 1 shows the variation of the obtained effective diffusion coefficient at $t = 30$, 45, and 60 minutes of drying. It was pointed out that the diffusion model seems not to be a good method to predict the interior moisture profile of food gel. Koptyug et al. [31] used MRI to study the diffusion of water in alumina pellets and showed that MRI can provide good information on the real-time variation of liquid content in the course of drying of porous solids. MRI is a good experimental method because it is able to measure moisture content at any point within a complex material. Additionally, it provides a quick, accurate, nondestructive method and therefore allows the evaluation of various drying models.

More recently, Guillard et al. [32] used the diffusion model to predict the moisture distribution in multicomponent heterogeneous food where components of high/low water activity are placed adjacent to one another. The calculation of moisture distributions compared well with the experimental results, and it was proposed that, in this special application, the model could be useful. Efremov [33] used another approach for the description of the drying kinetics of porous materials. The approach is based on the analytical solution of the diffusion equation (for one-dimensional isotropic diffusion) with a flux-type boundary condition in the form of mass flux. In this work, the drying kinetics (dimensionless moisture content versus time and drying rate) were determined by applying the Laplace transformation to express the mass flux. Porto and Lisbôa [34] developed a three-dimensional model based on the diffusion model with the constant diffusion coefficient in order to describe the drying process of a parallelepipedic oil shale particle. Lim et al. [35] introduced an equation derived from

a diffusion equation in which the diffusion coefficient is a function of space. Akpinar and Dincer used the diffusion model to investigate moisture transfer in a slab of potato [36] and in eggplant slices [37]. The works dealt with drying processes at different air temperatures and flow velocities. The influence of boundary conditions on drying process was investigated. The model is limited to the one-dimensional problem of an infinite slab. In their works, the thermophysical properties of the drying material are taken as constant, and the effect of heat transfer on the moisture loss is neglected.

The diffusion model can lead to an incorrect prediction and misinterpretation of the moisture distribution or of the drying behavior due to the fact that only moisture transport is considered and that the physical meaning of the diffusion coefficient is either lost (in the case of a constant) or it becomes a lumped parameter of all simultaneous effects (in the case of a variable). However, this model is still used as a simple way to describe drying in some cases.

2.2. Receding Front Theory. Different versions of the so-called receding front model were developed in order to obtain a better understanding and describe the influence of other mechanisms (capillarity, gravity, or external forces in gradients of pressure and temperature) on the motion of water during drying. According to this model, at the critical point (when the falling rate period starts), an evaporation front arises and gradually moves into the interior of the body [11]. The moving evaporation front divides the system into two zones: the wet and the dry zones as shown in Figure 2. For a hygroscopic material, the dry zone is called the sorption zone due to the adsorptive nature of moisture retention. In the dry zone, the free water content is zero and the main mechanism of moisture transfer is vapor flow. However, in this region, the movement of adsorbed water may also play an important role [38]. During drying, the position of the receding evaporation front varies with time.

The receding front model was first developed in the 1960s. A review of the development of the receding front model can be found in the work of Tsotsas [11]. The simplest version of the receding front model is a model where saturation S is 1 in the wet region and 0 in the dry region. In what follows, the model presented by Chen and Schmidt [38] is shown as one example. According to Chen and Schmidt, the set of one-dimensional equations describing the coupled heat and mass transfer can be written as (subscripts 1 and 2 denoting the wet and dry zones)

Wet zone $(0 < z < z_F(t))$:

$$\frac{\partial X_{fw}}{\partial t} = \frac{\partial}{\partial z}\left(\delta_1 \frac{\partial X_{fw}}{\partial z}\right),$$

$$\rho c_{p,w}\frac{\partial T_1}{\partial t} = \frac{\partial}{\partial z}\left(\lambda_{eff}\frac{\partial T_1}{\partial z}\right), \tag{5}$$

where δ_1 is the liquid transfer coefficient and $c_{p,w}$ is the specific heat capacity of water. The term X_{fw} is the moisture content of free water and λ_{eff} is the effective thermal conductivity, which is calculated by

$$\lambda_{eff} = \lambda_1 + \frac{\delta_v' \tilde{M}_v}{\tilde{R}T}\cdot\frac{\partial P_v^*(T)}{\partial T}\Delta h_v, \tag{6}$$

where λ_1 is the thermal conductivity of liquid, Δh_v is the evaporation enthalpy, $P_v^*(T)$ is the saturation vapor pressure, and δ_v' is the vapor transfer coefficient. The vapor transfer coefficient covers the contribution of both convective and diffusive flows:

$$\delta_v' = \delta_v\left(1 + \frac{(k_g K P_v/\eta)}{\delta_v + (k_g K/m\eta)\cdot(P_g - P_v)}\right), \tag{7}$$

where m is the ratio of air and vapor diffusion coefficient, k_g is the relative permeability of the gas phase, η is the dynamic viscosity, and δ_v is the vapor diffusion coefficient.

Dry or sorption zone $(z < z_F(t) < L)$:

$$\rho\frac{\partial X_{sorb}}{\partial t} = \rho\frac{\partial}{\partial z}\left(\delta_{sorb}\frac{\partial X_{sorb}}{\partial z}\right) + \frac{\partial}{\partial z}\left(\frac{\delta_v' \tilde{M}_v}{\tilde{R}T}\cdot\frac{\partial P_v}{\partial z}\right),$$

$$\rho c_{p,v}\frac{\partial T_2}{\partial t} = \frac{\partial}{\partial z}\left(\lambda_v\frac{\partial T_2}{\partial z}\right), \tag{8}$$

where $c_{p,v}$ is the specific heat capacity of vapor, λ_v is the thermal conductivity of vapor, X_{sorb} is the adsorbed water content, and δ_{sorb} is the adsorbed water transfer coefficient. For a nonhygroscopic material, X_{sorb} is zero and δ_{sorb} is negligible. $\tilde{M}_v$ denotes the molar mass of vapor, and P_v is the partial vapor pressure.

In addition to the above equations, the mass and heat transfer at the moving boundary must fulfill the following conditions:

$$\rho\delta_1\frac{\partial X_{fw}}{\partial z} = \rho\delta_{sorb}\frac{\partial X_{sorb}}{\partial z} + \frac{\delta_v'\tilde{M}_v}{\tilde{R}T}\cdot\frac{\partial P_v}{\partial z},$$

$$\lambda_{eff}\frac{\partial T_1}{\partial z} = \lambda_v\frac{\partial T_2}{\partial z} + \Delta h_v\frac{\delta_v'\tilde{M}_v}{\tilde{R}T}\cdot\frac{\partial P_v}{\partial z}, \tag{9}$$

$$T_1 = T_2; \quad X_{fw} = 0.$$

Sorption isotherm is applied in the model, and the surface boundary conditions are needed. For more details, refer [38].

A drawback of the receding front approach is that the diffusion equation is used instead of more fundamental concepts like capillary pressure, liquid pressure gradients, and permeability to describe the capillary activity in the wet zone. In addition, heat transfer is only described by an effective thermal conductivity. Difficulties appear in determining the boundary of the moving evaporation front and the coefficients for heat and mass transfer, which are the functions of dry and wet zones.

2.3. Drying Model of Philip and De Vries. Philip and De Vries [39] and De Vries [40–41] extended the previous treatment of the diffusion equations by including effects of capillary flow and vapor transport. In their work, the thermal energy equation was also incorporated into the set of the governing equations to describe the drying process. This set of equations was treated under the combination of moisture and temperature gradients. The obtained system consists of diffusion-like equations whose coefficients must be determined by experiment. The model is briefly presented below.

2.3.1. Liquid Water Transfer. Free liquid water movement is macroscopically described by Darcy's law:

$$\dot{\mathbf{m}}_w = -\rho_w\frac{K\cdot k_w}{\eta_w}(\nabla P_w - \nabla\Psi_w), \tag{10}$$

where Ψ_w is the gravity potential. By expressing the term ∇P_w as a function of X and T and by substituting this function into (10), the liquid water flux can be written as a combination of three components due to the moisture gradient, temperature gradient, and gravity [42].

$$\dot{\mathbf{m}}_w = -\delta_{wX}\nabla X - \delta_{wT}\nabla T + \rho_w\frac{K\cdot k_w}{\eta_w}\nabla\Psi_w, \tag{11}$$

where δ_{wX} and δ_{wT} are the isothermal and thermal diffusivities of water given by

$$\delta_{wX} = \rho_w\frac{K\cdot k_w}{\eta_w}\left(\frac{\partial P_w}{\partial X}\right),$$

$$\delta_{wT} = \rho_w\frac{K\cdot k_w}{\eta_w}\left(\frac{\partial P_w}{\partial T}\right). \tag{12}$$

2.3.2. Water Vapor Transfer. The transport of water vapor by molecular diffusion is described macroscopically by Fick's first law and by using the assumption of a steady diffusion in a closed system between an evaporation source and a condensation sink, a commonly used expression for the vapor flux in terms of moisture and temperature gradients:

$$\dot{\mathbf{m}}_v = -\delta_{vX}\nabla X - \delta_{vT}\nabla T, \tag{13}$$

where δ_{vX} and δ_{vT} are the isothermal and thermal diffusivities of vapor, respectively. These terms are expressed as [42]

$$\delta_{vX} = f(\psi)\delta_{va}\frac{P_g}{P_g - P_v} \cdot \frac{\tilde{M}_v g}{\tilde{R}T} \cdot \frac{\rho_v}{\rho_l} \cdot \frac{\partial P_w}{\partial X},$$

$$\delta_{vT} = f(\psi)\delta_{va}\frac{P_g}{P_g - P_v} \cdot \frac{\rho_v}{\rho_w} \cdot \frac{\zeta}{P_v^*} \cdot \frac{dP_v^*(T)}{dT}, \quad \zeta = \frac{(\nabla T)_{av}}{(\nabla T)}, \tag{14}$$

where $f(\psi)$ is a function of porosity and moisture content, δ_{va} is the diffusion coefficient of vapor in air, g is the gravitational acceleration, $P_v^*(T)$ is the saturation vapor pressure, $(\nabla T)_{av}$ is the average air temperature gradient, and ρ_v and ρ_w are the densities of vapor and liquid.

In addition to liquid and vapor transfer, Philip and De Vries assumed that the gas pressure can be treated as constant and the gas phase momentum equation can be ignored.

2.3.3. Mass and Heat Conservation Equations.
The partial differential equations of mass and energy are formulated as follows [9, 39]:

$$\frac{\partial X}{\partial t} = \nabla \cdot (\delta_T \nabla T) + \nabla \cdot (\delta_X \nabla X) + \nabla \cdot \left(\frac{K \cdot k_w}{\eta_w}\nabla\Psi_w\right),$$

$$(\rho C_p)\frac{\partial T}{\partial t} = \nabla \cdot (\lambda\nabla T) + \Delta h_v \nabla \cdot (\delta_{vX}\nabla X), \tag{15}$$

where $\delta_T = \delta_{wT} + \delta_{vT}$ is the overall thermal mass diffusivity, $\delta_X = \delta_{wX} + \delta_{vX}$ is the overall isothermal mass diffusivity, λ denotes the thermal conductivity, and ρC_p is the volumetric heat capacity of the moist porous medium. Note that convective energy terms are assumed negligible.

The theory of Philip and De Vries has become generally known and has been applied to porous media other than soil, which was chosen to investigate the heat and mass transfers by the authors. The major restriction of the theory is that it does not include the gradient of gas pressure, there is no convection contribution in heat equation, and the coefficients of the model are complicated.

2.4. Luikov's Theory.
Independent of Philip and De Vries's work, Luikov [43–45] investigated the heat and mass transfer during drying of capillary-porous bodies by employing the principles of irreversible thermodynamics. In this theory, the total moisture flux is assumed to be made up of three components: the first one is due to a gradient in moisture content, the second due to a gradient in temperature, and the last due to a gradient in the total pressure [46]:

$$\dot{\mathbf{m}}_m = -\rho_s\delta_m(\nabla X + \delta_T\nabla T + \delta_p\nabla P_g), \tag{16}$$

where $\dot{\mathbf{m}}_m$ is the total moisture flux, δ_m is the moisture diffusion coefficient, δ_T is the thermal gradient coefficient, and δ_p is the pressure gradient coefficient.

The conservation equations of Luikov's model are written in the following form [46]:

$$\rho_s\frac{\partial X}{\partial t} + \nabla \cdot \dot{\mathbf{m}}_m = 0,$$

$$\epsilon\,\rho_s\frac{\partial P_g}{\partial t} + \nabla \cdot \dot{\mathbf{m}}_g = \dot{M}_{ev},$$

$$(\rho C_p)\frac{\partial T}{\partial t} + \left(c_{p,w}\dot{\mathbf{m}}_w + c_{p,g}\dot{\mathbf{m}}_g\right) \cdot \nabla T = -\nabla \cdot \dot{\mathbf{q}} - \Delta h_v\dot{M}_{ev}, \tag{17}$$

where $\dot{\mathbf{q}} = -\lambda\nabla T$, λ is the thermal conductivity of the moist body, $\dot{\mathbf{m}}_w$ is calculated from Darcy's law (10), $\dot{\mathbf{m}}_g = -k_p\nabla P_g$ with k_p as the filtration coefficient, and $\dot{M}_{ev}$ is the mass rate of evaporation per unit volume:

$$\dot{M}_{ev} = \epsilon\,\rho_s\frac{\partial X}{\partial t}, \tag{18}$$

where ϵ is a dimensionless factor characterizing resistance to vapor diffusion in a body.

By using the above conservation equations, three interdependent partial differential equations involving variables X, T, and P can be obtained as

$$\frac{\partial X}{\partial t} = k_{11}\nabla^2 X + k_{12}\nabla^2 T + k_{13}\nabla^2 P_g,$$

$$\frac{\partial T}{\partial t} = k_{21}\nabla^2 X + k_{22}\nabla^2 T + k_{23}\nabla^2 P_g, \tag{19}$$

$$\frac{\partial P_g}{\partial t} = k_{31}\nabla^2 X + k_{32}\nabla^2 T + k_{33}\nabla^2 P_g,$$

where the kinetic coefficients k_{ij} depend not only on temperature and moisture content but also on material properties and drying conditions. For example,

$$k_{11} = \frac{k_m}{c_m\rho_s};$$

$$k_{12} = k_x \cdot k_{11}; \tag{20}$$

$$k_{13} = k_p \cdot k_{11},$$

where k_m is the coefficient of moisture conductivity, c_m is the moisture capacity, ρ_s is the density of the dry solid, and k_x is the thermogradient coefficient related to the moisture content difference. For more details on the computation of these kinetic coefficients, refer Luikov [45], Irudayaraj and Wu [47], and Lewis and Ferguson [48].

It is noted that under the assumption of constant gas pressure, Luikov's equations are similar to those proposed by Philip and De Vries. The biggest problem encountered in using Luikov's equations is the definition of the coefficient k_{ij}. In practice, it is often not possible to obtain these parameters to solve the full system of equations. However, Luikov's theory provides a well-established model in the treatment of simultaneous heat and mass transfer of the drying problem. The solution of Luikov's partial differential equations was studied numerically by Irudayaraj and Wu [47] and Lewis and Ferguson [48]. These equations are still commonly employed today and quite often solved by the finite element method.

2.5. Krischer's Theory. Krischer is also among the first researchers who had investigated the role of heat and mass transfer during drying of porous media. The research work of Krischer was and is still used today as a basis for much of the development in drying theory. In his work [49], which was first published in 1956, Krischer proposed a set of equations to describe the moisture transport for several geometries (plate, cylinder, and sphere). Krischer assumed that moisture transfer is controlled by the combined influence of capillary flow of liquid and diffusion of vapor.

In Krischer's model, the liquid flux is calculated from Darcy's law (10), and the vapor flux is written as

$$\dot{\mathbf{m}}_v = -\frac{\delta_{va}}{\mu} \cdot \frac{P_g}{P_g - P_v} \cdot \frac{\tilde{M}_v}{\tilde{R}T} \nabla P_v, \tag{21}$$

where μ (>1) is called the diffusion resistance factor and describes the decrease of the vapor flow in the considered material in comparison with that in a stagnant gas.

By using the theory of Krischer, Berger and Pei [50] included the sorption isotherm (empirically obtained) into the model as a coupling equation among liquid, vapor, and heat transfer. Based on Krischer's theory, Berger and Pei introduced two balanced equations for heat and mass (the gas pressure was taken as constant). In this model all phenomenological coefficients (e.g., liquid conductivity, vapor diffusivity, and thermal conductivity) are taken as constant. Heat transfer is assumed to take place only by conduction through the solid skeleton. The overall mass and heat balance equations proposed by Berger and Pei [50] are expressed in terms of moisture content and vapor pressure as follows:

$$\rho_s K^* \nabla^2 X + \delta_v \frac{\tilde{M}_v}{\tilde{R}T} \left[\frac{1}{\varepsilon_s} \left(\varepsilon_g - \varepsilon_w \right) \nabla^2 P_v - \frac{\rho_s}{\rho_w} \nabla X \nabla P_v \right]$$
$$= \rho_s \left(1 - \frac{1}{\rho_w} \cdot \frac{\tilde{M}_v P_v}{\tilde{R}T} \right) \frac{\partial X}{\partial t} + \frac{1}{\varepsilon_s} \cdot \frac{\tilde{M}_v}{\tilde{R}T} \left(\varepsilon_g - \varepsilon_w \right) \frac{\partial P_v}{\partial t},$$
$$\frac{\lambda_s}{\rho_s c_{p,s}} \nabla^2 T + \frac{\Delta h_v}{\rho_s c_{p,s}} \frac{\tilde{M}_v}{\tilde{R}T} \left\{ \delta_v \left[\frac{1}{\varepsilon_s} \left(\varepsilon_g - \varepsilon_w \right) \nabla^2 P_v - \frac{\rho_s}{\rho_w} \nabla X \nabla P_v \right] \right.$$
$$\left. - \frac{1}{\varepsilon_s} \left(\varepsilon_g - \varepsilon_w \right) \frac{\partial P_v}{\partial t} + \frac{\rho_s}{\rho_w} P_v \frac{\partial X}{\partial t} \right\} = \frac{\partial T}{\partial t}, \tag{22}$$

where δ_v is the vapor diffusivity; K^* is the liquid conductivity; ε_s, ε_w and ε_g are the volume fraction of solid, water, and gas, respectively.

The main difficulties encountered in using Krischer's model to predict the drying rate are the assumption of surface boundary conditions (Krischer postulated that "at the surface of the drying material, the corresponding equilibrium values of the dependent variables were reached instantaneously at the beginning of the drying process") and the application of the sorption isotherm for the whole range of moisture content [50]. Even though sorption isotherm is taken into account, the approach of Berger and Pei does not offer much innovation over Luikov's and Philip and De Vries' models [46]. In addition, as for the previous models, experimental tests are needed to ensure its validity.

3. Whitaker's Model and the Framework for the Numerical Simulation

In the late 1970s and early 1980s, Whitaker [51, 52] presented a set of equations to describe the simultaneous heat, mass, and momentum transfer in porous media. Based on the traditional conservation laws, the model proposed by Whitaker, an important milestone in the development of drying theory, incorporated all mechanisms for heat and mass transfer: liquid flow due to capillary forces, vapor and gas flow due to convection and diffusion, and internal evaporation of moisture and heat transfer by convection, diffusion, and conduction. By using the volume averaging method, the macroscopic differential equations were defined in terms of average field quantities.

Based on the model of Whitaker, a system of governing equations can be built to represent the drying process, in which the most important equations are the conservation equation for water in the liquid and the gas phases, the conservation equation for air in the gas phase, and the conservation equation of energy of the whole porous system under consideration. These equations can be solved numerically and form the framework for the numerical simulation of the drying process, in which simultaneous mass and heat transfer together with phase changes (vaporization) can be taken into account. We will briefly discuss these equations here.

The first equation governing the drying process is the conservation equation for water in both liquid and gas phases. This equation can be written as

$$\frac{\partial}{\partial t} \left(\rho_w \varepsilon_w + \varepsilon_g \rho_v \right) + \nabla \cdot \left(\rho_w \mathbf{v}_w + \rho_v \mathbf{v}_g \right) = \nabla \cdot \left[\rho_g \mathbf{D}_{eff} \cdot \nabla \left(\frac{\rho_v}{\rho_g} \right) \right], \tag{23}$$

where ρ_w, ρ_v, and ρ_g are the mass density of the liquid, vapor, and gas phases; ε_w and ε_g are the volume fractions of the liquid and gas phases; $\mathbf{v}_w$ and $\mathbf{v}_g$ are the velocities of the liquid and gas phases, and $\mathbf{D}_{eff}$ is the effective diffusivity tensor. The conservation equation for water states that the mass of water at each point inside the porous body is conserved.

The second equation governing the drying process is the conservation equation for air in the gas phase:

$$\frac{\partial}{\partial t} \left(\varepsilon_g \rho_a \right) + \nabla \cdot \left(\rho_a \mathbf{v}_g \right) = \nabla \cdot \left[\rho_g \mathbf{D}_{eff} \cdot \nabla \left(\frac{\rho_a}{\rho_g} \right) \right], \tag{24}$$

which also states that the mass of air at each point inside the porous body is conserved.

The third equation is the conservation equation of energy, which can be formulated as

$$\frac{\partial}{\partial t} \left(\varepsilon_s \rho_s h_s + \varepsilon_w \rho_w h_w + \varepsilon_g \rho_v h_v + \varepsilon_g \rho_a h_a \right)$$
$$+ \nabla \cdot \left[\rho_w h_w \mathbf{v}_w + \left(\rho_v h_v + \rho_a h_a \right) \mathbf{v}_g \right]$$
$$= \nabla \cdot \left[\rho_g h_a \mathbf{D}_{eff} \nabla \left(\frac{\rho_a}{\rho_g} \right) \right] + \nabla \cdot \left[\rho_g h_v \mathbf{D}_{eff} \nabla \left(\frac{\rho_v}{\rho_g} \right) \right]$$
$$+ \nabla \cdot \left(\boldsymbol{\lambda}_{eff} \nabla T \right). \tag{25}$$

Note that in the above system of equations, the velocity of water can be computed using the equation of motion for the liquid phase:

$$v_w = -\frac{\mathbf{K}\mathbf{k}_w}{\eta_w}\nabla P_w, \qquad (26)$$

and the velocity of air can be computed using the equation of motion for the gas phase:

$$\mathbf{v}_g = -\frac{\mathbf{K}\mathbf{k}_g}{\eta_g}\nabla P_g, \qquad (27)$$

where the dynamic viscosity of water η_w and of gas η_g are temperature dependent, $\mathbf{K}$ is the absolute permeability tensor, and $\mathbf{k}_w$ and $\mathbf{k}_g$ denote the relative permeability tensors of liquid and gas, respectively.

In addition to the above governing equations, the conditions for mass and heat transfer at the external drying surfaces of the porous medium must be specified. For example, it can be assumed that, at the external drying surfaces, the fluxes of mass and heat are described for convective drying by the boundary layer theory with Stefan correction:

$$\mathbf{J}_w \cdot \hat{\mathbf{n}} = \dot{\mathbf{m}}_v = \beta \frac{P_g \tilde{M}_v}{\tilde{R}T}\ln\left(\frac{P_g - P_{v,\infty}}{P_g - P_v}\right),$$

$$\mathbf{J}_e \cdot \hat{\mathbf{n}} = \dot{\mathbf{q}} + \Delta h_v \dot{\mathbf{m}}_v = \alpha(T - T_\infty) \qquad (28)$$

$$+ \Delta h_v \beta \frac{P_g \tilde{M}_v}{\tilde{R}T}\ln\left(\frac{P_g - P_{v,\infty}}{P_g - P_v}\right),$$

where $\mathbf{J}_w$ and $\mathbf{J}_e$ are the fluxes of water and heat respectively, $P_{v,\infty}$ and T_∞ are vapor pressure and temperature of bulk drying air, $\hat{\mathbf{n}}$ is the outward-pointing normal vector at the boundary surface, and β and α are mass and heat transfer coefficients. Additionally, we can assume that the gas pressure at the external drying surfaces is fixed at the pressure of the bulk drying air:

$$P_g = P_\infty. \qquad (29)$$

The advantage of Whitaker's model is that it offers a very good representation of the physical phenomena occurring in porous media during drying. However, the problem encountered in using Whitaker's model is the difficulty in determining its complicated transport coefficients, such as the effective diffusivity and permeabilities, which depend strongly on the material properties and structure. These parameters are either functions of moisture content or temperature or both moisture content and temperature. In addition, solving the coupled equations of heat and mass transfer, which are strongly nonlinear, requires very complicated numerical methods.

The theory of Whitaker was further developed and applied in drying analysis of various porous media, for example, in the drying analysis of sand by Whitaker and Chou [53], Hadley [54], Oliveira and Fernandes [55], and Puiggali et al. [56], analysis of glass beads by Quintard and Puiggali [57] and Kaviany and Mittal [58], analysis of sandstone by Wei et al. [59], analysis of porous insulators by Tien and Vafai [60], analysis of brick by Nasrallah and Perré [61], analysis of cellular materials by Crapiste et al. [62], analysis of wood by Spolek and Plumb [63], Michel et al. [64], Perré [65], and Lartigue et al. [66]. In these works, the model was usually quite successfully matched against experimental data.

Whitaker and Chou [53] simplified the theory to obtain two nonlinear equations for the distribution of saturation and temperature. In this work, the gas pressure was assumed as constant, the gas momentum equation was ignored, and a quasisteady state was applied. It is interesting to note that there is a resemblance of these two equations to the equations proposed by Luikov and by Philip and De Vries [46]. In this simplified case, the comparison between theory and experiment was made by Hougen et al. [15]. The important conclusion is that the gas phase momentum equation must be included in solving the comprehensive set of equations. Crapiste et al. [62] applied Whitaker's model to investigate the drying of cellular materials. To validate the model, a comparison of one-dimensional drying to the experimental drying of apple and potato was presented and a good agreement was found. Wei et al. [59] applied Whitaker's model to the drying of a cylinder of sandstone subjected to convective heating. The obtained partial differential equations in one dimension were solved by a three-point, two-level implicit finite difference method. The calculated results were compared with experimental results and showed a quite good agreement.

Ferguson [67] focused on a two-dimensional problem of the high-temperature drying of spruce. The numerical results highlighted the advantage of the discretization technique (control volume finite element method) in solving the problem with structured and unstructured meshes. A numerical investigation was conducted by Boukadida et al. [68] to study the convective drying of a slab of clay-brick. The work analyzed the influence of the properties of the surrounding drying agent (temperature, gas pressure, and vapor concentration) as well as the initial medium conditions (temperature and moisture content) on the drying process by considering several configurations. However, the full investigation of the effect of the boundary layer on the coupled heat and mass transfer still requires further work, as concluded by the authors. Silva [69], based on Whitaker's theory, presented a general model to describe the momentum, heat, and mass transfer in drying problems with moving boundary. By using the volume averaging method, a set of equations for multiphase systems was applied to porous media. Numerical results showed a good agreement with the experimental data of kaolin drying.

One of the most significant advances in developing Whitaker's theory as well as in modelling the drying of porous media comes from the works of Nasrallah and Perré [61] and Perré et al. [70]. In their work, the drying of two quite different porous media—clay-brick and softwood—was investigated. The most important advance in the work of Perré is the consideration of bound water [71–73]. By considering bound water, the driving potential for bound water migration was assumed to be proportional to the gradient in the bound moisture content. For the case of wood, Perré and his colleagues introduced two equations to calculate the transport of this kind of bound water

$$\frac{\partial \overline{\rho}_b}{\partial t} + \nabla \cdot (\overline{\rho_b \cdot \mathbf{v}_b}) = -\dot{m}_b,$$

$$\overline{\rho_b \cdot \mathbf{v}_b} = -\overline{\rho}_c \cdot \delta_b \cdot \nabla\left(\frac{\overline{\rho}_b}{\overline{\rho}_c}\right),$$

(30)

where $\dot{m}_b$ is the rate of bound water evaporation and the subscripts b and c denote bound water and cellulose matter, respectively. For wood, the diffusion coefficient of bound water δ_b is calculated in m^2/s from the following equation [72]:

$$\delta_b = \exp\left(-9.9 + 9.8 \cdot X_b - \frac{4300}{T}\right),$$

(31)

where X_b is moisture content of bound water and T is temperature (Kelvin).

In his work, Perré solved the one-dimensional problem of drying with three state variables (temperature, pressure, and moisture content). The control volume method was applied to solve the nonlinear partial differential equations. The mathematical schemes for equidistant and nonequidistant meshes were discussed [61]. The authors also investigated the sensitivity upon model parameters by numerically varying the effective diffusivity, effective thermal conductivity, intrinsic and relative permeabilities, and external drying conditions (heat and mass transfer coefficients).

With the rapid development of the computer technology, modern computers allow the simulation of drying not only in one dimension but also in two and three dimensions. Besides, numerical methods are also more efficient in obtaining accurate results and reducing the computational time. Among the advancements during the 1990s in the study of drying of porous media is the simulation of drying processes in two dimensions with unstructured meshes proposed by Perré [74] and Perré and Turner [75]. The first comprehensive three-dimensional drying model using structured meshes was introduced by Perré and Turner [76]. In this work, a homogeneous model, which employed the full set of conservation equations, was considered. A cube of light concrete (isotropic medium) and a board of wood (anisotropic medium) were chosen to investigate the influence of the number of exchange faces. Several simulation results for low- and high-temperature drying of softwood were presented and discussed. By comparing the different simulation results, the study showed that a three-dimensional model is required to describe correctly the drying behavior of porous media.

Concerning the heterogeneity of material properties, Perré [74] developed a heterogeneous drying model for wood. The variation of the material property information such as capillary pressure and absolute permeability was taken into account with the help of experiments [77, 78]. The material information of wood obtained from this work was later applied to a two-dimensional heterogeneous drying model [79]. In this work, the effects of material heterogeneity and local material direction on the heat and mass transport during drying were investigated. Two cases of

low- and high-temperature drying were considered. Following this direction, more recently, Truscott [80] and Truscott and Turner [81] developed a three-dimensional heterogeneous drying model for wood. The work considered the heterogeneity of the material properties, which vary within the transverse plane with respect to the position that defines the radial and tangential directions. Two nonlinear partial equations for moisture content and temperature (pressure was assumed as constant) were solved.

4. Numerical Simulation Based on Whitaker's Model

By solving the system (23)–(30) resulting from the model of Whitaker, the drying process in porous media can be simulated numerically taking into account complex mass and heat transport phenomena. Different numerical methods can be used to solve the above system of equations, for example, the finite element method, the finite difference method, or the control volume method.

In simple cases, the numerical solution can be obtained with relatively small computational effort. One of such examples can be found in [52], where numerical simulation was compared with experimental measurement [14] of a sand plate under isothermal drying conditions. In the work, the above system (23)–(30) was applied. The plate of sand has a thickness of 5.08 cm. The drying took place by air. The initial saturation was set to be 100%. One surface of the plate was considered impermeable. The other surface was in contact with air. Since the width and length of the plate are much larger than its thickness, the problem can be reduced to be 1-dimensional. The numerical solution was obtained with the help of the finite difference method. Note that during the experiment [14], the averaged saturation was determined at different time instants. Corresponding to these time instants, the saturation profiles were measured. The measurement delivered then the saturation profiles at different values of averaged saturation, namely, at 36%, 53%, 79%, and 89%. The numerical solution was then compared with the experimental result as shown in Figure 3. In Figure 3, the saturation profile is plotted as a function of the normalized distance from the impermeable surface of the plate. The comparison shows that the model of Whitaker delivers a reasonable result. The adequacy of Whitaker's model in modeling the drying process of porous media can also be seen by comparing this model with other models mentioned above. From Table 1, it can be observed that Whitaker's model offers the most complete picture of the different phenomena happening during the drying of a porous medium.

In this work, the control volume method is chosen to solve the model of Whitaker. The reason for using the control volume method lies in the fact that this method satisfies the conservation requirement of the basic physical quantities such as mass and energy at any discrete level. This means without the need to enforce this requirement by using additional constraints, the heat and mass flows across a boundary of a control volume element or over the

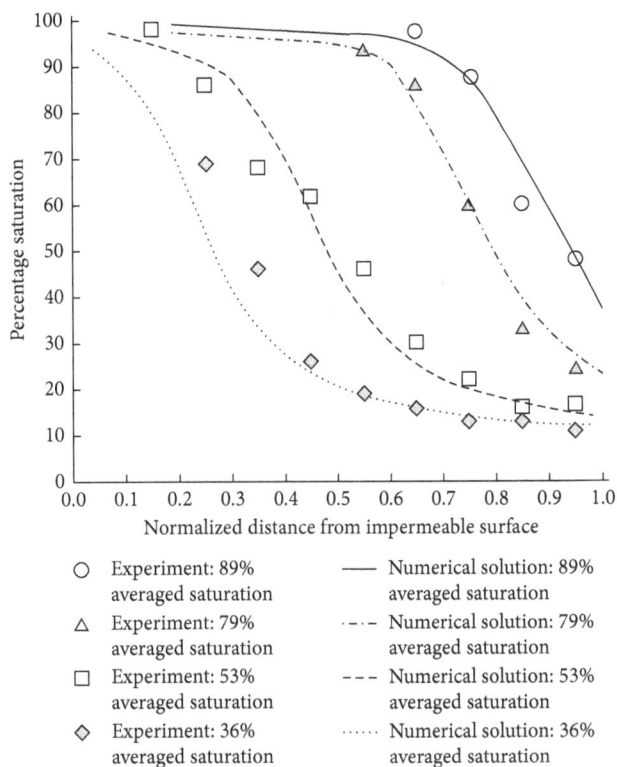

FIGURE 3: Comparison of numerical simulation and experimental values in drying of wet sand: saturation versus distance from the impermeable surface [52].

○	Experiment: 89% averaged saturation	—— Numerical solution: 89% averaged saturation
△	Experiment: 79% averaged saturation	·-·- Numerical solution: 79% averaged saturation
□	Experiment: 53% averaged saturation	--- Numerical solution: 53% averaged saturation
◇	Experiment: 36% averaged saturation	······ Numerical solution: 36% averaged saturation

boundary of the whole porous medium are automatically conserved. In this work, we will not discuss the details of the application of the control volume method in solving the system (23)–(30) and limit ourselves in presenting a numerical example in order to demonstrate the capability of the approach using the model of Whitaker. For more details on the approach and numerical implementation, refer [12].

We examine here the drying of a sphere of light concrete with radius $R = 2.5$ mm. The temperature of the sample at the start of the drying process is $T_0 = 20°C$, the moisture content before drying is $X_0 = 1$, and the pressure of air inside the sample is $P_0 = 1$ bar. In order to dry the sample, the sample is put in an oven with dried air. The temperature of the oven is $T_\infty = 80°C$ and the vapor pressure in the oven is $P_{v,\infty} = 0$. At the boundary of the sample, heat and mass transfer takes place during the drying process. For these two processes between the sample and the surrounding air of the oven, we assume that the heat transfer coefficient is $\alpha = 14.25$ W·m^{-2}·K^{-1} and mass transfer coefficient is $\beta = 0.015$ m·s^{-1}. Since the sample is symmetric and the boundary conditions applied to the sample can also be considered as symmetric, the drying problem of our sample can be solved by the control volume method in one-dimensional space. For other details concerning transport properties, refer [12].

As numerical results, the temporal evolution of moisture, temperature, and pressure for approximately every 0.5 mm in distance along the radial direction of the sample from the center to the boundary of the sample is shown in

Figures 4–6. In Figure 4, the dashed curve presents the average moisture content of the whole sample.

From the numerical results presented here, some important drying characteristics can be observed. Starting from a uniform initial moisture content $X_0 = 1$ (corresponding to saturation $S_0 = 62.5\%$) over the whole sample, after a short warm-up period, the moisture content is reduced at a constant rate (constant slope of the curves in Figure 4). This is called the first drying period (or constant rate drying period). During the first drying period (approximately 17.6 minutes), free water is brought to the surface by capillary forces where heat supplied by the convectional air is used for the rapid vaporization of water. Due to this reason, the sample remains at the wet bulb temperature of the drying air ($T_{wb} = 23.81°C$, Figure 5). The moisture gradient increases (relative permeability k_w decreases), and our analysis shows that the moisture profiles as the function of radius appear fairly flat. Within this period, the pressure stays constant at atmospheric pressure (Figure 6). As the drying process continues, the moisture content at the surface reaches the irreducible value $X_{irr} = 0.07$ (the average moisture content of the whole sample reaches the critical moisture content $X_{cr} = 0.1774$) and the second drying period commences. In the second drying period, the dominating forces are viscous forces. Our analysis shows that a front separating the regions of adsorbed water and free water recedes from the surface into the sample. This process is finished when the moisture content everywhere in the sample is below the irreducible value, that is, when all free water of the sample has been removed. During this period, heat transfer is almost unchanged (resistance in the sample is slightly increased), but mass transfer experiences an important additional resistance. Heat is used not only to evaporate water but also to increase the temperature of the sample. Therefore, the temperature of the sample starts to rise from the wet bulb temperature. The center of the sample stays cooler than the outside (Figure 5). This is due to the fact that the evaporation of water takes place not at the surface but at a place inside the sample (front). The free water region heats up until a new equilibrium is attained (if we assume a stationary front). At the front, heat is consumed for evaporation. As we can see from Figure 6, in the second drying period, an overpressure appears due to the Stefan effect. The pressure inside the sample increases to its maximum value, whereas the pressure at the surface always stays at the atmospheric pressure (1 bar). When the receding front has passed through the whole sample, the sample becomes dry and the entire porous medium is in the hygroscopic zone. The moisture content goes down to the equilibrium value. The temperature of solid gradually approaches the dry bulb temperature of the drying air ($T_\infty = 80°C$, Figure 5), and the pressure falls back to the atmospheric pressure (Figure 6).

By considering the numerical example presented here, it is easy to see that the implemented simulation framework based on the model of Whitaker can be used effectively in studying the complex drying process of porous media. Especially important is the incorporation of the simultaneous heat and mass transfer processes together with the evaporation process in the simulation.

TABLE 1: Comparison of different drying models.

		Diffusion theory	Receding front theory	Model of Philip and De Vries	Luikov's theory	Krischer's theory	Whitaker's model
Mass transport due to	Diffusion of air						Yes
	Convection of air						Yes
	Diffusion of vapor	Yes	Yes	Yes	Yes	Yes	Yes
	Convection of vapor		Yes	Yes	Yes	Yes	Yes
	Capillary force		Yes	Yes	Yes	Yes	Yes
Energy transport due to	Diffusion			Yes	Yes	Yes	Yes
	Convection				Yes	Yes	Yes
	Conduction		Yes	Yes	Yes	Yes	Yes

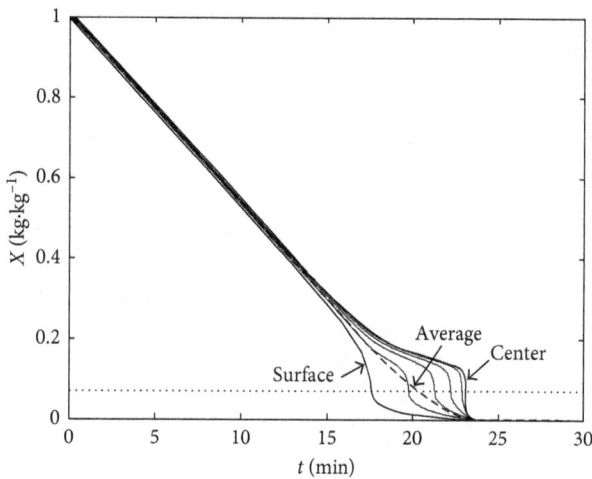

FIGURE 4: Drying of a sphere of light concrete: temporal evolution of moisture content for approximately every 0.5 mm in distance.

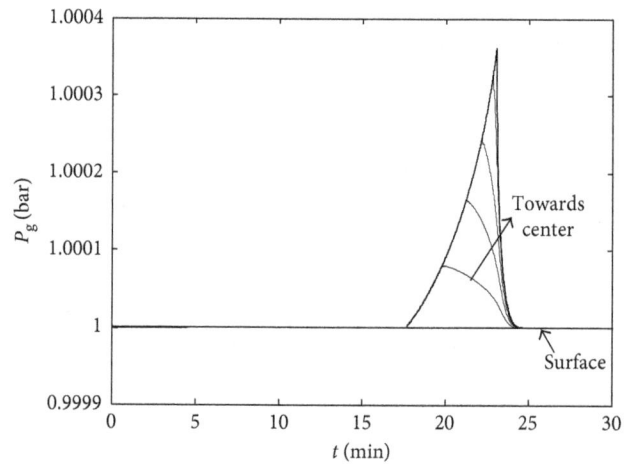

FIGURE 6: Drying of a sphere of light concrete: temporal evolution of pressure for approximately every 0.5 mm in distance along the radial direction.

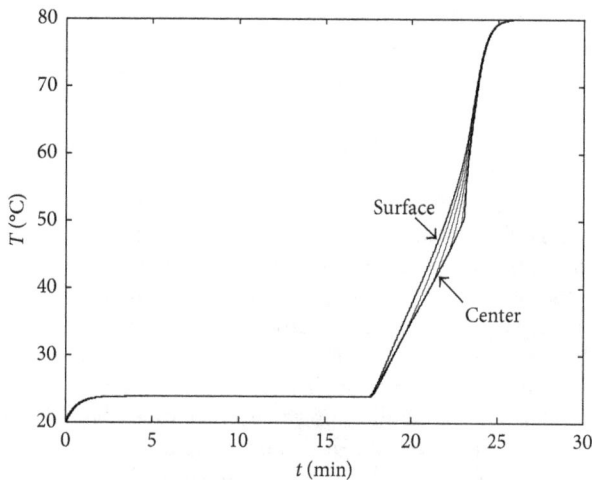

FIGURE 5: Drying of a sphere of light concrete: temporal evolution of temperature for approximately every 0.5 mm in distance.

5. Conclusion

In this work, a review of the development of some drying models, their application, and their restrictions is presented.

Among the most complex and modern models is the model developed by Whitaker. By using Whitaker's model, the basic laws of mass and heat transport can be formulated at macroscopic level. The result is a system of equations governing the drying process of porous media, which can be solved numerically using modern numerical methods, in particular the control volume method. Our numerical implementation of Whitaker's model is presented through a numerical example to show to capability of this model. The numerical example and the different characteristics of drying observed in our simulation results show that the implemented simulation framework can be used to study realistic drying processes in porous media.

Conflicts of Interest

The authors declare that there are no conflicts of interest regarding the publication of this paper.

References

[1] J. Azmir, Q. Hou, and A. Yua, "Discrete particle simulation of food grain drying in a fluidised bed," *Powder Technology*, vol. 323, pp. 238–249, 2018.

[2] A. A. Moghaddam, A. Kharaghani, E. Tsotsas, and M. Prat, "Kinematics in a slowly drying porous medium: reconciliation of pore network simulations and continuum modeling," *Physics of Fluids*, vol. 29, no. 2, p. 022102, 2017.

[3] L. K. Hiep, A. Kharaghani, C. Kirsch, and C. E. Tsotsas, "Discrete pore network modeling of superheated steam drying," *Drying Technology*, vol. 35, no. 13, pp. 1584–1601, 2017.

[4] K. Chen, P. Bachmann, A. Bueck, M. Jacob, and E. Tsotsas, "Experimental study and modeling of particle drying in a continuously-operated horizontal fluidized bed," *Particuology*, vol. 34, pp. 134–146, 2017.

[5] P. Krawczyk, "Numerical modeling of simultaneous heat and moisture transfer during sewage sludge drying in solar dryer," *Procedia Engineering*, vol. 157, pp. 230–237, 2016.

[6] I. N. Ramos, T. R. S. Brandao, and C. L. M. Silva, "Simulation of solar drying of grapes using an integrated heat and mass transfer model," *Renewable Energy*, vol. 81, pp. 896–902, 2015.

[7] P. S. Nasirabadi, M. Jabbari, and J. H. Hattel, "Numerical simulation of transient moisture transfer into an electronic enclosure," in *AIP Conference Proceedings*, vol. 1738, Rhodes, Greece, 2016.

[8] H. B. Fan, E. K. L Chan, C. K. Y. Wong, and M. M. F Yuen, "Moisture diffusion study in electronic packaging using molecular dynamic simulation," in *Proceedings of 56th Electronic Components and Technology Conference*, San Diego, CA, USA, 2006.

[9] M. Fortes and M. R. Okos, "Drying theories: their bases and limitations as applied to foods and grains," in *Advances in Drying 1*, A. S. Mujumdar, Ed., pp. 119–154, Hemisphere Publ. Corp., Washington, DC, USA, 1980.

[10] S. Bories, *Recent Advances in Modelisation of Coupled Heat and Mass Transfer in Capillary-Porous Bodies*, Drying '89, Hemisphere Publishing Corporation, New York, NY, USA, 1989.

[11] E. Tsotsas, "Measurement and modelling of intraparticle drying kinetics: a review," in *Drying'92, Proceedings of the 8th International Drying Symposium, Part A*, pp. 17–41, Montreal, QC, Canada, 1992.

[12] H. T. Vu, *Influence of Pore Size Distribution on Drying Behaviour of Porous Media by a Continuous Model*, Ph.D. thesis, University of Magdeburg, Magdeburg, Germany, 2016.

[13] T. K. Sherwood, "The drying of solid–I," *Industrial and Engineering Chemistry*, vol. 21, no. 1, pp. 12–16, 1929.

[14] N. H. Ceaglske and O. A. Hougen, "Drying of granular solids," *Industrial and Engineering Chemistry*, vol. 29, no. 7, pp. 805–813, 1937.

[15] O. A. Hougen, H. J. Cauley, and W. R. Marshall, *Limitations of Diffusion Equations in Drying*, American Institute of Chemical Engineers, Houston, TX, USA, 1939.

[16] M. Suzuki and S. Maeda, "An approximation of transient change of moisture distribution within porous material being dried," in *Proceedings of the first International Symposium on Drying (IDS '78)*, pp. 42–47, Montreal, QC, USA, 1978.

[17] G. D. Saravacos and G. S. Raouzeos, "Diffusivity of moisture in air-drying of raisins," in *Drying '86, Proceedings of the 5th International Drying Symposium*, vol. 2, pp. 487–491, Lyon, France, August 1986.

[18] M. Jaros, S. Cenkowski, D. S. Jayas, and S. Pabis, "A method of determination of the diffusion coefficient based on kernel moisture content and its temperature," *Drying Technology*, vol. 10, no. 1, pp. 213–222, 1992.

[19] M. Mourad, M. Hemati, and C. Laguerie, "A new correlation for the estimation of moisture diffusivity in corn kernels from drying kinetics," *Drying Technology*, vol. 14, no. 3-4, pp. 873–894, 1996.

[20] J. A. Ribeiro, A. W. Pereira, D. T. Oliveira, and M. A. S. Barrozo, "Determination of drying kinetic parameters of *Bixa orellana* seeds in thin layer," in *Drying'2002, Proceedings of the 13th International Drying Symposium (IDS'2002)*, vol. B, pp. 1293–1300, Beijing, China, August 2002.

[21] Z. Li and N. Kobayashi, "Determination of moisture diffusivity by thermo-gravimetic analysis under non-isothermal condition," *Drying Technology*, vol. 23, no. 6, pp. 1331–1342, 2005.

[22] J. Srikiatden and J. S. Roberts, "Measuring moisture diffusivity of potato and carrot (core and cortex) during convective hot air and isothermal drying," *Journal of Food Engineering*, vol. 74, no. 1, pp. 143–152, 2005.

[23] H. Koponen, *Moisture Diffusion Coefficients of Wood, Drying '87*, Hemisphere publ. corp., New York, NY, USA, 1987.

[24] A. A. J. Ketelaars, L. Pel, W. J. Coumans, and P. J. A. M. Kerkhof, "Drying kinetics: a comparison of diffusion coefficients from moisture concentration profiles and drying curves," *Chemical Engineering Science*, vol. 50, no. 7, pp. 1187–1191, 1995.

[25] L. Pel, H. Broken, and K. Kopinga, "Determination of moisture diffusivity in porous media using moisture concentration profiles," *International Journal of Heat and Mass Transfer*, vol. 39, no. 6, pp. 1273–1280, 1996.

[26] H. A. C. Thijssen and W. J. Coumans, "Short-cut calculation for non-isothermal drying of shrinking and non-shrinking particles and of hollow spheres containing an expanding central gas core," in *Drying '85: Selection of Papers from the 4th International Drying Symposium*, pp. 11–20, Kyoto, Japan, July1984.

[27] S. Blackband and P. Mansfield, "Diffusion in liquid-solid systems by NMR imaging," *Journal of Physics C: Solid State Physics*, vol. 19, no. 2, pp. L49–L52, 1986.

[28] G. W. Schrader and J. B. Litchfield, "Moisture profiles in a model food gel during drying: measurement using magnetic resonance imaging and evaluation of the Hckian model," *Drying Technology*, vol. 10, no. 2, pp. 295–332, 1992.

[29] L. Pel, A. A. J Ketelaars, O. C. G Adan, and A. A. Van Well, "Determination of moisture diffusivity in porous media using scanning neutron radiography," *International Journal of Heat and Mass Transfer*, vol. 36, no. 5, pp. 1261–1267, 1993.

[30] P. J. McDonald, T. Pritchard, and S. P. Roberts, "Diffusion of water at low saturation levels into sandstone rock plugs measured by broad line magnetic resonance profiling," *Journal of Colloid and Interface Science*, vol. 177, no. 2, pp. 439–445, 1996.

[31] I. V. Koptyug, S. I. Kabanikhin, K. T. Iskakov et al., "A quantitative NMR imaging study of mass transport in porous solids during drying," *Chemical Engineering Science*, vol. 55, no. 9, pp. 1559–1571, 2000.

[32] V. Guillard, B. Broyart, C. Bonazzi, S. Guilbert, and N. Gontard, "Water transfer in multi-component heterogeneous food: moisture distribution measurement and prediction," in *Drying'2002, Proceedings of the 13th International Drying Symposium (IDS'2002)*, vol. B, pp. 1155–1160, Beijing, China, August 2002.

[33] G. I. Efremov, "Drying kinetics derived from diffusion equation with flux-type boundary conditions," *Drying Technology*, vol. 20, no. 1, pp. 55–66, 2002.

[34] P. S. S. Porto and A. C. L. Lisbôa, "Drying modelling of a parallelepipedic oil shale particle," in *Drying 2004, Proceedings*

of the 14th International Drying Symposium (IDS 2004), vol. A, pp. 542–548, Campinas, SP, Brazil, August 2004.

[35] L. C. Lim, S. M. Tasirin, and W. R. W. Daud, "Derivation of new drying model from theoretical diffusion controlled drying period," in Drying 2004, Proceedings of the 14th International Drying Symposium (IDS 2004), vol. A, pp. 430–435, Campinas, SP, Brazil, August 2004.

[36] E. K. Akpinar and I. Dincer, "Application of moisture transfer models to solids drying," Proceedings of the Institution of Mechanical Engineers, Part A: Journal of Power and Energy, vol. 219, no. 3, pp. 235–244, 2005.

[37] E. K. Akpinar and I. Dincer, "Moisture transfer models for slabs drying," International Communications in Heat and Mass Transfer, vol. 32, no. 1-2, pp. 80–93, 2005.

[38] P. Chen and P. S. Schmidt, An Integral Model for Convective Drying of Hygroscopic and Non-Hygroscopic Materials, Drying '89, Hemisphere publ. corp., New York, NY, USA, 1990.

[39] J. R. Philip and D. A. De Vries, "Moisture movement in porous materials under temperature gradient," Transactions, American Geophysical Union, vol. 38, no. 2, pp. 222–232, 1957.

[40] D. A. De Vries, "Simultaneous transfer of heat and moisture in porous media," Transactions, American Geophysical Union, vol. 39, no. 5, pp. 909–916, 1958.

[41] D. A. De Vries, "Theory of heat and moisture transfer in porous media revisited," International Journal of Heat Mass Transfer, vol. 30, no. 7, pp. 1343–1350, 1987.

[42] J. Van der Kooi, Moisture Transport in Cellular Concrete Roofs, Ph.D. thesis, Uitgeverij Waltman-Delft, Delft, Netherlands, 1971.

[43] A. V. Luikov, "Application of irreversible thermodynamics methods to investigation of heat and mass transfer," International Journal of Heat and Mass Transfer, vol. 9, no. 2, pp. 139–152, 1966.

[44] A. V. Luikov, "Heat and mass transfer in capillary-porous bodies," Advances in Heat Transfer, vol. 1, pp. 123–184, 1964.

[45] A.V. Luikov, "Systems of differential equations of heat and mass transfer in capillary-porous bodies," International Journal of Heat and Mass Transfer, vol. 18, no. 1, pp. 1–14, 1975.

[46] I. W. Turner, The Modelling of Combined Microwave and Convective Drying of a Wet Porous Material, Ph.D. thesis, University of Queensland, Brisbane, QLD, Australia, 1991.

[47] J. Irudayaraj and Y. Wu, "Analysis and application of Luikov's heat, mass, and pressure transfer model to a capillary porous media," Drying Technology, vol. 14, no. 3-4, pp. 803–824, 1996.

[48] R.W. Lewis and W. J. Ferguson, "The effect of temperature and total gas pressure on the moisture content in a capillary porous body," International Journal for Numerical Methods in Engineering, vol. 29, no. 2, pp. 357–369, 1990.

[49] O. Krischer and W. Kast, Die wissenschaftlichen Grundlagen der Trocknungstechnik ter Band, dritte Auflage, Springer, Berlin, Germany, 1992.

[50] D. Berger and D. C. T. Pei, "Drying of hygroscopic capillary porous solids–A theoretical approach," International Journal of Heat and Mass Transfer, vol. 16, no. 2, pp. 293–302, 1973.

[51] S. Whitaker, "Simultaneous heat, mass, and momentum transfer in porous media: a theory of drying," Advances in Heat Transfer, vol. 13, pp. 119–203, 1977.

[52] S. Whitaker, "Heat and mass transfer in granular porous media," in Advances in Drying 1, A. S. Mujumdar, Ed., pp. 23–61, Hemisphere Publ. Corp., Washington, DC, USA, 1980.

[53] S. Whitaker and W. T.-H. Chou, "Drying granular porous media-theory and experiment," Drying Technology, vol. 1, no. 1, pp. 3–33, 1983.

[54] G. R. Hadley, "Numerical modeling of the drying of porous materials," in Drying'85, Proceedings of the 4th International Symposium on Drying, pp. 135–142, Kyoto, Japan, 1985.

[55] A. C. Oliveira and E. O. Fernandes, "Simulation of the convective drying of capillary-porous materials," in Drying '86, Proceedings of the 5th International Symposium on Drying, vol. 1, pp. 65–70, Cambridge, MA, USA, August 1986.

[56] J. R. Puiggali, M. Quintard, and S. Whitaker, "Drying granular porous media: gravitational effects in the isenthalpic regime and the role of diffusion models," Drying Technology, vol. 6, no. 4, pp. 601–629, 1988.

[57] M. Quintard and J.-R. Puiggali, "Numerical modelling of transport processes during the drying of a granular porous medium," Heat and Technology, vol. 4, no. 2, pp. 37–57, 1986.

[58] M. Kaviany and M. Mittal, "Funicular state in drying of a porous slab," International Journal of Heat and Mass Transfer, vol. 30, no. 7, pp. 1407–1418, 1987.

[59] C. K. Wei, H. T. Davis, E. A. Davis, and J. Gordon, "Heat and mass transfer in water-laden sandstone: convective heating," AIChE Journal, vol. 31, no. 8, pp. 1338–1348, 1985.

[60] H. C. Tien and K. Vafai, "A synthesis of infiltration effects on an insulation matrix," International Journal of Heat and Mass Transfer, vol. 33, no. 6, pp. 1263–1280, 1990.

[61] S. B. Nasrallah and P. Perré, "Detailed study of a model of heat and mass transfer during convective drying of porous media," International Journal of Heat and Mass Transfer, vol. 31, no. 5, pp. 957–967, 1988.

[62] G. H. Crapiste, S. Whitaker, and E. Rotstein, "Drying of cellular material-II. Experimental and numerical results," Chemical Engineering Science, vol. 43, no. 11, pp. 2929–2936, 1988.

[63] G. A. Spolek and O. A. Plumb, "A numerical model of heat and mass transport in wood during drying," in Drying'80, Proceedings of the Second International Symposium on Drying, vol. 2, pp. 84–92, New York, NY, USA, 1980.

[64] D. Michel, M. Quintard, and J.-R. Puiggali, Experimental and Numerical Study of Pine Wood Drying at Low Temperature, Drying '87, Hemisphere Publ. Corp., Washington, DC, USA, 1987.

[65] P. Perré, Le séchage convectif de bois résineux: choix, validation et utilisation d'un modèle, Ph.D. thesis, University Pasis VII, Paris, France, 1987.

[66] C. Lartigue, J. R. Puiggali, and M. Quintard, A Simplified Study of Moisture Transport and Shrinkage in Wood, Drying '89, Hemisphere Publ. Corp., Washington, DC, USA, 1990.

[67] W. J. Ferguson, "A control volume finite element numerical simulation of the high temperature drying of spruce," Drying Technology, vol. 13, no. 3, pp. 607–634, 1995.

[68] N. Boukadida, S. B. Nasrallah, and P. Perré, "Mechanism of two-dimensional heat and mass transfer during convective drying of porous media under different drying conditions," Drying Technology, vol. 18, no. 7, pp. 1367–1388, 2000.

[69] M. A. Silva, "A general model for moving boundary problems-application to drying of porous media," Drying Technology, vol. 18, no. 3, pp. 601–624, 2000.

[70] P. Perré, S. B. Nasrallah, and G. Arnaud, "A Theoretical study of drying: numerical simulations applied to clay-brick and softwood," in Drying '86, Proceedings of the 5th International Symposium on Drying, vol. 1, pp. 382–390, Cambridge, MA, USA, August 1986.

[71] P. Perré, J. P. Fohr, and G. Arnaud, A Model of Drying Applied to Softwoods: The Effect of Gaseous Pressure Below the Boiling Point, Drying '89, Hemisphere Publ. Corp., Washington, DC, USA, 1989.

[72] P. Perré and A. Degiovanni, "Simulation par volumes finis des transferts couplés en milieux poreux anisotropes: séchage du bois a basse et à haute temperature," *International Journal of Heat and Mass Transfer*, vol. 33, no. 11, pp. 2463–2478, 1990.

[73] P. Perré, M. Moser, and M. Martin, "Advances in transport phenomena during convective drying with superheated steam and moist air," *International Journal of Heat and Mass Transfer*, vol. 36, no. 11, pp. 2725–2746, 1993.

[74] P. Perré, "Image analysis, homogenization, numerical simulation and experiment as complementary tools to enlighten the relationship between wood anatomy and drying behavior," *Drying Technology*, vol. 15, no. 9, pp. 2211–2238, 1997.

[75] P. Perré and I. W. Turner, "Transpore: a generic heat and mass transfer computational model for understanding and visualising the drying of porous media," in *Drying'98, Proceedings of the 11th International Drying Symposium (IDS'98)*, vol. A, pp. 365–374, Halkidiki, Greece, August 1998.

[76] P. Perré and I. W. Turner, "A 3-D version of TransPore: a comprehensive heat and mass transfer computational model for simulating the drying of porous media," *International Journal of Heat and Mass Transfer*, vol. 42, no. 24, pp. 4501–4521, 1999.

[77] P. Perré and I. W. Turner, "Determination of the material property variations across the growth ring of softwood for use in a heterogeneous drying model. Part 1. Capillary pressure, tracheid model and absolute permeability," *Holzforschung*, vol. 55, no. 3, pp. 318–323, 2001.

[78] P. Perré and I. W. Turner, "Determination of the material property variations across the growth ring of softwood for use in a heterogeneous drying model. Part 2. Use of homogenization to predict bound liquid diffusivity and thermal conductivity," *Holzforschung*, vol. 55, no. 3, pp. 417–425, 2001.

[79] P. Perré and I. W. Turner, "A heterogeneous wood drying computational model that accounts for material property variation across growth rings," *Chemical Engineering Journal*, vol. 86, no. 1-2, pp. 117–131, 2002.

[80] S. L. Truscott, *A Heterogeneous Three-Dimensional Computational Model for Wood Drying*, Ph.D. thesis, University of Queensland, Brisbane, QLD, Australia, 2004.

[81] S. L. Truscott and I. W. Turner, "A heterogeneous three-dimensional computational model for wood drying," *Applied Mathematical Modelling*, vol. 29, no. 4, pp. 381–410, 2005.

Synthesis of CuNi/C and CuNi/γ-Al$_2$O$_3$ Catalysts for the Reverse Water Gas Shift Reaction

Maxime Lortie,[1,2] **Rima Isaifan,**[1,2] **Yun Liu,**[1] **and Sander Mommers**[1]

[1]*Chemical and Biological Engineering, University of Ottawa, Ottawa, ON, Canada K1N 6N5*
[2]*Centre for Catalysis Research and Innovation, University of Ottawa, Ottawa, ON, Canada K1N 6N5*

Correspondence should be addressed to Maxime Lortie; m.lortie90@gmail.com

Academic Editor: Donald L. Feke

A new polyol synthesis method is described in which CuNi nanoparticles of different Cu/Ni atomic ratios were supported on both carbon and gamma-alumina and compared with Pt catalysts using the reverse water gas shift, RWGS, reaction. All catalysts were highly selective for CO formation. The concentration of CH$_4$ was less than the detection limit. Cu was the most abundant metal on the CuNi alloy surfaces, as determined by X-ray photoelectron spectroscopy, XPS, measurements. Only one CuNi alloy catalyst, Cu$_{50}$Ni$_{50}$/C, appeared to be as thermally stable as the Pt/C catalysts. After three temperature cycles, from 400 to 700°C, the CO yield at 700°C obtained using the Cu$_{50}$Ni$_{50}$/C catalyst was comparable to that obtained using a Pt/C catalyst.

1. Introduction

The carbon dioxide hydrogenation reaction has been proposed for use with carbon capture technologies for the production of industrially viable chemicals, such as long chain hydrocarbons, methanol, formic acid, and carbon monoxide [1, 2]. When cofeeding CO$_2$ and H$_2$ over a hydrogenation catalyst, there are two main hydrogenation processes that can take place, the reverse water gas shift reaction

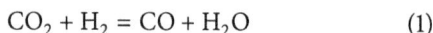

$$CO_2 + H_2 = CO + H_2O \qquad (1)$$

and the subsequent hydrogenation of CO to either hydrocarbons or alcohols, depending on the values of x, y, and z in

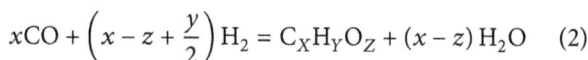

$$x\text{CO} + \left(x - z + \frac{y}{2}\right)\text{H}_2 = \text{C}_X\text{H}_Y\text{O}_Z + (x - z)\,\text{H}_2\text{O} \qquad (2)$$

One of the reactions in (2), the hydrogenation of CO to methane, is of particular interest in this study:

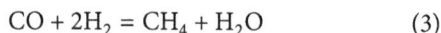

$$CO + 2H_2 = CH_4 + H_2O \qquad (3)$$

When CO is selectively formed via (1) and mixed with H$_2$, the resulting syngas can be a feed-stock for the Fischer Tropsch process [2] that produces liquid fuels. In contrast CH$_4$ formed via (3) is an undesirable by-product that is not convertible to liquid fuels in a Fischer Tropsch process.

Wang et al. [1] have recently reviewed catalysts for the RWGS reaction. They reported that noble metals have been studied and shown to be among the best catalysts for the RWGS reaction because they generally promote H$_2$ dissociation. Among noble metals used for the RWGS reaction, platinum (Pt) has received considerable attention. It was found to produce high CO yields [3–5]. In addition, Pekridis et al. [6] tested the electrokinetics of the RWGS reaction in solid oxide fuel cells containing a Pt/YSZ catalyst. The Pt/YSZ catalyst was found to be stable at high temperatures and the cell achieved a maximum power density of 9 mW/cm^2 [6]. In spite of the recognized performance of noble metals as catalysts for the RWGS reaction, their main drawback is their high cost that limits their commercial application.

Transition metals such as copper (Cu), nickel (Ni), and iron (Fe) are promising alternatives to noble metal catalysts for the RWGS reaction. Both copper and nickel based catalysts have shown good conversion for the WGS reaction as well as the RWGS reaction [7–12]. Chen et al. [7] investigated Cu nanoparticles ranging in size from 2.4 to 3.4 nm and found that the catalyst becomes unstable at higher temperatures [13]. These researchers [14] also added

Fe to Cu in an attempt to stabilize the catalyst. Although Fe alone had poor conversions, Fe stabilized the Cu catalyst for 120 h and caused an increase in conversion of approximately 7% at 600°C. On the other hand, the Cu catalyst without the iron stabilizer was deactivated rapidly and reached zero conversion after 120 hours.

Similar research was performed by Chen et al. [15] using Ni catalysts. Ni alone showed high selectivity towards methane. When they added potassium to a Ni/γ-Al$_2$O$_3$ catalyst they reported higher selectivity towards CO even though they did not notice an increase in CO$_2$ conversion. However with the potassium promoter they noticed the formation of coke.

Y. Liu and D. Liu [16] studied a Ni-Cu catalyst that was prepared by immersing gamma-alumina (γ-Al$_2$O$_3$) in an aqua ammonia solution of nickel nitrate and copper nitrate. Their catalysts were not selective in that they reported large yields of both CH$_4$ and CO. They interpreted their results as CO$_2$ being adsorbed on Cu and H$_2$ being adsorbed on Ni.

The polyol synthesis method has been used extensively in the past for the synthesis of metal particles from a metal salt precursor [Cu(NO$_3$)$_2$ or Ni(NO$_3$)$_2$ represented here as Me(NO$_3$)$_2$]. Bonet et al. [17, 18] indicated that the overall reactions at 180°C included the following reactions to form acetic acid, glycolaldehyde, and glycolic acid:

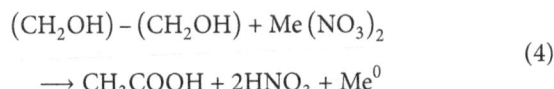

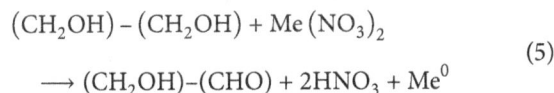

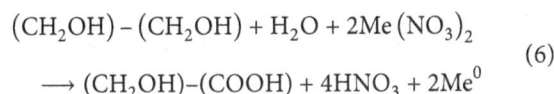

$$(CH_2OH) - (CH_2OH) + Me(NO_3)_2$$
$$\longrightarrow CH_3COOH + 2HNO_3 + Me^0 \tag{4}$$

$$(CH_2OH) - (CH_2OH) + Me(NO_3)_2$$
$$\longrightarrow (CH_2OH)-(CHO) + 2HNO_3 + Me^0 \tag{5}$$

$$(CH_2OH) - (CH_2OH) + H_2O + 2Me(NO_3)_2$$
$$\longrightarrow (CH_2OH)-(COOH) + 4HNO_3 + 2Me^0 \tag{6}$$

Bock et al. [19] found that oxalic acid, HOOC–COOH, is also formed and suggested that the majority of the metal is formed by the oxidation of ethylene glycol to glycolic acid. At the boiling point of ethylene glycol, 196°C, Poul et al. [17] indicated that the overall reactions included the formation of diacetyl:

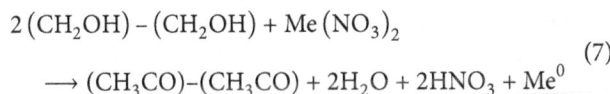

$$2(CH_2OH) - (CH_2OH) + Me(NO_3)_2$$
$$\longrightarrow (CH_3CO)-(CH_3CO) + 2H_2O + 2HNO_3 + Me^0 \tag{7}$$

where Me^0 represents either copper or nickel in the metallic state. Poul et al. [17] also commented on the reaction mechanism and indicated that intermediate solid phases (metal glycolates) precipitate before the metal powder is formed. Bonet et al. [18] indicated that oxidation products containing carboxylic acids act as stabilizers for metal colloid particles. Specifically they indicated that glycolate anions, the deprotonated form of glycolic acid, are good stabilizers for colloidal metal particles and that their concentration increases when the pH is greater than 6. In their work, they increased the pH by the addition of NaOH and reported

that PtRu bimetallic nanoparticle sizes decreased when the pH was increased. Their explanation of glycolate anions on the exterior of the metal particles preventing metal colloid agglomeration seems to be consistent with their nanoparticle size results. The polyol synthesis method has become known for its simplicity and accurate control of particle size [17, 18].

Other researchers [16, 18, 20] have also used the polyol synthesis method to obtain bimetallic particles. When alloying two metals together, the resulting reaction properties can often be enhanced compared to the pure metal. In the past both Cu and Ni have been alloyed with other metals to form alloys. For instance, Viau et al. [20] prepared Co-Ni and Fe-Ni particles using the polyol synthesis method.

Bonet et al. [18] used a polyol synthesis method to obtain CuNi particles. When nickel carbonate and copper carbonate were used at 140°C they obtained a CuNi powder composed of both a Ni rich CuNi solid solution and a Cu rich CuNi solid solution. When the carbonates were used at 196°C they obtained a CuNi powder composed of a Cu rich CuNi solid solution and a solid Ni metal phase. They noted that the reduction temperature for Cu is less than that for Ni. Their particles had a particle size of 140 nm. They did not report any reaction results.

In this work base metal CuNi nanoparticle catalysts were prepared by a new synthesis technique and used for the RWGS reaction. Cu was chosen because it is selective for the formation of CO [16], although it is unstable (sinters) at the higher temperatures where the equilibrium for the RWGS reaction is more favourable. Ni was chosen because it also produces CO [15] although it also can form unwanted by-products, CH$_4$, and coke.

The main objective of this investigation was to obtain physical characterization of the CuNi nanocatalyst and to use this catalyst in the RWGS reaction. The results obtained with the CuNi catalyst were then compared to Pt nanoparticle catalysts that were already synthesized and tested using a variety of reactions: ethylene oxidation [21], CO oxidation [22], and toluene oxidation [23]. These Pt catalysts are considered to be among the best catalysts for the RWGS reaction because they achieve reaction equilibrium at some conditions. Another purpose of the investigation was to determine if sintering of pure Cu could be prevented by the addition of Ni in the same way that it was prevented by the addition of Fe [14]. In what follows, we report CuNi catalyst compositions that promote CO formation and inhibit CH$_4$ formation at specific reaction conditions.

2. Experiment

2.1. Catalyst Preparation. CuNi nanoparticles were synthesised using a modified polyol technique. Nickel nitrate (Ni (NO$_3$)$_2$) (hexahydrate 99.999% metal basis, Alfa Aesar) was dissolved in ethylene glycol (anhydrous 99.8%, Sigma Aldrich). Next, an increase in pH to 11 was achieved by sodium hydroxide (NaOH) pellets (EM Science, ACS grade) to obtain the first solution. Separately, copper nitrate (Cu (NO$_3$)$_2$) (hexahydrate 99.999% metal basis, Alfa Aesar) was also dissolved in ethylene glycol. Its pH was also increased to

TABLE 1: Catalyst physical characteristics.

Catalyst	Metal loading (wt%)	XRD crystalline size (nm)	Typical particle size (nm)
Pt/C			2.8
Pt/γ-Al$_2$O$_3$	1	3.8	N/A
Cu$_{80}$Ni$_{20}$/C			64.4
Cu$_{80}$Ni$_{20}$/γ-Al$_2$O$_3$	10	30.2	N/A
Cu$_{50}$Ni$_{50}$/C			53.4
Cu$_{50}$Ni$_{50}$/γ-Al$_2$O$_3$	10	24	N/A
Cu$_{20}$Ni$_{80}$/C			41.1
Cu$_{20}$Ni$_{80}$/γ-Al$_2$O$_3$	10	16.7	N/A

11 using NaOH pellets to obtain a second solution. Solution 1 was then refluxed and stirred at 196°C. Once the temperature reached 196°C, the second solution, which was at room temperature, was added to the reflux. The mixture was left to reflux at 196°C for 30 minutes and then cooled. The colloidal particles were then stored in solution at room temperature.

A wet impregnation technique was used to deposit the colloidal particles on supports. The powdered support was first placed into a beaker. Then, a specific amount of colloidal particles was injected in the powder. A nominal amount of 10 wt% of CuNi was chosen. The mixture was sonicated for 1 hour and stirred for 24 hours. The catalyst was then centrifuged and washed with deionized water several times in order to remove any salts remaining from the synthesis procedure. The supports used were carbon black (Vulcan XC-72R, Cabot Corp., specific surface area of 254 m^2/g) and gamma-alumina (Alfa Aesar, specific surface area of 120 m^2/g). A freeze dryer was used to dry the catalyst. The catalyst was finely crushed prior to all experiments.

CuNi particles of three different compositions were prepared. The combined solution for each composition was prepared with a different ratio of Solution 1 to Solution 2. The ratio was selected to obtain CuNi colloidal particles of 80 wt% Cu/20 wt% Ni (nominally Cu$_{80}$Ni$_{20}$), 50 wt% Cu/50 wt% Ni (nominally Cu$_{50}$Ni$_{50}$), and 20 wt% Cu/80 wt% Ni (nominally Cu$_{20}$Ni$_{80}$).

Pt nanoparticles were synthesized using a different polyol synthesis method described here [22]. In short, PtCl$_4$ was dissolved in a 0.06 M NaOH solution of ethylene glycol and refluxed at 160°C for 3 hours. The nanoparticles were then deposited on C and γ-Al$_2$O$_3$ using the same deposition technique described previously. A nominal value of 1 wt% of Pt was chosen.

2.2. Physical Characterization. Supported Pt nanoparticles were analyzed using transmission electron microscopy (TEM) with a JEOL JEM 2100F FETEM operating at 200 kV. A particle size distribution was obtained using ImageJ software. More in-depth characterisation of the Pt nanoparticle is described elsewhere [20].

Scanning electron microscopy (SEM) was conducted on carbon supported CuNi nanoparticles using a JEOL model JSM-7500F field emission scanning electron microscope, FESEM, in both lower-secondary electron image, LEI, and compositional, COMPO, modes set at a distance of 8 mm

with an acceleration voltage of 5 kV. In addition, an energy-dispersive X-ray spectroscope (EDS) operating with the SEM was used to obtain a quantifiable amount of Cu and Ni in the CuNi particles.

CuNi colloidal particles were analyzed using X-ray diffraction (XRD) with a Rigaku Ultima IV diffractometer which used a Cu Kα X-ray (40 ma, 44 kV) operating with focused beam geometry and a divergence slit of 2/3 degree, a scan speed of 0.17 deg min^{-1}, and a scan step of 0.06 degrees were used while operating between 35° and 55°. Table 1 demonstrates the crystal sizes obtained through XRD.

2.3. Reaction Experiments. The performances of the supported CuNi catalysts were evaluated using the RWGS reaction. 50 mg of powdered catalyst was placed on a fritted quartz bed within a 35 mL quartz tube to act as a fixed bed reactor. A gas mixture of 1 kPa H$_2$ (Grade 4.0, Linde), 1 kPa CO$_2$ (Grade 3.0, Linde), and the balance He (Grade 4.7 Linde) flowed through the reactor at a total flow rate of 510 mL/min. The reaction was performed at atmospheric pressure using three consecutive temperature cycles. Each temperature cycle consisted of a series of experiments over the temperature range from 400°C to 700°C. Before each experiment, the temperature was held constant for 30 min. Although the same mass of catalyst was used in each experiment, the gas hourly space velocity (GHSV) was different because the supports had different bulk densities (288 g/L for the carbon support and 461 g/L for the alumina support). The GHSV values were 176000 h^{-1} and 282000 h^{-1}, respectively, for CuNi/C and CuNi/Al$_2$O$_3$ catalysts. The effluent was dehumidified by flowing through an adsorbent and was analyzed by flowing through a mass spectrometer (Ametek Proline DM 100) and a nondispersive infrared CO gas analyzer (Horiba VIA-510). Each set of experiments was repeated three times (24 hrs total) in order to examine reproducibility and stability. The yield of CO was calculated using the following formula:

$$\text{Yield of CO (\%)} = \frac{[\text{CO}]_{\text{OUT}}}{[\text{CO}_2]_{\text{IN}}} \times 100\%. \tag{8}$$

The mass spectrometer identified any by-products that were formed via side reactions such as CO methanation. The mass spectrometer indicated the presence of gases with a molecular weight of up to 50 atomic units and had a detection limit of 50 ppm.

3. Results and Discussion

The X-ray diffraction spectra of the CuNi nanoparticles are shown in Figure 1. The positions of the peaks for both pure Cu ($2\theta = 43.2$, ICSD Collection Code 53246) and pure Ni ($2\theta = 44.6$, ICSD Collection Code 43397) are shown as straight vertical lines in Figure 1. They are the X-ray reflections from the 111 crystal lattice planes. These peaks have corresponding lattice constants of 3.627 Å and 3.519 Å for pure Cu and pure Ni, respectively. The smaller peaks near 2θ values of 50.4 and 51.5 are the reflections from the 200 crystal lattice planes of Cu and Ni, respectively. No other species or oxides were identified.

There is a slight difference between the peak positions for the pure metals and the metals in the catalysts. Deviations exist because the catalysts are bimetallic solid solutions rather than pure metals. Because of these shifts, the lattice constants of each alloy are shifted. The catalyst that is nominally $Cu_{50}Ni_{50}$ in Figure 1(b) had a first 2θ peak position of 43.385 and a lattice constant of 3.612 Å which is slightly greater than the one for pure Cu. A second peak is observed near the first 2θ peak in Figure 1(b) which has a peak position of 44.346 and a lattice constant of 3.537 Å for Ni, which is slightly less than the one for pure Ni. These changes in lattice constants suggest that the two peaks represent a Cu rich alloy and a Ni rich alloy.

There is only one peak for the $Cu_{80}Ni_{20}$ catalyst in Figure 1(c). That means that all of the Ni was soluble in the Cu lattice. The single peak in Figure 1(c) is experimental evidence for a copper-rich CuNi solid solution in which the spacing between planes of the catalyst lattice is close to that of pure copper.

There are two peaks for the $Cu_{20}Ni_{80}$ catalyst in Figure 1(a). Because the first peak has a 2θ value at 43.394 and a lattice constant of 3.611 Å, close to that for pure Cu, the spacing between its planes will be similar to pure copper. Because the second peak has a 2θ value at 44.412 and a lattice constant of 3.533 Å, close to that for pure Ni, the spacing between its planes will be similar to pure nickel. Even though there is a large disparity between the bulk Cu content and the bulk Ni content of the catalyst, the two peaks appear to have similar areas. Therefore a substantial amount of Ni must be dissolved in the first Cu-like peak, and the first peak must represent a CuNi solid solution. Since the second peak has a lattice constant slightly different from pure Ni it will contain some Cu making it a Ni rich NiCu solid solution.

The observation of solid solutions is consistent with other works reported in the literature. Bonet et al. [18] synthesized CuNi particles using a similar technique. In their work they refluxed copper and nickel carbonates starting materials in ethylene glycol. After 39 hours at 140°C they observed the presence of a copper-rich solid solution, $Cu_{81}Ni_{19}$, and a Ni rich solid solution, $Ni_{86}Cu_{14}$.

The crystalline size of the synthesized nanoparticles increases with Cu content. A summary of their diameters (15–65 nm) can be found in Table 1. They were calculated from the XRD data using Scherrer's formula. The CuNi particles described by Poul et al. [17] had diameters of 250–400 nm. It is possible that the longer refluxing times and the

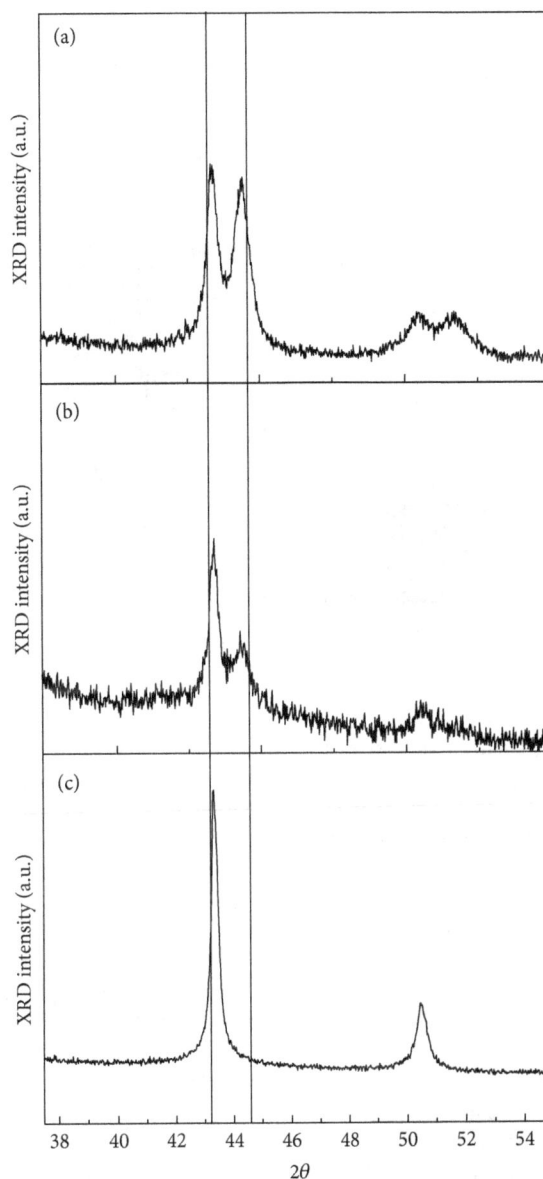

FIGURE 1: XRD spectra of colloidal: (a) $Cu_{20}Ni_{80}$, (b) $Cu_{50}Ni_{50}$, and (c) $Cu_{80}Ni_{20}$.

absence of NaOH may have provided more opportunity for agglomeration.

Pt/C nanoparticles characterisation can be found in [21]. In her work, Isaifan et al. demonstrate that the particles are mainly spherical with a reasonably narrow size distribution. The dispersion also appears to be relatively high. Numerical data derived from several TEM images indicated that a typical particle size for the Pt/C particles was 2.8 nm (Table 1).

An SEM image for the $Cu_{50}Ni_{50}$/C catalyst is shown in Figure 2. The particles appear to be generally spherical and to vary in size. The largest observed particle was approximately 100 nm. There are visible signs of agglomeration which is to be expected since no antiagglomerate such as polyvinylpyrrolidone (PVP) was used.

TABLE 2: Cu-Ni surface ratios obtained from XPS measurements.

	Theoretical bulk Cu/Ni ratio from nominal composition	Actual Surface Cu/Ni ratio from XPS	Percentage Increase
$Cu_{20}Ni_{80}$	0.25	1.2	380%
$Cu_{50}Ni_{50}$	1	2	100%
$Cu_{80}Ni_{20}$	4	4.2	5%

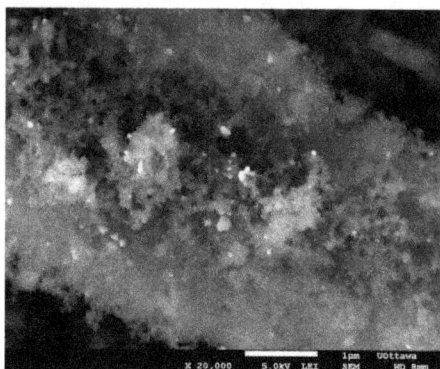

FIGURE 2: SEM image of a $Cu_{50}Ni_{50}/C$ catalyst.

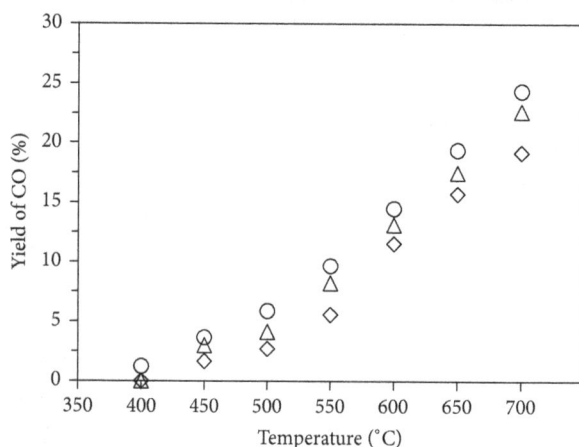

FIGURE 3: RWGS reaction at 1 atm, $P_{H2} = P_{CO2} = 1$ kPa, balance He, GHSV = 176000 h^{-1}, and 50 mg of catalyst: $Cu_{80}Ni_{20}/C$, 10 wt%, where $\bigcirc$ = 1st cycle, $\triangle$ = 2nd cycle, and $\diamond$ = 3rd cycle.

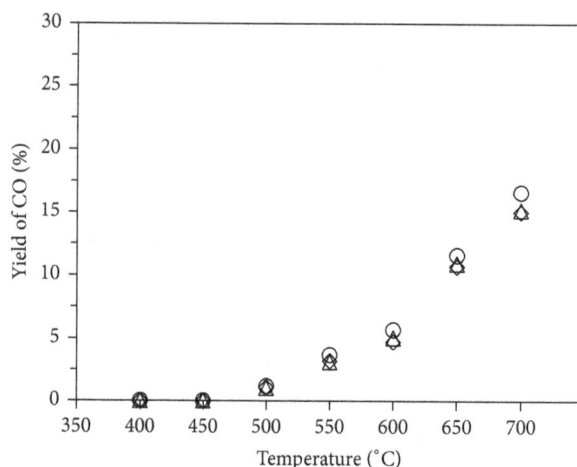

FIGURE 4: RWGS reaction at 1 atm, $P_{H2} = P_{CO2} = 1$ kPa, balance He, GHSV = 176000 h^{-1}, and 50 mg of catalyst: $Cu_{50}Ni_{50}/C$, 10 wt%, where $\bigcirc$ = 1st cycle, $\triangle$ = 2nd cycle, and $\diamond$ = 3rd cycle.

Typical particle sizes for all of the CuNi catalysts measured by SEM are listed in Table 1. In general the CuNi particle sizes are an order of magnitude larger than the Pt particle size. Since the CuNi metal loading, for example, 10 wt%, is an order of magnitude larger than the Pt loading, for example, 1 wt%, a greater extent of metal particle agglomeration might be expected for the CuNi particles.

An energy-dispersive X-ray spectroscopy, EDS, analysis was also performed on the catalyst. For example, the $Cu_{80}Ni_{20}/C$ catalyst had a measured composition of 82.7 wt% Cu and 17.3 wt% Ni. The EDS measurement was repeated at two different sites on the catalyst's surface with reproducible results. That EDS result is essentially the same as the XRD composition of $Cu_{82}Ni_{18}$ that was mentioned above. Both the EDS results and the XRD results were consistent with the $Cu_{80}Ni_{20}$ nominal composition of the synthesized particles. This indicates that the synthesis method was successful in obtaining the nominal Cu : Ni ratio that was intended.

XPS measurements were performed on the CuNi/C catalysts to determine their surface compositions. The results in Table 2 show that the Cu surface concentration was greater than that of the Cu bulk concentrations for all three of the CuNi/C catalysts. Furthermore, the surface concentration of the Cu always exceeded the surface concentration of the Ni, even for the nominal $Cu_{20}Ni_{80}/C$ catalyst. That result is consistent with the literature. An early report by van der Plank and Sachtler [24] indicated that Cu was the dominant species on the surface of CuNi alloys. Subsequently Watanabe et al. [25] provided definitive experimental data for the phenomenon. Later Sakurai et al. stated that the phenomenon had been conclusively shown [26].

In order to examine the physical changes of the catalyst, $Cu_{50}Ni_{50}/C$ was examined by SEM and XPS before and after exposure to high temperatures and reactants. There were no visible signs of additional agglomeration or any other physical

changes to the metal in comparison to the unreacted catalyst. This suggests that the catalyst is compositionally stable at temperatures of at least 700°C.

The reverse water gas shift reaction was performed using eight different catalysts. These results can be seen in Figures 3–10. In each figure the results for the 3 consecutive temperature cycles, over the temperature range from 400°C to 700°C, are shown. Some of the CuNi catalysts showed slight deactivation between the first and second cycles and also between the second and third cycles.

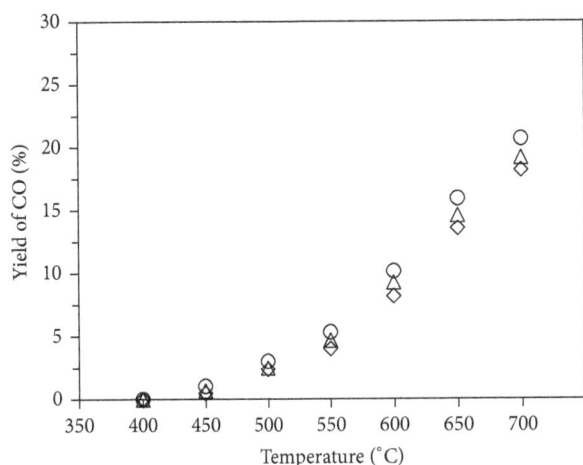

FIGURE 5: RWGS reaction at 1 atm, $P_{H2} = P_{CO2} = 1$ kPa, balance He, GHSV = 176000 h^{-1}, and 50 mg of catalyst: $Cu_{20}Ni_{80}$/C, 10 wt%, where ◯ = 1st cycle, △ = 2nd cycle, and ◇ = 3rd cycle.

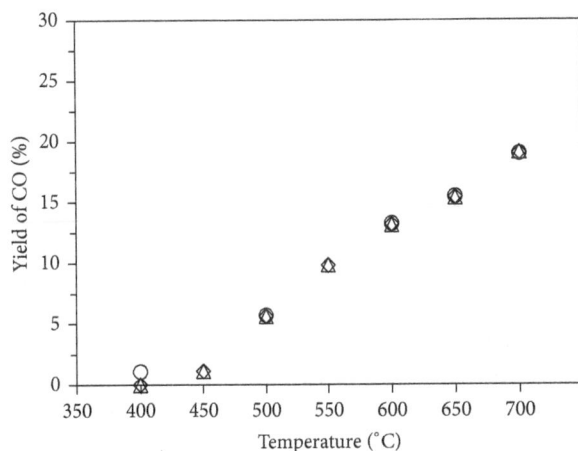

FIGURE 7: RWGS reaction at 1 atm, $P_{H2} = P_{CO2} = 1$ kPa, balance He, GHSV = 282000 h^{-1}, and 50 mg of catalyst: $Cu_{80}Ni_{20}$/γ-Al$_2$O$_3$, 10 wt%, where ◯ = 1st cycle, △ = 2nd cycle, and ◇ = 3rd cycle.

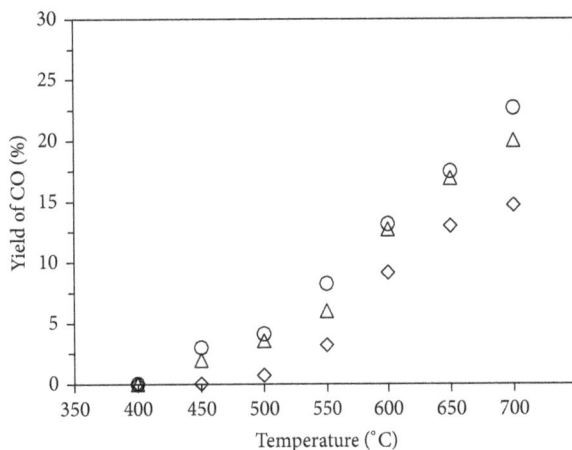

FIGURE 6: RWGS reaction at 1 atm, $P_{H2} = P_{CO2} = 1$ kPa, balance He, GHSV = 176000 h^{-1}, and 50 mg of catalyst: Pt/C, 1 wt%, where ◯ = 1st cycle, △ = 2nd cycle, and ◇ = 3rd cycle.

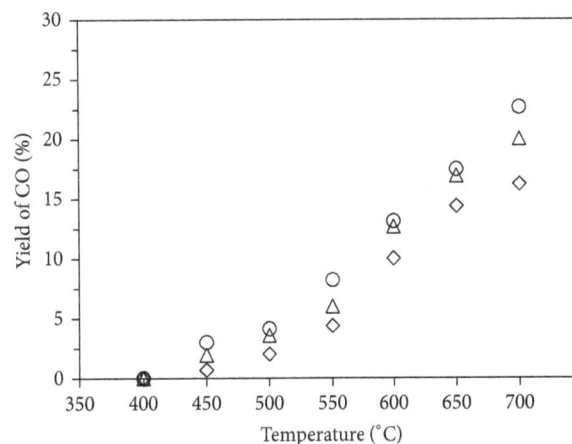

FIGURE 8: RWGS reaction at 1 atm, $P_{H2} = P_{CO2} = 1$ kPa, balance He, GHSV = 282000 h^{-1}, and 50 mg of catalyst: $Cu_{50}Ni_{50}$/γ-Al$_2$O$_3$, 10 wt%, where ◯ = 1st cycle, △ = 2nd cycle, and ◇ = 3rd cycle.

The catalysts supported on gamma-alumina, γ-Al$_2$O$_3$, are shown in Figures 7–10. In general the CO yields on the γ-Al$_2$O$_3$ supported catalyst were slightly greater than those on the carbon supported catalysts in Figures 3–6. For the first temperature cycle the CuNi metal γ-Al$_2$O$_3$ supported catalysts produced CO yields at 700°C that was 2–4% less than those produced by the Pt metal γ-Al$_2$O$_3$ supported catalyst. In addition, only the Pt metal γ-Al$_2$O$_3$ supported catalyst produced a nonzero CO yield at 400°C during the first cycle.

The only observable components in the gas stream entering the mass spectrometer were CO$_2$, H$_2$, CO, trace amounts of H$_2$O, and the carrier gas, He. These results indicate that CO was the main product having a typical concentration of 2000 ppm. Other products including CH$_4$ had concentrations less than the detection limit of the spectrometer, 50 ppm.

The absence of CH$_4$ in the products was a highly desirable result, since CH$_4$ is an undesirable by-product if syngas for a Fischer Tropsch process is the goal. Cu is known to favour

CO production while CH$_4$ is known to form on Ni catalysts [1]. Since some of the catalysts used in this work contained 80 wt% Ni the absence of CH$_4$ might be considered to be inconsistent with the literature [1]. The advantage of the CuNi alloys made using this particular polyol synthesis method is that more than one-half of the surface was composed of Cu, even for catalysts consisting of 80% Ni in bulk metal, as was shown by our XPS results. Perhaps the presence of sufficient Cu on the surface may allow CO to desorb before additional hydrogenation occurs to form CH$_4$.

The catalysts supported on carbon are shown in Figures 3–6. $Cu_{80}Ni_{20}$/C showed the highest yield among all catalysts during the first cycle. In addition, it was the only CuNi metal carbon supported catalyst that produced a CO yield at 400°C during the first cycle. It is well known that Cu alone has better performance for the RWGS reaction than Ni [1, 16, 27] since Ni tends to further hydrogenate CO to CH$_4$. The increased Cu content on the surface of the $Cu_{80}Ni_{20}$

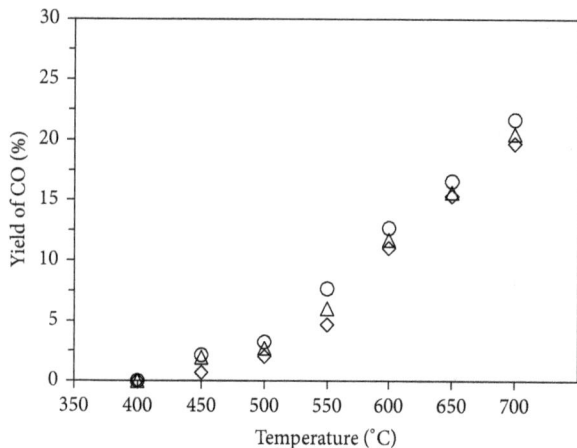

FIGURE 9: RWGS reaction at 1 atm, $P_{H2} = P_{CO2} = 1$ kPa, balance He, GHSV = 282000 h^{-1}, and 50 mg of catalyst: $Cu_{20}Ni_{80}/\gamma\text{-}Al_2O_3$, 10 wt%, where $\bigcirc$ = 1st cycle, $\triangle$ = 2nd cycle, and $\diamondsuit$ = 3rd cycle.

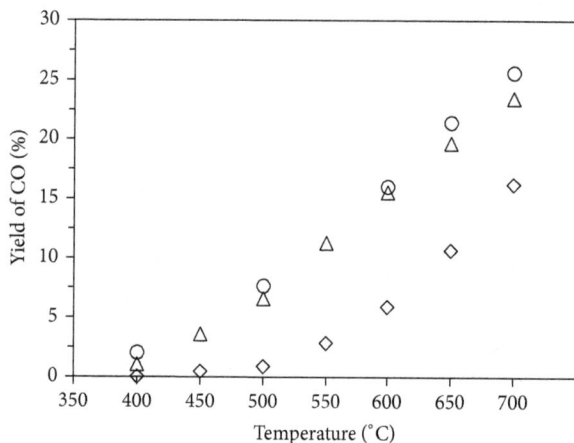

FIGURE 10: RWGS reaction at 1 atm, $P_{H2} = P_{CO2} = 1$ kPa, balance He, GHSV = 282000 h^{-1}, and 50 mg of catalyst: $Pt/\gamma\text{-}Al_2O_3$, 1 wt%, where $\bigcirc$ = 1st cycle, $\triangle$ = 2nd cycle, and $\diamondsuit$ = 3rd cycle.

catalyst seems to cause the increase in catalytic performance observed here. However, because of copper's instability at high temperatures [28], a reduction in performance was expected and is observed over time as shown in Figure 3. The yields of CO at 700°C using all the CuNi carbon supported catalysts differed from those using Pt by no more than 3%.

Deactivation was observed between the first and second temperature cycles for all of the CuNi catalysts supported on carbon. Virtually no deactivation was observed when using the Pt metal carbon supported catalyst. This suggests that the deactivation observed with the CuNi carbon supported catalyst may have been related to the CuNi metal and not to the catalyst support. The $Cu_{50}Ni_{50}/C$ catalyst in Figure 4 was different from the $Cu_{20}Ni_{80}/C$ and $Cu_{80}Ni_{20}/C$ catalysts in that no deactivation occurred between the second and third temperature cycles. This suggests that after sufficient time-on-stream the performance of the $Cu_{50}Ni_{50}/C$ may become

invariant with time and that it may become a thermally stable catalyst.

Furthermore, XPS experiments of the $Cu_{50}Ni_{50}/C$ showed no differences in carbon surface composition for both of the Cu and Ni atoms. These experiments suggest that coking did not occur. SEM images also showed no change in particle size either before or after testing. The sample seems to be compositionally stable throughout the experiments.

Deactivation was observed between the first and third temperature cycles for all the $\gamma\text{-}Al_2O_3$ supported catalysts, for both CuNi and Pt. The smallest extent of deactivation, 2%, was observed with the $Cu_{20}Ni_{80}\gamma\text{-}Al_2O_3$ supported catalyst. It is the CuNi catalyst with the smallest Cu content. In contrast the CO yields at 700°C for $Cu_{80}Ni_{20}$ and Pt $\gamma\text{-}Al_2O_3$ supported catalysts decreased by over 8% between the first and third temperature cycles. Conversely, no deactivation was observed for Pt metal carbon supported catalysts. This suggests that the $\gamma\text{-}Al_2O_3$ support may contribute to the deactivation. It is known [29] that as the temperature is increased above 500°C, gamma-alumina can be converted to other phases such as delta alumina, theta alumina, and alpha alumina. Alpha alumina has a much smaller surface area than gamma-alumina. The tendency of the $\gamma\text{-}Al_2O_3$ support to deactivate at high temperatures is consistent with literature data [14, 30].

Y. Liu and D. Liu [16] tested a similar catalyst having $C_{50}Ni_{50}/\gamma\text{-}Al_2O_3$ at a maximum temperature of 600°C. In their research, Y. Liu and D. Liu use a typical coimpregnation method where $\gamma\text{-}Al_2O_3$ is immersed in a solution of nickel nitrate and copper nitrate. They used twice the metal content (20 wt%) and much smaller GHSVs (much greater residence times in the reactor) and obtained CO_2 conversions that exceeded the ones being reported here. They used a catalyst preparation method that was different compared to the one used here. In fact, it did not seem to be possible to make a CuNi alloy using this technique when it was attempted in our laboratories. Perhaps that may explain why the CH_4 selectivity (e.g., 28.2%) in their work was much greater than in this work. Even though their CO_2 conversions were greater than the ones reported here, their finite selectivity to CH_4 caused their CO yields to be similar to the ones reported here. Y. Liu and D. Liu [16] noticed an increase in CH_4 when the Ni content of the CuNi catalysts was increased.

A comparison of the average CO yields for the carbon supported catalysts is shown in Figure 11. The CO yields obtained with $Cu_{50}Ni_{50}/C$ catalyst are smaller than those obtained with the other two CuNi catalysts. The $Cu_{50}Ni_{50}/C$ catalyst was the one that appeared to be compositionally stable according to the XPS results discussed previously. It was also the one that appeared to be the most thermally stable in Figure 4. Perhaps there is an association between minimum CO yield and thermal stability. In other words, perhaps the reaction sites on CuNi catalysts with the largest turnover frequencies are the ones that are the most thermally unstable. Thermal stability was further investigated and is discussed in another work [27]. Here, testing using $Cu_{50}Ni_{50}$ deposited on samarium-doped ceria showed no sign of deactivation when the catalyst is exposed to temperatures of 600–700°C for 48 consecutive hours. The high CO yields

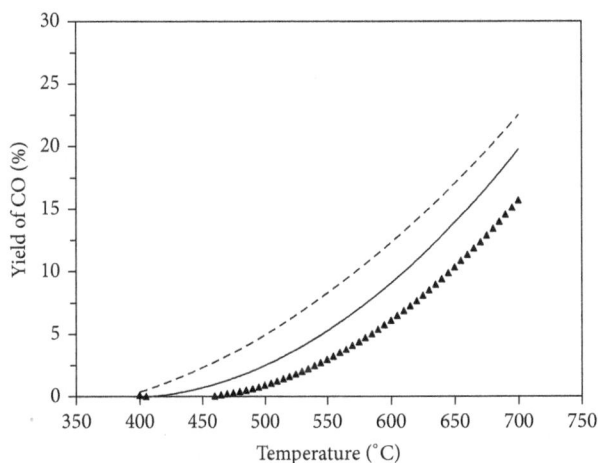

FIGURE 11: Average yield of CO for the reverse water gas shift reaction at 1 atm, $P_{H2} = P_{CO2} = 1$ kPa, balance He, GHSV $= 176000 \, h^{-1}$, and 50 mg of Cu_xNi_{1-x}/C catalyst: The solid line represents $Cu_{20}Ni_{80}/C$ catalysts, the triangles represent $Cu_{50}Ni_{50}/C$ catalysts, and the dashed line represents $Cu_{80}Ni_{20}/C$ catalysts.

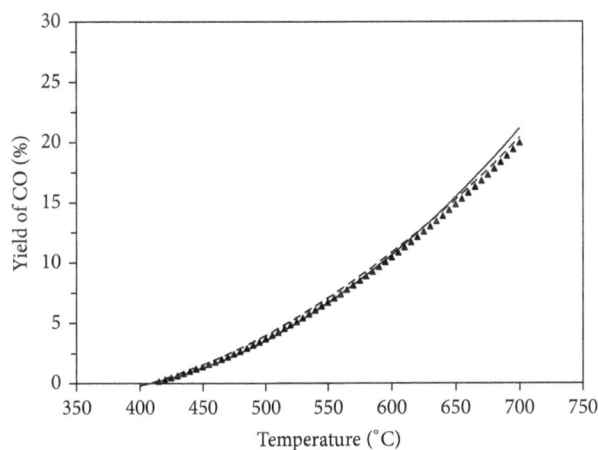

FIGURE 12: Average yield of CO for the reverse water gas shift reaction at 1 atm, $P_{H2} = P_{CO2} = 1$ kPa, balance He, GHSV $= 282000 \, h^{-1}$, and 50 mg of Cu_xNi_{1-x}/γ-Al_2O_3 catalyst: The solid line represents $Cu_{20}Ni_{80}/\gamma$-Al_2O_3 catalysts, the triangles represent $Cu_{50}Ni_{50}/\gamma$-Al_2O_3 catalysts, and the dashed line represents $Cu_{80}Ni_{20}/\gamma$-Al_2O_3 catalysts.

obtained using $Cu_{80}Ni_{20}/C$ are believed to be associated with a high Cu concentration on the surface of the catalyst as shown in the XPS results. As mentioned previously, Cu is known to obtain higher CO yields than Ni for the RWGS reaction [16].

A comparison of the average CO yields for the alumina supported catalysts shown in Figure 12 is almost identical regardless of catalyst composition. Nevertheless substantial catalyst deactivation was evident in Figures 7–10. That suggests that the deactivation caused by thermal transitions in alumina (loss of surface area, spinel formation) may have had more influence on catalyst performance than variations in CuNi catalyst composition. In spite of the similarity of the

results in Figure 12, the CO yields for the $Cu_{50}Ni_{50}/\gamma$-Al_2O_3 catalyst appear to be slightly less than the other two catalyst compositions. That is consistent with the observation in Figure 11 that the $Cu_{50}Ni_{50}/C$ catalyst also produced smaller yields than the other two CuNi catalysts.

As discussed above, some deactivation occurs when CuNi metal is supported on either carbon or γ-Al_2O_3. As a result, most catalysts composed of CuNi metal supported on either carbon or γ-Al_2O_3 will not satisfy one of the main objectives of this research, namely, obtaining a thermally stable catalyst capable of operating under high temperatures. The one exception discussed here is the $Cu_{50}Ni_{50}/C$ catalyst. Unless deactivation can be mitigated CuNi catalysts will not meet the requirements for an efficient, industrially viable catalyst.

4. Conclusion

The results of this investigation can be summarized by the following statements: The new polyol synthesis method permits both Cu and Ni metal salts to be reduced at the same time and at the same temperature, by first heating the Ni salt solution to 196°C and then adding the Cu salt solution that was at room temperature. In the past [18] a Ni rich surface was obtained because $Cu(NO_3)_2$ was reduced first at temperatures as low as 140°C followed by $Ni(NO_3)_2$ reduction as the solution continued to be heated to 196°C. Instead, a Cu rich surface is obtained which is ideal for the RWGS reaction because Cu has a higher selectivity towards CO than Ni. The CuNi alloy catalysts investigated in this work are similar to pure Cu catalysts in that they show selectivity for CO formation and the absence of CH_4 formation. The selectivity to CO was attributed to Cu being the most abundant metallic species on the surface of the catalyst, as determined by XPS measurements. Although some of the CuNi alloys show some deactivation, they are not nearly as thermally unstable as pure Cu (sintering) at the higher temperatures that are necessary for the equilibrium of the RWGS reaction to be thermodynamically favorable. Deactivation was observed in each case that an alumina catalyst support was used which was attributed to the instability of alumina at high temperatures. With one carbon supported catalyst, $Cu_{50}Ni_{50}/C$, there was no deactivation between the second and third temperature cycles and it appeared to be compositionally stable according to XPS and SEM results. Finally, CO yields at 700°C during the third temperature cycle of each CuNi catalyst were comparable to those with the Pt/C catalyst.

In conclusion, considering the difference in cost between CuNi alloys and Pt metal, these results suggest that more studies are warranted on the use of CuNi alloy catalysts for the RWGS reaction using the synthesis method described in this work.

Conflict of Interests

The authors declare that there is no conflict of interests regarding the publication of this paper.

Acknowledgments

The Natural Science and Engineering Research Council (NSERC) and Phoenix Canada Oil Company Limited are gratefully acknowledged for their financial support. The scientific contributions provided by Dr. Marten Ternan are also acknowledged.

References

[1] W. Wang, S. Wang, X. Ma, and J. Gong, "Recent advances in catalytic hydrogenation of carbon dioxide," *Chemical Society Reviews*, vol. 40, no. 7, pp. 3703–3727, 2011.

[2] P. Vibhatavata, J.-M. Borgard, M. Tabarant, D. Bianchi, and C. Mansilla, "Chemical recycling of carbon dioxide emissions from a cement plant into dimethyl ether, a case study of an integrated process in France using a Reverse Water Gas Shift (RWGS) step," *International Journal of Hydrogen Energy*, vol. 38, no. 15, pp. 6397–6405, 2013.

[3] S. S. Kim, K. H. Park, and S. C. Hong, "A study of the selectivity of the reverse water-gas-shift reaction over Pt/TiO$_2$ catalysts," *Fuel Processing Technology*, vol. 108, pp. 47–54, 2013.

[4] S. S. Kim, H. H. Lee, and S. C. Hong, "A study on the effect of support's reducibility on the reverse water-gas shift reaction over Pt catalysts," *Applied Catalysis A: General*, vol. 423-424, pp. 100–107, 2012.

[5] S. S. Kim, H. H. Lee, and S. C. Hong, "The effect of the morphological characteristics of TiO$_2$ supports on the reverse water-gas shift reaction over Pt/TiO$_2$ catalysts," *Applied Catalysis B: Environmental*, vol. 119-120, pp. 100–108, 2012.

[6] G. Pekridis, K. Kalimeri, N. Kaklidis et al., "Study of the reverse water gas shift (RWGS) reaction over Pt in a solid oxide fuel cell (SOFC) operating under open and closed-circuit conditions," *Catalysis Today*, vol. 127, no. 1–4, pp. 337–346, 2007.

[7] C. S. Chen, J. H. Wu, and T. W. Lai, "Carbon dioxide hydrogenation on Cu nanoparticles," *The Journal of Physical Chemistry C*, vol. 114, no. 35, pp. 15021–15028, 2010.

[8] C.-S. Chen, W.-H. Cheng, and S.-S. Lin, "Mechanism of CO formation in reverse water-gas shift reaction over Cu/Al$_2$O$_3$ catalyst," *Catalysis Letters*, vol. 68, no. 1-2, pp. 45–48, 2000.

[9] L. Wang, S. Zhang, and Y. Liu, "Reverse water gas shift reaction over Co-precipitated Ni-CeO$_2$ catalysts," *Journal of Rare Earths*, vol. 26, no. 1, pp. 66–70, 2008.

[10] F. S. Stone and D. Waller, "Cu-ZnO and Cu-ZnO/Al$_2$O$_3$ catalysts for the reverse water-gas shift reaction. The effect of the Cu/Zn ratio on precursor characteristics and on the activity of the derived catalysts," *Topics in Catalysis*, vol. 22, no. 3-4, pp. 305–318, 2003.

[11] J. Lin, *Supported Copper, Nickel and Copper-Nickel nanoparticle Catalyst for Low Temperature WGS Reaction*, University of Cincinnati, 2012.

[12] C. Chen, C. Ruan, Y. Zhan, X. Lin, Q. Zheng, and K. Wei, "The significant role of oxygen vacancy in Cu/ZrO$_2$ catalyst for enhancing water-gas-shift performance," *International Journal of Hydrogen Energy*, vol. 39, no. 1, pp. 317–324, 2014.

[13] C. S. Chen, W. H. Cheng, and S. S. Lin, "Study of reverse water gas shift reaction by TPD, TPR and CO$_2$ hydrogenation over potassium-promoted Cu/SiO$_2$ catalyst," *Applied Catalysis A: General*, vol. 238, no. 1, pp. 55–67, 2002.

[14] C.-S. Chen, W.-H. Cheng, and S.-S. Lin, "Study of iron-promoted Cu/SiO$_2$ catalyst on high temperature reverse water gas shift reaction," *Applied Catalysis A: General*, vol. 257, no. 1, pp. 97–106, 2004.

[15] C. S. Chen, J. H. Lin, J. H. You, and K. H. Yang, "Effects of potassium on Ni-K/Al$_2$O$_3$ catalysts in the synthesis of carbon nanofibers by catalytic hydrogenation of CO$_2$," *Journal of Physical Chemistry A*, vol. 114, no. 11, pp. 3773–3781, 2010.

[16] Y. Liu and D. Liu, "Study of bimetallic Cu-Ni/γ-Al$_2$O$_3$ catalysts for carbon dioxide hydrogenation," *International Journal of Hydrogen Energy*, vol. 24, no. 4, pp. 351–354, 1999.

[17] L. Poul, N. Jouini, and F. Fiévet, "Layered hydroxide metal acetates (metal = zinc, cobalt, and nickel): elaboration via hydrolysis in polyol medium and comparative study," *Chemistry of Materials*, vol. 12, no. 10, pp. 3123–3132, 2000.

[18] F. Bonet, S. Grugeon, L. Dupont, R. Herrera Urbina, C. Guéry, and J. M. Tarascon, "Synthesis and characterization of bimetallic Ni-Cu particles," *Journal of Solid State Chemistry*, vol. 172, no. 1, pp. 111–115, 2003.

[19] C. Bock, C. Paquet, M. Couillard, G. A. Botton, and B. R. MacDougall, "Size-selected synthesis of PtRu nano-catalysts: reaction and size control mechanism," *Journal of the American Chemical Society*, vol. 126, no. 25, pp. 8028–8037, 2004.

[20] G. Viau, F. Fiévet, and F. Fiévet-Vincent, "Nucleation and growth of bimetallic CoNi and FeNi monodisperse particles prepared in polyols," *Solid State Ionics*, vol. 84, no. 3-4, pp. 259–270, 1996.

[21] R. J. Isaifan, S. Ntais, and E. A. Baranova, "Particle size effect on catalytic activity of carbon-supported Pt nanoparticles for complete ethylene oxidation," *Applied Catalysis A: General*, vol. 464-465, pp. 87–94, 2013.

[22] R. J. Isaifan, H. A. E. Dole, E. Obeid, L. Lizarraga, E. A. Baranova, and P. Vernoux, "Catalytic CO oxidation over Pt nanoparticles prepared from the polyol reduction method supported on Yttria-Stabilized Zirconia," *ECS Transactions*, vol. 35, no. 28, pp. 43–57, 2011.

[23] H. A. E. Dole, R. J. Isaifan, F. M. Sapountzi et al., "Low temperature toluene oxidation over Pt nanoparticles supported on yttria stabilized-zirconia," *Catalysis Letters*, vol. 143, no. 10, pp. 996–1002, 2013.

[24] P. van der Plank and W. M. H. Sachtler, "Surface composition of equilibrated copper-nickel alloy films," *Journal of Catalysis*, vol. 7, no. 3, pp. 300–303, 1967.

[25] K. Watanabe, M. Hashiba, and T. Yamashina, "A quantitative analysis of surface segregation and in-depth profile of copper-nickel alloys," *Surface Science*, vol. 61, no. 2, pp. 483–490, 1976.

[26] T. Sakurai, T. Hashizume, A. Jimbo, A. Sakai, and S. Hyodo, "New result in surface segregation of Ni-Cu binary alloys," *Physical Review Letters*, vol. 55, no. 5, pp. 514–517, 1985.

[27] M. Lortie, *Reverse water gas shift reaction over supported Cu-Ni nanoparticle catalysts, chapter 3 [M.S. dissertation]*, University of Ottawa, 2014.

[28] M. V. Twigg and M. S. Spencer, "Deactivation of supported copper metal catalysts for hydrogenation reactions," *Applied Catalysis A: General*, vol. 212, pp. 161–174, 2001.

[29] R. K. Oberlander, "Aluminas for catalysts: their preparation and properties," in *Applied Industrial Catalysis*, vol. 3, p. 67, Academic Press, Orlando, Fla, USA, 1984.

[30] J. Hu, K. P. Brooks, J. D. Holladay, D. T. Howe, and T. M. Simon, "Catalyst development for microchannel reactors for martian in situ propellant production," *Catalysis Today*, vol. 125, no. 1-2, pp. 103–110, 2007.

Product Characterization and Kinetics of Biomass Pyrolysis in a Three-Zone Free-Fall Reactor

Natthaya Punsuwan and Chaiyot Tangsathitkulchai

School of Chemical Engineering, Institute of Engineering, Suranaree University of Technology, Nakhon Ratchasima 30000, Thailand

Correspondence should be addressed to Chaiyot Tangsathitkulchai; chaiyot@sut.ac.th

Academic Editor: Deepak Kunzru

Pyrolysis of biomass including palm shell, palm kernel, and cassava pulp residue was studied in a laboratory free-fall reactor with three separated hot zones. The effects of pyrolysis temperature (250–1050°C) and particle size (0.18–1.55 mm) on the distribution and properties of pyrolysis products were investigated. A higher pyrolysis temperature and smaller particle size increased the gas yield but decreased the char yield. Cassava pulp residue gave more volatiles and less char than those of palm kernel and palm shell. The derived solid product (char) gave a high calorific value of 29.87 MJ/kg and a reasonably high BET surface area of 200 m^2/g. The biooil from palm shell is less attractive to use as a direct fuel, due to its high water contents, low calorific value, and high acidity. On gas composition, carbon monoxide was the dominant component in the gas product. A pyrolysis model for biomass pyrolysis in the free-fall reactor was developed, based on solving the proposed two-parallel reactions kinetic model and equations of particle motion, which gave excellent prediction of char yields for all biomass precursors under all pyrolysis conditions studied.

1. Introduction

Pyrolysis is a viable thermal process for efficient and economical conversion of biomass into alternative energy in forms of solid char, liquid biooil, and combustible gases [1]. Based on the pyrolysis time and heating rate, the pyrolysis process can be classified into conventional and fast or flash pyrolysis. Conventional pyrolysis may also be termed "slow pyrolysis." This type of pyrolysis is defined as the one which occurs under a slow heating rate (less than 10°C/s), slow heat transfer rate in the reaction zone, and relatively long mean residence time [2]. Normally, conventional pyrolysis has been used mainly for charcoal production. Conversely, fast or flash pyrolysis is a thermal decomposition process that occurs at a high heating rate and short mean residence time. Heating rate of flash pyrolysis is around 100°C/s, or even 10,000°C/s, and the mean residence time is normally less than 2 seconds [2]. Flash pyrolysis process generally produces 45–75 wt% of liquid, 15–25 wt% of solid, and 10–20 wt% of noncondensable gases, depending on the feedstock used and pyrolysis conditions.

Generally, solid, liquid, and gas produced from pyrolysis are of prime interest for use as a primary fuel. However, a number of researchers [3, 4] have directed attention towards investigating the effect of pyrolysis conditions on maximizing the yield of pyrolytic liquids which show promise as a substitute for petroleum based fuel. Since the liquid product or biooil from the pyrolysis process is composed of a large number of oxygenated compounds, its fuel properties do not conform to the standards of petroleum based fuels. It is therefore necessary to upgrade the crude oil along with some kinds of pretreatment before it can be fully utilized in any combustion systems [5]. Due to the numerous utility of pyrolysis products and in particular the potential application of biooil, the present work is thus concentrated on the flash pyrolysis of biomass which is capable of producing a high proportion of liquid product.

In this work, a laboratory free-fall reactor with a central heated zone is proposed for the study of biomass flash pyrolysis. A free-fall reactor has been widely used in laboratory studies on flash pyrolysis due to the following advantages: it provides high heating rate, determinations of mass balance and mean residence time are simple and straightforward [6], the mean residence time can be moderately controlled [6], and the kinetic parameters can be conveniently examined from the pyrolysis results.

Due to a large number of complex reactions involved in the pyrolysis process, different model approaches have been proposed for the analysis of pyrolysis kinetics and mechanism such as the global kinetics model [7], the Brodi-Shafizadeh model [8], the two-parallel reactions model [9], and the three pseudocomponents model [10]. In this study, the two-parallel reactions model was adopted and used to describe the pyrolysis mechanism. According to Luangkiattikhun et al. (2008) [9], the prediction of pyrolysis kinetics by the two-parallel reactions model gave excellent fitting with the experimental results from TGA for all palm solid wastes under the pyrolysis conditions investigated. However, as opposed to studying pyrolysis in the static mode, the analysis of pyrolysis in a free-fall reactor requires the consideration of relevant forces acting on the falling particles. Also, the particle velocity and drag force can be affected by particle properties and properties of the flowing fluid which are changing along the pyrolysis reactor. In summary, the present work aims to study the distribution and properties of pyrolysis products from three types of biomass wastes, namely, palm shell, palm kernel cake, and cassava pulp residue in a three-zone free-fall reactor as well as the kinetic modeling of the pyrolysis process.

2. Experimental Section

2.1. Material Characterization. Three types of biomass were studied in a laboratory free-fall pyrolysis reactor, including palm shell (PS), cassava pulp residue (CPR), and palm kernel (PK). These raw materials were milled and sieved to obtain average particle sizes in the range from 0.18 to 1.55 mm for palm shell and 0.28 mm for palm kernel and cassava pulp residue. Each sample was dried in an electric oven at 110°C for 24 h to remove the excess moisture and kept in a desiccator for subsequent characterization and pyrolysis experiments.

Compositions of the biomass precursors were characterized by means of proximate and ultimate analyses. Proximate analysis was performed according to ASTM method to determine the weight percentage of moisture (ASTM D2867-95), volatile matter (ASTM D5832-95), ash content (ASTM D2866-94), and fixed carbon (by difference). Ultimate analysis, determined with a CHNS/O analyzer (Perkin Elmer PE2400 series), gives information on weight % of carbon, nitrogen, hydrogen, sulfur, and oxygen (by difference). True and apparent densities of the test biomasses and char product were measured with a helium pycnometer (Accu Pyc 1330, Micromeritics). In addition, thermal decomposition of biomass precursors was determined with a thermogravimetric analyzer (SDT 2960 DSC-TGA model, TA Instruments), using nitrogen (99.99% purity) as an inert carrier gas.

2.2. Flash Pyrolysis in a Free-Fall Reactor. A standard vertical tube furnace supplied by Carbolite, UK, was employed for the heating of a free-fall pyrolysis reactor. The attached alumina tube coming with the furnace has an internal diameter of 0.1 m and an overall length of 1.10 m. It has a centrally heated zone of 0.20 m in length and the upper and lower nonheated sections of 0.45 m long for each section. The free-fall reactor used for the pyrolysis study, which was fabricated from

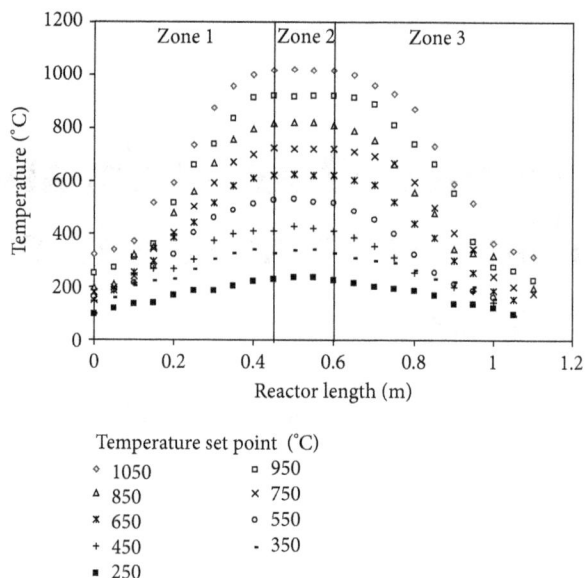

FIGURE 1: Temperature profiles in a free-fall reactor.

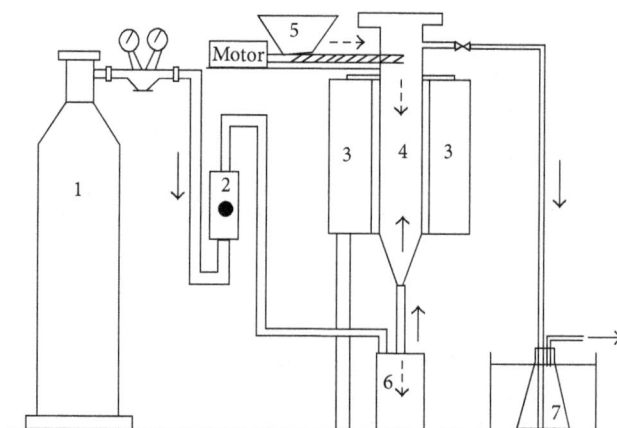

(1) Nitrogen tank cylinder   (7) Condenser
(2) Flow meter   - - - Solid
(3) High temperature vertical tube furnace   —— Gas
(4) Stainless steel free-fall reactor
(5) Raw material feeder
(6) Solid product collector

FIGURE 2: Schematic diagram of the free-fall reactor unit.

a stainless steel pipe of 0.04 m I.D. and has the same length as the furnace, was inserted inside the alumina tube. Figure 1 shows the temperature profile inside the reactor for various setting temperatures of the central hot zone, starting from the top of reactor for zone I to be called preheating zone, followed by zone II (heated zone) and zone III (cooling zone), respectively. Figure 2 shows the schematic diagram of the free-fall reactor system and Table 1 lists the operating conditions used.

For each experimental run, the furnace was turned on and nitrogen gas was admitted into the reactor from the bottom and flew upward at the rate of 200 cm³/min, corresponding

TABLE 1: Experimental conditions for biomass pyrolysis study in a free-fall reactor.

Biomass	Temperature (°C)	Inert gas flow rate (cm³/min)	Average particle size (mm)	Biomass feed rate (g/min)
PS	250–1050	200	0.18–1.55	0.6
PK	250–1050	200	0.28	0.6
CPR	250–1050	200	0.28	0.6

PS: palm shell, PK: palm kernel, CPR: cassava pulp residue.

to the flow velocity of 2.65×10^{-3} m/s. When the desired pyrolysis temperature was reached, a batch of 50 g of biomass sample was continuously fed from the top into the reactor at a rate of 0.6 g/min by a calibrated screw feeder. The char produced dropped to the bottom part of the reactor and it was collected in a solid collector. The pyrolysis vapor flew out the reactor and passed through a condenser where the liquid product and entrained solid were collected. The condenser operated at −10°C using a temperature-controlled bath filled with a glycerine-water mixture. At the completion of each run, the solid and liquid products left in the condenser were separated and weighed. The total yields of liquid and char products were calculated based on the total amount of biomass fed into the system. The gas yield was estimated by mass balance knowing the total yields of both the solid and liquid products. The apparent density of final char product (ρ_p) was estimated from the total weight of collected char and volume of a 50 g batch of biomass feed calculated from the apparent density of the feed particles, assuming that there is no change in the char particle volume during the course of biomass pyrolysis.

2.3. Analysis of Pyrolysis Products. Liquid products collected from palm shell pyrolysis were analyzed for physicochemical properties. Water in the biooil was removed by isotropic distillation with toluene [11]. A chemical analysis of the dewatered biooil was determined by gas chromatography (Varian CP-3800) equipped with a capillary coated with VF-5MS (30 m × 0.39 mm and 0.25 μm film thickness). In addition, the biooil derived from biomass pyrolysis was also analyzed for its calorific value (ASTM D240-92), density (Gay-Lussac bottle), viscosity (ASTM D445-96), and water by the Karl Fischer Titration.

The solid product derived from pyrolysis of palm shell was characterized for elemental analysis (CHNS/O analyzer, Perkin Elmer PE2400 series II). Proximate analysis, consisting of moisture content (ASTM D2867-95), volatile content (ASTM D5832-95), ash content (ASTM D2866-94), and fixed carbon content was also determined according to ASTM procedures. Furthermore, the true density and porous properties of char product were analyzed by using a helium pycnometer (Accupyc 1330 Micromeritics) and Autosorb-iQ (Quantachrome), respectively.

The gas product was characterized by a gas chromatograph (Varian CP-3800) equipped with a thermal conductivity detector (TCD). A capillary coated with CP-Carboplot (27.5 m × 0.53 mm and 0.25 μm film thickness) was used for the analysis of CO_2, CO, H_2, and CH_4.

3. Modeling of Biomass Pyrolysis in a Free-Fall Reactor

For the modeling of biomass pyrolysis, it is assumed that no particle shrinkage occurs; that is, diameter of a biomass particle remains constant during the course of pyrolysis reaction and that there is no heat and mass transfer resistances across the pyrolyzed particle. The following equations form the basis of pyrolysis modeling for a three-zone free-fall reactor studied in this work.

3.1. Equation of Particle Motion. The equation of motion for a spherical char particle settling in a flowing fluid is derived based on the conservation of momentum which states that the rate of change of momentum equals the net forces acting on the particle due to gravity, buoyancy, and drag forces. Mathematically,

$$\frac{\pi}{6} d_p^3 \rho_p \frac{d\left(v_p - v_g\right)}{dt} = \frac{1}{6} \pi d_p^3 \left(\rho_p - \rho_g\right) g \\ - \frac{1}{8} \pi \rho_g d_p^2 C_D \left(v_p - v_g\right), \tag{1}$$

or after rearranging,

$$\frac{dv_p}{dt} = \frac{\left(\rho_p - \rho_g\right) g}{\rho_p} - \frac{3}{4} \frac{C_D \rho_g \left(v_p - v_g\right)^2}{d_p \rho_p}, \tag{2}$$

where v_p = particle velocity, ρ_p = particle apparent density (weight of particle/volume of solid and void), d_p = particle diameter, v_g = gas velocity, ρ_g = gas density, C_D = drag coefficient, a function of Reynolds number = $18/\text{Re}_p^{0.6}$, $1 \leq \text{Re}_p \leq 1000$, and Re_p = particle Reynolds number = $d_p(v_p - v_g)\rho_g/\mu_g$.

3.2. Pyrolysis Kinetic Equations. The two-parallel reaction kinetic model [9] was adopted in this work to describe the conversion of pyrolyzed particles along the vertical distance of the reactor. It is based on the assumption that a biomass consists of two major components, with weight fractions a and b, that decompose independently into volatile and char products, according to the following rate equations:

$$\frac{d\rho_1}{dt} = -A_1 \exp\left[\frac{E_1}{RT}\right] \rho_1^{n1},$$

$$\frac{d\rho_2}{dt} = -A_2 \exp\left[\frac{E_2}{RT}\right]\rho_2^{n2},$$

$$\rho_p = \rho_1 + \rho_2,$$

$$(3)$$

where ρ, A, E, n, R, and T are density of biomass component, the preexponential factor, activation energy, order of reaction, gas constant, and absolute temperature, respectively. Subscripts 1 and 2 refer to component 1 and 2 of the biomass, respectively. ρ_1 and ρ_2 are equal to $\rho_p a$ and $\rho_p b$, respectively, and ρ_p represents the apparent density of a char particle. The temperature zones inside the reactor are displayed in Figure 1 which indicates that the reactor can be divided from the top of reactor into three consecutive regions: (i) preheating zone, (ii) heated zone, and (iii) cooling zone. The above kinetic expressions for the pyrolysis of the first and second components in the three hot zones of the reactor can now be written in terms of reactor length (L) by employing the relation, $v_p - v_g = dL/dt$, to obtain the following.

(1) Preheating zone (Zone I):

1st component:

$$\frac{d\rho_1}{\rho_1^{n_1}} = \int_0^{0.45} -\frac{A_1 \exp\left[E_1/RT_1(L)\right]}{\left(v_p - v_g\right)}dL, \qquad (4)$$

2nd component:

$$\frac{d\rho_2}{\rho_2^{n_2}} = \int_0^{0.45} -\frac{A_2 \exp\left[E_2/RT_1(L)\right]}{\left(v_p - v_g\right)}dL. \qquad (5)$$

(2) Heated zone (Zone II):

1st component:

$$\frac{d\rho_1}{\rho_1^{n_1}} = \int_{0.45}^{0.60} -\frac{A_1 \exp\left[E_1/RT_2\right]}{\left(v_p - v_g\right)}dL, \qquad (6)$$

2nd component:

$$\frac{d\rho_2}{\rho_2^{n_2}} = \int_{0.45}^{0.60} -\frac{A_2 \exp\left[E_2/RT_2\right]}{\left(v_p - v_g\right)}dL. \qquad (7)$$

(3) Cooling zone (Zone III):

1st component:

$$\frac{d\rho_1}{\rho_1^{n_1}} = \int_{0.60}^{1.10} -\frac{A_1 \exp\left[E_1/RT_3(L)\right]}{\left(v_p - v_g\right)}dL, \qquad (8)$$

2nd component:

$$\frac{d\rho_2}{\rho_2^{n_2}} = \int_{0.60}^{1.10} -\frac{A_2 \exp\left[E_2/RT_3(L)\right]}{\left(v_p - v_g\right)}dL. \qquad (9)$$

The initial weight fractions of the first and second components are designated as a and b and the sum of the two numbers equal to unity. The apparent density of each biomass component (ρ_1 and ρ_2) leaving one zone will be the initial density for the next adjacent zone. To simplify the calculation, fluid properties and fluid velocity are assumed to be constant at the average temperature of that zone. $T_1(L)$ and $T_3(L)$ are the temperature profiles of the preheating and cooling zone of the reactor, respectively. The estimation of these temperatures at various positions along the reactor is obtained directly from the reactor temperature profile as displayed in Figure 1.

The calculation scheme commences by first setting the initial values of kinetic parameters A_1, A_2, E_1, E_2, n_1, n_2, a and b, and ρ_p, C_D, Re_p, and v_p for the condition at the reactor inlet. The inlet particle velocity was estimated by the equation, $v_{p,\text{initial}} = [(2g/27)(\rho_p/\rho_f - 1)]^{5/7}D_p^{8/7}(\rho_f/\mu_f)^{3/7}$. Then, the computation is performed by numerical method to obtain the values of ρ_p and v_p as a function of reactor length which include those at the reactor exit, using (4)–(9). The calculation is repeated to obtain an optimum set of kinetic parameters, using the nonlinear least square (NLS) algorithm to minimize the objective function defined as O.F. $= \sum_{i=1}^{N}(\rho_{\text{cal},i} - \rho_{\text{exp},i})^2$, where $\rho_{\text{exp},i}$ and $\rho_{\text{cal},i}$ represent the experimental and calculated char product density (char density at the reactor exit). Subscript i denotes the discrete values of ρ, and N is the number of data points used in the model fitting. Next, the calculated char yield is computed from the calculated char density by the following equation:

$$\text{char yield (\%wt)} = 100\left(\frac{\rho_p}{\rho_{\text{feed}}}\right), \qquad (10)$$

where ρ_p and ρ_{feed} are the product char density and the feed particle density, respectively.

4. Results and Discussion

4.1. Biomass Properties. Table 2 presents the proximate and ultimate analysis of the biomass precursors. It is seen that the biomass samples contain more than 70% of volatile matters with cassava pulp residue having the highest volatile content. The high volatile matter found in the samples suggests the high potential of these residues for energy production by pyrolysis [12]; high levels of volatile matter result in more liquid and gas fuel to be obtained from the pyrolysis process. The fixed carbon content is the carbon found in the biomass that is left after volatile matters are driven off and it is used as an estimate of the amount of solid product left after the pyrolysis of biomass [13]. The fixed carbon reported in Table 2 indicates that the solid yield of pyrolysis from palm shell is higher than those of palm kernel and cassava pulp residue. Table 2 indicates that the carbon content of biomass precursor varies from 43.08 to 49.34%. Generally, higher carbon content leads to a higher heating value of solid combustion [14], thus making these biomasses to be good precursors for energy production. The results also indicate that the sulfur content in the biomass precursors is much lower than most typical fossil fuels (bituminous coal 0.5–1.5%, typical distillate oil 0.2–1.2%) [15]. This suggests that

TABLE 2: Proximate and ultimate analysis of oil-palm shell, oil-palm kernel, and cassava pulp residue and particle true densities.

Biomass	Ultimate analysis[1], (%wt)					Proximate analysis[2], (%wt)			Solid density[3], (g/cm^3)	
	C	H	N	S	O	Volatile	Fixed carbon	Ash	True density	Apparent density
PS	49.34	6.00	0.78	0.33	43.55	72.56	25.97	1.47	1.43	1.28
PK	43.84	6.13	3.11	0.06	46.86	79.68	16.78	3.54	1.40	1.30
CPR	43.08	6.10	0.40	0.07	50.36	83.86	12.12	4.02	1.37	1.13

[1] Dry and ash-free basis.
[2] Dry basis.
[3] Particle size of 1.55 mm.

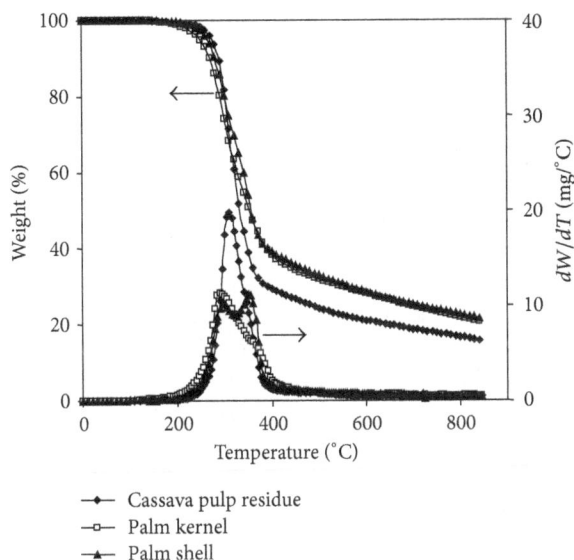

FIGURE 3: TG and DTG data of palm shell, palm kernel, and cassava pulp residue.

thermal decomposition of biomass should give lower sulfur emission than fossil fuels. The reason that biomass fuels are almost devoid of sulfur and coupled with its low ash content has made biomass a highly desirable fuel from the standpoint of pollution control cost. In addition, the composition of biomass ash which contains alkaline metals (e.g., Na, K, Mg, and Ca) can react with the released sulfur dioxide as well [16]. Overall, the use of biomass as a fuel creates less environmental problems. Table 2 also shows the true and apparent densities of the test biomasses.

4.2. Thermal Behavior of Biomass. In this work, TGA is used as an analytical technique to study the thermal decomposition behavior of biomass sample in an inert atmosphere of N_2. The results of TGA analysis are displayed in Figure 3 which show the weight loss curves (TG) and derivative thermogravimetric (DTG) evolution profiles as a function of heating temperature. As can be observed from TGA results, the cassava pulp residue started to decompose first followed by palm kernel and palm shell. Thermal decomposition of the biomass precursors started at approximately 250°C, possibly by the liberation of inherent moisture. Then it was followed by a major loss of weight where the main devolatilization occurs over the temperature range from 250 to 400°C. Further

observation indicates that cassava pulp residue and palm kernel showed one DTG peak while palm shell exhibited a two peak characteristic. This is attributed to the differences in the cellulosic composition of the biomass and the pyrolysis behavior of each biomass component. Typically, hemicellulose decomposition occurs over the temperature of 200–260°C [17], cellulose between 240 and 350°C [18], and lignin decomposes when heated over the wider range of 280–500°C [17].

4.3. Pyrolysis in a Free-Fall Reactor. On studying the effect of pyrolysis temperature and particle size, the experiments were conducted on palm shell with nine different pyrolysis temperatures (heated zone temperature) and six different particle size ranges under a fixed sweep gas flow rate of 200 cm^3/min. For the effect of biomass type, the experiments were performed with different precursors under a fixed sweep gas flow rate of 200 cm^3/min and particle size of 0.28 mm.

Figure 4 shows the final solid, liquid, and gas yields derived from palm shell pyrolysis expressed as a fraction of the initial sample weight for different particle sizes and pyrolysis temperatures. For all particle sizes, the solid yields continuously decreased as the temperature was increased (see Figure 4(a)). At a high pyrolysis temperature, the solid yields tended to become constant approaching the value of 15% except for the particle size greater than 0.36 mm and the highest conversion is achieved for the smallest size particles. Also, the thermal decomposition of a small particle size started at the lower pyrolysis temperature than that of a larger particle size. Moreover, at a fixed pyrolysis temperature, the solid yield increased with increasing in particle size. This can be explained by the fact that a larger size particle has a greater temperature gradient due to the longer heat diffusional path. This effect leads to a lower average particle temperature and hence giving less solid conversion by the pyrolysis reaction. As a result of this behavior, to cover a wide range of solid yield or solid conversion for large particle size, it is necessary to extend the length of heated zone of the reactor to allow sufficient time for increased solid decomposition.

Figure 4(b) shows the effect of palm shell particle size and pyrolysis temperature on the yields of liquid product. The liquid yield increased with the increasing of pyrolysis temperature and passed through a maximum at temperature around 650°C for particle size of 0.18 mm and at 750°C for particle size in the range of 0.23–0.36 mm. For particle size larger than 0.36 mm, the liquid yield tended to increase continuously with increasing pyrolysis temperature. The decrease

in higher yield of gaseous products. The increase in gaseous products is believed to be predominantly due to secondary cracking of the pyrolysis vapors at higher temperatures [19]. At high pyrolysis temperatures both the rate of primary pyrolysis and the rate of thermal cracking of tar to gaseous products are expectedly high. Further, a smaller particle is expected to produce higher gas yield because of the higher heat up rate and heat flux as compared to the larger particles. This observation agrees with the work of Wei et al. (2006) [20] who studied the effect of particle size on product distribution from pyrolysis of pine sawdust and apricot stone in a free-fall reactor at 800°C. They reported that the decrease of biomass particle size contributed to an increase in the gas yields.

Figure 5(a) compares the solid yields from pyrolysis of palm shell, palm kernel, and cassava pulp residue at various pyrolysis temperatures. The cassava pulp residue started to decompose at a lower pyrolysis temperature than palm kernel and palm shell. These results are also consistent with the results from the TGA analysis (Figure 2). It was also found that the thermal decomposition behavior of biomass was consistent with the amount of volatile matter content. Biomass with a high volatile matter can be easily decomposed by heating than that with lower volatile content. Therefore, cassava pulp residue which has the highest amount of volatile matter (83.86 wt%) could decompose more rapidly than palm kernel (79.62 wt%) and palm shell (72.56 wt%). The yield of liquid product (Figure 5(b)) was found to increase with pyrolysis temperature to give a maximum value at around 400°C for cassava pulp residue and at 520°C for palm shell and palm kernel, and then it decreased with increasing in pyrolysis temperature. The decrease in liquid yields and the corresponding increase in gas yields above the optimum temperature (see Figure 5(c)) are probably due to secondary cracking of the pyrolysis vapor at relatively high temperatures [21]. Furthermore, the secondary decomposition of the char at higher temperatures may as well give additional noncondensable gaseous product [22]. Figure 5(c) shows that the gas yield increased over the whole temperature range and the pyrolysis of cassava pulp residue gave higher gas product than those of palm kernel and palm shell. This is probably due to the differences in cellulosic components of these biomasses. The cellulose and hemicellulose components of biomass are mainly responsible for the volatile portion of the pyrolysis products while lignin is the main contributor to the formation of char [21].

Typical characteristics of char, liquid, and gas product derived from palm shell pyrolysis at temperature of 650°C, N_2 flow rate of 200 cm³/min, particle size of 0.28 mm, and feed rate of 0.6 g/min are listed in Table 3. Proximate analysis indicates that the main component in char is fixed carbon. It may be that during pyrolysis, hydrogen and oxygen were consumed via reactions of dehydrogenation and deoxygenation to produce CO_2, CO, H_2, H_2O [23]. This should cause the decrease in the molar ratio of H/C and O/C in char. Calorific value of the derived char is 29.87 MJ which is comparable to that of coal (~35 MJ) [24]. Porous characteristics of char indicates that the derived char has reasonable BET surface

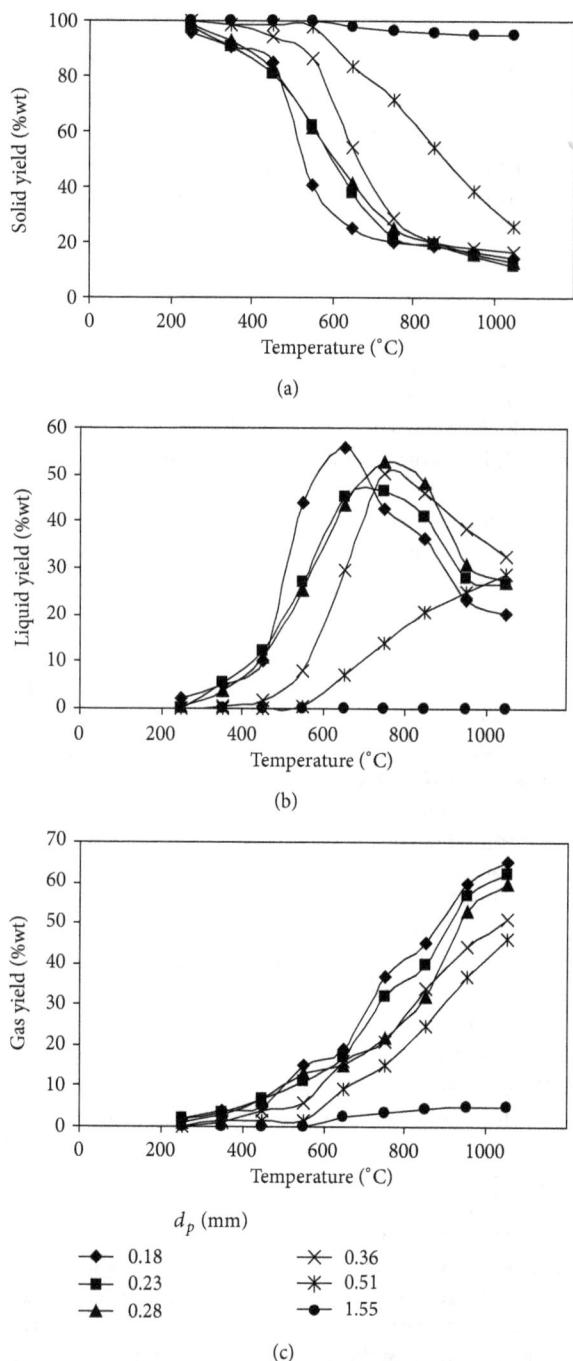

FIGURE 4: Effect of particle size and pyrolysis temperature on (a) solid yield, (b) liquid yield, and (c) gas yield (palm shell feed rate 0.6 g/min and sweep gas flow rate 200 cm³/min).

in liquid yield and the corresponding increase in gas yield (see Figure 4(c)) above the optimum pyrolysis temperature are probably caused by the decomposition of some liquid vapors in the gas product. It is also noted that there was no liquid product being produced from the pyrolysis of the largest size of 1.55 mm at the temperature below 1000°C.

From Figure 4(c), it is seen that higher pyrolysis temperature and small particle size led to more volatilization resulting

TABLE 3: Properties of chars, liquid, and gas products derived from palm shell pyrolysis at temperature of 650°C, N_2 flow rate of 200 cm^3/min, particle size of 0.28 mm, and feed rate of 0.6 g/min.

Properties	Value
Char product	
Proximate analysis (dry basis) (wt%)	
Volatile	20.22
Fixed carbon	69.97
Ash	9.82
Elemental analysis (dry and ash—free basis) (wt%)	
C	59.28
H	1.67
O	37.52
N	1.53
H/C (mole ratio)	0.20
O/C (mole ratio)	0.47
Calorific value (MJ/kg)	29.87
Porous characteristics	
BET surface area (m^2/g)	200
Micropore area (m^2/g)	182
Average pore size (nm)	2.03
Total pore volume (cm^3/g)	0.111
Micropore volume (cm^3/g)	0.091
Liquid product	
Viscosity at 40° (cP)	3.5
Density at 30°C (g/cm^3)	1.16
pH	2.67
Calorific value (MJ/kg)	5.8 (35.3*)
Water contents (wt %)	51
n-Octane, C_8 (wt%)	2.542
Phenol (wt%)	1.987
Furfural (wt%)	0.876
Benzene (wt%)	0.386
Toluene (wt%)	0.712
Xylene (wt%)	0.364
Styrene (wt%)	0.128
Gas product	
CO (%)	18.13
CO_2 (%)	10.10
CH_4 (%)	4.21
H_2 (%)	1.67

*Water removed.

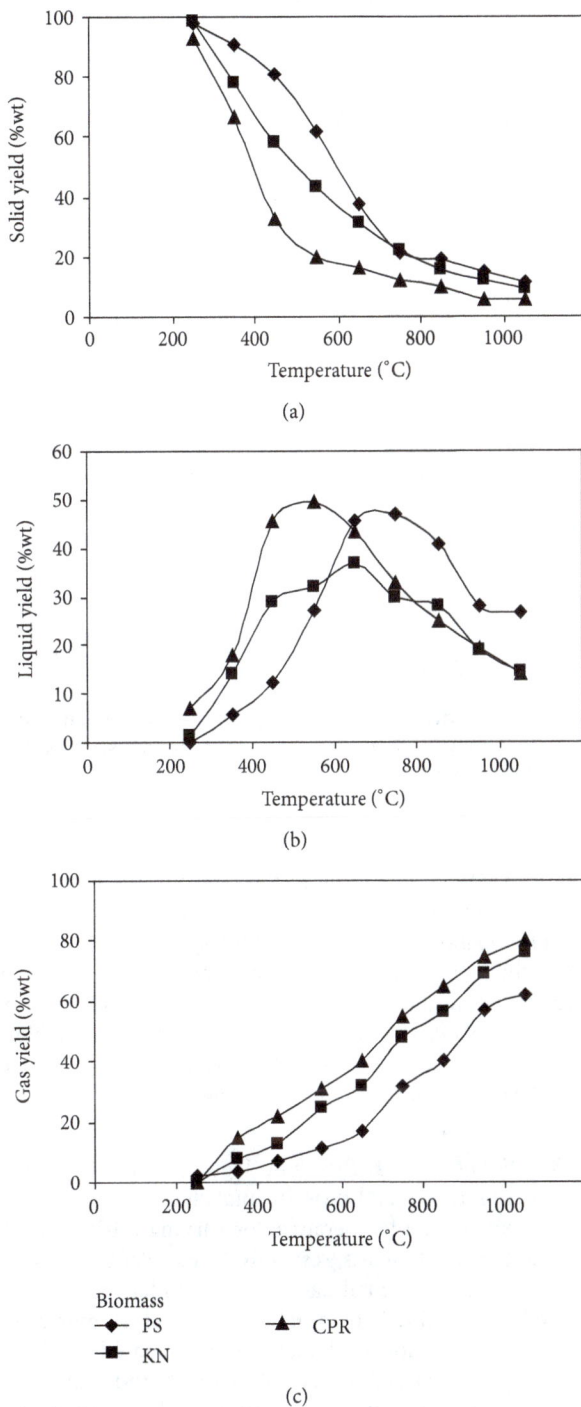

FIGURE 5: Effect of biomass type and pyrolysis temperature on (a) solid yield, (b) liquid yield, and (c) gas yield (biomass feed rate 0.6 g/min, sweep gas flow rate 200 mL/min, and particle size of 0.28 mm).

area of around 200 m^2/g and contains mainly microspores which account for about 80% of the total pore volume.

Crude biooil displayed opaque dark color and gave the single phase chemical solution. Density of crude biooil (1.16 g/cm^3) is denser than that of diesel fuel (0.78 g/cm^3) and heavy fuel oil (0.85 g/cm^3). The crude biooil shows relatively low viscosity characteristic of 3.5 cP at 40°C. The low viscosity is associated with the high value of water content (51 wt%) present in the crude biooil. Generally, the water content in biooil is mainly derived from the decomposition of lignin-derived materials [25]. The crude biooil shows a high acid level; pH at room temperature is about 2.67. It has been reported that biooil with low pH is very corrosive to the metals [26]. The corrosiveness of biooils is more severe when the water content is high and also when biooils are used at a high temperature [27]. In addition, removing of water from the crude biooil gave rise to an increase in calorific value from

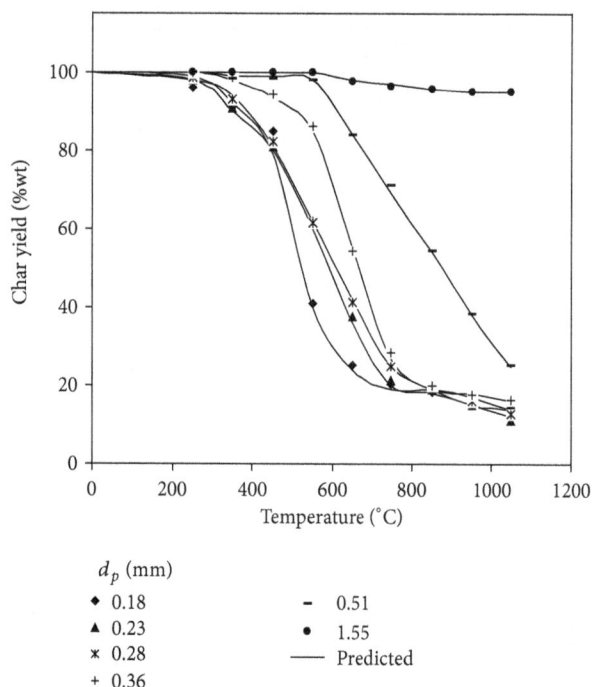

FIGURE 6: Char yields for palm shell pyrolysis in a free-fall reactor at different pyrolysis temperatures for various particle sizes (N_2 flow rate is 200 cm^3/min).

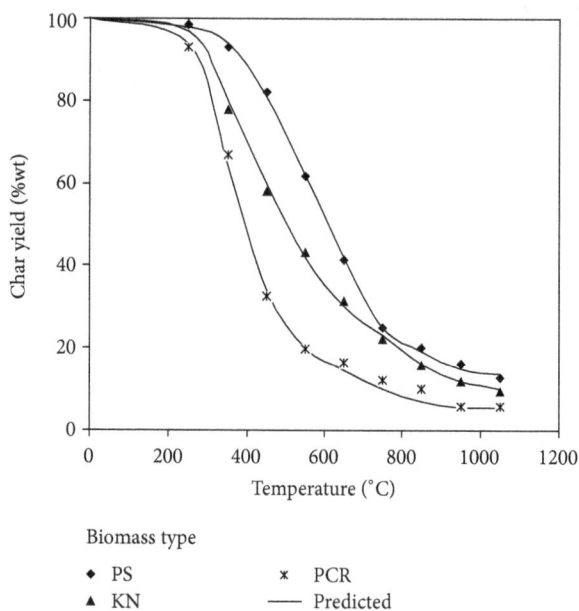

FIGURE 7: Char yields of biomass pyrolysis for different pyrolysis temperatures (particle size is 0.28 mm and N_2 flow rate is 200 cm^3/min).

5.8 to 35.3 MJ/kg. According to these results, the removal of water from biooil still gives calorific value lower than that of commercial diesel oil fuel (45.0 MJ/kg) [28]. GC analysis results show that the derived biooil provides a source of potential chemical feedstock such as octane, phenol, furfural, benzene, toluene, xylene, and styrene. As shown in Table 3, the gas products consist of CO, CO_2, CH_4, and H_2 with CO being the dominant components among the others.

4.4. Kinetics of Biomass Pyrolysis. Figures 6 and 7 compare the model predicted and experimental char yields as a function of heated zone temperature for varying initial particle sizes and types of biomass, respectively. Excellent agreement between the experimental data and the model prediction is discernible. The kinetic parameters (E, A and n), maximum error, and correlation coefficient (R^2) estimated from the model are summarized in Table 4. It was found that for all conditions, the values of the correlation coefficient are higher than 0.99 and with the maximum error of estimate being less than 7.2%. This proves that the pyrolysis model proposed in this study can be used to describe the thermal decomposition behaviors of the test biomasses in a free-fall reactor within acceptable accuracy.

As observed from Table 4, the first component appears to have higher activation energy (E_1) than that of the second component (E_2). This indicates that the first component requires a larger amount of energy to initiate the pyrolysis reaction as compared to the second component. Also, the first biomass component has higher value of the preexponential factor. In addition, the larger value of b in comparison with

that of a signifies that the decomposition of biomass precursor should be contributed predominantly by the second component. For this work, the lignocellulosic component of biomass has been analyzed by the Kurschner-Hoffer method for holocellulose extraction and the Klason method for lignin determination [29]. It has been found that palm shell contains 45.42% cellulose, 21.74% hemicellulose, and 17.64% lignin; palm kernel contains 30.4% cellulose, 25.12% hemicellulose, and 15.72% lignin; and cassava pulp residue contains 15.6% cellulose, 4.6% hemicellulose, and 2.2% lignin. Based solely on these data, it could be inferred that the first component (smaller weight fraction) is probably contributed by lignin, and the second component (larger weight fraction) by both the cellulose and hemicellulose.

It is further observed that, for palm shell pyrolysis, the kinetic parameters (n, A and E) appear to vary with particle size but showing no definite trend. In principle, the kinetic parameters should be constant and independent of particle size if the pyrolysis process is indeed a true kinetic control. This observed inconsistency probably results from the influence of heat and mass transfer limitations which are not taken into account by the present pyrolysis model.

The variation of char yield of the three biomass precursors with pyrolysis temperature is illustrated in Figure 7. Over the pyrolysis temperature range studied, the char yield of cassava pulp residue was lower than those of palm kernel and palm shell. This indicates that cassava pulp residue should decompose more rapidly than both palm kernel and palm shell. This result is consistent with the much lower values of activation energies (E_1 and E_2) of cassava pulp residue compared to those of palm shell and palm kernel pyrolysis, as shown in Table 4.

TABLE 4: Estimated kinetic parameters of biomass pyrolysis in a free-fall reactor.

Biomass	Size (mm)	a	A_1 (s^{-1})	E_1 (kJ/mol)	n_1	b	A_2 (s^{-1})	E_2 (kJ/mol)	n_2	Max. error (%)	Correlation coefficient (R^2)
PS	0.18		2.73×10^4	556	1.00		8.63×10^4	123	3.00	7.16	0.9958
	0.23		2.75×10^4	500	1.30		7.90×10^4	124	2.95	6.95	0.9955
	0.28	0.14	1.02×10^5	650	3.00	0.86	2.00×10^4	129	2.80	6.40	0.9997
	0.36		5.50×10^5	473	6.50		1.88×10^3	150	2.94	5.79	0.9976
	0.51		5.80×10^5	600	6.40		1.09×10^4	152	6.5	3.13	0.9939
	1.55		6.30×10^4	640	4.30		1.90×10^4	149	3.00	0.62	0.9961
PK	0.28	0.18	7.56×10^4	630	9.20	0.82	1.26×10^4	107	2.98	6.27	0.9925
CPR	0.28	0.08	3.48×10^4	520	9.70	0.90	4.90×10^4	100	3.10	5.17	0.9993

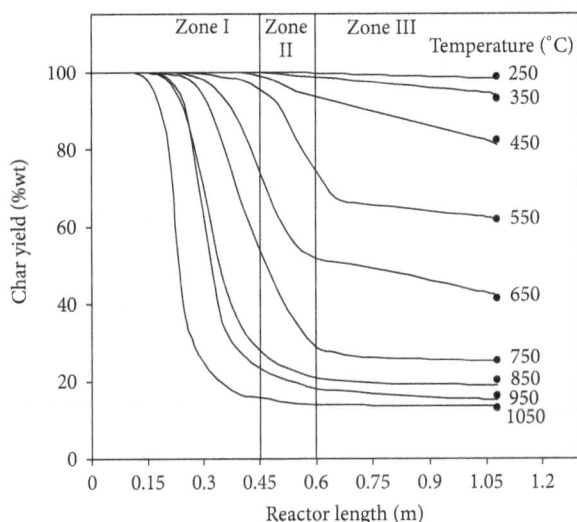

FIGURE 8: Char yields calculated from the pyrolysis model for palm shell at the average particle size of 0.28 mm and N$_2$ flow rate 200 cm^3/min (line is by simulation and dot represents experimental data).

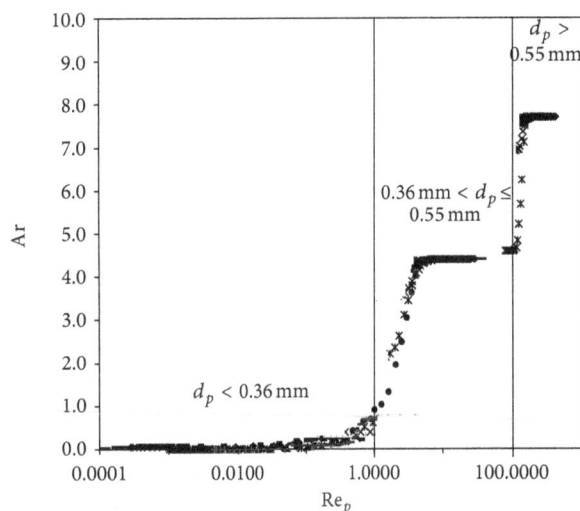

FIGURE 9: The relationship between particle Reynolds number and Archimedes number for biomass pyrolysis in a three-zone free-fall reactor.

Based on the obtained kinetic parameters, the predicted profiles of char yield from palm shell pyrolysis versus the length of the free-fall reactor are calculated and presented in Figure 8. As seen, palm shell (0.28 mm in size) starts to decompose in the first zone and appears to reach the completion of devolatilization in the second zone of the reactor for pyrolysis above 650°C. Also, the decomposition of biomass precursor does not occur at a relatively low pyrolysis temperature of 250°C at which only 1% reduction in the char yield is observed.

The relationship between the flow regime and the characteristics of char product as influenced by the pyrolysis process can be further examined by plotting particle Reynolds number ($Re_p = d_p(v_p - v_g)\rho_g/\mu_g$) versus Archimedes number ($Ar = g\rho_g(\rho_p - \rho_g)d_p^3/\mu_f^2$), as delineated in Figure 9. The pyrolysis behavior of biomass with reference to particle motion can be divided into three separate regimes. For small particle size ($d_p < 0.36$ mm), the relatively low values of Re_p and Ar are found in this zone. This indicates a large reduction of char particle density which in turn affects the velocity of biomass particle. The values of Ar and Re_p in this zone are lower than unity which indicate that viscous force acting on

the particle is much larger than gravity force, thus rendering the particle to move slowly. The slow moving particles will experience a long residence time in the reactor, resulting in the high weight loss of particles. In the intermediate zone for $d_p = 0.36$–0.55 mm, the Archimedes number rises continuously with approximately fivefold increase for about an order of magnitude increase of Reynolds number from 1 to about 10, and remains substantially constant up to the Reynolds number of 100. In the last zone for $d_p > 0.55$ mm, the Archimedes number increases markedly over a small increase of Reynolds number before finally attaining a constant value of around 7.5. The extent of pyrolysis reaction for a relatively large size particle is much reduced by the rapid settling of particles that does not allow sufficient time for complete devolatilization to occur.

5. Conclusions

The pyrolysis temperature, particle size, and type of biomass precursor had a significant effect on the yields and properties of pyrolysis products in a three-zone free-fall reactor. It was observed that the thermal decomposition of biomass did not occur when the biomass particle was larger than 1.55 mm due to a relatively short central heated zone of 20 cm in length.

For the biooil product, its yield increased continuously with the increase in pyrolysis temperature and passed through a maximum when an optimum temperature was reached. Typical biooil from palm shell pyrolysis is less attractive to be used as a direct fuel, due to its high water content, low calorific value, and a high acid level (pH = 2.67). The derived solid product (char) had reasonably high calorific value 29.87 MJ/kg comparable to that of coal (~35 MJ/kg) and showed reasonable BET surface area of about 200 m^2/g. The gas product contained CO, CO_2, CH_4, and H_2 with CO being the dominant component. The pyrolysis model developed for the free-fall reactor in the present study predicted extremely well the char yields under all pyrolysis conditions studied.

Conflict of Interests

The authors declare that there is no conflict of interests regarding the publication of this paper.

Acknowledgment

This research was funded by the Ministry of Science and Technology of Thailand.

References

[1] P. McKendry, "Energy production from biomass (part 2): conversion technologies," *Bioresource Technology*, vol. 83, no. 1, pp. 47–54, 2002.

[2] A. K. Jain, S. K. Sharma, and D. Singh, *Reaction Kinetics of Paddy Husk Thermal Decomposition*, Energy Research Center Panjab University, Panjab, India, 1996.

[3] C. Acıkgoz, O. Onay, and O. M. Kockar, "Fast pyrolysis of linseed: product yields and compositions," *Journal of Analytical and Applied Pyrolysis*, vol. 71, no. 2, pp. 417–429, 2004.

[4] W. T. Tsai, M. K. Lee, and Y. M. Chang, "Fast pyrolysis of rice husk: product yields and compositions," *Bioresource Technology*, vol. 98, no. 1, pp. 22–28, 2007.

[5] Natural Resources Management and Environment Department, "Integrated Energy System in China—The Cold Northeastern Region Experience," http://www.fao.org/docrep/T4470E/T4470E00.htm.

[6] J. Lehto, "Determination of kinetic parameters for Finnish milled peat using drop tube reactor and optical measurement techniques," *Fuel*, vol. 86, no. 12-13, pp. 1656–1663, 2007.

[7] M. J. Safi, I. M. Mishra, and B. Prasad, "Global degradation kinetics of pine needles in air," *Thermochimica Acta*, vol. 412, no. 1-2, pp. 155–162, 2004.

[8] N. Nugranad, *Pyrolysis of biomass [Ph.D. thesis]*, University of Leeds Department of Fuel and Enegy, 1997.

[9] P. Luangkiattikhun, C. Tangsathitkulchai, and M. Tangsathitkulchai, "Non-isothermal thermogravimetric analysis of oil-palm solid wastes," *Bioresource Technology*, vol. 99, no. 5, pp. 986–997, 2008.

[10] S. Hu, A. Jess, and M. Xu, "Kinetic study of Chinese biomass slow pyrolysis: comparison of different kinetic models," *Fuel*, vol. 86, no. 17-18, pp. 2778–2788, 2007.

[11] E. G. Baker and D. C. Elliott, "Catalytic hydrotreating of biomass—derived oil," in *Pyrolysis Oils From Viomass*, American Chemical Society, 1988.

[12] Q. Wu, J. Dai, Y. Shiraiwa, G. Sheng, and J. Fu, "A renewable energy source—hydrocarbon gases resulting from pyrolysis of the marine nanoplanktonic alga Emiliania huxleyi," *Journal of Applied Phycology*, vol. 11, no. 2, pp. 137–142, 1999.

[13] M. Hutagalung, "Understanding coal analysis," Majari Magazine, 2008, http://majarimagazine.com/2008/06/understanding-coal-sample-analysis/.

[14] A. Demirbas, "Combustion characteristics of different biomass fuels," *Progress in Energy and Combustion Science*, vol. 30, no. 2, pp. 219–230, 2004.

[15] B. R. Miller, "Structure of Wood," http://www.fpl.fs.fed.us/document/fplgtr/fplgtr113/ch02.pdf.

[16] M. V. Dagaonkar, A. A. C. M. Beenackers, and V. G. Pangarkar, "Enhancement of gas-liquid mass transfer by small reactive particles at realistically high mass transfer coefficients: absorption of sulfur dioxide into aqueous slurries of $Ca(OH)_2$ and $Mg(OH)_2$ particles," *Chemical Engineering Journal*, vol. 81, no. 1-3, pp. 203–212, 2001.

[17] M. J. Prins, K. J. Ptasinski, and F. J. J. G. Janssen, "Torrefaction of wood. Part 1. Weight loss kinetics," *Journal of Analytical and Applied Pyrolysis*, vol. 77, no. 1, pp. 28–34, 2006.

[18] J. B. Wooten, J. I. Seeman, and M. R. Hajaligol, "Observation and characterization of cellulose pyrolysis intermediates by 13C CPMAS NMR. A new mechanistic model," *Energy and Fuels*, vol. 18, no. 1, pp. 1–15, 2004.

[19] D. Mohan, C. U. Pittman Jr., and P. H. Steele, "Pyrolysis of wood/biomass for bio-oil: a critical review," *Energy and Fuels*, vol. 20, no. 3, pp. 848–889, 2006.

[20] L. Wei, S. Xu, L. Zhang et al., "Characteristics of fast pyrolysis of biomass in a free fall reactor," *Fuel Processing Technology*, vol. 87, no. 10, pp. 863–871, 2006.

[21] O. Onay and O. M. Koçkar, "Pyrolysis of rapeseed in a free fall reactor for production of bio-oil," *Fuel*, vol. 85, no. 12-13, pp. 1921–1928, 2006.

[22] A. H. Patrick and T. Williams, "Influence of temperature on the products from the flash pyrolysis of biomass," *Fuel*, vol. 75, no. 9, pp. 1051–1059, 1996.

[23] B. B. Uzun, A. E. Pütün, and E. Pütün, "Fast pyrolysis of soybean cake: product yields and compositions," *Bioresource Technology*, vol. 97, pp. 569–576, 2006.

[24] E. Jorjani, J. C. Hower, S. Chehreh Chelgani, M. A. Shirazi, and S. Mesroghli, "Studies of relationship between petrography and elemental analysis with grindability for Kentucky coals," *Fuel*, vol. 87, no. 6, pp. 707–713, 2008.

[25] F. Abnisa, W. M. A. Wan Daud, W. N. W. Husin, and J. Sahu, "Utilization possibilities of palm shell as a source of biomass energy in Malaysia by producing bio-oil in pyrolysis process," *Biomass and Bioenergy*, vol. 35, no. 5, pp. 1863–1872, 2011.

[26] E. Hu, Y. Xu, X. Hu, L. Pan, and S. Jiang, "Corrosion behaviors of metals in biodiesel from rapeseed oil and methanol," *Renewable Energy*, vol. 37, no. 1, pp. 371–378, 2012.

[27] S. Czernik, *Environment Health and Safety in Fast Pyrolysis of Biomass*, vol. 1, CPL Press, Newbury, UK, 1999.

[28] H. E. Saleh, "The preparation and shock tube investigation of comparative ignition delays using blends of diesel fuel with biodiesel of cottonseed oil," *Fuel*, vol. 90, no. 1, pp. 421–429, 2011.

[29] C. D. Blasi, G. Signorelli, C. D. Russo, and G. Rea, "Product distribution from pyrolysis of wood and agricultural residues," *Industrial and Engineering Chemistry Research*, vol. 38, no. 6, pp. 2216–2224, 1999.

Permissions

The contributors of this book come from diverse backgrounds, making this book a truly international effort. This book will bring forth new frontiers with its revolutionizing research information and detailed analysis of the nascent developments around the world.

We would like to thank all the contributing authors for lending their expertise to make the book truly unique. They have played a crucial role in the development of this book. Without their invaluable contributions this book wouldn't have been possible. They have made vital efforts to compile up to date information on the varied aspects of this subject to make this book a valuable addition to the collection of many professionals and students.

This book was conceptualized with the vision of imparting up-to-date information and advanced data in this field. To ensure the same, a matchless editorial board was set up. Every individual on the board went through rigorous rounds of assessment to prove their worth. After which they invested a large part of their time researching and compiling the most relevant data for our readers.

The editorial board has been involved in producing this book since its inception. They have spent rigorous hours researching and exploring the diverse topics which have resulted in the successful publishing of this book. They have passed on their knowledge of decades through this book. To expedite this challenging task, the publisher supported the team at every step. A small team of assistant editors was also appointed to further simplify the editing procedure and attain best results for the readers.

Apart from the editorial board, the designing team has also invested a significant amount of their time in understanding the subject and creating the most relevant covers. They scrutinized every image to scout for the most suitable representation of the subject and create an appropriate cover for the book.

The publishing team has been an ardent support to the editorial, designing and production team. Their endless efforts to recruit the best for this project, has resulted in the accomplishment of this book. They are a veteran in the field of academics and their pool of knowledge is as vast as their experience in printing. Their expertise and guidance has proved useful at every step. Their uncompromising quality standards have made this book an exceptional effort. Their encouragement from time to time has been an inspiration for everyone.

The publisher and the editorial board hope that this book will prove to be a valuable piece of knowledge for researchers, students, practitioners and scholars across the globe.

List of Contributors

Liza A. Dosso, Carlos R. Vera and Javier M. Grau
Instituto de Investigaciones en Catálisis y Petroquímica "Ing. JoséMiguel Parera" (INCAPE), FIQ, UNL-CONICET, CCT CONICET Santa Fe, "Dr. Alberto Cassano," Colec. Ruta Nac. No. 168, KM 0, Paraje El Pozo, S3000AOJ Santa Fe, Argentina

Victor J. Law and Denis P. Dowling
School of Mechanical and Materials Engineering, University College Dublin, Belfield D04 V1W8, Dublin 4, Ireland

Nicolás Carrara, Carolina Betti and Cecilia R. Lederhos
Instituto de Investigaciones en Catálisis y Petroquímica (INCAPE) (FIQ-UNL, CONICET), Colectora Ruta Nac., No. 168 Km 0, Pje El Pozo, 3000 Santa Fe, Argentina

Laura Gutierrez and Mónica E. Quiroga
Instituto de Investigaciones en Catálisis y Petroquímica (INCAPE) (FIQ-UNL, CONICET), Colectora Ruta Nac., No. 168 Km 0, Pje El Pozo, 3000 Santa Fe, Argentina
Facultad de Ingeniería Química, Universidad Nacional del Litoral, Santiago del Estero 2654, 3000 Santa Fe, Argentina

FernandoColoma-Pascual and María Cristina Almansa
Servicios Técnicos de Investigación, Facultad de Ciencias, Universidad de Alicante, Apartado 99, 03080 Alicante, Spain

Cristian Miranda
Laboratorio de Investigación en Cat´alisis y Procesos (LICAP), Universidad del Valle, Ciudad Universitaria Mel´endez, Calle 13 # 100-00, Cali, Colombia

K. S. Makarevich, A. V. Zaitsev, O. I. Kaminsky, E. A. Kirichenko and I. A. Astapov
Institute of Materials Khabarovsk Scientific Center, Far Eastern Branch of the Russian Academy of Sciences, Khabarovsk, Russia

Basma Khoualdia, Samia Ben-Ali and Ahmed Hannachi
University of Gabes, National School of Engineers of Gabes, Laboratory of Research Process Engineering and Industrial Systems, LR11ES54, St. Omar Ibn El Khattab 6029, Gabes, Tunisia

Mahmoud A. Mohsin
Department of Chemistry, University of Sharjah, Sharjah, UAE

Tahir Abdulrehman and Yousef Haik
College of Science and Engineering, Hamad Bin KhalifaUniversity, Doha, Qatar

Handoko Darmokoesoemo, Suyanto Suyanto and Leo Satya Anggara
Department of Chemistry, Faculty of Science and Technology, Airlangga University, Surabaya 60115, Indonesia

Andrew Nosakhare Amenaghawon
Department of Chemical Engineering, Faculty of Engineering, University of Benin, PMB 1154, Ugbowo, Benin City, Edo State, Nigeria

Heri Septya Kusuma
Department of Chemical Engineering, Faculty of Industrial Technology, Institut Teknologi Sepuluh Nopember, Surabaya 60111, Indonesia

Javier Paul Montalvo Andia
Instituto de Energía y Medio Ambiente, Universidad Cat´olica San Pablo, Arequipa, Peru

Lidia Yokoyama
Escola de Química, Departamento de Processos Inorgûnicos, Universidade Federal do Rio de Janeiro, Rio de Janeiro, RJ, Brazil

Luiz Alberto Cesar Teixeira
Departamento de Engenharia Química e de Materiais, Pontifícia Universidade Católica do Rio de Janeiro, Rio de Janeiro, RJ, Brazil

Misri Gozan, Jabosar Ronggur Hamonangan Panjaitan and Dewi Tristantini
Chemical Engineering Department, Faculty of Engineering, Universitas Indonesia, Depok 16424, Indonesia

Rizal Alamsyah
Centre for Agro-Based Industry, Bogor 16122, Indonesia

Young Je Yoo
School of Chemical & Biological Engineering, Seoul National University, Seoul, Republic of Korea

Fei Chang
Institute of Comprehensive Utilization of Plant Resources, Kaili University, Kaili 556011, China

Chen Yan
An Shun City People's Hospital, People's Hospital Republic of China, An Shun 561000, China

Quan Zhou
Pharmaceutical and Bioengineering College, Hunan Chemical Vocational Technology College, Zhuzhou, Hunan 412000, China

Elizabeth Henao and Germán Quintana
Grupo Pulpa y Papel, Facultad de Ingeniería Química, Universidad Pontificia Bolivariana, Sede Central Medelln, Circular 1 No. 70-01, Medellín, Colombia

Ezequiel Delgado and Héctor Contreras
Departamento de Madera, Celulosa y Papel (DMCyP), Universidad de Guadalajara, Km. 15.5 Carretera Guadalajara-Nogales Las Agujas, 45020 Zapopan, JAL, Mexico

Xiaofang Liu and Rui Wang
Guizhou Engineering Research Center for Fruit Processing, Food and Pharmaceutical Engineering Institute, Guiyang University, Guiyang 550005, China

M. E. Becerra
Departamento de Física y Química, Facultad de Ciencias Exactas y Naturales, Universidad Nacional de Colombia, Sede Manizales 170003, Colombia
Departamento de Ingeniería Química, Facultad de Ingeniería y Arquitectura, Universidad Nacional de Colombia, Sede Manizales 170003, Colombia
Laboratorio de Materiales Nanoestructurados y Funcionales, Facultad de Ciencias Exactas y Naturales, Universidad Nacional de Colombia, Sede Manizales 170003, Colombia

Grupo de Investigación en Procesos Químicos, Catalíticos y Biotecnológicos, Universidad Nacional de Colombia, Sede Manizales 170003, Colombia
Departamento Química, Facultad de Ciencias Exactas y Naturales, Universidad de Caldas, Manizales 17003, Colombia

O. Giraldo
Departamento de Física y Química, Facultad de Ciencias Exactas y Naturales, Universidad Nacional de Colombia, Sede Manizales 170003, Colombia
Laboratorio de Materiales Nanoestructurados y Funcionales, Facultad de Ciencias Exactas y Naturales, Universidad Nacional de Colombia, Sede Manizales 170003, Colombia
Grupo de Investigación en Procesos Químicos, Catalíticos y Biotecnológicos, Universidad Nacional de Colombia, Sede Manizales 170003, Colombia

A. M. Suarez
Laboratorio de Materiales Nanoestructurados y Funcionales, Facultad de Ciencias Exactas y Naturales, Universidad Nacional de Colombia, Sede Manizales 170003, Colombia
Departamento Química, Facultad de Ciencias Exactas y Naturales, Universidad de Caldas, Manizales 17003, Colombia

N. P. Arias
Grupo de Investigación en Procesos Químicos, Catalíticos y Biotecnológicos, Universidad Nacional de Colombia, Sede Manizales 170003, Colombia
Facultad de Ciencias e Ingeniería, Universidad de Boyacá, Sogamoso, Colombia

Ditpon Kotatha and Supitcha Rungrodnimitchai
Department of Chemical Engineering, Faculty of Engineering, Thammasat University, Khlong Luang, Pathum Thani 12120, Thailand

Qazi Mohammed Omar and Javeed A. Awan
Institute of Chemical Engineering and Technology, Faculty of Engineering and Technology, University of the Punjab, Lahore, Pakistan

Jean-Noël Jaubert
Universitède Lorraine, Ecole Nationale Supèrieure des Industries Chimiques, Laboratoire Rèactions et Gènie des Procèdès (UMRCNRS 7274), 1 rue Grandville, 54000 Nancy, France

Endalkachew Chanie Mengistie
Bahir Dar Institute of Technology (BiT), Faculty of Chemical and Food Engineering, Bahir Dar University, 1920 Bahir Dar, Ethiopia

Jean-François Lahitte
Laboratoire de Genie Chimique, Universite Paul
Sabatier, 118 route de Narbonne, Toulouse, France

Hong Thai Vu
Department of Chemical Process Equipment,
School of Chemical Technology, Hanoi University
of Science and Technology, Hanoi, Vietnam

Evangelos Tsotsas
Chair of \$ermal Process Engineering, Faculty of
Process and Systems Engineering, Otto-von-Guericke-
University Magdeburg, Magdeburg, Germany

Maxime Lortie and Rima Isaifan
Chemical and Biological Engineering, University of
Ottawa, Ottawa, ON, Canada K1N 6N5
Centre for Catalysis Research and Innovation,
University of Ottawa, Ottawa, ON, Canada K1N 6N5

Yun Liu and Sander Mommers
Chemical and Biological Engineering, University of
Ottawa, Ottawa, ON, Canada K1N 6N5

Natthaya Punsuwan and Chaiyot Tangsathitkulchai
School of Chemical Engineering, Institute of
Engineering, Suranaree University of Technology,
Nakhon Ratchasima 30000, Thailand

Index